Periodic Table

1	2	3	4	5	6	7	8	9	10	11	12	13	14	15	16	17	18
																	2 **He** 4.003
3 **Li** 6.941	4 **Be** 9.012											5 **B** 10.81	6 **C** 12.011	7 **N** 14.007	8 **O** 15.9994	9 **F** 19.00	10 **Ne** 20.179
11 **Na** 22.990	12 **Mg** 24.31											13 **Al** 26.98	14 **Si** 28.09	15 **P** 30.974	16 **S** 32.064	17 **Cl** 35.453	18 **Ar** 39.948
19 **K** 39.102	20 **Ca** 40.08	21 **Sc** 44.96	22 **Ti** 47.88	23 **V** 50.94	24 **Cr** 52.00	25 **Mn** 54.94	26 **Fe** 55.85	27 **Co** 58.93	28 **Ni** 58.69	29 **Cu** 63.55	30 **Zn** 65.38	31 **Ga** 69.72	32 **Ge** 75.59	33 **As** 74.92	34 **Se** 78.96	35 **Br** 79.90	36 **Kr** 83.80
37 **Rb** 85.47	38 **Sr** 87.62	39 **Y** 88.906	40 **Zr** 91.22	41 **Nb** 92.91	42 **Mo** 95.94	43 **Tc** (98)	44 **Ru** 101.1	45 **Rh** 102.905	46 **Pd** 106.4	47 **Ag** 107.870	48 **Cd** 112.41	49 **In** 114.82	50 **Sn** 118.69	51 **Sb** 121.75	52 **Te** 127.60	53 **I** 126.90	54 **Xe** 131.29
55 **Cs** 132.905	56 **Ba** 137.33	57–71 **Rare Earths**	72 **Hf** 178.49	73 **Ta** 180.95	74 **W** 183.85	75 **Re** 186.2	76 **Os** 190.2	77 **Ir** 192.2	78 **Pt** 195.09	79 **Au** 196.97	80 **Hg** 200.59	81 **Tl** 204.37	82 **Pb** 207.19	83 **Bi** 208.98	84 **Po** (210)	85 **At** (210)	86 **Rn** (222)
87 **Fr** (223)	88 **Ra** (226)	89–103 **Actinides**	104 **Rf** (261)	105 **Du** (260)	106 **Sg** (263)	107 **Bh** (262)	108 **Hs** (265)	109 **Mt** (266)	110 — (273)	111 — (?)	112 — (277)						

Rare Earths (Lanthanides)

57 **La** 138.91	58 **Ce** 140.12	59 **Pr** 140.91	60 **Nd** 144.24	61 **Pm** (147)	62 **Sm** 150.36	63 **Eu** 152.0	64 **Gd** 157.25	65 **Tb** 158.92	66 **Dy** 162.50	67 **Ho** 164.93	68 **Er** 167.26	69 **Tm** 168.93	70 **Yb** 173.04	71 **Lu** 174.97

Actinides

89 **Ac** 227.03	90 **Th** 232.04	91 **Pa** 231.04	92 **U** 238.03	93 **Np** 237.05	94 **Pu** (244)	95 **Am** (243)	96 **Cm** (247)	97 **Bk** (247)	98 **Cf** (251)	99 **Es** (252)	100 **Fm** (257)	101 **Md** (258)	102 **No** (259)	103 **Lr** (260)

The 1–18 group designation has been recommended by the International Union of Pure and Applied Chemistry (IUPAC).

ENVIRONMENTAL CHEMISTRY

THIRD EDITION

Colin Baird
University of Western Ontario

Michael Cann
University of Scranton

W. H. Freeman and Company

New York

ACQUISITIONS EDITOR: JESSICA FIORILLO

PUBLISHER: SUSAN FINNEMORE BRENNAN

MARKETING MANAGER: MARK SANTEE

PROJECT EDITOR: JANE O'NEILL

MEDIA EDITOR: VICTORIA ANDERSON

ASSOCIATE EDITOR: AMY SHAFFER

TEXT AND COVER DESIGNER: VICTORIA TOMASELLI

PHOTO EDITOR: VIKII WONG

ILLUSTRATIONS: NUGRAPHIC DESIGN, INC.

ILLUSTRATION COORDINATOR: SHAWN CHURCHMAN

PRODUCTION COORDINATOR: SUSAN WEIN

COMPOSITION: PROGRESSIVE INFORMATION TECHNOLOGIES

PRINTING AND BINDING: RR DONNELLEY & SONS COMPANY

Library of Congress Cataloging-in-Publication Data
Baird, Colin.
　　Environmental chemistry / Colin Baird, Michael Cann.–3rd ed.
　　　p. cm.
　　Includes bibliographical references and index.
　　ISBN 0-7167-4877-0 (EAN: 9780716748779)
　　1. Environmental chemistry.　　I. Cann, Michael C. II. Title.

　　TD192.B35 2004
　　628.5'01'54–dc22　　　　　　　　　　　　　　　　2004047245
　　EAN: 9780716748779

Printed in the United States of America

Second printing
W. H. Freeman and Company
41 Madison Avenue
New York, NY 10010
Houndmills, Basingstoke RG21 6XS, England

www.whfreeman.com

BRIEF CONTENTS

CONTENTS

PART II   TOXIC ORGANIC
           CHEMICALS   305

CHAPTER 7   Pesticides   307

Background   307

DDT   313

The Accumulation of Organochlorines
    in Biological Systems   318

Other Organochlorine Insecticides   323

Principles of Toxicology   327

The Distribution of Environmental
    Pollutants   334

Organophosphate and Carbamate
    Insecticides   337

PREFACE

To the Student

There are many definitions of environmental chemistry. To some, it is solely the chemistry of Earth's natural processes in air, water, and soil. More commonly, as in this book, environmental chemistry is concerned principally with the chemical aspects of problems that humankind has created in the natural environment. Part of this infringement on the natural chemistry of our planet has been a result of the activities of our everyday lives. In addition, chemists, through the products that they create and the processes by which they make these products, have also had a significant impact on the chemistry of the environment.

Chemistry has played a major role in the advancement of society and in making our lives longer, healthier, more comfortable, and more enjoyable. The effects of human-made chemicals are ubiquitous and in many instances quite positive. Without chemistry there would be no pharmaceutical drugs, no computers, no automobiles, no TVs, no DVDs, no lights, no synthetic fibers. However, along with all the positive advances that result from chemistry, copious amounts of toxic and corrosive chemicals have been produced and dispersed into the environment. Historically, chemists as a group have not always paid enough attention to the environmental consequences of their activities.

But it is not just the chemical industry, or even industry as a whole, that has emitted troublesome substances into the air, water, and soil. The fantastic increase since the Industrial Revolution in population and affluence has overloaded our atmosphere with carbon dioxide and toxic air pollutants, our waters with sewage, and our soil with garbage. We are exceeding the planet's natural capacity to cope with waste, and in many cases we do not know the consequences of these actions. As a character in Margaret Atwood's novel *Oryx and Crake* states, "The whole world is now one vast uncontrolled experiment."

During your journey through the chapters in this text, you will see that scientists do have a good handle on many environmental chemistry problems and have suggested ways—although sometimes very expensive ones—to keep us from inheriting the whirlwind of uncontrolled experiments on the planet. Chemists have also become more aware of the contributions of their own profession and industry in creating pollution and have developed the concept of *green chemistry* to help minimize their environmental footprint in the future. To illustrate these efforts, case studies of their initiatives have been included in the text. However, as a prelude to these studies, in the Introduction to Green Chemistry we discuss some of the history of environmental regulations—especially in the United States—and the principles and an illustrative application of the green chemistry movement that has developed.

Although the science underlying environmental problems is often maddeningly complex, its central aspects can usually be understood and appreciated with only introductory chemistry as background preparation. However, students who have not had some introduction to organic chemistry are encouraged to work through the Appendix, particularly before tackling Chapters 7 and 8.

To the Instructor

Environmental Chemistry, Third Edition, has been revised and updated in line with comments and suggestions made by a variety of readers and reviewers of the second edition. The number of end-of-chapter Additional Problems has been increased, and we have included more higher-level problems, partly as a result of contributions from various users and especially from Dr. Brian Wagner of the University of Prince Edward Island. As in previous editions, the background required to solve both in-text and end-of-chapter problems is either developed in the text or would have been covered previously in an introductory chemistry course. Where appropriate, hints are given to start students on the solution.

New to This Edition
Green Chemistry

The most important single addition to the third edition is the inclusion of introductory material and 12 case studies integrated into the text on the emerging area of green chemistry. These have been written by Professor Michael Cann of the University of Scranton. Mike has been extensively involved in the teaching of green chemistry, as coauthor of *Real-World Cases in Green Chemistry*, published by the American Chemical Society in Washington, D.C.

The green chemistry cases discussed within the text are:

• The concept of atom economy, which helps to determine the efficiency of a reaction, and the application of this concept to the synthesis of ibuprofen (see the Introduction to Green Chemistry).

• The replacement of CFCs and hydrocarbon blowing agents with carbon dioxide (Chapter 1).

• New pesticides that elicit natural defenses of plants against disease (Chapter 1) and that are specific to target organisms (Chapter 7); a termite control method that drastically reduces the amount of the pesticide that is required (Chapter 7).

- A new method for dry cleaning clothes that eliminates the use of organic solvents (Chapter 2).

- A new method for making computer chips that significantly reduces waste, pollution, and resource use (Chapter 4).

- A new polymer that is made from corn, that biodegrades, and that can be used to produce cups, bags, containers for foods, carpets, and fibers for clothing (Chapter 5).

- A new way of bleaching paper that has the potential to produce less toxic waste and reduce energy requirements (Chapter 8).

- A new method of producing cotton that significantly reduces water pollution, water use, and energy requirements (Chapter 9).

- A biodegradable chelating agent that has the potential to replace phosphates in detergents (Chapter 10).

- The removal of lead from automobile paint (Chapter 11).

- The removal of arsenic and chromium from pressure-treated wood (Chapter 11).

- The production of a biodegradable dispersing polymer from renewable resources (Chapter 12).

Other Topics

Generally speaking, all material on contemporary environmental problems in the second edition of the text has been updated to reflect the newest discoveries and analyses. The following additions and chapter reorganizations have also been made, in part to produce a better balance among the areas of air, water, and soil chemistry, and in part to include more mathematically oriented analyses of environmental issues:

- A steady-state analysis of atmospheric reactions, including the Chapman mechanism for ozone depletion, has been added to the stratospheric chemistry chapter (now Chapter 1).

- The detailed reaction mechanism sections, discussing the sequence of chemical reaction steps in both the troposphere and stratosphere, have been pulled together to form a separate chapter (Chapter 3).

- Subsections on sick building syndrome and on the hazardous air pollutant benzene have been added to the air pollution chapter (now Chapter 2).

- The concept of critical load in acid rain and a discussion of the origin and health effects of atmospheric smoke have been added to the air pollution chapter (Chapter 2).

• The Six Cities air pollution case study has been extended by more up-to-date research and incorporated into the main text of the air pollution chapter (Chapter 2).

• A new chapter, 13, covers and expands on all the nuclear chemistry— including radioactivity from radon gas and nuclear energy—of the previous edition.

• The role of volcanoes and variation in the Sun's output in affecting global air temperatures has been added to the greenhouse effect chapter (Chapter 4). The concept of residence time has been moved here from the heavy metals chapter and expanded.

• The various signs in nature that global warming is under way are discussed, and the effects of atmospheric aerosols are further elucidated (Chapter 4).

• The predictions concerning future global warming have been updated by summarizing the new IPCC report (2001) in Chapter 5. Predictions concerning future energy supplies and use have also been updated, including topics such as renewable energy and the hydrogen society (Chapter 6).

• The chapter on toxic organic chemicals in the second edition has been expanded somewhat and divided in two to make it more manageable, now covering pesticides in Chapter 7 and nonpesticide toxic organic substances in Chapter 8.

• A new section on calculating the distribution of pollutants through the various compartments of the environment has been added to Chapter 8. The precautionary principle is also introduced in Chapter 8.

• The role of PCBs in affecting mental development has been moved from the introductory chapter to Chapter 8 and updated.

• Conclusions about the carcinogenic and other health effects of dioxins and related compounds have been updated and expanded as a result of the EPA Dioxin 2000 Report (Chapter 8).

• New sections have been added in Chapter 8 on new environmental organic concerns: insect repellants, PBDEs, and perfluorinated sulfonates.

• A new subsection that introduces and applies pE–pH diagrams has been added to the natural waters chapter (now Chapter 9), and a new subsection on pollution of water by perchlorate salts has also been added.

• All the material on groundwater from the water chemistry chapters has been consolidated into the water purification chapter (now Chapter 10). Similarly, material on nitrates, nitrites, and nitrosamines in water and food has been transferred to Chapter 10 to consolidate material on nitrate pollution.

- Chapter 10 on water purification has been reorganized. It now begins with the purification and disinfection of water, with added material on new membrane technologies. The terms *disinfection by-product* and *maximum contaminant level goal* are now explicitly introduced and discussed. The treatment of sewage by conventional techniques has been placed just before the treatment of wastewater by conventional and modern techniques to improve the flow of material into later chapters.

- Because heavy-metal contamination is predominantly a water pollution problem that is best covered later, it has been moved to a new position (now Chapter 11). Pollution by chromium has been added, and the material on arsenic has been expanded and updated to reflect the growing importance of this element in drinking water in both the United States and Asia.

- The chapter on contaminated soils and sediments and the treatment of solid waste (now Chapter 12) has been reorganized so that it moves from the least to the most toxic. The chapter now begins with the various possible treatments for garbage, including landfills (extended coverage), incineration, and recycling, and then moves on to the natural chemistry of soil, with additional material on acidity and salinity. Material on mine tailings is also new.

- Nuclear chemistry, Chapter 13, consolidates radioactivity and radon coverage from Chapter 3 and nuclear energy from Chapter 5 of the second edition. New material has been added on the operation of fission power reactors, nuclear waste disposal, and fusion reactors and their environmental problems.

- New subsections have been added on the health effects of ionizing radiation, on depleted uranium, on the accidents at Chernobyl and Three Mile Island, and on dirty bombs (Chapter 13).

Supplements

The *Solutions Manual* (0-7167-1169-9) includes worked solutions to almost all problems (other than Review Questions, which are designed to direct students back to descriptive material within each chapter).

Students and instructors interested in pursuing specific topics in more detail should consult the Further Reading section at the end of each chapter and the Websites of Interest that are given for each chapter on the textbook's website, **www.whfreeman.com/envchem3e/**

To All Readers of the Text

The authors are happy to receive comments and suggestions about the content of this book from instructors and students at their home addresses: Colin Baird at cbaird@uwo.ca and Michael Cann at cannm1@uofs.edu.

Acknowledgments

The authors wish to express their gratitude and appreciation to a number of people who in various ways have contributed to this third edition of the book:

To Professor Thomas Chasteen of Sam Houston State University for masterfully writing the Environmental Instrumental Analysis boxes. These boxes bring in the view of a scientist working in this area and add much to the book.

To Professor Brian D. Wagner of the University of Prince Edward Island, for supplying some of the Additional Problems found in this edition.

To the students and instructors who have used previous editions of the text and, via their reviews and e-mails, have pointed out sections and problems that needed clarifying or extending.

To the W. H. Freeman and Company editors of the three editions—Deborah Allen, Michelle Russel Julet, and Jessica Fiorillo, and their editorial assistants Guy Copes and Jenness Crawford—for their encouragement, ideas, insightful suggestions, patience, and organizational abilities. To Margaret Comaskey for her careful copyediting and suggestions; Vicki Tomaselli for her crisp, clean design; Vikii Wong for researching the photographs so diligently; and Jane O'Neill for skillfully "quarterbacking" the book's production.

Colin Baird wishes to express his thanks:

To Ron Martin and Martin Stillman, his colleagues at the University of Western Ontario, who used the first two editions and have made valuable suggestions for improvement, and to his colleagues at Western and elsewhere who supplied information or answered queries on various subjects: Myra Gordon, Duncan Hunter, Roland Haines, Edgar Warnhoff, Marguerite Kane, Currie Palmer, Rob Lipson, Dave Shoesmith, Geoff Rayner-Canham, and Chris Willis. Thanks also to Nancy Russell for research assistance and to Chip Weseloh of Environment Canada for supplying data.

To his secretaries through the years, Sandy McCaw, Darlene Hagen, Diana Timmermans, Elizabeth Moreau, Shannon Woodhouse, Wendy Smith, and Judy Purves, for their brave attempts to decipher his writing and for dealing with the always-urgent problems that authors seem to have.

To his daughter, Jenny; son-in-law, Jim; and step-daughters, Zora and Cara.

Mike Cann wishes to express his thanks:

To his students (especially Marc Connelly) and fellow faculty at the University of Scranton, who have made valuable suggestions and contributions to his understanding of green chemistry and environmental chemistry.

To Joe Breen, who was one of the pioneers of green chemistry and one of the founders of the Green Chemistry Institute.

To Paul Anastas (White House Office of Science and Technology Policy) and Tracy Williamson (U.S. Environmental Protection Agency), whose boundless energy and enthusiasm for green chemistry are contagious.

To his loving wife, Cynthia, who has graciously and enthusiastically endured countless discussions of green chemistry and environmental chemistry.

To his children, Holly and Geoffrey, and his grandchildren, McKenna, Alexia, Alan Joshua, and Samantha, who, along with future generations, will reap the rewards of sustainable chemistry.

Both authors wish to express thanks to the reviewers of the third edition of the text for their helpful comments and suggestions:

Sheikh Ahmed, *Alderson-Broaddus College*
Floyd Beckford, *Lyon College*
Suzanne Bell, *Eastern Washington University*
David Neal Boehnke, *Jacksonville University*
Kathy M. Carvalho-Knighton, *University of South Florida, St. Petersburg*
Paul A. Flowers, *University of North Carolina, Pembroke*
James R. Gaier, *Manchester College*
Andreas Gebauer, *University of Alabama, Huntsville*
Marcia L. Gillette, *Indiana University, Kokomo*
Joel Gohdes, *Pacific University*
James Gordon, *Central Methodist College*
Michael P. Greene, *Copiague High School*
Robert Guy, *Dalhousie University*
Eric Hardegree, *Abilene Christian University*
Alton Hassell, *Baylor University*
Munir Humayan, *University of Chicago*
William C. Kaska, *University of California, Santa Barbara*
Dietmar Kennepohl, *Athabasca University*
Gregory S. Kowalczyk, *Southern Connecticut State University*
Anthony Lagalante, *Pennsylvania State University*
Leroy E. Laverman, *University of California, Santa Barbara*
David Lehmpuhl, *University of Southern Colorado*
Danuta Leszczynska, *Florida State University*
Susan Libes, *Coastal Carolina University*

Walter Loveland, *Oregon State University*
John Manock, *University of North Carolina, Wilmington*
Rusty Myers, *Alaska Pacific University*
Jodi Paar, *Pacific University*
Barrie Peake, *University of Otago*
Harry E. Pence, *State University of New York, Oneonta*
Jimmie M. Purser, *Millsaps College*
John R. Reinfelder, *Rutgers University, Cook College*
Thomas Rhyne, *Appalachian State University*
Kathryn Rowberg, *Purdue University, Calumet*
Jean Smolen, *Saint Joseph's University*
Brian D. Wagner, *University of Prince Edward Island*
Randall Wanke, *Augustana College*
Robert Wei, *Cleveland State University*
James J. Worman, *Rochester Institute of Technology*
Catherine Woytowicz, *George Washington University*
Z. Diane Xie, *University of Utah*

Introduction to Green Chemistry

A Brief History of Environmental Regulation

In the United States, many environmental disasters came to a head in the 1960s and 1970s. In 1962 the deleterious effects of the insecticide DDT were brought to the forefront by Rachel Carson in her seminal book, *Silent Spring*. In 1969 the Cayahoga River, which runs through Cleveland, Ohio, was so polluted with industrial waste that it caught fire. The Love Canal neighborhood in Niagara Falls, New York, was built on the site of a chemical dump and in the mid-1970s, during an especially rainy season, toxic waste began to ooze into the basements of homes, and drums of waste chemicals surfaced. The U.S. government purchased the land and cordoned off the entire Love Canal neighborhood. These distressing events were brought into the homes of Americans on the nightly news and, along with other environmental disasters, they became rallying points for environmental reform.

This era saw the creation of the U.S. Environmental Protection Agency (EPA) in 1970, the celebration of the first Earth Day, also in 1970, and a mushrooming number of environmental laws. Before 1960 there were approximately 20 environmental laws in the United States, and there are now more than 120. Most of the earliest of these were focused on conservation or setting land aside from development. The focus of environmental laws changed dramatically starting in the 1960s. Some of the most familiar U.S. environmental acts include the Clean Air Act (1970) and the Clean Water Act (originally known as the Federal Water Pollution Control Act Amendments of 1972). One of the major provisions of these acts was to set up pollution control programs. In effect, these programs attempted to control the release of toxic and other harmful chemicals into the environment. The Comprehensive Environmental Response, Compensation and Liability

If mankind is to survive, we shall require a substantially new manner of thinking.

Albert Einstein

Act (also known as the Superfund Act) set up a procedure and provided funds for cleaning up toxic waste sites. These acts thus focused on dealing with pollutants after they were produced, and they are known as "end-of-the-pipe solutions" and "command and control laws."

The risk due to a hazardous substance is a function of the exposure to and the hazard from the substance:

$$\text{risk} = f(\text{exposure} \times \text{hazard})$$

The end-of-the-pipe laws attempt to control risk by preventing exposure to these substances. However, exposure controls inevitably fail, which points out the weakness of these laws. The Pollution Prevention Act of 1990 is the only U.S. environmental act that focuses on the paradigm of *prevention* of pollution at the source: if hazardous substances are not used or produced, then their risk is eliminated. There is also no need to worry about controlling exposure, controlling dispersion into the environment, or cleaning up hazardous chemicals.

The U.S. Pollution Prevention Act of 1990 set the stage for **green chemistry.** Green chemistry became a formal focus of the U.S. EPA in 1991 and became part of a new direction set by the EPA, by which the agency worked with and encouraged companies to voluntarily find ways to reduce the environmental consequences of their activities. While at the EPA, Paul Anastas (now at the White House Office of Science and Technology Policy) and John Warner (University of Massachusetts–Boston) defined *green chemistry* as *the design of chemical products and processes that reduce or eliminate the use and generation of hazardous substances.* Moreover, green chemistry seeks to

- reduce waste (especially toxic waste),
- reduce the consumption of resources and ideally use renewable resources, and
- reduce energy consumption.

Anastas and Warner also formulated the Twelve Principles of Green Chemistry. These principles provide guidelines for chemists in assessing the environmental impact of their work.

The Twelve Principles of Green Chemistry

1. It is better to **prevent waste** than to treat or clean up waste after it is formed.

2. Synthetic methods should be designed to **maximize the incorporation of all materials** used in the process into the final product.

3. Wherever practicable, synthetic methodologies should be designed to use and generate **substances that possess little or no toxicity** to human health and the environment.

4. Chemical products should be designed to **preserve efficacy of function while reducing toxicity.**

5. The use of **auxiliary substances** (e.g., solvents, separation agents, etc.) **should be made unnecessary** whenever possible and innocuous when used.

6. **Energy requirements** should be recognized for their **environmental and economic impacts and should be minimized.** Synthetic methods should be conducted at ambient temperature and pressure.

7. A raw material **feedstock should be renewable** rather than depleting whenever technically and economically practical.

8. Unnecessary **derivatization** (blocking group, protection/deprotection, temporary modification of physical/chemical processes) should be **avoided** whenever possible.

9. **Catalytic reagents** (as selective as possible) are superior to stoichiometric reagents.

10. **Chemical products** should be designed so that at the end of their function they **do not persist in the environment** and break down into innocuous degradation products.

11. Analytical methodologies need to be further developed to allow for **real-time, in-process** monitoring and control prior to the formation of hazardous substances.

12. Substances and the form of a substance used in a chemical process should be chosen so as to **minimize the potential for chemical accidents,** including releases, explosions, and fires.

In most of the chapters in the text, real-world examples of green chemistry are discussed. During these discussions, you should keep in mind the Twelve Principles of Green Chemistry and decide which of them are met by the particular example. Although we won't consider all the principles at this point, a brief discussion of some of them is beneficial.

• Principle 1 is the heart of green chemistry and places the emphasis on the prevention of pollution at the source rather than cleaning up waste after it has been produced.

• Principles 2–5, 7–10, and 12 focus on the materials used in the production of chemicals and the products that are formed.

• In a chemical synthesis, in addition to the desired product, unwanted by-products are often formed and these compounds are usually discarded as waste. Principle 2 encourages chemists to look for synthetic routes that maximize the production of the desired product and at the same time minimize the production of unwanted by-products (see the synthesis of ibuprofen discussed later).

• Principles 3 and 4 stress that the toxicity of materials and products should be kept to a minimum. As we will see in later discussions of green chemistry, Principle 4 is often met when new pesticides are designed with reduced toxicity to nontarget organisms.

• During the course of a synthesis, chemists employ not only the compounds that are actually involved in the reaction (reactants) but also auxiliary substances such as solvents (to dissolve the reactants and to purify the products) and agents used to separate and dry the products. These materials are usually used in much larger quantities than the reactants, and they contribute a great deal to the waste produced during a chemical synthesis. When they are designing a synthesis, Principle 5 reminds chemists to consider ways to minimize the use of these auxiliary substances.

• Many organic chemicals are produced from petroleum, which is a non-renewable resource. Principle 7 urges chemists to consider ways to produce chemicals from renewable resources such as plant material (biomass).

• As we will see in Chapter 7, DDT is an effective pesticide. However, a major environmental problem is its stability in the natural environment. DDT degrades only slowly. Although it has been banned in most developed countries since the 1970s (in the United States since 1972), it can still be found in the environment, particularly in the fatty tissues of animals. Principle 10 stresses the need to consider the lifetime of chemicals in the environment and the need to focus on materials (such as pesticides) that degrade rapidly in the environment to harmless substances.

• Many chemical reactions require heating or cooling or a pressure higher or lower than atmospheric pressure, or both. Performing reactions at other than ambient temperature and pressure requires energy, and Principle 6 reminds chemists of these considerations when designing a synthesis.

Presidential Green Chemistry Challenge Awards

To recognize outstanding examples of green chemistry, the **Presidential Green Chemistry Challenge Awards** were established in 1996 by the U.S. EPA. Generally five awards are given each year at a ceremony held at the

National Academy of Sciences in Washington, D.C. The awards are given in the following three categories:

1.   The use of alternative synthetic pathways for green chemistry, in areas such as

- catalysis and biocatalysis,

- natural processes, such as photochemistry and biomimetic synthesis, and

- alternative feedstocks that are more innocuous and renewable (e.g., biomass).

2.   The use of alternative reaction conditions for green chemistry, such as

- solvents that have a reduced impact on human health and the environment and

- increased selectivity and reduced wastes and emissions.

3.   The design of safer chemicals that are, for example,

- less toxic than current alternatives and

- inherently safer with regard to accident potential.

Real-World Examples of Green Chemistry

To introduce you to the important and exciting world of green chemistry, you will encounter real-world cases of green chemistry throughout this book. These examples are winners of Presidential Green Chemistry Challenge Awards. As you explore these examples, it will become apparent that green chemistry is very important in mitigating the ecological footprint of chemical processes in the air, water, and soil.

We begin our journey into this important topic by briefly exploring how green chemistry can be applied to the synthesis of *ibuprofen*, an important everyday drug. In this discussion we will see how the redesign of a chemical synthesis can eliminate a great deal of waste and pollution and reduce the amount of resources required.

Before discussing the synthesis of ibuprofen, we must first look at the concept of **atom economy.** This concept was developed by Barry Trost of Stanford University and won a Presidential Green Chemistry Challenge Award in 1998. Atom economy focuses our attention on green chemistry Principle 2 by asking *how many of the atoms of the reactants are incorporated into the final desired product and how many are wasted.* As we will see in our discussion of the synthesis of ibuprofen, when chemists synthesize a compound, not all the atoms of the reactants are used in the desired product.

Many of these atoms may end up in unwanted products (by-products) that are in many instances considered waste. These waste by-products may be toxic and can cause considerable environmental damage if not disposed of properly. In the past, waste products from chemical and other processes have not been disposed of properly, and environmental disasters such as Love Canal have occurred.

Before we take on the synthesis of ibuprofen, let us look at a simple illustration of the concept of atom economy using the production of the desired compound 1-bromobutane (compound 4) from 1-butanol (compound 1).

$$H_3C—CH_2—CH_2—CH_2—OH + Na—Br + H_2SO_4 \longrightarrow$$
$$ 1 2 3$$
$$H_3C—CH_2—CH_2—CH_2—Br + NaHSO_4 + H_2O$$
$$ 4 5 6$$

If we inspect this reaction, we find that not only the desired product is formed, but also the unwanted by-products sodium hydrogen sulfate and water, compounds 5 and 6. On the left side of this reaction we have printed all the atoms of the reactants that are used in the desired product in green and the remainder of the atoms (which become part of our waste by-products) in black. If we add up all the green atoms on the left side of the reaction, we get 4 C, 9 H, and 1 Br (reflecting the molecular formula of the desired product, 1-bromobutane). The mass of these atoms collectively is 137 amu (atomic mass units), the molecular mass of 1-bromobutane. Adding up *all* the atoms of the reactants gives 4 C, 12 H, 5 O, 1 Br, 1 Na, and 1 S, and the total mass of all these atoms is 275 amu. If we take the mass of the atoms that are used, divide by the mass of all the atoms, and multiply by 100, we obtain the % **atom economy,** here 50%. Thus we see that half of the mass of all the atoms of the reactants is wasted and only half is actually incorporated into the desired product.

$$\% \text{ atom economy} = \frac{\text{(molecular mass of atoms used)}}{\text{(molecular mass of all reactants)}} \times 100$$

$$= \frac{137}{275} \times 100 = 50\%$$

This is one method of assessing the efficiency of a reaction. Armed with this information, a chemist may want to explore other methods of producing 1-bromobutane that have a greater % atom economy. We will now see how the concept of atom economy can be applied to the preparation of ibuprofen.

Ibuprofen is a common analgesic and anti-inflammatory drug found in brand-name products such as *Advil*, *Motrin*, and *Medipren*. The first commercial synthesis of ibuprofen was by the Boots Company PLC of Nottingham, England. This synthesis, which has been used since the 1960s, is shown in Figure In-1. Although a detailed discussion of the chemistry of the synthesis is beyond the scope of this book, we can calculate its atom economy and obtain some idea of the waste produced. In Figure In-1 the atoms printed in

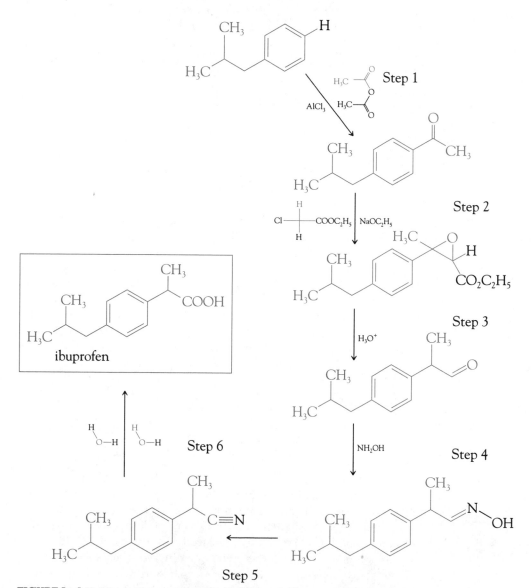

FIGURE In-1 The Boots Company synthesis of ibuprofen. [Source: M. C. Cann and M. E. Connelly, *Real-World Cases in Green Chemistry* (Washington, DC: American Chemical Society, 2000).]

green are those that are incorporated into the final desired product, ibuprofen, while those in black end up in waste by-products. We can inspect the structures of each of the reactants and determine that the total of all the atoms in the reactants is 20 C, 42 H, 1 N, 10 O, 1 Cl, and 1 Na. The masses of all these atoms added together total 514.5 amu. We can also determine the number of atoms of the reactants that are used in the ibuprofen (the atoms shown in green): 13 C, 18 H, and 2 O (the molecular formula of ibuprofen). These atoms have a total mass of 206.0 amu (the molecular mass of the ibuprofen).

$$\% \text{ atom economy} = \frac{(\text{molecular mass atoms used})}{(\text{molecular mass of all reactants})} \times 100$$

$$= \frac{206.0}{514.5} \times 100 = 40\%$$

Only 40% of the mass of all the atoms of the reactants in this synthesis end up in the ibuprofen and 60% are wasted. Because more than 30 million pounds of ibuprofen are produced each year, if we produced all the ibuprofen by this synthesis, there would be more than 35 million pounds of unwanted waste produced, just from the poor atom economy of this synthesis.

A new synthesis (Figure In-2) of ibuprofen was developed by the BHC Company (a joint venture of the Boots Company PLC and Hoechst Celanese Corporation), which won a Presidential Green Chemistry Challenge Award in 1997. This synthesis has only three steps as opposed to the six-step Boots synthesis and is less wasteful in many ways. One of the most obvious ways is in the increased atom economy. The mass of all the atoms of the reactants in this synthesis is 266.0 amu (13 C, 22 H, 4 O; note that the HF, Raney nickel, and Pd are only used in catalytic amounts and thus do not contribute to the atom economy) while the atoms used (printed in green) again weigh 206.0 amu. This yields a % atom economy of 77%:

$$\% \text{ atom economy} = \frac{(\text{molecular mass atoms used})}{(\text{molecular mass of all reactants})} \times 100$$

$$= \frac{206.0}{266.0} \times 100 = 77\%$$

A by-product of the acetic anhydride used in step 1 is acetic acid, which is isolated and used, increasing the atom economy of this synthesis to more than 99%. Additional environmental advantages of the BHC synthesis include the elimination of auxiliary materials (Principle 5), such as solvents and the aluminum chloride promoter (replaced with the catalyst HF, Principle 9), and higher yields. Thus the green chemistry of the BHC Company synthesis lowers the environmental impact for the synthesis of ibuprofen by lowering the consumption of reactants and auxiliary substances while simultaneously reducing the waste.

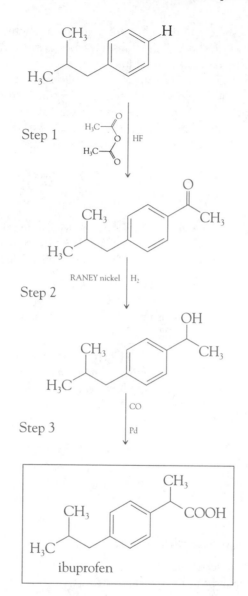

FIGURE In-2 The BHC Company synthesis of ibuprofen. [Source: M. C. Cann and M. E. Connelly, *Real-World Cases in Green Chemistry* (Washington, DC: American Chemical Society, 2000).]

Other improved syntheses that are winners of Presidential Green Chemistry Awards include the pesticide *Roundup*, the antiviral agent *Cytovene*, and the active ingredient in the antidepressant *Zoloft*.

Sustainability has been defined as *meeting the needs of the present generation without compromising the ability of future generations to meet their needs.* This popular definition was coined by the Brundtland Commission of the United Nations in 1987. Green chemistry provides a paradigm for reducing both the consumption of resources and the production of waste, thus moving toward sustainability. One of the primary considerations in the manufacture of chemicals must be the environmental impact of the chemical and the process by which it is produced. Sustainable chemistry must become part of the psyche of not only chemists and scientists, but also business leaders and policymakers. With this in mind, real-world examples of green chemistry have been incorporated throughout this text to expose you (our future scientists, business leaders, and policymakers) to sustainable chemistry.

Further Reading

1. P. T. Anastas and J. C. Warner, *Green Chemistry Theory and Practice* (New York: Oxford University Press, 1998).

2. M. C. Cann and M. E. Connelly, *Real-World Cases in Green Chemistry* (Washington, DC: American Chemical Society, 2000).

3. M. C. Cann, "Bringing State of the Art, Applied, Novel, Green Chemistry to the Classroom, by Employing the Presidential Green Chemistry Challenge Awards, *Journal of Chemical Education* 76 (1999): 1639–1641.

4. M. C. Cann, "Greening the Chemistry Curriculum at the University of Scranton," *Green Chemistry* 3 (2001): G23–G25.

5. M. A. Ryan and M. Tinnesand, eds, *Introduction to Green Chemistry* (Washington, DC: American Chemical Society, 2002).

6. M. Kirchhoff and M. A. Ryan, eds, *Greener Approaches to Undergraduate Chemistry Experiments* (Washington, DC: American Chemical Society, 2002).

7. World Commission on Environment and Development, *Our Common Future* [The Bruntland Report] (New York: Oxford University Press, 1987).

Websites of Interest

1. EPA "Green Chemistry": http://www.epa.gov/greenchemistry/index.html

2. The Green Chemistry Institute of the American Chemical Society website: http://www.chemistry.org/portal/a/c/s/1/acsdisplay.html?DOC=greenchemistry institute\index.html

3. University of Scranton "Green Chemistry": http://academic.scranton.edu/faculty/CANNM1/greenchemistry.html

4. American Chemical Society Educational Activities: http://www.chemistry.org/portal/Chemistry?PID=acsdisplay.html&DOC= education\greenchem\index.html

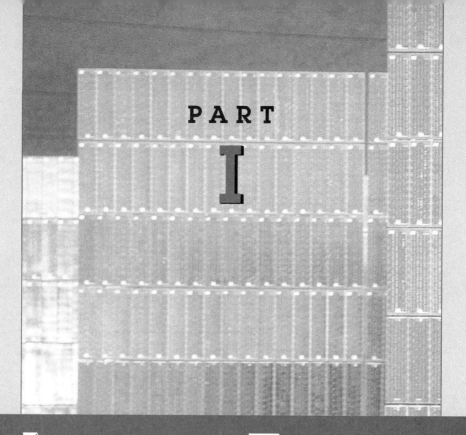

PART
I

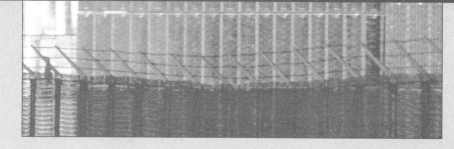

AIR AND ENERGY

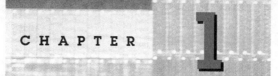

CHAPTER 1

Stratospheric Chemistry: The Ozone Layer

Introduction

The **ozone layer** is a region of the atmosphere that is called "Earth's natural sunscreen" because it filters out harmful ultraviolet (UV) rays from sunlight before they can reach the surface of our planet and cause damage to humans and other life forms. Any substantial reduction in the amount of this ozone would threaten life as we know it. Consequently, the appearance in the mid-1980s of a large "hole" in the ozone layer over Antarctica represented a major environmental crisis.

A young girl applies sunscreen to protect her skin against UV rays from the Sun. (Lowell George/CORBIS)

Dobson Units for Overhead Ozone

Ozone, O_3, is a gas that is present in small concentrations throughout the atmosphere. The total amount of atmospheric ozone that lies over a given point is measured in terms of **Dobson units** (DU). One Dobson unit is equivalent to a 0.01-mm (0.001-cm) thickness of pure ozone at the density it would possess if it were brought to ground level (1 atm) pressure and 0°C temperature. Some authors use milliatmosphere centimeter (matm cm) rather than the equivalent Dobson unit for the amount of overhead ozone: 1 matm cm = 1 DU (see Additional Problem 1).

On average, this total overhead ozone at temperate latitudes amounts to about 350 DU; thus if all the ozone were to be brought down to ground level, the layer of pure ozone would be only 3.5 mm thick. Because of stratospheric winds, ozone is transported from tropical regions, where most of it is produced, toward polar regions. Thus, ironically, the closer to the equator you live, the less the total amount of ozone that protects you from ultraviolet light. Ozone concentrations in the tropics usually average 250 DU, whereas those in subpolar regions average 450 DU, except, of course, when holes appear in the ozone layer over such areas. There is some natural variation of ozone concentration with season, with the highest levels in the early spring and the lowest in the fall, as illustrated by the black curve in Figure 1-1.

The Annual Ozone Hole Above Antarctica

The Antarctic ozone hole was discovered by Dr. Joe C. Farman and his colleagues in the British Antarctic Survey. They had been recording ozone levels over this region since 1957. Their data indicated that the total amounts of ozone each October had been gradually falling, with precipitous declines beginning in the late 1970s. This is illustrated in Figure 1-2, where the September–October minimum daily amount of overhead ozone is plotted against year. The period from September to November corresponds to the spring season at the South Pole and follows a period of very cold 24-hour nights common to polar winters. By the mid-1980s the springtime loss in ozone at some altitudes over Antarctica was complete and resulted in a loss of more than 50% of the total overhead amount. It is therefore appropriate

FIGURE 1-1 The seasonal dependence of total overhead ozone (black curve) and the January 1979–March 1994 depletions in ozone (green curve) for the 35–60°N region. [Source: Redrawn from R. D. Bojkov, L. Bishop, and V. E. Fioletov, "Total Ozone Trends from Quality-Controlled Ground-Based Data," *Journal of Geophysical Research* 100 (1995): 25867–25876.]

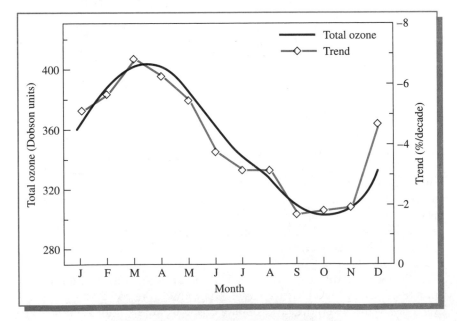

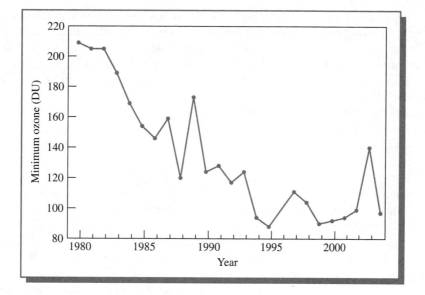

FIGURE 1-2 Minimum concentration of total overhead ozone above Antarctica each spring in recent years. [Source: NASA website http://toms.gsfc.nasa.gov/multi/min_ozone.jpg]

to speak of a "hole" in the ozone layer that now appears each spring over the Antarctic and lasts for several months. The average geographic area covered by the ozone hole has increased substantially since it began, and in most recent years it has been comparable in size to that of the North American continent (see Figure 1-3).

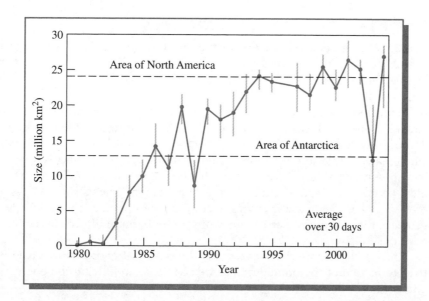

FIGURE 1-3 The variation with year of *the* size of the Antarctic ozone hole, measured as the average area within the 220-DU contour line. [Source: NASA website http://toms.gsfc.nasa.gov/multi/oz_hole_area.jpg]

Initially it was not clear whether the hole was due to a natural phenomenon involving meteorological forces or to a chemical mechanism involving air pollutants. In the latter possibility, the suspect chemical was chlorine, produced mainly from gases that are released into the air in large quantities as a consequence of their use, for example, in air conditioners. Scientists had predicted that the chlorine would destroy ozone, but only to a small extent and only after several decades had elapsed. The discovery of the Antarctic ozone hole came as a complete surprise to everyone.

To discover why the hole formed each spring, an American emergency ozone research expedition headed by Dr. Susan Solomon of the National Oceanic and Atmospheric Administration in Boulder, Colorado, went to the Antarctic in late winter in August 1986. Using the Moon as a light source, Solomon and her co-workers were able to identify from the specific wavelengths of light absorbed by atmospheric gases that certain molecules were present in the air far above their heads. As a result of that research and subsequent investigations, it is known that the hole indeed does occur as a result of chlorine pollution. Furthermore, it has been predicted that the hole will continue to reappear each spring until about the middle of this century and that a corresponding hole may one day appear above the Arctic region.

As a consequence of these discoveries, governments worldwide moved quickly to legislate a phase-out in production of the chemicals responsible so that the situation does not become much worse by the development of even more severe ozone depletion over populated areas than is already the case, with the corresponding threat to the health of humans and other organisms that this increase would bring.

Ozone Depletion in Temperate Areas

Ozone is being depleted not just in the air above Antarctica but to some extent worldwide. The extent of the nonpolar depletion is illustrated in Figure 1-4, which shows the changes that occurred in total ozone concentrations from 1979 to 2002 for nonpolar regions. After accounting for the expected seasonal variations in ozone levels, the average loss at these latitudes amounted to about 3% over the 1980s. For reasons that scientists do not fully understand, there were no net additional losses over the 1990s. The greatest depletion in northern mid-latitude regions (35–60°N) occurs in the winter-spring. There is some depletion in the summertime, when most people's exposure to sunlight is at a maximum. The worldwide loss of ozone has become a major environmental concern, since it results in less protection to life at the Earth's surface from the harmful ultraviolet component of sunlight.

In this chapter we shall investigate the chemical processes involved in the production of the ozone layer, as well as those underlying the phenomenon of ozone depletion.

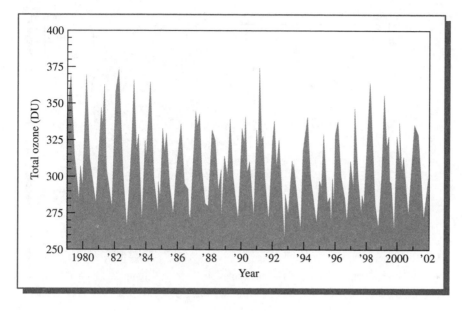

FIGURE 1-4 Overhead ozone variations over mid-latitude regions. [Source: R. A. Kerr, "A Brighter Outlook for Good Ozone," *Science* 297 (2002): 1623.]

Regions of the Atmosphere

The main components (ignoring the normally ever-present but variable water vapor) of an unpolluted version of the Earth's atmosphere are **diatomic nitrogen** (N_2, about 78% of the molecules), **diatomic oxygen** (O_2, about 21%), *argon* (Ar, about 1%), and *carbon dioxide* (CO_2, presently about 0.04%). (The names of chemicals important to a chapter are printed in bold, along with their formulas, when they are introduced. The names of chemicals less important in the present context are printed in italic.) This mixture of chemicals seems unreactive in the lower atmosphere, even at temperatures or sunlight intensities well beyond those naturally encountered at the Earth's surface.

The lack of noticeable reactivity in the atmosphere is deceptive. In fact, many environmentally important chemical processes occur in air, whether clean or polluted. In the next two chapters these reactions will be explored in detail. In Chapters 2 and 3, reactions that occur in the **tropo-sphere,** the region of the sky that extends from ground level to about 15 kilometers altitude and contains 85% of the atmosphere's mass, are discussed. In this chapter we will consider processes in the **stratosphere,** the portion of the atmosphere from approximately 15 to 50 km (9–30 miles) altitude that lies just above the troposphere. The chemical reactions to be considered are vitally important to the continuing health of the ozone layer, which is found in the bottom half of the stratosphere. The ozone concentrations and the average temperatures at altitudes up to 50 km in the Earth's atmosphere are shown in Figure 1-5.

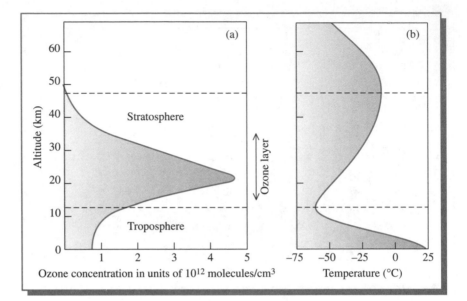

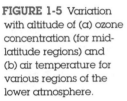

FIGURE 1-5 Variation with altitude of (a) ozone concentration (for mid-latitude regions) and (b) air temperature for various regions of the lower atmosphere.

The stratosphere is defined as the region that lies between the altitudes where the temperature trends display reversals: the bottom of the stratosphere occurs where the temperature first stops decreasing with height and begins to increase, and the top of the stratosphere is the altitude where the temperature stops increasing with height and begins to decrease. The exact altitude at which the troposphere ends and the stratosphere begins varies with season and with latitude.

Environmental Concentration Units for Atmospheric Gases

Two types of concentration scales are commonly used for gases present in air. For *absolute* concentrations, the most common scale is the number of **molecules per cubic centimeter** of air. The variation in the concentration of ozone on the molecules per cubic centimeter scale with altitude is illustrated in Figure1-5a. Absolute concentrations are sometimes expressed in terms of the **partial pressure** of the gas, which is expressed in units of atmospheres or kilopascals or bars. According to the **ideal gas law** ($PV = nRT$), partial pressure is directly proportional to the molar concentration n/V, and hence to the molecular concentration per unit volume, when different gases or components of a mixture are compared at the same Kelvin temperature T.

Relative concentrations are usually based on the chemists' familiar **mole fraction** scale (called *mixing ratios* by physicists), which is also a molecule fraction scale. Because the concentrations for many constituents are so small, atmospheric and environmental scientists often re-express the mole or molecule fraction as a *parts per* _____ value. Thus a concentration of 100 molecules of a gas such as carbon dioxide dispersed in 1 million (10^6) molecules of air would be expressed as 100 parts per million, i.e., 100 ppm, rather than as a molecule or mole fraction of 0.0001. Similarly, ppb and ppt stand for parts per billion (one in 10^9) and parts per trillion (one in 10^{12}).

It is important to emphasize that for *gases*, these relative concentration units express the number of *molecules* of a pollutant (i.e., the solute in chemists' language) that are present in one million or billion or trillion *molecules* of air. Since, according to the ideal gas law, the volume of a gas is proportional to the number of molecules it contains, the "parts per" scales also represent the *volume* a pollutant gas would occupy, compared to that of the stated *volume of air*, if the pollutant were to be isolated and compressed until its pressure equaled that of the air. In order to emphasize that the concentration scale is based upon molecules or volumes rather than upon mass, a v (for volume) is sometimes shown as part of the unit, e.g., 100 ppm$_v$ or 100 ppmv.

The Physics and Chemistry of the Ozone Layer

Absorption of Light by Molecules

The chemistry of ozone depletion, and of many other processes in the stratosphere, is driven by energy associated with light from the Sun. For this reason, we begin by investigating the relationship between light absorption by molecules and the resulting activation, or energizing, of the molecules that enables them to react chemically.

An object that we perceive as black in color absorbs light at all wavelengths of the visible spectrum, which runs from about 400 nm (violet light) to about 750 nm (red light); note that one nanometer (nm) equals 10^{-9} meter. Substances differ enormously in their propensity to absorb light of a given wavelength because of differences in the energy levels of their electrons. Diatomic molecular oxygen, O_2, does not absorb visible light very readily but does absorb some types of **ultraviolet** (UV) light, which is light having wavelengths between about 50 and 400 nm. The most environmentally relevant portion of the electromagnetic spectrum is illustrated in Figure 1-6. Notice that the UV region begins at the violet edge of the visible region, hence the name *ultraviolet*. The division of the UV region into components will be discussed later in this chapter. At the other end of the spectrum, beyond the red portion of the visible region, lies **infrared** light, which will become important to us when we discuss the greenhouse effect in Chapter 4.

An **absorption spectrum** such as that illustrated in Figure 1-7 is a graphical representation that shows the relative fraction of light that is absorbed by a given type of molecule as a function of wavelength. Here, the efficient light-absorbing behavior of O_2 molecules for the UV region

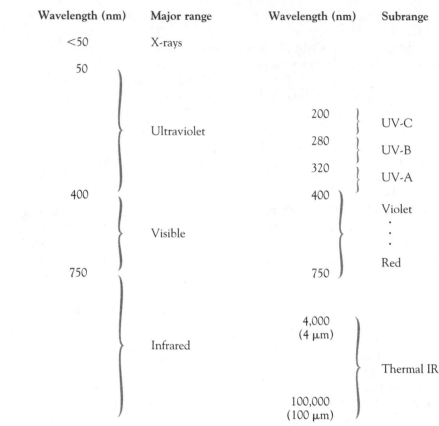

Wavelength (nm)	Major range	Wavelength (nm)	Subrange
<50	X-rays		
50			
	Ultraviolet	200	UV-C
		280	UV-B
		320	UV-A
400		400	Violet
	Visible		•
			•
			•
750		750	Red
	Infrared	4,000 (4 μm)	Thermal IR
		100,000 (100 μm)	

FIGURE 1-6 The electromagnetic spectrum. The ranges of greatest environmental interest in this book are shown.

between 70 and 250 nm is shown; some minuscule amount of absorption continues beyond 250 nm, but in an ever-decreasing fashion (not shown). Notice that the fraction of light absorbed by O_2 (given on a logarithmic scale in Figure 1-7) varies quite dramatically with wavelength. This sort of selective absorption behavior is observed for all atoms and molecules, although the specific regions of strong absorption and of zero absorption vary widely, depending upon the structure of the species and the energy levels of their electrons.

Filtering of Sunlight's UV Component by Atmospheric O_2 and O_3

As a result of these absorption characteristics, the O_2 gas that lies *above* the stratosphere filters from sunlight most of the UV light from 120 to 220 nm; the remainder of the light in this range is filtered by the O_2 in the stratosphere. Ultraviolet light that has wavelengths shorter than 120 nm is

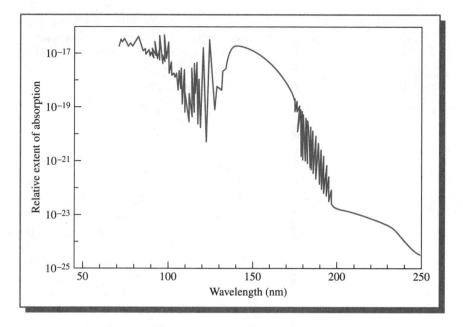

FIGURE 1-7 Absorption spectrum of O_2. [Source: T. E. Graedel and P. J. Crutzen, *Atmospheric Change: An Earth System Perspective* (New York: W. H. Freeman, 1993).]

filtered in and above the stratosphere by O_2 and other constituents of air such as N_2. Thus no UV light having wavelengths shorter than 220 nm reaches the Earth's surface. This screening protects our skin and eyes, and in fact protects all biological life, from extensive damage by this part of the Sun's output.

Diatomic oxygen also filters some, but not all, of sunlight's UV in the 220- to 240-nm range. The ultraviolet light in the whole 220- to 320-nm range is filtered from sunlight mainly by ozone molecules, O_3, that are spread through the middle and lower stratosphere. The absorption spectrum of ozone in this wavelength region is shown in Figure 1-8. Since its molecular constitution, and thus its set of energy levels, is different from that of diatomic oxygen, its light absorption characteristics also are quite different.

Ozone, aided to some extent by O_2 at the shorter wavelengths, filters out all of the Sun's ultraviolet light in the 220- to 290-nm range, which overlaps the 200- to 280-nm region known as **UV-C** (see Figure 1-6). However, ozone can absorb only a fraction of the Sun's UV light in the 290- to 320-nm range, since, as you can infer from Figure1-8b, its inherent ability to absorb light of such wavelengths is quite limited. The remaining amount of the sunlight of such wavelengths, 10 to 30% depending upon latitude, penetrates the atmosphere to the Earth's surface. Thus ozone is not *completely* effective in shielding us from light in the **UV-B** region, defined as that which lies from 280 to 320 nm (although different authors vary slightly on the limits for this range). Since the absorption by

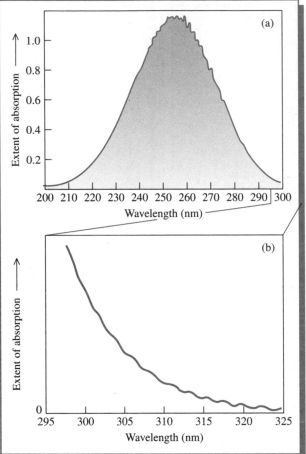

ozone falls off in an almost exponential manner with wavelength in this region (see Figure 1-8b), the fraction of UV-B that reaches the troposphere increases with increasing wavelength.

Because neither ozone nor any other constituent of the clean atmosphere absorbs significantly in the **UV-A** range, i.e., 320–400 nm, most of this, the least biologically harmful type of ultraviolet light, does penetrate to the Earth's surface. (Nitrogen dioxide gas does absorb UV-A light but is present in such small concentration in clean air that its net absorption of sunlight is quite small.)

The net effect of diatomic oxygen and ozone in screening the troposphere from the UV component of sunlight is illustrated in Figure 1-9. The curve at the left corresponds to the intensity of light received outside the Earth's atmosphere, whereas the curve at the right corresponds to the light that is transmitted to the troposphere (and thus to the surface). The vertical separation between the curves at each wavelength corresponds to the amount of sunlight that is absorbed in the stratosphere and outer regions of the atmosphere.

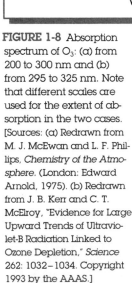

FIGURE 1-8 Absorption spectrum of O_3: (a) from 200 to 300 nm and (b) from 295 to 325 nm. Note that different scales are used for the extent of absorption in the two cases. [Sources: (a) Redrawn from M. J. McEwan and L. F. Phillips, *Chemistry of the Atmosphere*. (London: Edward Arnold, 1975). (b) Redrawn from J. B. Kerr and C. T. McElroy, "Evidence for Large Upward Trends of Ultraviolet-B Radiation Linked to Ozone Depletion," *Science* 262: 1032–1034. Copyright 1993 by the AAAS.]

Biological Consequences of Ozone Depletion

A reduction in stratospheric ozone concentration allows more UV-B light to penetrate to the Earth's surface. A 1% decrease in overhead ozone is predicted to result in a 2% increase in UV-B intensity at ground level. This increase in UV-B is the principal environmental concern about ozone depletion, since it leads to detrimental consequences to many life forms, including humans. Exposure to UV-B causes human skin to sunburn and suntan; overexposure can lead to skin cancer, the most prevalent form of cancer. Increasing amounts of UV-B may also adversely affect the human immune system and the growth of some plants and animals.

Most biological effects of sunlight arise because UV-B can be absorbed by DNA molecules, which then may undergo damaging reactions. By comparing the variation in wavelength of UV-B light of differing intensity arriving at the Earth's surface with the absorption characteristics of DNA as shown in Figure 1-10, it can be concluded that the major detrimental effects of sunlight absorption will occur at about 300 nm. Indeed, in light-skinned people, the skin shows maximum UV absorption from sunlight at about 300 nm.

Most skin cancers in humans are due to overexposure to UV-B in sunlight, so any decrease in ozone is eventually expected to yield an increase in the incidence of this disease. Fortunately, the great majority of skin cancer cases are not the often-fatal (25% mortality rate) **malignant melanoma,** but rather one of the slowly spreading types that can be treated and that collectively affect about one in four Americans at some point in their lives. The plot in Figure 1-11, which is based on health data from eight countries at different latitudes and that therefore receive different amounts of ground-level UV, shows that the rise in the incidence of nonmelanoma skin cancer with exposure to UV is exponential, since the logarithm of the incidence is linearly related to the UV intensity. For example, the skin cancer rate in Europe is only about half that in the United States.

The incidence of the malignant melanoma form of skin cancer, which affects about one in one hundred Americans, is thought to be related to short periods of very high UV exposure, particularly early in life. Particularly susceptible are fair-skinned, fair-haired, freckled people who burn easily and who have moles with irregular shapes or colors. The incidence of malignant melanoma is also related to latitude. White males living in sunny climates such as Florida or Texas are twice as likely to die from this disease than those in the more northerly states, although part of this increase is as likely due to different patterns of personal behavior, such as choice of clothing, as it is to increased UV-B content in the sunlight. Curiously, indoor workers—who have intermittent exposure to the Sun—are more susceptible than are tanned outdoor workers! The lag period between first exposure and the appearance of melanoma is 15–25 years. If malignant melanoma is not treated early, it can spread via the bloodstream to body organs such as the brain and the liver.

The use of sunscreens that block UV-B, but not UV-A, may actually lead to an increase in melanoma skin cancer, since sunscreen usage allows people to expose their skin to sunlight for prolonged periods without burning. The

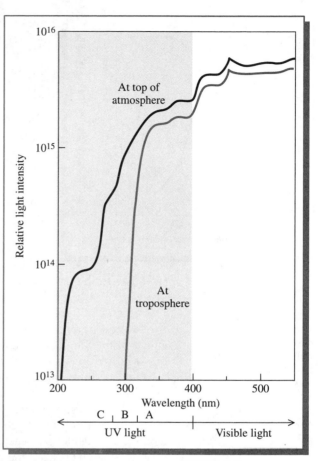

FIGURE 1-9 The intensity of sunlight in the UV and in part of the visible region measured outside the atmosphere and in the troposphere. [Source: W. L. Chameides and D. D. Davis, *Chemical & Engineering News* (4 October 1982): 38–52. Copyright 1982 by the American Chemical Society. Reprinted with permission.]

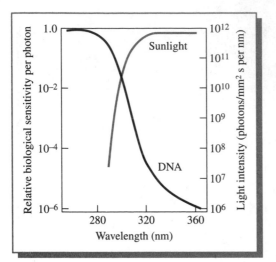

FIGURE 1-10 The absorption spectrum for DNA and the intensity of sunlight at ground level versus wavelength. The degree of absorption of light energy by DNA reflects its biological sensitivity to a given wavelength. [Source: Adapted from R. B. Setlow, *Proceedings of the National Academy of Science USA* 71 (1974): 3363–3366.]

substances used in sunscreen lotions either reflect or scatter sunlight (e.g., zinc oxide, titanium oxide) or absorb its UV component before it can reach the skin. Sunscreens are chosen that do not undergo an irreversible chemical reaction when they absorb sunlight, because this would quickly reduce the effectiveness of the application and because the reaction products could be toxic to the skin.

It is predicted that there will be a 1–2% increase in malignant skin cancer incidence for each 1% decrease in overhead ozone. For people who live at a latitude of about 45°N (e.g., the northern United States and southern Canada), there is eventually expected to be about an 11% increase in one type (basal cell carcinoma) of nonmalignant skin cancer and about a 22% increase in the other (squamous cell) type, due to the 6.6% decrease in stratospheric ozone that occurred between 1979 and 1992 over those regions.

Because of the long time lag (30–40 years) between exposure to UV and the subsequent manifestation of nonmalignant skin cancers, it is unlikely that effects from ozone depletion are observable as yet. The rise in skin cancer that has occurred in many areas of the world is probably due instead to greater amounts of time spent by people outdoors in the Sun over the past few decades. For example, the incidence of skin cancer among residents of Queensland, Australia, most of whom are light-skinned, rose to about 75% of the population as lifestyle changes increased their exposure to sunlight years before ozone depletion began. As a consequence of its experience with skin cancer, Australia has led the world in public health awareness of the need for protection from ultraviolet exposure.

In addition to skin cancer, UV exposure has been linked to several other human medical conditions. The front of the eye is the one part of the human anatomy where ultraviolet light can penetrate the human body. However, the cornea and lens filter out about 99% of the UV from light before it reaches the retina. Over time, the UV-B absorbed by the cornea and lens produces highly reactive molecules called free radicals that attack the structural molecules and can produce cataracts. Indeed, there is some evidence that increased UV-B levels give rise to an increased incidence of eye cataracts, particularly among the nonelderly. A 10% increase in UV-B is predicted to result in about 6% more cataracts among 50-year-olds. Increased UV-B exposure also leads to a suppression of the human immune system, probably with a resulting increase in the incidence of infectious diseases, although this has not yet been extensively researched.

However, sunlight does have some positive effects on human health. Vitamin D, which is synthesized from precursor chemicals by the absorption of UV by the skin, is an anticancer agent. Some controversial recent research indicates that moderate exposure to the Sun can reduce the incidence of several types of cancer.

Humans are not the only organisms affected by ultraviolet light. It is speculated that increases in UV-B exposure can interfere with the efficiency of photosynthesis, and plants may respond by producing less leaf, seed, and fruit. All organisms that live in the first 5 m or so below the surface in bodies of clear water would also experience increased UV-B exposure arising from ozone depletion and may be at risk. It is feared that production of the microscopic plants called phytoplankton near the surface of seawater may be at significant risk from increased UV-B; this would affect the marine food chain for which it forms the base. Experiments indicate that there is a complex interrelationship between plant production and UV-B intensity, since the latter also affects the survival of insects that feed off the plants. The recent worldwide drop in the population of frogs and other amphibians as well as the increase in deformities in frogs that do survive have been linked to increasing levels of UV. In particular, the UV-induced mortality of amphibian eggs laid in the springtime in shallow waters may be combining with habitat destruction to cause this decline.

In summary, the fraction of UV-B in sunlight that does reach the Earth's surface can cause increased skin cancer, both malignant and nonmalignant, in humans and may also lead to increases in infection. The effect of UV-B on plants and amphibians is also detrimental.

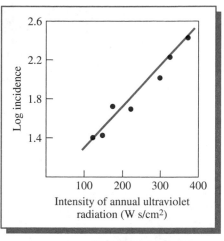

FIGURE 1-11 Incidence (logarithmic scale) for nonmelanoma skin cancer per 100,000 males versus annual UV light intensity, using data from various countries. [Source: Redrawn from D. Gordon and H. Silverstone, in R. Andrade et al., *Cancer of the Skin* (Philadelphia: W. B. Saunders, 1976), pp. 405–434.]

Variation in Light's Energy with Wavelength

As Albert Einstein was the first to realize, light can be considered not only a wave phenomenon but also to have particle-like properties in that it is absorbed (or emitted) by matter only in finite packets, now called **photons.** The quantity of energy, E, associated with each photon is related to the frequency, ν, and the wavelength, λ, of the light by the formulas

$$E = h\nu \quad \text{or} \quad E = hc/\lambda \quad \text{since} \quad \lambda\nu = c$$

Here h is Planck's constant (6.626218×10^{-34} J s) and c is the speed of light (2.997925×10^8 m/s). From the equation, it follows that *the shorter the wavelength of the light, the greater the energy it transfers to matter when absorbed.* Ultraviolet light is high in energy content, visible light is of intermediate energy, and infrared light is low in energy. Furthermore, UV-C is higher in energy than UV-B, which in turn is more energetic than is UV-A.

For convenience, the product hc in the equation above can be evaluated on a molar basis to yield a simple formula relating the energy absorbed by 1 mole of matter when each molecule in it absorbs one photon of a particular wavelength of light. If the wavelength is expressed in nanometers, the value of hc per mole is 119,627 kJ nm/mol, so the equation becomes

$$E = 119,627/\lambda \qquad \text{where } E \text{ is in kJ/mol if } \lambda \text{ is in nm}$$

The photon energies for light in the UV and visible regions are of the same order of magnitude as the enthalpy (heat) changes, ΔH°, of chemical reactions, including those in which atoms dissociate from molecules. For example, dissociation of molecular oxygen into its monatomic form requires an enthalpy change of 495 kJ/mol:

$$O_2 \longrightarrow 2\,O \qquad \Delta H^\circ = 495 \text{ kJ/mol}$$

Recall that ΔH° stands for the enthalpy change determined under standard conditions. To a good approximation, for a dissociation reaction ΔH° is equal to the energy required to drive the reaction. Since all the energy has to be supplied by one photon per molecule (see below), the corresponding wavelength for the light is

$$\lambda = 119,627\ \frac{\text{kJ nm}}{\text{mol}}\ \bigg/\ 495\ \frac{\text{kJ}}{\text{mol}} = 241 \text{ nm}$$

Thus any O_2 molecule that absorbs a photon from light of wavelength 241 nm or shorter has sufficient excess energy to dissociate:

$$O_2 + \text{UV photon } (\lambda < 241 \text{ nm}) \longrightarrow 2\,O$$

If energy in the form of light initiates a reaction, it is called a **photochemical reaction.** The oxygen molecule in the above reaction is variously said to be *photochemically dissociated* or *photochemically decomposed* or to have undergone *photolysis*.

Molecules that absorb light (usually in the ultraviolet or visible region) immediately undergo a change in the organization of their electrons. They are said to exist temporarily in an electronically **excited state,** and to denote this, their formulas are followed by a superscript asterisk (*). However, molecules generally do not remain in the excited state, and therefore do not retain the excess energy provided by the photon, for very long. Within a tiny fraction of a second they must either use the energy to react photochemically or return to their **ground state**—the lowest energy (most stable) arrangement of the electrons. They quickly return to the ground state either by themselves emitting a photon or by converting the excess energy into heat that becomes shared among several neighboring molecules as a result of collisions (i.e., molecules must "use it or lose it").

Consequently, molecules normally are not able to accumulate energy from several photons until they receive sufficient energy to react; all the excess energy required to drive a reaction usually must come from a single photon. Therefore light of wavelength less than 241 nm can result in the dissociation of O_2 molecules, but light of longer wavelength does not contain enough energy to promote the reaction at all, even though certain wavelengths of such light can be absorbed by the molecule (see Figure 1-7). In the case of an O_2 molecule, the energy from a photon of wavelength greater than 241 nm can, if absorbed temporarily, raise the molecules to an excited state, but the energy is rapidly converted to an increase in the energy of motion of the molecule and those that surround it.

$$O_2 + \text{photon } (\lambda > 241 \text{ nm}) \longrightarrow O_2{}^* \longrightarrow O_2 + \text{heat}$$

$$O_2 + \text{photon } (\lambda < 241 \text{ nm}) \longrightarrow O_2{}^* \longrightarrow 2\,O \quad \text{or} \quad O_2 + \text{heat}$$

PROBLEM 1-1

What is the energy, in kilojoules per mole, associated with photons having the following wavelengths? What is the significance of each of these wavelengths? [Hint: See Figure 1-6.]

(a) 280 nm   (b) 400 nm   (c) 750 nm   (d) 4000 nm

PROBLEM 1-2

The $\Delta H°$ for the decomposition of ozone into O_2 and atomic oxygen is $+105$ kJ/mol:

$$O_3 \longrightarrow O_2 + O$$

What is the longest wavelength of light that could dissociate ozone in this manner? By reference to Figure 1-6, determine the region of sunlight (UV, visible, or infrared) in which this wavelength falls.

PROBLEM 1-3

Using the enthalpy of formation information given below, calculate the maximum wavelength that can dissociate NO_2 to NO and atomic oxygen. Recalculate the wavelength if the reaction is to result in the complete dissociation into free atoms (i.e., $N + 2\,O$). Is light of these wavelengths available in sunlight? [Hint: Recall from introductory chemistry that, for any reaction, the enthalpy change equals the sum of the enthalpies of formation, $\Delta H_f^°$, of the products minus those of the reactants.]

$\Delta H_f^°$ values (kJ/mol): NO_2: $+ 33.2$; NO: $+ 90.2$; N: $+ 472.7$; O: $+ 249.2$

Of course, in order that a sufficiently energetic photon supply the energy to drive a reaction, it must be absorbed by the molecule. As you can infer from the examples of the absorption spectra of O_2 and O_3 (Figures 1-7 and 1-8), there are many wavelength regions in which molecules simply do not absorb significant amounts of light. Thus, for example, because ozone molecules do not absorb visible light near 400 nm, shining light of this wavelength on them does not cause them to decompose, even though 400-nm photons carry sufficient energy to dissociate them to atomic and molecular oxygen (see Problem 1-2). Furthermore, as discussed, the fact that molecules of a substance absorb photons of a certain wavelength and that such photons are sufficiently energetic to drive a reaction does not mean that the reaction will necessarily occur; the photon energy can be diverted by the molecule into other processes undergone by the excited state. Thus the availability of light with sufficient photon energy is a necessary but not a sufficient condition for reaction to occur.

Creation of Ozone in the Stratosphere

In this section, the formation of ozone in the stratosphere and its destruction by noncatalytic processes are analyzed. As we shall see, the formation reaction generates sufficient heat to determine the temperature in this region of the atmosphere. *Above* the stratosphere, the air is very thin and the concentration of molecules is so low that most oxygen exists in atomic form, having been dissociated from O_2 molecules by UV-C photons from sunlight. The eventual collision of oxygen atoms with each other leads to the re-formation of O_2 molecules, which subsequently dissociate photochemically again when more sunlight is absorbed.

In the stratosphere itself, the intensity of the UV-C light is much lower since much of it is filtered by the oxygen that lies above. In addition, since the air is denser than it is higher up, the molecular oxygen concentration is much higher in the stratosphere. For this combination of reasons, most stratospheric oxygen exists as O_2 rather than as atomic oxygen. Because the concentration of O_2 molecules is relatively large and the concentration of atomic oxygen is so small, the most likely fate of the stratospheric oxygen atoms created by the photochemical decomposition of O_2 is *not* their mutual collision to re-form O_2 molecules. Rather, they are more likely at such altitudes to collide and react with undissociated, intact diatomic oxygen molecules, an event that results in the production of ozone:

$$O + O_2 \longrightarrow O_3 + \text{heat}$$

Indeed, this reaction is the source of all the ozone in the stratosphere. During daylight hours, ozone is constantly being formed by this process, the rate of which depends upon the amount of UV light and therefore on the concentration of oxygen atoms and molecules at a given altitude.

At the bottom of the stratosphere, the abundance of O_2 is much greater than at the top because air density increases progressively as one approaches the surface. However, relatively little of the oxygen at this level is dissociated, and thus little ozone is formed because almost all the high-energy UV has been filtered from sunlight before it descends to this altitude. For this reason the ozone layer does not extend much below the stratosphere. Indeed, the ozone present in the lower stratosphere is largely formed at higher altitudes and transported there. In contrast, at the top of the stratosphere, the UV-C intensity is greater but the air is thin and therefore relatively little ozone is produced, since the oxygen atoms collide and react with each other rather than with the small number of intact O_2 molecules. Consequently, the production of ozone reaches a maximum where the product of UV-C intensity and O_2 concentration is maximum. The maximum density of ozone occurs lower—at about 25 km over tropical areas, 21 km over mid-latitudes, and 18 km over subarctic regions—since much of it is transported downward after its production. Collectively, most of the ozone is located in the region between 15 and 35 km, i.e., the lower and middle stratosphere, known informally as the **ozone layer** (see Figure 1-5a).

A third molecule, which we will designate as M, such as N_2 or H_2O or even another O_2 molecule, is required to carry away the heat energy generated in the collision between atomic oxygen and O_2 that produces ozone. Thus the reaction is written more realistically as

$$O + O_2 + M \longrightarrow O_3 + M + heat$$

The release of heat by this reaction results in the temperature of the stratosphere as a whole being *higher* than the air that lies below or above it, as indicated in Figure 1-5b.

Notice from Figure 1-5b that within the stratosphere the air at a given altitude is cooler than that which lies above it. The general name for this phenomenon is a **temperature inversion.** Because cool air is denser than hot air, it does not rise spontaneously due to the force of gravity; consequently, vertical mixing of air in the stratosphere is a very slow process compared to mixing in the troposphere. The air in this region therefore is *stratified*— hence the name *strato*sphere.

PROBLEM 1-4

Given that the total concentration of molecules in air decreases with increasing altitude, would you expect the *relative* concentration of ozone, on the ppb scale, to peak at a higher or a lower altitude or the same altitude compared to the peak for the absolute concentration of the gas?

Destruction of Stratospheric Ozone

The results for Problem 1-2 show that photons of light in the visible range and even in portions of the infrared range of sunlight possess sufficient energy to split an oxygen atom from a molecule of O_3. However, such photons are not efficiently absorbed by ozone molecules and consequently their dissociation by such light is not important, except in the lower stratosphere where little UV penetrates. As we have seen previously, ozone does efficiently absorb UV light with wavelengths shorter than 320 nm, and the excited state thereby produced does undergo a dissociation reaction. Thus absorption of a UV-C or UV-B photon by an ozone molecule in the stratosphere results in the decomposition of that molecule. This reaction accounts for much of the ozone destruction in the middle and upper stratosphere:

$$O_3 + UV \text{ photon } (\lambda < 320 \text{ nm}) \longrightarrow O_2^* + O^*$$

The oxygen atoms produced in the reaction of ozone with UV light have an electron configuration that differs from that of atoms with the lowest energy and therefore exist in an electronically excited state; the oxygen molecules also are produced in an excited state.

PROBLEM 1-5

By reference to the information in Problem 1-2, calculate the longest wavelength of light that decomposes ozone to O^* and O_2^*, given the following thermochemical data:

$$O \longrightarrow O^* \quad \Delta H^\circ = 190 \text{ kJ/mol}$$
$$O_2 \longrightarrow O_2^* \quad \Delta H^\circ = 95 \text{ kJ/mol}$$

[Hint: Express the overall reaction as a sum of simpler reactions and add the ΔH° values according to Hess' law.]

Most oxygen atoms produced in the stratosphere by photochemical decomposition of ozone or of O_2 subsequently react with intact O_2 molecules to re-form ozone. However, some of the oxygen atoms react instead with intact ozone molecules and in the process destroy them, since they are converted to O_2:

$$O_3 + O \longrightarrow 2 O_2$$

In effect, the unbonded oxygen atom takes one oxygen atom from the ozone molecule. This reaction is inherently inefficient, since, although it is an exothermic reaction, its activation energy is 17 kJ/mol, a sizable one for atmospheric reactions to overcome. Consequently, few collisions between O_3 and O occur with sufficient energy to result in reaction.

To summarize the processes, ozone in the stratosphere is constantly being formed, decomposed, and re-formed during daylight hours by a series of reactions that proceed simultaneously, though at very different rates depending upon altitude. Ozone is produced in the stratosphere because

there is adequate UV-C from sunlight to dissociate some O_2 molecules and thereby produce
oxygen atoms, most of which collide with
other O_2 molecules and form ozone. The
ozone gas filters UV-B and UV-C from sunlight but is destroyed temporarily by this
process or by reaction with oxygen atoms. The

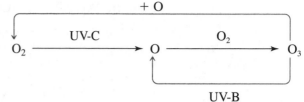

FIGURE 1-12 The
Chapman mechanism.

average lifetime of an ozone molecule at an altitude of 30 km is about half
an hour, whereas it is months in the lower stratosphere.

   Ozone is not formed below the stratosphere due to a lack of the UV-C
required to produce the O atoms necessary to form O_3, because this type of
sunlight has been absorbed by O_2 and O_3 in the stratosphere. Above the
stratosphere, oxygen atoms predominate and usually collide with other
O atoms to re-form O_2 molecules.

   The ozone production and destruction processes discussed constitute the
so-called **Chapman mechanism** (or cycle), shown in Figure 1-12. Recall that
the series of simple reaction steps that document how an overall chemical
process, such as ozone production and destruction, occurs at the molecular
level is called a **reaction mechanism.** Recently an additional mechanism that
could produce ozone in the high stratosphere has been proposed, but it has not
been accepted by all atmospheric scientists as contributing significant amounts
of ozone; it is the subject of Additional Problem 3 at the end of the chapter.

   Even in the ozone-layer portion of the stratosphere, O_3 is not the gas
of greatest abundance or even the dominant oxygen-containing species; its
relative concentration never exceeds 10 ppm. Thus the term *ozone layer*
is something of a misnomer. Nevertheless, this tiny concentration of ozone
is sufficient to filter all the remaining UV-C and much of the UV-B from
sunlight before it reaches the lower atmosphere. Perhaps the alternative
name *ozone screen* is more appropriate than ozone layer.

The Steady State in Atmospheric Reactions

It is not uncommon to find that the concentration of a substance, natural or
synthetic, in some compartment of the environment or in an organism does
not change much with time. This does not necessarily mean that there are
no inputs or sinks of the substance. More often, the concentration does not
vary much with time because the input rate and the rate at which the substance decays or is eliminated from some compartment in the environment
have become equal: we say that the substance has achieved a **steady state.**
Equilibrium is a special case of the steady state; it arises when the decay
process is the exact opposite of the input process.

   In the material to follow, we discover the implications of the steady
state in common situations involving reactive substances such as those
found in the ozone layer.

The Steady-State Approximation

If we know the nature of the creation and destruction reaction steps for a reactive substance, we can often algebraically derive a useful equation for its steady-state concentration.

As a simple example, consider the formation and destruction of oxygen atoms *above* the stratosphere. As mentioned previously, the atoms are formed by the photochemical dissociation of molecules of diatomic oxygen:

$$O_2 \longrightarrow 2\,O \tag{i}$$

The atoms re-form diatomic oxygen when two of them collide simultaneously with a third molecule, M, which can carry away most of the energy released by the newly formed O_2 molecule:

$$O + O + M \longrightarrow O_2 + M \tag{ii}$$

Recall from introductory chemistry that the rates of the individual steps in reaction mechanisms can be calculated from the concentrations of the reactants and from the **rate constant,** k, for the step. Thus the rate of reaction (i) equals $k_i\,[O_2]$. The rate constant k_i here incorporates the intensity of the light impinging upon the molecular oxygen. Thus, since two O atoms are formed for each O_2 molecule that dissociates,

$$\text{rate of formation of O atoms} = 2\,k_i\,[O_2]$$

The rate of destruction of oxygen atoms by reaction (ii) is

$$\text{rate of destruction of O atoms} = 2\,k_{ii}[O]^2\,[M]$$

where we square the oxygen atom concentration because two of them are involved as reactants in the step.

The net rate of change of the O atom concentration with time equals the rate of its formation minus the rate of its destruction:

$$\text{rate of change of } [O] = 2\,k_i\,[O_2] - 2\,k_{ii}[O]^2\,[M]$$

When atomic oxygen is at a steady state, this net rate must be zero; thus the right-hand side of this equation must also be zero. As a consequence, it follows that

$$k_{ii}\,[O]^2\,[M] = k_i\,[O_2]$$

By rearrangement of this equation, we obtain a relationship between the steady-state concentrations of O and of O_2:

$$[O]^2_{ss}\,/[O_2]_{ss} = k_i/(k_{ii}[M])$$

We see now why the ratio of oxygen atoms to diatomic molecules increases as we go higher and higher above the stratosphere: it is because the air pressure drops, therefore [M] also drops, so the O_2 re-formation rate decreases.

Steady-State Analysis of the Chapman Mechanism

After this introduction, we now are ready to apply the steady-state analysis to the Chapman mechanism. The four reactions of concern are shown again below. Notice that the recombination of O atoms, i.e., reaction (ii), is not included because its rate in the mid- and low-stratosphere is not competitive with other reactions, since the oxygen atom concentration is small there.

$$O_2 \longrightarrow 2\,O \tag{1}$$

$$O + O_2 + M \longrightarrow O_3 + M \tag{2}$$

$$O_3 \longrightarrow O_2 + O \tag{3}$$

$$O_3 + O \longrightarrow 2\,O_2 \tag{4}$$

Noting that O is produced or consumed in all four reactions, we obtain four terms in its overall rate expression and assume it is in a steady state:

$$\text{rate of change of } [O] = 2\,\text{rate}_1 - \text{rate}_2 + \text{rate}_3 - \text{rate}_4 = 0 \tag{A}$$

Other useful information about concentrations can be obtained by considering the steady-state expression for the ozone concentration:

$$\text{rate of change of } [O_3] = \text{rate}_2 - \text{rate}_3 - \text{rate}_4 = 0 \tag{B}$$

If we add the expressions for the rates of change in [O] and in [O_3], i.e., equations (A) and (B), we find that the rates for reactions (2) and (3) cancel, and we obtain

$$2\,\text{rate}_1 - 2\,\text{rate}_4 = 0$$

Using the expressions for these two rates in terms of reactant concentrations, we find

$$2\,k_1[O_2] - 2\,k_4[O_3][O] = 0$$

or

$$[O_3][O] = k_1[O_2]/k_4 \tag{C}$$

Another useful expression can be obtained by subtracting equation (B) from (A). We obtain

$$2\,\text{rate}_1 - 2\,\text{rate}_2 + 2\,\text{rate}_3 = 0$$

which by rearrangement and cancellation becomes

$$\text{rate}_3 = \text{rate}_2 - \text{rate}_1$$

It is known from experiment that rate_2 (and rate_3) are much larger than rate_1, so the latter can be neglected here, giving simply

$$\text{rate}_3 = \text{rate}_2$$

Using the expressions for these two reaction rates in terms of the concentrations of their reactants,

$$k_3[O_3] = k_2[O][O_2][M]$$

Rearranging this equation, we can solve for the ratio of ozone to atomic oxygen:

$$[O_3]/[O] = k_2[O_2][M]/k_3 \tag{D}$$

Equations (C) and (D) give us two equations in the two unknowns, $[O]$ and $[O_3]$. Multiplying their left sides and equating the result to the product of their right sides eliminates $[O]$ and leaves us with an equation for the ozone concentration:

$$[O_3]^2 = [O_2]^2[M] k_1 k_2/k_3 k_4$$

or, taking the square root of both sides, we obtain an expression for the steady-state concentration of ozone in terms of the diatomic oxygen concentration:

$$[O_3]_{ss}/[O_2]_{ss} = [M]^{0.5}(k_1 k_2/k_3 k_4)^{0.5} \tag{E}$$

Thus the steady-state ratio of ozone to diatomic oxygen depends on the square root of the air density through $[M]$. The ratio is also proportional to the square root of the product of the rate constants for the reactions (1) and (2) in which atomic oxygen and then ozone are produced, and inversely proportional to the square root of the product of the ozone destruction reaction rate constants. Substitution of numerical values for the rate constants k and $[M]$ into equation (E) predicts the correct order of magnitude for the ozone/diatomic oxygen ratio, i.e., about 10^{-4} in the mid-stratosphere. Ozone never is the main oxygen-containing species in the atmosphere, not even in the "ozone layer."

Equation (E) predicts that the concentration of ozone relative to diatomic oxygen should fall slowly as we climb in the atmosphere, given that it is proportional to the square root of the air density, through the $[M]$ dependence. This occurs because the formation reaction of ozone, through step (2), will slow down as $[M]$ declines. This decline with increasing altitude is observed in the upper stratosphere and above. Below about 35 km, however, the more important change in the terms of equation (E) involves k_1, and consequently the $[O_3]/[O_2]$ ratio is not simply proportional to $[M]^{0.5}$.

The rate constant k_1 incorporates the intensity of sunlight capable of dissociating diatomic oxygen into its atoms. Since the UV-C sunlight required ($\lambda < 242$ nm) is successively filtered by absorption as the light descends toward the Earth's surface, the value of k_1 declines especially rapidly in the low stratosphere and below. Thus the concentration of ozone predicted by applying the steady-state analysis to the Chapman mechanism successfully predicts that the ozone concentration will peak in the stratosphere. However, as discussed, the actual peak of ozone concentration (~25 km above the equator) occurs rather lower in the stratosphere than the altitude of maximum production (~40 km) because horizontal air movement transports ozone downward.

Substitution of equation (E) into (C) allows us to deduce an expression for the steady-state concentration of free oxygen atoms:

$$[O]_{ss} = (k_1 k_3 / k_2 k_4)^{0.5} / [M]^{0.5}$$

Thus the concentration of atomic oxygen is predicted to increase with altitude as [M] declines—as in our previous analysis for the upper atmosphere—and as k_1 and k_3 increase, since UV light intensity increases with increasing altitude. Indeed, atomic oxygen dominates over ozone at high altitudes, whereas below about 50 km, ozone is always dominant. These trends are illustrated in Figure 1-13.

The production of ozone through reaction (2) is critically dependent on the supply of free oxygen atoms in reaction (1). The rate of oxygen atom production, in turn, is highly dependent on the intensity of UV-C sunlight. As we have noted, this intensity falls sharply as we descend through the stratosphere. The UV-C light intensity also depends strongly on latitude,

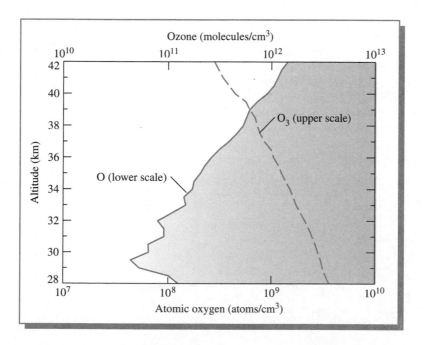

FIGURE 1-13 Concentrations of atomic oxygen and ozone versus altitude on December 2, 1977, at Palestine, Texas (32°N). (World Meteorological Organization, 1981). Note different scales for the concentrations. [Source: J. N. Seinfeld and S. N. Pandis, *Atmospheric Chemistry and Physics* (New York: Wiley-Interscience, 1998).]

FIGURE 1-14 Comparison of stratospheric ozone concentrations as a function of altitude as predicted by the Chapman mechanism and as observed over Panama (9°N) on November 13, 1970. [Source: Adapted from J. N. Seinfeld and S. N. Pardis, *Atmospheric Chemistry and Physics* (New York: Wiley-Interscience, 1998).]

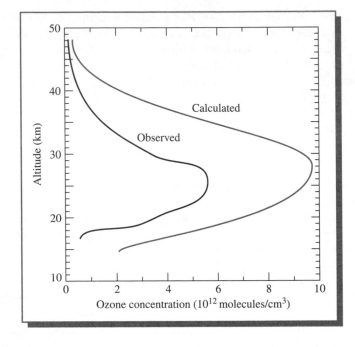

being strongest over the equator and declining continuously toward the poles. Thus ozone production is greatest over the equator.

The variation of ozone with altitude calculated by equation (E) is compared to the experimental trend in Figure 1-14. The qualitative behavior of the predicted variation is correct, but the predicted amount of ozone exceeds the observed—by about a factor of two near the peak concentration. Scientists eventually found that they had underestimated the rate of the ozone destruction reaction (4) by about a factor of four, since there are catalysts in the stratosphere that greatly speed up the overall reaction. These reactions are discussed in the next section.

PROBLEM 1-6

Consider the following three-step mechanism for the production and destruction of excited oxygen atoms, O*, in the atmosphere:

$$O_2 \xrightarrow{\text{light}} O + O^*$$

$$O^* + M \longrightarrow O + M$$

$$O^* + H_2O \longrightarrow 2\,OH$$

Develop an expression for the steady-state concentration of O* in terms of the concentrations of the other chemicals involved.

Catalytic Processes of Ozone Destruction

In the early 1960s it was realized that there are mechanisms for ozone destruction in the stratosphere in addition to the processes described in the Chapman mechanism. In particular, there exist a number of atomic and molecular species, designated in general as X, that react efficiently with ozone by abstracting (removing) an oxygen atom from it:

$$X + O_3 \longrightarrow XO + O_2$$

In those regions of the stratosphere where the atomic oxygen concentration is appreciable, the XO molecules react subsequently with oxygen atoms to produce O_2 and to re-form X:

$$XO + O \longrightarrow X + O_2$$

The **overall reaction** corresponding to this reaction mechanism is obtained by algebraically summing the successive steps that occur in air over and over again an equal number of times. In the case of the additional steps of the mechanism, the *reactants* in the two steps are added together and become the reactants of the overall reaction, and similarly for the products of the two steps:

$$X + O_3 + XO + O \longrightarrow XO + O_2 + X + O_2$$

Molecules that are common to both sides of the reaction equation, in this case X and XO, are then canceled and the common terms collected, yielding the balanced overall reaction:

$$O_3 + O \longrightarrow 2\,O_2 \quad \text{overall reaction}$$

Thus the species X are **catalysts** for ozone destruction in the stratosphere, since they speed up a reaction (here, between O_3 and O) but are eventually re-formed intact and are able to begin the cycle again—with, in this case, the destruction of further ozone molecules.

As previously discussed (Chapman cycle), this overall reaction can occur as a simple collision between an ozone molecule and an oxygen atom even in the absence of a catalyst, but almost all such direct collisions are ineffective in producing a reaction. The X catalysts greatly increase the efficiency of this reaction, i.e., they effectively increase the value of k_4 in equation (E) and thus decrease the steady-state concentration of ozone. All the environmental concerns about ozone depletion arise from the fact that we are inadvertently increasing the stratospheric concentrations of several X catalysts by the release at ground levels of certain gases, especially those containing chlorine. Such an increase in the catalyst concentration leads to a reduction in the concentration of ozone in the stratosphere by the mechanism shown earlier and by one to be discussed later.

Most ozone destruction by the catalytic mechanism (i.e., the combination of sequential steps) just described, hereafter designated **Mechanism I,** occurs in the middle and upper stratosphere, where the ozone concentration is low to start with. Chemically, all the X catalysts are **free radicals,** which are atoms or molecules containing an odd number of electrons. As a consequence of the odd number, one electron is not paired with another of opposite spin character (as occurs for all the electrons in almost all stable molecules). Free radicals are usually very reactive, since there is a driving force for their unpaired electron to pair with one of the opposite spin, even if it is located in a different molecule. The determination of the appropriate bonding structure for simple free radicals is described in Chapter 3.

An analysis of which free radical reactions are feasible in air and which are not is given in Box 1-1. The general conclusions from the analysis are that

- exothermic reactions involving one or more free radicals as reactant will have a very small activation energy and thus be fast, whereas

- endothermic reactions involving free radicals will have activation energies approximately equal to their endothermicities and thus be slower the greater the amount of heat required to drive the process.

Catalytic Destruction of Ozone by Nitric Oxide

The catalytic destruction of ozone occurs even in a "clean" atmosphere (one unpolluted by artificial contaminants) since small amounts of the X catalysts have always been present in the stratosphere. One important natural version of X—i.e., one of the species responsible for catalytic ozone destruction in a nonpolluted stratosphere—is the free-radical molecule **nitric oxide,** NO. It is produced when molecules of **nitrous oxide,** N_2O, rise from the troposphere to the stratosphere, where they may eventually collide with an excited oxygen atom produced by photochemical decomposition of ozone. Most of these collisions will yield $N_2 + O_2$ as products, but a few of them result in the reaction

$$N_2O + O^* \longrightarrow 2\,NO$$

We can ignore the possibility that NO produced in the troposphere will migrate to the stratosphere. As explained in Chapter 2, the gas is efficiently oxidized to nitric acid, which is soluble in rain and therefore readily washed out of the tropospheric air, before this process can occur.

The NO molecules that are the products of this reaction catalytically destroy ozone by extracting an oxygen atom from ozone and forming **nitrogen dioxide,** NO_2, i.e., they act as X in Mechanism I:

$$NO + O_3 \longrightarrow NO_2 + O_2$$
$$\underline{NO_2 + O \longrightarrow NO + O_2}$$
$$O_3 + O \longrightarrow 2\,O_2 \quad \text{overall}$$

BOX 1-1 | The Rates of Free-Radical Reactions

The rate of a given chemical reaction is affected by a number of parameters, most notably the magnitude of the activation energy required before the reaction can occur. Thus reactions with appreciable activation energies are inherently very slow processes and can often be ignored compared to alternative, faster processes for the chemicals involved. In gas-phase reactions involving simple free radicals as reactants, the activation energy exceeds that imposed by their endothermicity by only a small amount. Thus we can assume, conversely, that all exothermic free-radical reactions will have only a small activation energy (Figure 1a). Therefore, exothermic free-radical reactions usually are fast (providing, of course, the reactants exist in reasonable concentrations in the atmosphere). An example of an exothermic radical reaction with a small energy barrier is

$$Cl + O_3 \longrightarrow ClO + O_2$$

The activation energy here is only 2 kJ/mol.

Reactions involving the combining of two free radicals generally are exothermic, since a new bond is formed, so they too proceed quickly with little activation energy, provided that the radical concentrations are high enough that the reactants do in fact collide with each other at a fast rate.

In contrast, endothermic reactions in the atmosphere will be much slower since the activation barrier must of necessity be much larger (see Figure 1b). At atmospheric temperatures, few if any collisions between the molecules would have energy sufficient to overcome this large barrier and allow reaction to occur. An example is the endothermic reaction:

$$OH + HF \longrightarrow H_2O + F$$

(continued on p. 30)

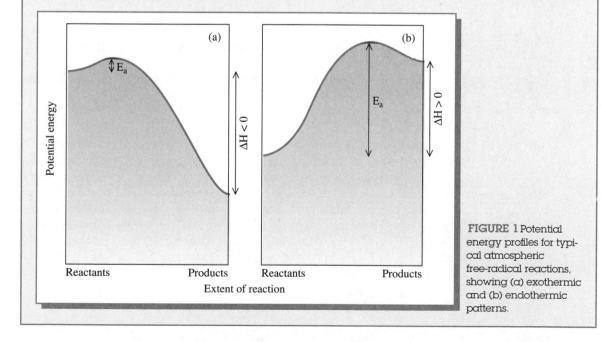

FIGURE 1 Potential energy profiles for typical atmospheric free-radical reactions, showing (a) exothermic and (b) endothermic patterns.

BOX 1-1   **The Rates of Free-Radical Reactions** *(continued)*

Its activation energy must be at least equal to its $\Delta H^{\circ} = +69$ kJ/mol, and consequently the reaction would be so very slow at stratospheric temperatures that we can ignore it completely.

PROBLEM 1

Draw an energy profile diagram, i.e., one similar to Figure 1b, for the abstraction from water of a hydrogen atom by ground-state atomic oxygen, given that the reaction is endothermic by about 69 kJ/mol. On the same diagram, show the energy profile for the reaction of O* with H_2O to give the same products, given that O* lies above ground-state atomic oxygen (O) by 190 kJ/mol. From these curves, predict why abstraction by O* occurs quickly but that by O is extremely slow in the atmosphere.

PROBLEM 1-7

Not all XO molecules such as NO_2 survive long enough to react with oxygen atoms; some are photochemically decomposed to X and atomic oxygen, which then reacts with O_2 to re-form ozone. Write out the three steps (including one for ozone destruction) for this process and add them together to deduce the net reaction. Does this sequence destroy ozone overall, or is it a *null cycle*, which involves no chemical change but simply a transformation of one type of energy to another?

Since there is one nitric oxide molecule and one ozone molecule in the reaction step that produces nitrogen dioxide and molecular oxygen, the rate of the step equals its rate constant times the concentration of nitric oxide times that of ozone:

$$\text{rate} = k\,[NO]\,[O_3]$$

The units for k and those for the concentrations must be consistent. For example, if concentrations are given in molecules per cubic centimeter for the two gases, then the rate of the reaction will be expressed in molecules per cubic centimeter per unit time, so the units for k should be the reciprocal of molecules per cubic centimeter times time. In this case it has been determined experimentally that, at $-30°C$, the rate constant k has the value 6×10^{-15} with units of cm^3/molecules s. Thus if the concentration of nitric oxide is 1×10^9 molecules/cm^3 and that of ozone is 5×10^{12} molecules/cm^3, then by substitution we obtain

$$\text{rate} = 6 \times 10^{-15} \text{ cm}^3/\text{molecules s} \times 1 \times 10^9 \text{ molecules/cm}^3$$
$$\times 5 \times 10^{12} \text{ molecules/cm}^3$$
$$= 3 \times 10^7 \text{ molecules/cm}^3 \text{ s}$$

Another important X catalyst in the stratosphere is the **hydroxyl free radical,** OH. It originates from the reaction of excited oxygen atoms, O^*, with water or with *methane*, CH_4, molecules:

$$O^* + CH_4 \longrightarrow OH + CH_3$$

The methane originates at the Earth's surface, and a small fraction survives sufficiently long to migrate to the stratosphere.

PROBLEM 1-8

Write out the two-step mechanism by which the hydroxyl free radical catalytically destroys ozone by Mechanism I. By adding the steps together, deduce the overall reaction.

PROBLEM 1-9

By analogy with its reaction with methane, write a balanced equation for the reaction by which O^* produces OH from water vapor.

PROBLEM 1-10

At an altitude of about 35 km, the average concentrations of O^* and of CH_4 are approximately 100 and 1×10^{11} molecules/cm^3, respectively, and the rate constant k for their reaction is approximately 3×10^{-10} cm^3/molecules s. Calculate the rate of destruction of methane in molecules per second per cubic centimeter and in grams per year per cubic centimeter under these conditions. [Hint: Recall that the rate law for a simple process is its rate constant k times the product of the concentrations of its reactant concentrations.]

Destruction of Ozone Without Atomic Oxygen by Other Mechanisms

A factor that minimizes the catalyzed gas-phase destruction of ozone by Mechanism I is the requirement for atomic oxygen to complete the cycle by reacting with XO to permit the regeneration of the X catalyst in a usable form:

$$XO + O \longrightarrow X + O_2$$

In the region where most ozone is found, namely the lower stratosphere (15–25 km altitude), the concentration of oxygen atoms is very low because little UV-C penetrates down this far and because the O_2 concentration there is so high that few oxygen atoms survive for long before being converted to ozone. Thus the gas-phase destruction of ozone by reactions that require atomic oxygen is sluggish in the lower stratosphere. Most ozone loss in the lower stratosphere occurs instead via the overall reaction

$$2\,O_3 \longrightarrow 3\,O_2$$

In the nonpolluted atmosphere, the most important mechanism for this process is that catalyzed by OH and the **hydroperoxy radical,** HOO, both of which react with ozone in a two-step sequence:

$$OH + O_3 \longrightarrow HOO + O_2$$

$$HOO + O_3 \longrightarrow OH + 2O_2$$

Notice that the second step of this mechanism is not identical to that involving the HOO in Mechanism I, but nevertheless it does regenerate the hydroxyl radical, OH.

There is a another general catalytic mechanism, henceforth designated **Mechanism II,** that depletes ozone in the lower stratosphere, particularly when the concentrations of the X catalysts are relatively high. First, two ozone molecules are destroyed by the catalysts discussed previously and by the same initial reaction:

$$X + O_3 \longrightarrow XO + O_2$$
$$X' + O_3 \longrightarrow X'O + O_2$$

We have used X$'$ to symbolize the catalyst in the second equation to indicate that it need not be chemically identical to X, the one in the first equation. In the steps that follow, the molecules XO and X$'$O that have added an oxygen atom react with each other. As a consequence, the catalysts X and X$'$ are ultimately regenerated, usually after the combined, but unstable, molecule XOOX$'$ has formed and been decomposed by either heat or light:

$$XO + X'O \longrightarrow [XOOX'] \longrightarrow X + X' + O_2$$

(By convention in chemistry, a species shown in square brackets is one with a transient existence.) We shall see several examples of catalytic Mechanism II in operation in the ozone holes and in the mid-latitude lower stratospheres. Mechanisms I and II are summarized in Figure 1-15.

FIGURE 1-15 Summary of catalytic ozone destruction by Mechanisms I and II.

Mechanism I

$$X + O_3 \rightarrow XO + O_2$$
$$\underline{XO + O \rightarrow X + O_2}$$
$$O_3 + O \rightarrow 2\,O_2 \qquad \text{overall}$$

Mechanism II

$$X + O_3 \rightarrow XO + O_2$$
$$X' + O_3 \rightarrow X'O + O_2$$
$$\underline{XO + X'O \rightarrow \rightarrow X + X' + O_2}$$
$$2\,O_3 \rightarrow 3\,O_2 \qquad \text{overall}$$

Finally, we note that while the rate of production of ozone from oxygen depends only upon the concentrations of O_2 and O_3 and on the intensity of UV light at a given altitude, what determines the rate of ozone destruction is somewhat more complex. The rate of ozone decomposition by UV-B or by catalysts depends upon ozone's concentration multiplied by either the sunlight intensity or the catalyst concentration. In general, the concentration of ozone will rise until the net rate of destruction just meets the rate of production and then will remain constant at this steady-state level as long as the intensity of sunlight remains the same. If, however, the rate of destruction is temporarily increased by the introduction of additional molecules of a catalyst, the steady-state concentration of ozone must then decrease to a new, lower value at which the rates of formation and destruction are again equal. However, it should be clear from the discussion that, due to its constant re-formation reactions, atmospheric ozone cannot be permanently and totally destroyed, no matter how great the level of catalyst. It should also be realized that any decrease in the concentration of ozone at higher altitudes allows more UV penetration to lower altitudes, which produces more ozone there; thus there is some "self-healing" of total ozone loss.

Atomic Chlorine and Bromine as X Catalysts

The decomposition of synthetic chlorine-containing gases in the stratosphere over the last few decades has generated a substantial amount of **atomic chlorine,** Cl, in this region. As the stratospheric chlorine concentration increases, so does the potential for ozone destruction, since the free radical Cl is an efficient X catalyst.

However, synthetic gases are not the only suppliers of chlorine to the ozone layer. There always has been some chlorine in the stratosphere as a result of the slow upward migration of the **methyl chloride** gas, CH_3Cl (also called *chloromethane*), produced at the Earth's surface, mainly in the oceans as a result of the interaction of chloride ion with decaying vegetation. Recently another large source of methyl chloride, from tropical plants, has been discovered; this may be the missing source of the compound for which scientists had been searching.

Only a portion of the methyl chloride gas is destroyed in the troposphere. When intact molecules of it reach the stratosphere, they are photochemically decomposed by UV-C or attacked by OH radicals. In either case, atomic chlorine, Cl, is eventually produced.

$$CH_3Cl \xrightarrow{\text{UV-C}} Cl + CH_3$$

or

$$OH + CH_3Cl \longrightarrow Cl + \text{other products}$$

Chlorine atoms are efficient X catalysts for ozone destruction by Mechanism I:

$$Cl + O_3 \longrightarrow ClO + O_2$$

$$ClO + O \longrightarrow Cl + O_2$$

$$O_3 + O \longrightarrow 2\,O_2 \quad \text{overall}$$

Each chlorine atom can catalytically destroy many tens of thousands of ozone molecules in this manner. At any given time, however, the great majority of stratospheric chlorine normally exists not as Cl, nor as the free radical **chlorine monoxide,** ClO, but as a form that is not a free radical and that is inactive as a catalyst for ozone destruction. The two main **catalytically inactive** (or *reservoir*) molecules containing chlorine in the stratosphere are **hydrogen chloride** gas, HCl, and **chlorine nitrate** gas, $ClONO_2$.

The chlorine nitrate is formed by the combination of chlorine monoxide and nitrogen dioxide; after a few days or hours, a given $ClONO_2$ molecule is photochemically decomposed back to its components, and thus the catalytically active ClO is re-formed.

$$ClO + NO_2 \underset{\text{sunlight}}{\rightleftharpoons} ClONO_2$$

However, under normal circumstances, more chlorine exists as $ClONO_2$ than as ClO at any given time. (Processes similar to this reaction occur for several other constituents of the stratosphere; as we shall see at the end of Chapter 3, the reactions are easily systematized, thereby greatly reducing the number of processes that have to be learned.)

The other catalytically inactive form of chlorine, HCl, is formed when atomic chlorine abstracts a hydrogen atom from a molecule of stratospheric methane:

$$Cl + CH_4 \longrightarrow HCl + CH_3$$

This reaction is slightly endothermic, so its activation energy is nonzero, and it therefore proceeds at a slow but significant rate (see Box 1-1). The *methyl free radical*, CH_3, does not operate like the X catalysts since it combines with an oxygen molecule and is degraded eventually to carbon dioxide by reactions discussed in Chapter 3. Eventually, each HCl molecule is reconverted to the active form, i.e., chlorine atoms, by reaction with the hydroxyl radical:

$$OH + HCl \longrightarrow H_2O + Cl$$

Again, usually much more chlorine exists as HCl than as atomic chlorine at any given time under normal conditions.

When the first predictions concerning stratospheric ozone depletion were made in the 1970s, it was not realized that about 99% of stratospheric chlorine usually is tied up in the inactive forms. When the existence of inactive chlorine was discovered in the early 1980s, the predicted amounts of stratospheric ozone loss in the future were lowered appreciably. As we shall see, however, inactive chlorine can temporarily become activated and massively destroy ozone, a discovery that was not made until the late 1980s.

Although there has always been some chlorine in the stratosphere due to the natural release of CH_3Cl from the surface, in recent decades this has been completely overshadowed by much larger amounts of chlorine produced from synthetic chlorine-containing gaseous compounds that are released into air during their production or use. Most of these substances are chlorofluoro-carbons (CFCs); their nature, production, usage, and replacements will be discussed in detail later in this chapter.

As with methyl chloride, large quantities of **methyl bromide,** CH_3Br, are also produced naturally and some of it eventually reaches the stratosphere, where it is decomposed photochemically to yield atomic bromine. Like chlorine, bromine atoms can catalytically destroy ozone by Mechanism I:

$$Br + O_3 \longrightarrow BrO + O_2$$
$$BrO + O \longrightarrow Br + O_2$$

In contrast to chlorine, almost all the bromine in the stratosphere remains in the active free-radical forms Br and BrO, since the inactive forms, **hydrogen bromide,** HBr, and **bromine nitrate,** $BrONO_2$, are efficiently decomposed photochemically by sunlight. In addition, the formation of HBr from the attack of atomic bromine on methane is a slower reaction than is the analogous process involving atomic chlorine since it is much more endothermic and therefore has a higher activation energy:

$$Br + CH_4 \longrightarrow HBr + CH_3$$

A lower percentage of stratospheric bromine exists in inactive form than does chlorine because of the slower speed of this reaction and because of the efficiency of the photochemical decomposition reactions. For that reason, stratospheric bromine is more efficient at destroying ozone than is chlorine (by a factor of 40 to 50), but there is much less of it in the stratosphere, so overall it is less important.

When molecules such as HCl and HBr eventually diffuse from the stratosphere back into the upper troposphere, they dissolve in water droplets and are subsequently carried to lower altitudes and are transported to the ground by rain. Thus, although the lifetime of chlorine and bromine in the stratosphere is long, it is not infinite, and the catalysts are eventually removed. However, the average chlorine atom destroys about 10,000 molecules of ozone before it is removed!

The Ozone Hole and Other Sites of Ozone Depletion

As discussed previously, scientists discovered in 1985 that stratospheric ozone over Antarctica is reduced by about 50% for several months each year, due mainly to the action of chlorine. An episode of this sort, during which there is said to be a hole in the ozone layer, can occur from September to early November, corresponding to spring at the South Pole. The hole has been appearing since about 1979, as was shown in Figure 1-2, which illustrates the variation in the minimum September–October ozone concentrations at the Halley Bay research station in the Antarctic as a function of year. Extensive research in the late 1980s led to an understanding of the chemistry of this phenomenon.

The Activation of Catalytically Inactive Chlorine

The ozone hole occurs as a result of special polar winter weather conditions in the lower stratosphere, where ozone concentrations usually are highest, that temporarily convert all the chlorine that is stored in the catalytically inactive forms HCl and $ClONO_2$ into the active forms Cl and ClO. Consequently, the high concentration of active chlorine causes a large, though temporary, annual depletion of ozone.

The conversion of inactive to active chlorine occurs at the surface of particles formed by a solution of water, **sulfuric acid,** H_2SO_4, and **nitric acid,** HNO_3, the latter being formed by combination of OH with NO_2 gas. The same conversion reactions could potentially occur in the gas phase but are so slow there as to be of negligible importance; they become rapid only when they occur on the surfaces of cold particles.

In most parts of the world, even in winter, the stratosphere is cloudless. Condensation of water vapor into liquid droplets or solid crystals that would constitute clouds doesn't normally occur in the stratosphere since the concentration of water in that region is exceedingly small, although there are always small liquid droplets consisting largely of sulfuric acid present, as well as some solid sulfate particles. However, the temperature in the lower stratosphere drops so low ($-80°C$) over the South Pole in the sunless winter months that condensation does occur. The usual stratospheric warming mechanism—the release of heat by the $O_2 + O$ reaction—is absent because of the lack of production of atomic oxygen from O_2 and O_3 when there is total darkness. In turn, because the polar stratosphere becomes so cold during the total darkness at midwinter, the air pressure drops, since it is proportional to the Kelvin temperature, according to the ideal gas law, $PV = nRT$. This pressure phenomenon, in combination with the Earth's rotation, produces a **vortex,** a whirling mass of air in which wind speeds can exceed 300 km (180 miles) per hour. Since matter cannot penetrate the vortex, the air inside it is isolated and remains very cold for many months. At the South Pole the vortex is sustained well into the springtime (October). (The vortex around the North Pole usually breaks down in February or early March, before much sunlight returns to the area, but recently there have been exceptions to this generalization, as discussed later.)

The particles produced by condensation of the gases within the vortex form **polar stratospheric clouds,** or PSCs. As the temperature drops, the first particles to form are small ones containing water and sulfuric and nitric acids. Scientists are no longer sure if these so-called **Type 1** particles are crystalline solids, or a mixture of solid and liquid, or supercooled liquids, but since they are conventionally referred to as crystals, we shall follow this practice. Indeed, some of the solid ones correspond in composition to *nitric acid trihydrate*, $HNO_3 \cdot 3\ H_2O$. When the air temperature drops a few degrees further, below $-80°C$, a different, larger type of crystal—consisting mainly of frozen water ice and perhaps also nitric acid—also forms; they are called **Type 2** crystals.

Chemical reactions that lead ultimately to ozone destruction occur in a thin aqueous layer present at the surface of the PSC ice crystals. Upon contact, gaseous $ClONO_2$ reacts at the surface with water molecules to produce **hypochlorous acid,** HOCl:

$$ClONO_2(g) + H_2O(aq) \longrightarrow HOCl(aq) + HNO_3(aq)$$

Also in the aqueous layer, gaseous hydrogen chloride dissolves and forms ions:

$$HCl(g) \xrightarrow{\substack{\text{aqueous} \\ \text{layer}}} H^+(aq) + Cl^-(aq)$$

Reaction of the two forms of dissolved chlorine produces molecular chlorine, which escapes to the surrounding air:

$$Cl^-(aq) + HOCl(aq) \longrightarrow Cl_2(g) + OH^-(aq)$$

This process is illustrated schematically in Figure 1-16. Overall, the process corresponds to the net reaction

$$HCl(g) + ClONO_2(g) \longrightarrow Cl_2(g) + HNO_3(aq)$$

since the ions H^+ and OH^- re-form water. Similar reactions probably also occur on the surface of solid particles.

During the dark winter months, molecular chlorine accumulates within the vortex in the lower stratosphere and eventually becomes the predominant chlorine-containing gas. Once a little sunlight reappears in the very early Antarctic spring, or the air mass moves to the edge of the vortex where there is some sunlight, the Cl_2 molecules are decomposed by the light into atomic chlorine:

$$Cl_2 + \text{sunlight} \longrightarrow 2\ Cl$$

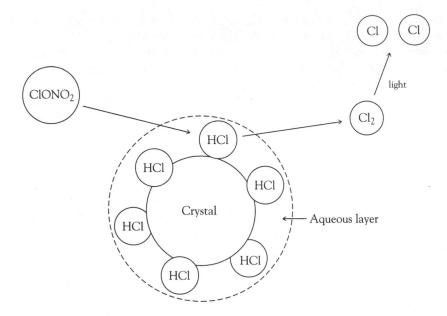

FIGURE 1-16 A scheme illustrating the production of molecular chlorine from inactive forms of chlorine in the winter and spring stratosphere in polar regions.

Similarly, any gaseous HOCl molecules released from the surface of the crystals undergo photolysis to hydroxyl radicals and atomic chlorine:

$$HOCl + sunlight \longrightarrow OH + Cl$$

Massive destruction of ozone by atomic chlorine then ensues.

PROBLEM 1-11

Given that diatomic chlorine gas is the stablest form of the element and that the ΔH_f° value for atomic chlorine is $+121.7$ kJ/mol, calculate the maximum wavelength of light that can dissociate diatomic chlorine into the monatomic form. Does such a wavelength correspond to light in the visible, the UV-A, or the UV-B region?

Since stratospheric temperatures above the Antarctic remain below $-80°C$ even in the early spring, the crystals persist for months. Any of the Cl that is converted back to HCl by the reaction with methane is subsequently reconverted to Cl_2 on the crystals and then back to Cl by sunlight. Inactivation of chlorine monoxide by conversion to chlorine nitrate does not occur, since all the NO_2 necessary for this reaction is temporarily bound as nitric acid in the crystals. The Type 2 crystals, and those Type 1 crystals that have grown large, move downward under the influence of gravity into the upper troposphere, thereby removing NO_2 from the lower stratosphere over the South Pole and further preventing the deactivation of chlorine. This

denitrification of the lower stratosphere extends the life of the Antarctic ozone hole and increases the ozone depletion.

Only when the PSCs and the vortex have vanished does chlorine return predominantly to the inactive forms. The liberation of nitric acid from the remaining Type 1 crystals into the gas phase results in its conversion to NO_2 by the action of sunlight:

$$HNO_3 + UV \longrightarrow NO_2 + OH$$

More importantly, air containing normal amounts of NO_2 mixes with polar air once the vortex breaks down in late spring. The nitrogen dioxide quickly combines with chlorine monoxide to form the catalytically inactive chlorine nitrate. Consequently, the catalytic destruction cycles largely cease operation and the ozone concentration builds back up toward its normal level a few weeks after the PSCs have disappeared and the vortex has ceased. Thus the ozone hole closes for another year, although the ozone levels nowadays never quite return to their natural levels, even in the fall. However, before the ozone levels build back up in the spring, some of the ozone-poor air mass can move away from the Antarctic and mix with surrounding air, temporarily lowering the stratospheric ozone concentrations in adjoining geographic regions, such as Australia, New Zealand, and the southern portions of South America.

Reactions That Create the Ozone Hole

In the lower stratosphere—the region where the PSCs form and chlorine is activated—the concentration of free oxygen atoms is small; few atoms are produced there on account of the scarcity of the UV-C light that is required to dissociate O_2. Furthermore, any atomic oxygen atoms produced in this way immediately collide with the abundant O_2 molecules to form O_3. Thus ozone destruction mechanisms based on the $O_3 + O \rightarrow 2\,O_2$ reaction, even when catalyzed, are not important here.

Rather, most of the ozone destruction in the ozone hole occurs via Mechanism II, with both X and X′ being atomic chlorine and with the overall reaction being $2\,O_3 \rightarrow 3\,O_2$. Thus the sequence starts with the reaction of chlorine with ozone:

$$\textit{step 1:} \quad Cl + O_3 \longrightarrow ClO + O_2$$

In Figure 1-17 the experimental ClO and O_3 concentrations are plotted as a function of latitude for part of the Southern Hemisphere during the spring of 1987. As anticipated, the two species display opposing trends, i.e., they anticorrelate very closely. At sufficient distances away from the South Pole (90°S), the concentration of ozone is relatively high and that of ClO is low, since chlorine is mainly tied up in inactive forms. However, as one travels closer to the Pole and enters the vortex region, the concentration of ClO suddenly becomes high and simultaneously that of O_3 falls off sharply: most of the chlorine has been activated and most of the ozone has consequently been destroyed. The latitude

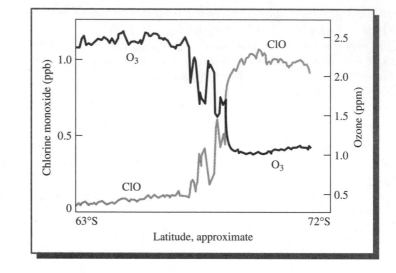

FIGURE 1-17 Stratospheric ozone and chlorine monoxide concentrations versus latitude near the South Pole on September 16, 1987. [Source: Reprinted with permission from P. S. Zurer, *Chemical & Engineering News* (30 May 1988): 16. Copyright 1988 by the American Chemical Society.]

at which the concentrations both change sharply marks the beginning of the ozone hole, which continues through to the region above the South Pole.

In the next reaction in the sequence, two ClO free radicals, produced in two separate step 1 events, combine temporarily to form a nonradical dimer, *dichloroperoxide*, ClOOCl (or Cl_2O_2):

> **step 2a:** $2\,ClO \longrightarrow Cl{-}O{-}O{-}Cl$

The rate of this reaction becomes important to ozone loss under these conditions because the chlorine monoxide concentration rises steeply due to the activation of the chlorine. Once the intensity of sunlight has risen to an appreciable amount in the Antarctic spring, the dichloroperoxide molecule ClOOCl absorbs UV and splits off one chlorine atom. The resulting ClOO free radical is unstable, so it subsequently decomposes (in about a day), releasing the other chlorine atom:

> **step 2b:** $ClOOCl + UV \text{ light} \longrightarrow ClOO + Cl$

> **step 2c:** $ClOO \longrightarrow O_2 + Cl$

Adding steps 2a, 2b, and 2c, we see that the net result is the conversion of two ClO molecules to atomic chlorine via the intermediacy of the dimer ClOOCl, which corresponds to the second stage of Mechanism II:

$$2\,ClO \longrightarrow [ClOOCl] \overset{\text{light}}{\longrightarrow} \longrightarrow 2\,Cl + O_2$$

By these processes ClO returns to the ozone-destroying form of chlorine, Cl. If we add this reaction to two times step 1, we obtain the overall reaction

$$2\,O_3 \longrightarrow 3\,O_2$$

Thus a complete catalytic ozone destruction cycle exists in the lower stratosphere under these special weather conditions, i.e., when a vortex is present. The cycle also requires very cold temperatures, since under warmer conditions the dimer ClOOCl is unstable and reverts back to two ClO molecules before it can undergo photolysis, thereby short-circuiting any ozone destruction. Before appreciable sunlight becomes available in the early spring, most of the chlorine exists as ClO and Cl_2O_2, since step 2b requires fairly intense light levels; such an atmosphere is said to be primed for ozone destruction.

About three-quarters of the ozone destruction in the Antarctic ozone hole occurs by this mechanism, in which chlorine is the only catalyst. This ozone destruction cycle contributes greatly to the creation of the ozone hole. Each chlorine destroys about 50 ozone molecules per day during the spring. The slow step in the mechanism is step 2a, which is the combination of two ClO molecules. Since the rate law for step 2a is second order in ClO concentration (i.e., its rate is proportional to the square of the ClO concentration), it proceeds at a substantial rate; hence the destruction of ozone is significant only when the ClO concentration is high. The sudden appearance of the ozone hole is consistent with the quadratic rather than linear dependence of ozone destruction on chlorine concentration by the Cl_2O_2 mechanism. Let us hope that there are not many more environmental problems whose effects will display such non-linear behavior and which would similarly surprise us!

A minor route for ozone destruction in the ozone hole involves Mechanism II with bromine as X′ and chlorine as X (or vice versa). Thus two ozone molecules are destroyed, one by a chlorine atom and one by a bromine atom. The ClO and BrO free-radical molecules produced in these processes then collide with each other and rearrange their atoms eventually to yield O_2 and atomic chlorine and bromine.

PROBLEM 1-12

Write out the mechanism by which one bromine atom and one chlorine atom catalytically destroy ozone, as in the process described above. Add the steps to determine the overall reaction.

PROBLEM 1-13

Suppose that the concentration of chlorine continues to rise in the stratosphere, but that the relative increase in bromine does not rise proportionately. Will the dominant mechanism involving dichloroperoxide or the chlorine plus bromine mechanism of Problem 1-12 become relatively more important as the destroyer of ozone in the Antarctic spring as a result of the increase?

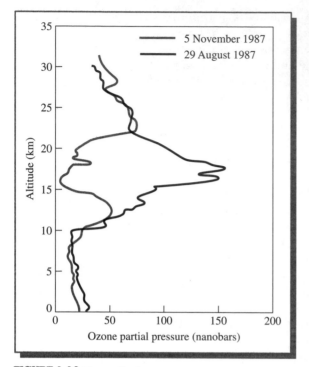

5 November 1987
29 August 1987

Altitude (km)

Ozone partial pressure (nanobars)

FIGURE 1-18 The vertical distribution of ozone over McMurdo, Antarctica, in late winter (August 29) and mid-spring (November 5) of 1987. [Source: Redrawn from B. J. Johnson, T. Deschler, and R. A. Thompson, *Geophysical Research Letters* 19 (1992): 1105–1108. Copyright by the American Geophysical Union.]

PROBLEM 1-14

Why is the mechanism involving dichloroperoxide of negligible importance in the destruction of ozone, compared to that which proceeds by ClO + O, in the upper levels of the stratosphere?

In the lower stratosphere above Antarctica, an ozone destruction rate of about 2% per day occurs each September due to the combined effects of the various catalytic reaction sequences. As a result, by early October almost all the ozone is wiped out between altitudes of 15 and 20 km, just the region in which its concentration normally is highest over the Pole. This result is illustrated in Figure 1-18, which shows the measured partial pressure of ozone as a function of altitude over the Antarctic in 1987 before (black line) and during (green line) the seasonal appearance of the hole.

In summary, the special vortex weather conditions in the lower stratosphere above the Antarctic in winter cause denitrification and lead to the conversion of inactive chlorine into Cl_2 and HOCl. These two compounds produce atomic chlorine when sunlight appears. The chlorine atoms efficiently destroy ozone via Mechanism II. Once the vortex disappears in the late spring, the ice particles on which the activation of chlorine compounds occurs disappear, the chlorine returns to inactive forms, and the hole heals.

The Size of the Antarctic Ozone Hole

Because (as will be explained later) the stratospheric concentration of chlorine continued to increase until the end of the twentieth century, the extent of Antarctic ozone depletion increased from the early 1980s until the late 1990s. There are several relevant measures of the extent of this depletion.

• One measure is the *surface area* covered by low ozone; Figure 1-3 shows the area that lies within the 220-DU contour line for the mid-September to mid-October period as a function of year. This area grew rapidly and approximately linearly from 1981 through 1985; the linear growth with lesser slope that occurred from the middle 1980s to the early 1990s seems now to have leveled off. The ozone hole now covers almost all the area of the vortex.

• Similarly, the linear decrease in the *minimum amount* of overhead ozone in the spring that occured from 1978 to 1985 was replaced subsequently by a slower decline since that time and may now have leveled off (see Figure 1-2; the very deep minima for 1993 and 1994 were due to the effects of the eruption of Mount Pinatubo, as explained later).

• The average *length of time* that ozone depletion occurs has also increased in recent years. Some reduction in ozone levels is now usually seen both in mid-winter (at least in the outer portions of the continent where there is some sunshine at that time) and in the summer as well as the spring, and, indeed, there is now some persistence of the depletion from one year to the next.

• The *vertical region* over which almost total ozone depletion occurs, 12–22 km, has not increased since the mid-1990s.

In 2002 the polar vortex was weaker and smaller than at any time over the previous decade, owing to unusual winds that mixed air from low to higher altitudes. As a result, the 2002 Antarctic ozone hole was smaller and of shorter duration than any for many years (see Figures 1-2 and 1-3). The hole in 2003 was similar in size and strength to those of the years preceding 2002.

The various reactions that lead to catalytic ozone destruction by atomic chlorine by various mechanisms are summarized in Figure 1-19. A possible

Ozone destruction step

$$O_3 + Cl \longrightarrow O_2 + ClO$$

Atomic chlorine reconstitution

Mid-stratosphere

$$ClO + O \longrightarrow Cl + O_2$$

Ozone hole/low stratosphere

$$2\,ClO \longrightarrow ClOOCl$$

$$ClOOCl + UV \longrightarrow ClOO + Cl$$

$$ClOO \longrightarrow Cl + O_2$$

Inactivation of chlorine

$$Cl + CH_4 \longrightarrow HCl + CH_3$$

$$ClO + NO_2 \longrightarrow ClONO_2$$

Activation of chlorine on particle surfaces

$$HCl(g) \xrightarrow{H_2O} H^+(aq) + Cl^-(aq)$$

$$H_2O(aq) + ClONO_2(g) \longrightarrow HOCl(aq) + HNO_3(aq)$$

$$Cl^-(aq) + HOCl(aq) \longrightarrow Cl_2(g) + OH^-(aq)$$

$$Cl_2(g) + sunlight \longrightarrow 2\,Cl(g)$$

$$H^+(aq) + OH^-(aq) \longrightarrow H_2O(aq)$$

FIGURE 1-19 A summary of the main ozone destruction reaction cycles operating in the Antarctic ozone hole.

additional mechanism by which ozone in the late spring (i.e., post-PSC) could be destroyed by large concentrations of chlorine nitrate is explored in Additional Problem 7.

Stratospheric Ozone Destruction over the Arctic Region

Given the similarity in climate, it is perhaps surprising that an ozone hole above the Arctic did not start to form at the same time as the one in the Antarctic. Episodes of partial springtime ozone depletion over the Arctic region have occurred several times since the mid-1990s. The phenomenon is less severe than in Antarctica: the reasons for this are that the stratospheric temperature over the Arctic does not fall as low or for as long and that air circulation to surrounding areas is not as limited. Consequently, the polar stratospheric clouds form less frequently over the Arctic and do not last as long. In the past, only the smaller (Type 1) crystals were formed; these are not large enough to fall out of the stratosphere and thereby denitrify it. How-ever, during the polar night, the chlorine nitrate and hydrogen chloride do react on the surface of the Type 1 particles to produce molecular chlorine, which then dissociates to atomic chlorine and, by reaction with an ozone molecule, becomes chlorine monoxide, as illustrated in Figure 1-20.

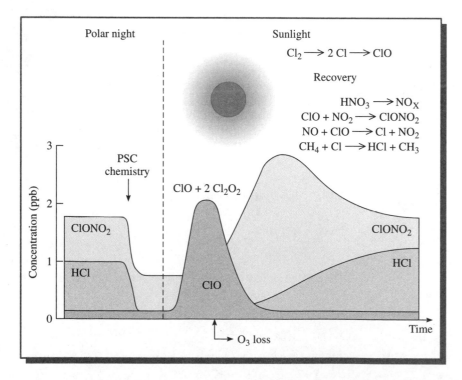

FIGURE 1-20 The evolution of stratospheric chlorine chemistry with time above the Arctic in winter and spring. [Source: Redrawn from C. R. Webster et al., *Science* 261 (1993): 1130.]

Before the mid-1990s, the vortex containing the cold air mass above the Arctic broke up by late winter; therefore NO_2-containing air mixed with vortex air before much sunlight returned to the polar region in the spring. Since the stratospheric air temperature usually rose above $-80°C$ by early March, the nitric acid in the particles was converted back to gaseous nitrogen dioxide before the intense spring sunlight could drive the Cl_2O_2 mechanism. Due to increases in NO_2 from both these sources, the activated chlorine was mostly transformed back to $ClONO_2$ before it could destroy much ozone (Figure 1-20). Thus the total extent of ozone destruction over the Arctic area was much less than that over the Antarctic.

Unfortunately, there were ominous signs in the mid-1990s that springtime conditions above the Arctic were changing for the worse, with the result that ozone depletion there accelerated in the lower stratosphere. The Arctic vortex in the winter and spring of 1995–1996 was exceptionally cold and persistent, resulting in chlorine-catalyzed losses of ozone amounting to about 130 DU as late as mid-April; the total overhead column of ozone at that date was about 200 DU. Large nitric acid–containing particles were formed and persisted long enough to fall out of the stratosphere, thereby denitrifying certain regions. A portion of the vortex passed across Great Britain during that March, producing record lows of 195 DU in northern Scotland. The magnitude of ozone depletion above the Arctic in the 1995–1996 winter was about the same as that observed over the South Pole in the early 1980s. Consequently, some atmospheric scientists have stated that an Arctic ozone hole formed in March of 1996. Since depletion of overhead ozone is never 100% complete, the definition of what conditions constitute a hole is somewhat arbitrary. Perhaps we should adopt the phrase of folk singer Joni Mitchell and refer to ulcerated ozone instead! The cold winter of 1999–2000 also saw considerable ozone loss—of about 100 DU compared to the mean in the 1980s—and 60% denitrification at the vortex core in the lower stratosphere.

In the winter of 1996–1997 the vortex above the Arctic was unusually strong and persisted into late March, i.e., well into the sunlit period, due to colder temperatures in the lower stratosphere. As a consequence, the usual springtime dynamic resupply of ozone from higher altitudes failed to occur, resulting in a record minimum in overhead ozone. However, the *chemical* loss of ozone was not as serious in the 1996–1997 winter as it was in 1995–1996, amounting to only about 50 DU.

For reasons that will be explained in Chapter 4, both the depletion of ozone and the increase in carbon dioxide levels themselves cool the stratosphere, and this will lead to even more depletion if cooling occurs in the springtime and thereby extends the period in which PSCs remain. Also, the often-irregular shape of the Arctic vortex means that there are frequent occasions when an arm of it passes over a sunlit area in late winter (before the bulk of the vortex is illuminated); temporary ozone depletion occurs within such arms. Some scientists predict that the recovery with time from ozone depletion will be slower in the Arctic than the Antarctic because of the cooling effects of CO_2 and O_3. Research reported in 2003 predicts that loss of H_2 to the atmosphere in a future hydrogen economy (discussed in Chapter 6) will produce more water in the stratosphere, thereby cooling it and producing more PSCs for longer periods over wider areas in both the Arctic and Antarctic.

The winters of 1995–1996 and 1996–1997 did not foreshadow a sustained period of longer stratospheric winters having late vortex destruction and increasing amounts of ozone depletion. Four of the five years following 1997 (all except 1999–2000) produced minimal springtime ozone losses above the Arctic. Although low ozone is likely to occur again during some winters over the next decade, current theoretical models predict that the Arctic hole will never rival in magnitude of ozone depletion that of the Antarctic. Students interested in following year-to-year developments of this phenomenon can consult the press releases, etc. on the NASA website listed at this textbook's site, www.whfreeman.com/envchem.

Notice in Figure 1-20 that, although HCl is converted completely in the PSCs, the $ClONO_2$, which is present in excess, is not completely eliminated in the stratosphere above the North Pole. Once the PSCs disappear as air temperatures rise, chlorine nitrate initially dominates since it forms rapidly from ClO and nitrogen dioxide. The reaction of atomic chlorine with methane is a slower process; thus the HCl concentration is slower to rise.

The chemistry underlying mid-latitude losses in stratospheric ozone is discussed in Box 1-2. A systemization of the various atmospheric chemical reactions discussed in this chapter is given in Chapter 3, after the corresponding reactions in the troposphere have been discussed.

Increases in UV at Ground Level

Experimentally, the amount of UV-B reaching ground level increases by a factor of three to six in the Antarctic during the early part of the spring because of the appearance of the ozone hole. Biologically, the most dangerous UV doses under hole conditions occur in the late spring (November and December), when the Sun is higher in the sky than in earlier months and low overhead ozone values still prevail. Abnormally high UV levels have also been detected in southern Argentina when ozone-depleted stratospheric air from the Antarctic traveled over the area.

Increases in ground-level UV-B intensity have also been measured in the spring months in mid-latitude regions in North America, Europe, and New Zealand. Calculations indicate that the extent of UV increases since the 1980s over mid- and high-latitude regions amounts to 6–14%. The most definitive experimental evidence comes from New Zealand, where long-term summertime increases in UV-B (but, as expected, not in UV-A) amounted to 12% by 1998–1999. The situation over mid-latitudes is complicated by two

BOX 1-2 | The Chemistry Behind Mid-Latitude Decreases in Stratospheric Ozone

As noted earlier, there was a worldwide decrease of several percent in the steady-state ozone concentration in the stratosphere over nonpolar areas during the 1980s and an additional short-term major decrease from 1992 to 1994 (Figure 1-4). In Figure 1-1, the variation of total overhead ozone with season (black curve) and the average trend in each month (green curve) are plotted for these regions. Notice that the extent of depletion closely mirrors the total ozone concentration for any given month; the greatest depletion occurs in the March–April period and the least in the early fall.

Scientists have had a harder time tracking down the source of the mid-latitude ozone depletion than that over polar regions. As in Antarctica, almost all the ozone loss in nonpolar regions occurs in the lower stratosphere. Some scientists have speculated that reactions leading to ozone destruction could occur not only on ice crystals but also on the surfaces of other particles present in the lower stratosphere. They suggested that the reactions could occur on cold liquid droplets consisting mainly of sulfuric acid that occur naturally in the lower stratosphere at all latitudes. The liquid droplets would have to be cold enough for them to take up significant amounts of gaseous HCl, or no net reaction would take place. There always exists a small background amount of the acid, due to the oxidation of the naturally occurring gas *carbonyl sulfide*, COS, some of which survives long enough to reach the stratosphere. However, the dominant though erratic source of the H_2SO_4 at these altitudes is direct injection into the stratosphere of sulfur dioxide gas emitted from volcanoes, followed by its oxidation to the

acid. Indeed, the steep decline in ozone in 1992–1993 followed the June 1991 massive eruption of Mt. Pinatubo in the Phillipines, and measurable ozone depletion was noted for several years after the eruption of El Chichon in Mexico in 1982. Note the dips significantly below the trend for the ozone levels in Figure 1-4; both of these periods temporarily increased the concentration of sulfuric acid droplets in the lower stratosphere.

The other relevant reaction that takes place on the surface of the sulfuric acid droplets results in some denitrification of stratospheric air. In the gas-phase steps of the sequence, ozone itself converts some nitrogen dioxide, NO_2, to *nitrogen trioxide*, NO_3, which then combines with other NO_2 molecules to form *dinitrogen pentoxide*, N_2O_5:

$$NO_2 + O_3 \longrightarrow NO_3 + O_2$$
$$NO_2 + NO_3 \longrightarrow N_2O_5$$

These gas-phase processes normally are reversible and do not remove much NO_2 from the air, but in the presence of high levels of liquid droplets, a conversion of N_2O_5 to nitric acid occurs instead:

$$N_2O_5 + H_2O\text{(droplets)} \xrightarrow{\substack{H_2SO_4 \\ \text{droplets}}} 2\,HNO_3$$

By this mechanism, much of the NO_2 that normally would be available to tie up chlorine monoxide as the nitrate $ClONO_2$ becomes unavailable for this purpose; hence a greater proportion of the chlorine atoms occur in the catalytically active form and destroy ozone. It should be realized, however, that even in the absence of particles, some NO_2 is converted to

(*continued on p. 48*)

BOX 1-2 | The Chemistry Behind Mid-Latitude Decreases in Stratospheric Ozone (continued)

nitric acid as a result of its reaction with the hydroxyl radical. This nitric acid eventually undergoes photochemical decomposition in daylight hours to reverse this reaction and to produce species that are catalytically active in ozone destruction.

In the mid-latitude lower stratosphere the most important catalytic ozone destruction reactions involving halogens employ Mechanism II, with X being atomic chlorine or bromine and X′ being the hydroxyl radical:

$$Cl + O_3 \longrightarrow ClO + O_2$$

$$OH + O_3 \longrightarrow HOO + O_2$$

$$ClO + HOO \longrightarrow HOCl + O_2$$

$$HOCl \xrightarrow{\text{sunlight}} OH + Cl$$

and similarly for the case where bromine replaces chlorine. The reaction sequence involving collision of ClO with BrO discussed for the Antarctic ozone hole is also operative here.

PROBLEM 1

Deduce the overall reaction equation for the reaction sequence shown above.

This mechanism explains why, in the current high-chlorine lower stratosphere, large volcanic eruptions can deplete mid-latitude stratospheric ozone for a few years, but it does not account for the overall trend of decreasing ozone in the last two decades. Some of the decrease is probably due to the mechanism operating on the background concentration of sulfuric acid particles in the lower stratosphere; its magnitude would have increased continuously in this time period since the chlorine levels were continuously increasing. Chlorine and bromine increases combined resulted in about a 4% decline in mid-latitude ozone levels in the 1979–1995 period. However, much of the gradual decline over mid-latitudes is believed to be due to other factors, such as springtime dilution of ozone-depleted polar air and its transport out of the polar regions, changes in the solar cycle, and both natural and anthropogenic changes in the pattern of atmospheric transport and temperatures.

facts: some UV-B is absorbed by the ground-level ozone produced by pollution reactions (as explained in Chapter 2), thereby masking any changes in UV-B due to small amounts of stratospheric ozone depletion, and records of UV received at the Earth's surface were started only in the 1990s.

The Chemicals That Cause Ozone Destruction

The recent increase in levels of stratospheric chlorine and bromine is due primarily to the release into the atmosphere of organic compounds containing chlorine and bromine that are **anthropogenic,** that is to say, they are man-made. These anthropogenic contributions to stratospheric halogen levels completely overshadow the natural input. The compounds at fault are those that do not have a **sink**—i.e., a natural removal process such as dissolution in rain or oxidation by atmospheric gases—in the troposphere. After a few years of traveling in the troposphere, they begin to diffuse into the stratosphere, where eventually they undergo photochemical decomposition by UV-C from sunlight, thereupon releasing their halogen atoms.

The variation in the total concentration of stratospheric chlorine and bromine atoms, expressed as the equivalent of chlorine in terms of ozone destruction power, measured over the last quarter-century and projected to the middle of the twenty-first century, is illustrated by the topmost curve in Figure 1-21. The peak chlorine equivalent concentration of about 3.8 ppb occurs in 1999 and is almost four times as great as the "natural" level that is due to methyl chloride and methyl bromide releases from the sea. The Antarctic ozone hole first appeared when the chlorine concentration reached about 2 ppb (dashed horizontal line).

CFC Decomposition Increases Atmospheric Chlorine

As is clear from inspection of Figure 1-21, the recent increase in stratospheric chlorine is due primarily to the use and release of **chlorofluorocarbons,** compounds containing only chlorine, fluorine, and carbon, which are commonly called **CFCs.** In the 1980s, about 1 million tonnes (i.e., metric tons, 1000 kg each) of CFCs were released annually to the atmosphere. These compounds are nontoxic, nonflammable, nonreactive, and have useful condensation

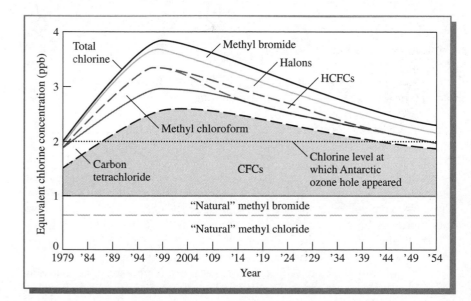

FIGURE 1-21 Actual and projected concentration of stratospheric chlorine versus time, showing the contributions of various gases. Note: Ozone-depleting effects of bromine atoms in halons and methyl bromide have been converted to their chlorine equivalents. [Source: DuPont.]

BOX 1-3   Formulas and Codes for Carbon Compounds (Including CFCs)

Since carbon is a Group 14 element and a nonmetal, its atoms each form four covalent bonds in most stable molecules. The simplest molecule of carbon is methane, CH_4, which is emitted into air predominantly as a result of biological decay processes. Stable compounds also are formed when any or all of the four hydrogen atoms are replaced by halogen atoms. Examples are CH_2Cl_2, CF_2Cl_2, and CHF_2Cl.

Carbon atoms can combine with each other to form extensive chains. The simplest chain has two carbon atoms held together by a single bond. Each carbon in such a unit forms three other bonds in order to bring the total bonds for each carbon to four. If all six atoms joined to the carbons are hydrogen, the gaseous compound C_2H_6, ethane, results. (Further background concerning organic chemistry is presented in the Appendix.) The formulas for such compounds can be written for each carbon separately in order that the atoms bonded to each carbon be shown. Thus ethane can also be written as CH_3CH_3, as $CH_3—CH_3$, or $H_3C—CH_3$ to show the carbon–carbon bond. Molecules containing only carbon and hydrogen, and in which all the linkages are single bonds, are called alkanes. All compounds with one or more of the hydrogens in an alkane replaced by halogens are stable species. Examples include $CH_3—CFCl_2$ and $CH_2F—CF_3$.

The chemical formula for an individual CFC, such as CFC-11, can be deduced by adding 90 to its code number, 11. The numbers in the sum, 101, correspond to the number of carbon, hydrogen, and fluorine atoms present in one molecule. Therefore, CFC-11 contains one carbon, zero hydrogens, and one fluorine. Since the total number of noncarbon atoms in substituted alkanes adds up to $2n + 2$, where n is the number of carbons, the number of chlorine atoms can be found by subtracting from $2n + 2$ the number of hydrogen plus fluorine atoms. Thus for CFC-11, $2n + 2 = 2(1) + 2 = 4$, so the number of chlorine atoms is $4 - (0 + 1) = 3$, and its formula is $CFCl_3$.

PROBLEM 1

Deduce the formulas for the compounds with the following code numbers:
(a) 12   (b) 113   (c) 123   (d) 134

PROBLEM 2

Deduce the code numbers for each of the following compounds:
(a) CH_3CCl_3 (b) CCl_4 (c) CH_3CFCl_2

properties (suiting them for use as coolants, for example); because of these favorable characteristics, they found a multitude of uses. There were several CFCs of commercial importance. We shall refer to them by their commercial code numbers 11, 12, etc., the significance of which is discussed in Box 1-3. We shall also return to the topic of CFCs in Chapter 4, since they act as greenhouse gases while in the troposphere and in the stratosphere.

CFC-12, which is pure CF_2Cl_2, is a gas at room temperature but is readily liquefied under pressure. Beginning in 1930 it was used as the circulating fluid in refrigerators, replacing the toxic gases *ammonia* and *sulfur dioxide*. Until recently it was also employed extensively in automobile air conditioners, from which much of it was released to the atmosphere during use and servicing. Today special equipment is used to capture the CFCs (and their modern replacements) when air conditioners in cars are serviced.

After World War II it was discovered that vaporizing CFC-12 from the liquid state could create bubbles in rigid plastic foams. The tiny bubbles of embedded CF_2Cl_2 make these products good thermal insulators, because the gas is a poor conductor of heat. In such rigid foams, the CFC-12 is trapped for decades before its release to the atmosphere occurs. By contrast, in the formation of foam sheets, the CFC-12 is immediately released.

The compound $CFCl_3$, called *CFC-11,* is a liquid that boils near room temperature. CFC-11 was used to blow holes in soft foam products and to make the rigid urethane foam products used to insulate refrigerators, freezers, and some buildings.

Both CFC-11 and CFC-12 were employed extensively as the propellants in aerosol spray cans. Because of concern for the ozone layer, this use was essentially eliminated in the late 1970s (except for a few crucial medical applications) by the United States, Canada, Norway, and Sweden. However, utilization of CFCs in spray cans continued elsewhere in the world. In the late 1980s this usage amounted to about one-fifth of worldwide CFC consumption and was the largest single source of CFC released to the atmosphere. Gases such as butane, often combined with a flame suppressant, have now replaced CFCs in aerosol spray cans.

The other CFC of major environmental concern was $CF_2Cl—CFCl_2$, called *CFC-113.* It was used extensively to clean residues from electronic circuit boards after their fabrication, to the extent of about 2 kg/m^2! Many major manufacturers have changed their fabrication processes so that no cleaning liquid of any type now has to be used.

CFCs have no tropospheric sink, so all of their molecules eventually rise to the stratosphere. In contrast to intuitive expectation, this vertical transport in the atmosphere is *not* affected by the fact that the mass of such molecules is greater than the average molecular mass of nitrogen and oxygen in air, because the differential force of gravity is much less than the forces due to the constant collisions of other molecules, which randomize the directions of even heavy molecules.

The CFC molecules eventually migrate to the middle and upper parts of the stratosphere where there is sufficient unfiltered UV-C from sunlight to photochemically decompose them, thereby releasing chlorine atoms. CFCs do not absorb sunlight with wavelengths greater than 290 nm and generally require wavelengths of 220 nm or less for photolysis. The CFCs must rise to the mid-stratosphere before decomposing, since UV-C does not penetrate to lower altitudes. Because vertical motion in the stratosphere is slow, their atmospheric lifetimes are long: 60 years on average for CFC-11 molecules, 105 years for CFC-12. CFC-11 is photochemically decomposed at lower altitudes than is CFC-12 and is therefore more able to destroy ozone at low stratospheric altitudes where the concentration of O_3 is greatest. It is because of the long stratospheric lifetimes of CFCs that the chlorine concentration in Figure 1-21 falls so slowly with time.

PROBLEM 1-15

Reactions of the type

$$OH + CF_2Cl_2 \longrightarrow HOF + CFCl_2$$

are conceivable tropospheric sinks for CFCs. Can you deduce why they don't occur, given that C—F bonds are much stronger than O—F?

The theory that CFCs cause most of the stratospheric ozone depletion was questioned in the past by some commentators because natural sources, especially seawater and volcanoes, spew much more chlorine into the atmosphere than do CFCs. They concluded that ozone depletion must therefore be due to natural, rather than artificial, causes.

However, the fact that invalidates this argument is that the natural sources emit almost all their chlorine into the troposphere rather than into the stratosphere. The sodium chloride emitted into ground-level air over oceans and the HCl emitted into the high troposphere and the very low stratosphere by volcanoes are both water soluble; hence they are rained out before they can rise to levels of the stratosphere where they could destroy ozone. Thus the total stratospheric chlorine introduced by natural processes is a small fraction of that which originates from CFCs. The finding that the concentrations of HF, a chemical that has no natural sources in the stratosphere, are equal to those expected *solely* as a result of the decomposition of CFCs confirms these substances as the source of stratospheric chlorine.

Other Chlorine-Containing, Ozone-Depleting Substances

Another widely used carbon–chlorine compound that lacks a tropospheric sink—although some of it ends up dissolving in ocean waters—is **carbon tetrachloride,** CCl_4, which also is photochemically decomposed in the stratosphere. Like CFCs, then, it is classified as an **ozone-depleting substance** (ODS). Commercially, carbon tetrachloride was used as a solvent and as an intermediate in the manufacture of CFC-11 and CFC-12; during production, some was lost to the atmosphere. Its use as a dry-cleaning solvent was discontinued in most developed countries some decades ago, but it was used until recently in many other countries. Because of its relatively long atmospheric lifetime (26 years), it will continue to make a significant contribution to stratospheric chlorine for several more decades (see Figure 1-21).

Methyl chloroform, CH_3—CCl_3, or *1,1,1-trichloroethane*, was produced in large quantities and used in metal cleaning in such a way that much of it was released into the atmosphere. Although about half of it is removed from the troposphere by reaction with the hydroxyl radical, the remainder survives long enough to migrate to the stratosphere. Because its average lifetime is only five years and its production has been largely phased out, its concentration in the atmosphere has declined rapidly since the 1990s. According to Figure 1-21, the contribution of methyl chloroform to stratospheric chlorine was substantial in the 1990s but by 2010 will become negligible.

Green Chemistry: The Replacement of CFC and Hydrocarbon Blowing Agents with Carbon Dioxide in Producing Foam Polystyrene

Polystyrene is a common polymer that is used to make many everyday items. This polymer varies in appearance from a rigid, solid plastic to foam polystyrene. Rigid plastic polystyrene is used in disposable silverware; audiocassette, CD, and DVD cases; and appliance casings. Foam polystyrene is utilized as insulation in coolers and houses, foam cups, meat and poultry trays, egg cartons, and in some countries it is still used in fast-food containers. Globally, about 10 million tons of polystyrene are produced on an annual basis, with approximately half being used to produce the foam form.

As indicated previously, CFCs have been employed as blowing agents for rigid plastic foams, and foam polystyrene is no exception. Low-molecular-weight hydrocarbons, such as pentane, have also been used as blowing agents; although these compounds do not deplete the ozone layer, they do contribute to ground-level smog when they are emitted into the atmosphere, as we will see in Chapter 3. Low-molecular-weight hydrocarbons are also very flammable and reduce worker safety.

The search for a replacement for CFC and hydrocarbon blowing agents led the Dow Chemical Company of Midland, Michigan, to develop a process employing 100% carbon dioxide as a blowing agent for polystyrene foam sheets. For this discovery Dow was the recipient of a Presidential Green Chemistry Challenge Award in 1996. Nonetheless, we will see in Chapter 4 that carbon dioxide is a greenhouse gas and thus contributes to the environmental problem of global warming, so we might wonder whether we are trading one environmental problem for another. However, waste carbon dioxide from other processes (natural gas production and the preparation of ammonia) that would otherwise be emitted into the atmosphere can be captured and used as a blowing agent. In addition, we will see that CFCs not only dramatically affect the ozone layer, but they also are greenhouse gases and are significantly more potent from this perspective than is carbon dioxide.

Dow Chemical found an added advantage to the polystyrene foam sheets made with carbon dioxide in that they remained flexible for a much longer time than those made with CFCs. This results in less breakage during use and a longer shelf life. In addition, foam sheets made with CFCs had to be degassed of the CFCs prior to recycling them, while carbon dioxide rapidly escapes from the polystyrene, leaving a sheet composed of 95% air and 5% polystyrene within a few days.

CFC Replacements

Compounds such as CFCs and CCl_4 have no tropospheric sinks because they do not undergo any of the normal removal processes: they are not soluble in water and thus they are not rained out from air; they are not attacked by the hydroxyl radical or any other atmospheric gases and so do not decompose; and they are not photochemically dissociated by either visible or UV-A light.

The compounds being implemented as the direct replacements for CFCs all contain hydrogen atoms bonded to carbon. Consequently a majority (though not necessarily 100%) of the molecules will be removed from the troposphere by a sequence of reactions that begins with hydrogen abstraction by OH:

$$OH + H-\overset{|}{\underset{|}{C}}- \longrightarrow H_2O + \text{C-centered free radical} \longrightarrow$$

$$CO_2 \text{ and HCl eventually}$$

Reactions of this type are discussed in more detail in Chapter 3. Because methyl chloride, methyl bromide, and methyl chloroform each contain hydrogen atoms, a fraction of such molecules are removed in the troposphere before they have a chance to rise to the stratosphere.

The *temporary* replacements for CFCs employed in the 1990s and the early years of the twenty-first century contain hydrogen, chlorine, fluorine, and carbon; they are called **HCFCs, hydrochlorofluorocarbons.** One important example is CHF_2Cl, the gas called *HCFC-22* (or just *CFC-22*), named and coded according to the same scheme used for CFCs. It is now employed in most domestic air conditioners and in some refrigerators and freezers, and has found some use in replacing CFC-11 in blowing foams such as those used in food containers. Since it contains a hydrogen atom and thus is mainly removed from the air before it can rise to the stratosphere, its long-term ozone-reducing potential is small—only 5% of that of CFC-11. This advantage is offset, however, by the fact that HCFC-22 decomposes to release chlorine more quickly than does CFC-11, so its *short-term* potential for ozone destruction is greater. Specifically, HCFC-22 results in 15% as much ozone destruction as does CFC-11 in the first 15 years after release.

But because most HCFC-22 is destroyed within a few decades after its release, it is responsible for almost *no long-term* ozone destruction, thus yielding an overall 5% value. However, most concerns about stratospheric ozone destruction center on the next few decades, before substantial reduction of stratospheric chlorine occurs from the phase-out of CFCs. Notice the contribution of HCFCs to the curve in Figure 1-21. They should be significant only from the late 1990s until about 2030. By 2000, HCFCs represented 6% of tropospheric chlorine, and their concentration was increasing linearly with time.

Reliance exclusively on HCFCs as CFC replacements would eventually lead to a renewed buildup of stratospheric chlorine, because the volume of HCFC consumption would presumably rise with increasing world population and affluence. Products that are entirely free of chlorine, and therefore pose no hazard to stratospheric ozone, will be the ultimate replacements for CFCs and HCFCs.

Hydrofluorocarbons, HFCs, substances that contain hydrogen, fluorine, and carbon, are the main long-term replacements for CFCs and HCFCs. The compound CH_2F—CF_3, called *HFC-134a,* has an atmospheric lifetime of several decades before finally succumbing to OH attack. HFC-134a, rather than CFC-12, now is used as the working fluid in refrigerators and some air conditioners, including those in automobiles. All HFCs eventually react to form hydrogen fluoride, HF. Unfortunately, one atmospheric degradation pathway for HFC-134a, and for several HCFCs as well, produces **trifluoroacetic acid,** TFA, CF_3—COOH, as an intermediate, which is then removed from the air by rainfall. Some scientists worry that TFA represents an environmental hazard to wetlands since it will accumulate in aquatic plants and could inhibit their growth. However, some of the TFA in the environment arises from the degradation under heating of polymers such as Teflon, not from CFC replacements. Polyfluorocarboxylic acids, of which the acid form of TFA is an example, have been used in certain commercial products but are now being phased out.

Another environmental concern about HFCs considers their accumulation in air after their inadvertent release during use. While present in the trosposphere, before they are destroyed, HFCs contribute to global warming by enhancing the greenhouse effect, a topic discussed in detail in Chapter 4. Outside of North America, industries usually use cyclopentane or isobutane rather than an HFC in refrigerant applications. Such hydrocarbons have a much shorter lifetime in air than HFCs. Some environmentalists hope that developing countries follow the hydrocarbon rather than the HFC route when they start to manufacture goods requiring coolants. In this context, it should be noted that fully fluorinated compounds are unsuitable replacements for CFCs because they have no tropospheric or stratospheric sinks and, if released into the air, they would contribute to global warming for very long periods of time.

Bromine-Containing, Ozone-Depleting Substances

Halon chemicals are bromine-containing, hydrogen-free substances such as CF_3Br and CF_2BrCl. Because they have no tropospheric sinks, they eventually rise to the stratosphere. There they are photochemically decomposed, with the release of atomic bromine (and chlorine, if present), which, as we have already discussed, is an efficient X catalyst for ozone destruction. Thus halons also are ozone-depleting substances. Bromine from halons will continue to account for a significant fraction of the ozone-destroying potential of stratospheric halogen catalysts for decades to come (Figure 1-21).

Halons operate to quell fires by releasing atomic bromine, which combines with the free radicals in the fire to form inert products. The halons release their bromine atoms even at only moderately high temperatures, since the C—Br bond is relatively weak. Since they are nontoxic and leave no residues upon evaporation, halons are very useful for fighting fires, particularly in inhabited, enclosed spaces, such as military aircraft, and spaces housing electronic equipment, such as computer centers. The substitution of other chemicals for halons in the testing of fire extinguishers drastically reduces halon emissions to the atmosphere, since only a minority of the releases are from the fighting of actual fires. Fine sprays of water can be substituted for halons in fighting many fires.

Some molecules of another commercial bromine-containing compound, methyl bromide, CH_3Br, eventually make their way to the stratosphere where each one is photodissociated to release atomic bromine. Methyl bromide is produced synthetically for use as a soil fumigant, as discussed below, and on that account its release into the troposphere has been increasing. Methyl bromide is also taken up by the oceans and by soils; overall, its average lifetime in air is about eight months. A sizable fraction of atmospheric CH_3Br is due to release during fires involving grass, crops, and trees. The fraction of methyl bromide emissions from its industrial production lies in the 10–40% range; scientists still find it difficult to quantify all the sources, natural and synthetic, emitting this chemical into the atmosphere.

Fluorine atoms are liberated in the stratosphere as a result of the decomposition of CFCs, halons, and other compounds, including HCFCs and HFCs. In principle, the fluorine atoms could catalytically destroy ozone (see Problem 1-17). However, the reaction of atomic fluorine with methane and other hydrogen-containing molecules in the stratosphere is rapid and produces HF, a very stable molecule. Because the H—F bond is much stronger than the O—H bond, the reactivation of fluorine by the attack of the hydroxyl radical on hydrogen fluoride molecules is very endothermic; consequently its activation energy is high and the reaction is extremely slow at atmospheric temperatures (see Box 1-1). Thus fluorine is quickly and permanently deactivated before it can destroy any significant amount of ozone.

PROBLEM 1-16

The free radical CF_3O is produced during the decomposition of HCF-134a. Show the sequence of reactions by which it could destroy ozone, acting as an X catalyst in a manner reminiscent of OH. (Note that it is too short-lived to actually destroy much ozone.)

PROBLEM 1-17

(a) Write the set of reactions by which atomic fluorine could operate as an X catalyst by Mechanisms I and II in the destruction of ozone. (b) An alternative to the second step of Mechanism I in the case of X = F is the reaction of FO with ozone to give atomic fluorine and two molecules of oxygen. Write out this mechanism and deduce its overall reaction. (c) Given that the mechanism in part (b) is similar to one discussed previously for X = OH, develop a generalized mechanism for the schemes in terms of X rather than F or OH.

International Agreements That Restrict ODSs

In contrast to almost all other environmental problems, such as global warming (Chapter 4), international agreement on remedies for stratospheric ozone depletion was obtained and successfully implemented in a fairly short period of time. As discussed, the use of CFCs in most aerosol products was banned in the late 1970s in North America and some Scandinavian countries. This decision was taken on the basis of predictions made by Sherwood Rowland and Mario Molina, chemists at the University of California, Irvine, concerning the effect of chlorine on the thickness of the ozone layer. There was no experimental indication of any depletion at the time of their prediction. Rowland and Molina, together with the German chemist Paul Crutzen, were jointly awarded the Nobel Prize in Chemistry in 1995 to honor their work in researching the science underlying ozone depletion.

The growing awareness of the seriousness of chlorine build-up in the atmosphere led to international agreements to phase out CFC production in the world. The breakthrough came at a conference in Montreal, Canada, in 1987 that gave rise to the **Montreal Protocol;** this agreement was strengthened at several follow-up conferences. As a result of this international agreement, all ozone-depleting chemicals are now destined for phase-out in all nations. All legal CFC *production* in developed countries ended in 1995. Developing countries have been allowed until 2010 to reach the same goal. Similarly, production of carbon tetrachloride and methyl chloroform has been phased out. Developed countries have agreed to end production of HCFCs by 2030, and developing countries by 2040, with no increases allowed after 2015.

Halon production was halted in developed countries in 1994 by the terms of the Montreal Protocol. However, use of existing stocks continues, as do releases from fire-fighting equipment. In addition, in the 1990s China and Korea—which, as developing countries, have until 2010 to terminate production—increased their production of these chemicals. For these reasons, the atmospheric concentrations of these substances continued to rise.

Methyl bromide has now been added to the list of ozone-depleting substances that will be banned. Developed countries are required to phase out its use by 2005, and developing countries by 2015, with possible exceptions if replacements cannot be found for essential uses. Developing countries have agreed that they will henceforth restrict their use of methyl bromide to the levels they employed in the 1990s.

Methyl bromide is sometimes used as the pesticide of last resort, when other alternatives fail. It has been widely used to sterilize soil—to control insects, nematodes, fungi, and plant diseases—by injecting it deep into soil covered temporarily by tarps before planting crops such as tomatoes, strawberries, grapes, tobacco, and flowers. Methyl bromide is also used to fumigate some crops such as dried fruit and nuts after harvest. The United States uses about 40% of current world production, mainly to fumigate soils and to kill insects in grain silos and mills. Currently practicable alternatives to the use of methyl bromide in these applications and to control termites in buildings are apparently less effective and more costly and time-consuming, although there has been some recent success in substituting methyl iodide for it. In some applications, *phosphine*, PH_3, or *carbonyl sulfide*, COS, or *sulfuryl fluoride*, SO_2F_2, have been used. All these substances, including methyl bromide, are quite toxic.

The 2005 phase-out of methyl bromide use in silos has been opposed by some in the North American milling industry on the basis that their competitors in developing countries are not required to follow suit. Indeed, in early 2003, the U.S. government applied to the United Nations for a two-year exemption from a complete phase-out in 2005 since it claimed that no safe, effective, and economically viable alternatives had been found for all uses. Developing countries have agreed that after 2002 they will restrict their use of methyl bromide to the levels employed in the mid-1990s. It is also used to treat insects found in wood products from developing countries.

As a direct result of the implementation of the gradual phase-out of ozone-depleting substances, the tropospheric concentration of chlorine peaked in 1993–1994 and had declined by about 5% by 2000. Much of the initial drop was due to the phase-out of methyl chloroform, which has a short atmospheric lifetime. The concentration of CFCs, especially CFC-12, is slow to drop because they were used in many applications such as foams and cooling devices that only slowly emit them to the atmosphere. The stratospheric chlorine equivalent level was predicted to have peaked, at less than 4 ppb, at the turn of the century, with a gradual decline predicted thereafter (see Figure 1-21). Observations in 2000 indicated that the actual chlorine content in the stratosphere had peaked, but the bromine abundance was still increasing. The slowness in the decline of the stratospheric chlorine level is due to

- the long time it takes for molecules to rise to the middle or upper stratosphere and to then absorb a photon and dissociate to atomic chlorine,

- the slowness of the removal of chlorine and bromine from the stratosphere, and

- the continued input of some chlorine and bromine into the atmosphere.

Because ozone is formed (and destroyed) in rapid natural processes, its level responds very quickly to a change in stratospheric chlorine concentration. Thus the Antarctic ozone hole probably will not continue to appear after the middle of the twenty-first century, i.e., once the chlorine equivalent concentration is reduced back to the 2-ppb level it had in the years before the hole began to form (Figure 1-21). Without the Montreal Protocol agreements, catastrophic increases in chlorine, to many times the present level, would have occurred, particularly since CFC usage and atmospheric release in developing countries would have increased dramatically. A further doubling of stratospheric chlorine levels would probably have led to the formation of a substantial ozone hole each spring over the Arctic region. And with the increase in ozone depletion would have come a catastrophic increase in skin cancers.

It has been pointed out that if bromine, rather than chlorine, had been used to make CFC-like refrigerants and blowing agents, stratospheric ozone depletion would be much worse, since atom for atom, bromine is much more effective because much more of it is normally in the catalytically active form.

PROBLEM 1-18

Given that their C—H bond energies are lower than that of CH_4, can you rationalize why ethane, C_2H_6, or propane, C_3H_8, is a better choice than methane to inactivate atomic chlorine in the stratosphere?

PROBLEM 1-19

No controls on the release of CH_3Cl, CH_2Cl_2, or $CHCl_3$ have been proposed. What does that imply about their atmospheric lifetimes, compared to those for CFCs, CCl_4, and methyl chloroform?

Clearly the stratosphere is a complicated system, and predictions concerning ozone depletion can be made only by sophisticated analyses that consider the effects of many reactions occurring simultaneously. It is not easy even to predict whether a given condition, such as nitric oxide emissions from airplane flights in this region, will result in an increase or a decrease in the amount of overhead ozone (see Additional Problem 10)! Only by extensive research over the last two decades have scientists been able to make semi-quantitative predictions regarding the causes of ozone loss.

Green Chemistry: Harpin Technology—Eliciting Nature's Own Defenses Against Diseases

Earlier in the chapter we learned that methyl bromide is used as a soil fumigant and some of these molecules find their way into the stratosphere, where they are involved in the destruction of the ozone layer. An interesting development, which offers an alternative to methyl bromide, is known as *harpin technology*. This technology was developed by EDEN Bioscience Corporation in Bothell, Washington, for which it was awarded a Presidential Green Chemistry Challenge Award in 2001.

Harpin is a naturally occurring bacterial protein that was isolated from the bacterium *Erwinia amylovora* at Cornell University. When applied to the stems and leaves of plants, harpin elicits the plant's natural defense mechanisms to diseases caused by bacteria, viruses, nematodes, and fungi. Hypersensitive response (HR), which is induced by harpin, is an initial defense by plants to invading pathogens that results in cell death at the point of infection, and the dead cells surrounding the infection act as a physical barrier to the spread of the pathogen. In addition, the dead cells may release compounds that are lethal to the pathogen.

Pests often build up immunity to pesticides. However, since harpin does not directly affect the pest, it is unlikely that immunity will occur with harpin. In addition to using traditional pesticides to control infestations in plants, more recently a second approach to this problem has been to develop genetically altered plants. The DNA in such plants is altered to provide the plant with a means to ward off various pests. Although this approach is often quite successful, it is not without its critics, especially in Europe, where genetically altered plants face serious restrictions. In contrast, harpin has no effect on the plant's DNA: it simply activates defenses that are innate to the plant.

Traditional pesticides are generally made by chemists employing lengthy chemical syntheses, which invariably create large quantities of often toxic waste. In addition, the compounds (chemical feedstocks) from which the pesticides are produced are derived from petroleum. Approximately 2.7% of all petroleum is used to produce chemical feedstocks; thus the production of these compounds is in part responsible for the depletion of this nonrenewable resource. In contrast, harpin is made from a genetically altered benign laboratory strain of the *Escherichia coli* bacteria through a fermentation process. After the fermentation is complete, the bacteria are destroyed and the harpin protein is extracted. Most of the wastes are biodegradable. Thus the production of harpin produces only nontoxic biodegradable wastes and does not require petroleum.

Harpin has very low toxicity. In addition, it is applied at a level of 0.002–0.06 kg/acre, which represents an approximately 70% reduction in quantity when compared to conventional pesticides. Harpin is rapidly decomposed by UV light and microorganisms, which is in part responsible for its lack of contamination and build-up in soil, water, and organisms and the fact that it leaves no residue in foods.

An added benefit of harpin is that it also acts as a plant growth stimulant. Harpin is thought to aid in photosynthesis and nutrient uptake, resulting in increased biomass, early flowering, and enhanced fruit yields. Harpin is sold as a 3% solution in a product called *Messinger*.

Review Questions

The questions listed below, and the comparable ones in succeeding chapters, are designed to test your knowledge mainly of some of the *factual* material presented in the chapter. Those involving green chemistry are given after this section.

The problems within the chapter and the more elaborate Additional Problems that follow these questions are designed to test your problem-solving abilities.

1. Which three gases constitute most of the Earth's atmosphere?

2. What is a *Dobson unit*? How is it used?

3. If the overhead ozone concentration at a point above the Earth's surface is 250 DU, what is the equivalent thickness in millimeters of pure ozone at 1.0 atm pressure ?

4. What range of altitudes constitutes the troposphere? the stratosphere?

5. What is the wavelength range for visible light? Does ultraviolet light have a shorter or longer wavelength than visible light?

6. Which atmospheric gas is primarily responsible for filtering sunlight in the 120- to 220-nm region? Which, if any, gas absorbs most of the Sun's rays in the 220- to 320-nm region? Which absorbs primarily in the 320- to 400-nm region?

7. What is the name given to the finite packets of light absorbed by matter?

8. What are the equations relating photon energy E to light's frequency ν and wavelength λ?

9. What is meant by the expression *photochemically dissociated* as applied to stratospheric O_2?

10. Write the equation for the chemical reaction by which ozone is formed in the stratosphere. What are the sources for the different forms of oxygen used here as reactants?

11. Write the two reactions that, aside from the catalyzed reactions, contribute most significantly to ozone destruction in the stratosphere.

12. What is meant by the phrase *excited state* as applied to an atom or molecule? Symbolically, how is an excited state signified?

13. Explain why the phrase *ozone layer* is a misnomer.

14. Define the term *free radical*, and give two examples relevant to stratospheric chemistry.

15. What are the two steps and the overall reaction by which X species such as Cl catalytically destroy ozone in the middle and upper stratosphere?

16. What is meant by the term *steady state* as applied to the concentration of ozone in the stratosphere? What is meant by the phrase *the steady-state approximation* for a substance?

17. Explain why, atom for atom, stratospheric bromine destroys more ozone than does chlorine.

18. Describe the process by which chlorine becomes activated in the Antarctic ozone hole phenomenon.

19. What is the principal four-step mechanism by which chlorine destroys ozone in the spring over Antarctica?

20. Explain why full-scale ozone holes have not yet been observed over the Arctic.

21. What are two effects on human health that scientists believe will result from ozone depletion?

22. Define the term *CFC* and write the formulas for the two most common CFCs. Give one use for each CFC you mention.

23. Define what is meant by a tropospheric *sink*.

24. Explain what HCFCs are and state what sort of reaction provides a tropospheric sink for them. Is their destruction in the troposphere 100% complete? Why are HCFCs not considered to be a suitable long-term replacements for CFCs?

25. What types of chemicals are proposed as long-term replacements for CFCs?

26. Chemically, what are *halons*? What was their main use?

27. What gases are being phased out according to the Montreal Protocol agreements?

28. Explain why ozone destruction via the reaction of O_3 with atomic oxygen does not occur to a significant degree in the lower stratosphere.

29. Write the mechanism and overall reaction by which ozone destruction occurs in the lower stratosphere by a catalytic process that does not involve chlorine or nitric oxide.

 # Green Chemistry Questions

1. The development of carbon dioxide as a blowing agent for foam polystyrene won a Presidential Green Chemistry Challenge Award.

(a) Which of the three focus areas (see the Introduction to Green Chemistry) for these awards does this one best fit into?

(b) List two of the twelve principles of green chemistry (see the Introduction to Green Chemistry) that are addressed by the green chemistry of the carbon dioxide process.

2. What environmental advantages does the use of carbon dioxide as a blowing agent have over the use of CFCs and hydrocarbons?

3. Does the carbon dioxide that is used as a blowing agent contribute to global warming?

4. The development of harpin won a Presidential Green Chemistry Challenge Award.

(a) Which of the three focus areas (see the Introduction to Green Chemistry) for these awards does this one best fit into?

(b) List four of the twelve principles of green chemistry (see the Introduction to Green Chemistry) that are addressed by the green chemistry in the use of harpin.

5. Why is there little concern that pests will develop immunity to harpin?

6. Why is harpin not expected to accumulate in the environment?

Additional Problems

1. **(a)** Prove that the number of moles of overhead ozone over a unit area on the Earth's surface is proportional to the height of the layer, as specified in the definition of Dobson units, and that 1 DU is equal to 1 matm cm. **(b)** Calculate the total mass of ozone that is present in the atmosphere if the average overhead amount is 350 DU and given that the radius of the Earth is about 5000 km. [Hints: The volume of a sphere, which you can approximate the Earth to be, is $4\pi r^3/3$. You may assume that ozone behaves as an ideal gas.]

2. The rate constants for the reactions of atomic chlorine and of hydroxyl radical with ozone are given by $3 \times 10^{-11} e^{-250/T}$ and $2 \times 10^{-12} e^{-940/T}$, where T is the Kelvin temperature. Calculate the ratio of the rates of ozone destruction by these catalysts at 20 km, given that at this altitude the average concentration of OH is about 100 times that of Cl and that the temperature is about $-50°C$. Calculate the rate constant for ozone destruction by chlorine under conditions in the Antarctic ozone hole when the temperature is about $-80°C$ and the concentration of atomic chlorine increases by a factor of 100 to about 4×10^5 molecules/cm^3.

3. It has been proposed by A. M. Wodkte and co-workers at the University of California, Santa Barbara, that an additional mechanism could exist for the creation of ozone in the high stratosphere. The sequence begins with the creation of (vibrationally) excited O_2 and ground-state atomic oxygen from the absorption of photons with wavelengths less than 243 nm. The O_2* reacts with a ground-state O_2 molecule to produce ozone and another atom of oxygen. What is the net reaction from these two steps? What do you predict is the fate of the two oxygen atoms, and what would be the overall reaction once this fate is included?

4. The Arrhenius equation (see your introductory chemistry textbook) relates reaction rates to temperature via the activation energy. Calculate the ratio of the rates at $-30°C$ (a typical stratospheric temperature) for two reactions having the same Arrhenius A factor and initial concentrations, one which is endothermic and has an activation energy of 30 kJ/mol and the other which is exothermic with an activation energy of 3 kJ/mol.

5. Using a steady-state analysis for $d[X]/dt$, determine an expression for the ratio of the concentration of XO to that of X for the two steps of Mechanism I.

6. Perform a steady-state analysis for $d[Cl]/dt$ and for $d[ClO]/dt$ in Mechanism II, i.e.,

$$Cl_2 \longrightarrow 2\ Cl \qquad (1)$$
$$Cl + O_3 \longrightarrow ClO + O_2 \qquad (2)$$
$$2\ ClO \longrightarrow 2\ Cl + O_2 \qquad (3)$$
$$ClO + NO_2 \longrightarrow ClONO_2 \qquad (4)$$

Obtain expressions for the steady-state concentrations of Cl and ClO, and hence for the rate of destruction of ozone.

7. J. A. Pyle at the University of Cambridge in England has proposed that ozone destruction in the late spring over northern latitudes may be initiated in the lower stratosphere by the photolysis of $ClONO_2$ to Cl and NO_3, followed by photolysis of the latter to NO and O_2. Deduce a catalytic ozone destruction cycle, requiring no atomic oxygen, that incorporates these reactions. What is the overall reaction? (Note that this mechanism is especially important in the spring in the polar ozone holes since initially the active chlorine is converted almost exclusively to $ClONO_2$ rather than predominantly to HCl because the former is much faster to re-form than is the latter.)

8. Deduce possible reaction step(s), none of which involve photolysis, for Mechanism II that follow(s) the $X + O_3 \rightarrow XO + O_2$ step so that the sum of all the mechanism's steps does not destroy or create any ozone.

9. When Mechanism II for ozone destruction operates with X = Cl and X′ = Br, the radicals ClO and BrO react to re-form atomic chlorine and bromine (see Problem 1-12). A fraction of the latter process proceeds by the intermediate formation of BrCl, which undergoes photolysis in daylight. At night, however, all the bromine eventually ends up as BrCl, which does not decompose and restart the mechanism until dawn. Deduce why all the bromine exists as BrCl at night, even though only a fraction of the ClO with BrO collisions yields this product.

10. Under conditions of low oxygen atom concentration, the radical HOO can react reversibly with NO_2 to produce a molecule of $HOONO_2$:

$$HOO + NO_2 \longrightarrow HOONO_2$$

(a) Deduce why the addition of nitrogen oxides to the lower stratosphere could lead to an *increase* in the steady-state ozone concentration as a consequence of this reaction. **(b)** Deduce how the addition of nitrogen oxides to the middle and upper stratosphere could *decrease* the ozone concentration there as a consequence of other reactions. **(c)** Given the information stated in parts (a) and (b), in what regions of the stratosphere should supersonic transport airplanes fly if they emit substantial amounts of nitrogen oxides in their exhaust? **(d)** How will your answer to part (c) be affected if you take into account the substantial amounts of sulfate particles emitted by the engines of such planes?

11. **(a)** What CFC number would be appropriate for methyl chloroform, CH_3CCl_3? **(b)** Show that 134 is the appropriate label for CH_2FCF_3. Why is an a or b designation also required to uniquely characterize the latter compound?

12. The chlorine dimer mechanism is not implicated in significant ozone destruction in the lower stratosphere at mid-latitudes, even when the particle concentration becomes enhanced by volcanoes. Deduce two reasons why this mechanism is not important under these conditions.

13. As discussed in the text, the seriousness of the Antarctic ozone hole in any given year can be analyzed in terms of (a) the minimum ozone concentration reached, (b) the average October (or mid-September to mid-October) ozone concentration, (c) the geographic area that the hole covers, or (d) the number of days or weeks that very low ozone values are recorded. By consulting several of the websites listed for this chapter at www.whfreeman.com/envchem/, construct up-to-date graphs of the variation in each of these parameters. Is the size of the annual hole showing signs yet of decreasing, according to any of these parameters?

14. Absorbance is defined as $\log(I_0/I)$, the logarithm of the inverse of the fraction of light passing through a sample. (I_0 is the light intensity incident on the sample and I is the light intensity that passes through the sample.) Absorbance depends on the concentration of molecules in the sample that absorb the light and the path length of the light through the sample. This is described by the Beer–Lambert law: $A = \varepsilon c l$, where ε is the molar absorptivity, c is the concentration in moles per liter, and l is the path length. Using the Beer–Lambert law, predict the increase in intensity of 300-nm light at the Earth's surface that would result from a 1% decrease in the concentration of ozone in the stratosphere. An estimate of the current absorbance at 300 nm can be obtained from Figure 1-9, which shows a decrease in light intensity of about two orders of magnitude at 300 nm, comparing the top of the atmosphere to the troposphere; thus assume $I/I_0 = 0.01$. Note that the answer obtained will be different from the 2% increase given in the text of the chapter, since that was for overall increased intensity throughout the UV-B, not just at 300 nm.

15. Consider the following set of compounds: $CFCl_3$, $CHFCl_2$, CF_3Cl, and CHF_3. Assuming that equal numbers of moles of each were released into the troposphere at ground level, rank these four compounds in terms of their potential to catalytically destroy ozone in the stratosphere. Explain your ranking.

16. In the chapter, hydrocarbons such as isobutane were mentioned as CFC replacements for refrigerant applications. Isobutane and other hydrocarbons are also being used increasingly as CFC replacements as the propellants in aerosol cans.

What are the advantages of using hydrocarbons instead of CFCs or HCFCs as aerosol propellants? What is the major disadvantage? What extra type of agent would need to be added to aerosol cans using hydrocarbon propellants to make them safer?

Further Reading

1. M. J. Molina and F. S. Rowland, "Stratospheric Sink for Chlorofluoromethanes: Chlorine Atom-Catalyzed Destruction of Ozone," *Nature* 249 (1974): 810–812. [The original article concerning CFC destruction of ozone]

2. S. Solomon, "Stratospheric Ozone Depletion: A Review of Concepts and History," *Journal of Geophysics* 37 (1999): 275–316.

3. (a) A. E. Waibel et al., "Arctic Ozone Loss Due to Denitrification," *Science* 283 (1999): 2064–2068. (b) G. Walker, "The Hole Story," *New Scientist* (25 March 2000): 24–28.

4. S. A. Montzka et al., "Present and Future Trends in the Atmospheric Burden of Ozone-Depleting Halogens," *Nature* 398 (1999): 690–693.

5. R. McKenzie, B. Connor, and G. Bodeker, "Increased Summertime UV Radiation in New Zealand in Response to Ozone Loss." *Science* 285 (1999): 1709–1711.

6. T. K. Tromp et al., "Potential Environmental Impact of a Hydrogen Economy on the Stratosphere," *Science* 300 (2003): 1740.

7. (a) S. Madronich and F. R. de Gruijl, "Skin Cancer and UV Radiation," *Nature* 366 (1993): 23 (b) J.-S. Taylor, "DNA, Sunlight, and Skin Cancer" *Journal of Chemical Education* 67 (1990): 835–841.

8. C. Biever, "Bring Me Sunshine," *New Scientist* (9 August 2003): 30–33.

Websites of Interest

See the listings on the website associated with Chapter 1 of this textbook: www.whfreeman.com/envchem3e/. Several of them show trends and up-to-date satellite data (often as color contour diagrams, some with animation) concerning stratospheric ozone depletion in polar regions.

2

Ground-Level Air Pollution— Outdoors and Indoors

Introduction

The best-known example of air pollution is the smog that occurs in many cities throughout the world. The reactants that produce the most common type of smog are mainly emissions from cars and electric power plants, although in rural areas some of the ingredients are supplied by emissions from forests. The operation of motor vehicles produces more air pollution than does any other single human activity. The most obvious manifestation of such smog is a yellowish-brownish-gray haze that is due to the presence in air of small water droplets containing products of chemical reactions that occur among pollutants in air. This haze, familiar to most of us who live in urban areas, now extends periodically to once-pristine areas such as the Grand Canyon in Arizona. Smog often has an unpleasant odor due to some of its gaseous components. More seriously, the intermediates and final products of the reactions in smog affect human health and can cause damage to plants, animals, and some materials.

It is a historical characteristic that once an undeveloped country starts industrial development, its air quality worsens significantly. The situation continues to deteriorate until a significant degree of affluence is attained, at which point emission controls are enacted and enforced, and the air begins to clear. Thus, although the quality of air is now improving with

*this most excellent canopy,
  the air,
look you, this excellent
  o'erhanging
firmament, this majestic roof
  fretted
with golden fire, why,
  it appears
no other thing to me than a foul
  and pestilent congregation of
  vapours*

William Shakespeare, Hamlet

The smoke from this petroleum refinery fire contains toxic, polluting gases and particles. (Brand X Pictures)

time in most developed countries, it is worsening in the larger cities of developing countries. Mexico City is generally considered to have the worst urban air pollution in the world at present.

One of the most important features of the Earth's atmosphere is that it is an *oxidizing* environment, a phenomenon due to the large concentration of **diatomic oxygen,** O_2, that it contains. Almost all the gases that are released into the air, whether they are "natural" substances or "pollutants," are eventually completely oxidized in air, and the end products are subsequently deposited on the Earth's surface. The oxidation reactions are vital to the cleansing of the air.

In this chapter, the pollution of tropospheric air is examined. The detrimental effects on animals, plants, and materials of polluted air—including the air we encounter indoors—and methods by which air pollution can be combated are described. As background, we begin the chapter by discussing the concentration units by which gases in the atmosphere are reported and the constitution and chemical reactivity of clean air.

Concentration Units for Atmospheric Pollutants

There is no consensus regarding the appropriate units by which to express concentrations of substances in air. In Chapter 1, ratios involving numbers of molecules—the "parts per" system—were emphasized as a measure. Other measures are often also encountered and will be used in this chapter:

Molecules of a gas per cubic centimeter of air, molecules/cm^3

Micrograms of a substance per cubic meter of air, $\mu g/m^3$

Moles of a gas per liter of air, mol/L

Given the lack of a consensus on a single appropriate scale, it is important to be able to convert gas concentrations from one set of units to another. This form of manipulation is discussed in Box 2-1.

The Chemical Fate of Trace Gases in Clean Air

In addition to the well-known stable constituents of the atmosphere (N_2, O_2, Ar, CO_2, H_2O), the troposphere contains a number of gases that are present in only trace concentrations, since they have efficient sinks and thus do not accumulate.

From biological and volcanic sources the atmosphere regularly receives inputs of the partially oxidized gases **carbon monoxide,** CO, and **sulfur dioxide,** SO_2, and of several gases that are simple compounds of hydrogen,

BOX 2-1 | The Interconversion of Gas Concentrations

Since the number of moles of a substance is proportional to the number of the molecules of it (Avogadro's number, 6.02×10^{23}, is the proportionality constant) and since the partial pressure of a gas is proportional to the number of moles of it, a concentration, for example, of 2 ppm for any pollutant gas present in air means

2 molecules of the pollutant in 1 million molecules of air

2 moles of the pollutant in 1 million moles of air

2×10^{-6} atm partial pressure of pollutants per 1 atm total air pressure

2 L of pollutant in 1 million liters of air (when the partial pressures and temperatures of pollutant and air have been adjusted to be equal)

Let us convert a concentration of 2 ppm to its value in molecules (of pollutant) per cubic centimeter (cm^3) of air for conditions of 1 atm total air pressure and 25°C. Since the value of the numerator, 2 molecules, in the new concentration scale is the same as in the original, all we need to do is establish the volume, in cubic centimeters, that 1 million molecules of air occupy. This volume is easy to evaluate using the ideal gas law ($PV = nRT$), since we know that

$$P = 1.0 \text{ atm}$$

$$T = 25 + 273 = 298 \text{ K}$$

$$n = (10^6 \text{ molecules})/$$
$$(6.02 \times 10^{23} \text{ molecules/mol})$$
$$= 1.66 \times 10^{-18} \text{ mol}$$

and the gas constant $R = 0.082$ L atm/mol K.

Now $PV = nRT$, so

$$V = nRT/P$$
$$= 1.66 \times 10^{-18} \text{ mol}$$
$$\times 0.082 \text{ L atm/mol K}$$
$$\times 298 \text{ K}/1 \text{ atm}$$
$$= 4.06 \times 10^{-17} \text{ L}$$

Since 1 L = 1000 cm^3, then V = 4.06×10^{-14} cm^3. Since 2 molecules of pollutant occupy 4.06×10^{-14} cm^3, it follows that the concentration in the new units is 2.0 molecules/4.06×10^{-14} cm^3, or 4.9×10^{13} molecules/cm^3.

In general, the most straightforward strategy to use to change the value of a concentration a/b from one scale to its value p/q on another is to independently convert the units of the numerator a to the units of the numerator p (both of which involve only the pollutant) and then convert the denominator b to its new value q (both of which involve the total air sample).

To convert a value in ppm or molecules/cm^3 to mol/L, we must change the molecules of pollutant to the number of moles of pollutant; for a pollutant concentration, again of 2 ppm, we can write

moles of pollutant

$$= (2 \text{ molecules} \times 1 \text{ mol})/$$
$$(6.02 \times 10^{23} \text{ molecules})$$
$$= 3.3 \times 10^{-24} \text{ mol}$$

Thus the molarity is (3.3×10^{-24} mol)/(4.06×10^{-17} L), or 8.2×10^{-8} M.

An alternative way to approach these conversions is to use the definition that 2 ppm means 2 L of pollutant per 1 million liters of air, and to find the number of moles and molecules of pollutant contained in a volume of 2 L at the stated pressure and temperature.

A unit often used to express concentrations in polluted air is micrograms per cubic meter, i.e., $\mu g/m^3$. If the pollutant is a pure substance, we can interconvert such values into the molarity and the "parts per" scales, provided that the pollutant's molar mass is known.

Consider as an example the conversion of 320 $\mu g/m^3$ to the ppb scale if the pollutant is SO_2, the total air pressure is 1.0 atm, and the temperature is 27°C. Initially the concentration is

$$\frac{320 \ \mu g \text{ of } SO_2}{1 \ m^3 \text{ of air}}$$

First we convert the numerator from grams of SO_2 to moles, since from there we can obtain the number of molecules of SO_2:

$$320 \times 10^{-6} g \ SO_2 \times \frac{1 \ mol \ SO_2}{64.1 \ g \ SO_2}$$

$$\times \frac{6.02 \times 10^{23} \text{ molecules of } SO_2}{1 \ mol \ SO_2}$$

$$= 3.01 \times 10^{18} \text{ molecules of } SO_2$$

Then using the ideal gas law we can change the volume of air to moles and then molecules, using $1 \ L = 1 \ dm^3 = (0.1 \ m)^3$:

$$n = PV/RT = 1.0 \ atm \times 1.0 \ m^3$$

$$\times \frac{\dfrac{1 \ L}{(0.1 \ m)^3} \Big/ 0.082 \ L \ atm}{mole \ K \times 300 \ K}$$

$$= 40.7 \ mol$$

Now 40.7 mol $\times$ 6.02 $\times$ 10^{23} molecules/mol = 2.45 $\times$ 10^{25} molecules, or 2.45 $\times$ 10^{16} billion molecules of air.

Thus the SO_2 concentration is

$$\frac{3.01 \times 10^{18} \text{ molecules of } SO_2}{2.45 \times 10^{16} \text{ billion molecules of air}}$$

$$= 123 \ ppb$$

Note that the conversion of moles to molecules was not strictly necessary, as Avogadro's number cancels from numerator and denominator. As stated previously, ppb refers to the ratio of the number of moles as well as to the ratio of the number of molecules.

It is vital in all interconversions to distinguish between quantities associated with the pollutant and those of air.

PROBLEM 1

Convert a concentration of 32 ppb for any pollutant to its value on

(a) the ppm scale,

(b) the molecules per cm^3 scale, and

(c) the molarity scale.

Assume 25°C and a total pressure of 1.0 atm.

PROBLEM 2

Convert a concentration of 6.0 $\times$ 10^{14} molecules/cm^3 to the ppm scale and to the moles per liter (molarity) scale. Assume 25°C and 1.0 atm total air pressure.

PROBLEM 3

Convert a concentration of 40 ppb of ozone, O_3, into

(a) the number of molecules per cm^3, and

(b) micrograms per m^3.

(continued on p. 70)

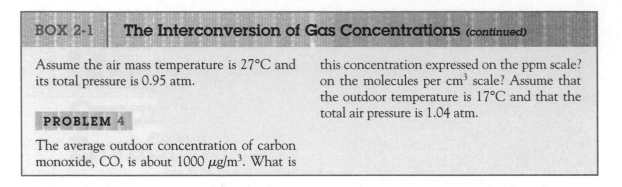

BOX 2-1 | The Interconversion of Gas Concentrations *(continued)*

Assume the air mass temperature is 27°C and its total pressure is 0.95 atm.

PROBLEM 4

The average outdoor concentration of carbon monoxide, CO, is about 1000 $\mu g/m^3$. What is this concentration expressed on the ppm scale? on the molecules per cm^3 scale? Assume that the outdoor temperature is 17°C and that the total air pressure is 1.04 atm.

some of whose atoms are in a highly reduced form (e.g., H_2S, NH_3); the most important of these "natural" substances are listed in Table 2-1.

Although most of these natural gases are gradually oxidized in air, none of them react directly with diatomic oxygen molecules. Rather, their reactions begin when they are attacked by the **hydroxyl free radical,** OH, even though the concentration of this species in air is exceedingly small, about a million molecules per cubic centimeter on average (see Problem 2-1). In clean tropospheric air, as in the stratosphere, the hydroxyl radical is produced when a small fraction of the excited oxygen atoms resulting from the photo-

TABLE 2-1	Some Important Gases Emitted into the Atmosphere from Natural Sources		
Formula	**Name**	**Main Natural Source**	**Atmospheric Lifetime**
NH_3	Ammonia	Anaerobic biological decay	Days
H_2S	Hydrogen sulfide	Anaerobic biological decay	Days
HCl	Hydrogen chloride	Anaerobic biological decay, volcanoes	
SO_2	Sulfur dioxide	Volcanoes	Days
NO	Nitric oxide	Lightning	Days
CO	Carbon monoxide	Fires; CH_4 oxidation	Months
CH_4	Methane	Anaerobic biological decay	Years
CH_3Cl	Methyl chloride	Oceans	Years
CH_3Br	Methyl bromide	Oceans	Years
CH_3I	Methyl iodide	Oceans	

chemical decomposition of the trace amounts of atmospheric ozone react with gaseous water to abstract one hydrogen atom from each H_2O molecule:

$$O_3 \xrightarrow{\text{UV-B}} O_2 + O^*$$

$$O^* + H_2O \longrightarrow 2\,OH$$

The average tropospheric lifetime of a given hydroxyl radical is only about one second, since it reacts quickly with one or another of the many atmospheric gases. Because the lifetime of hydroxyl radicals is short and sunlight is required to form more of them, the OH concentration drops quickly at night. Recall, from Problem 1 of Box 1-1, that because the corresponding reaction involving unexcited atomic oxygen is endothermic, its activation energy is high and consequently it occurs far too slowly to be a significant source of atmospheric OH.

PROBLEM 2-1

In one study, the concentration of OH in air at the time was found to be 8.7×10^6 molecules per cubic centimeter. Calculate its molar concentration and its concentration in parts per trillion, assuming that the total air pressure is 1.0 atm and the temperature is 15°C.

PROBLEM 2-2

For the concentration of hydroxyl radical associated with Problem 2-1 and a carbon monoxide concentration of 20 ppm, calculate the rate of its reaction with atmospheric carbon monoxide at 30°C, given that the rate constant for the process is $5 \times 10^{-13}\,e^{-300/T}$ cm^3/molecule s.

The hydroxyl free radical is reactive toward a wide variety of other molecules, including the hydrides of carbon, nitrogen, and sulfur listed in Table 2-1 and many molecules containing multiple bonds (double or triple bonds), including CO and SO_2. Although suspected for decades of playing a pivotal role in air chemistry, the presence of OH in the troposphere was confirmed only recently, since its concentration is so very small. The great importance of the hydroxyl radical to tropospheric chemistry arises because it, not O_2, initiates the oxidation of *all* the gases in Table 2-1, other than HCl. Without OH and its related reactive species HOO, most of these gases would not be efficiently removed from the troposphere, nor would most pollutant gases such as the unburned hydrocarbons emitted from vehicles. Indeed, OH has been called the "tropospheric vacuum cleaner" or "detergent." The reactions that it initiates correspond to a flameless, ambient-temperature "burning" of the reduced gases of the lower atmosphere. If these gases were to accumulate, the atmospheric composition would be quite different, as would the forms of life that were viable on Earth. Interestingly, hydroxyl is unreactive to molecular oxygen—in contrast to the behavior of many other free radicals—and to molecular nitrogen, so it survives long enough to react with many other species.

An example of the reactions initiated by the hydroxyl radical is the net oxidation of **methane** gas, CH_4, into the completely oxidized product **carbon dioxide,** CO_2:

$$CH_4 + 2\,O_2 \xrightarrow{\text{OH catalyst}} CO_2 + 2\,H_2O$$

As we shall see in Chapter 3, this overall reaction occurs by a sequence of steps, the first of which involves the reaction of hydroxyl radical with methane and the next-to-last of which involves the reaction of OH with carbon monoxide. Indeed, this pair of reactions accounts for the fate of most hydroxyl radicals in a clean atmosphere. However, since a new hydroxyl radical is also produced eventually in these multistep reactions, it is acting as a catalyst. Since the OH is originally produced from O_3, the case can be made that it is really ozone that causes the oxidation of most atmospheric gases. Diatomic molecular oxygen reacts with some of the free-radical species produced by OH reactions, so it does appear in the overall equation as the substance that oxidizes the reactants.

The *hydrogen halides* (HF, HCl, HBr) and fully oxidized gases such as carbon dioxide are relatively unreactive (from the oxidation–reduction point of view) in the troposphere because no further oxidation is possible with them; they eventually are deposited on the Earth's surface, often as a result of dissolving in raindrops.

Urban Ozone: The Photochemical Smog Process

The Origin and Occurrence of Smog

Many urban centers in the world undergo episodes of air pollution during which relatively high levels of ground-level **ozone,** O_3—an undesirable constituent of air if present in appreciable concentrations at low altitudes in the air that we breathe—are produced as a result of the light-induced chemical reaction of pollutants. This phenomenon is called **photochemical smog** and is sometimes characterized as "an ozone layer in the wrong place," to contrast it with the stratospheric ozone depletion problems discussed in Chapter 1. The word *smog* is a combination of *smoke* and *fog*. The process of smog formation involves hundreds of different reactions, involving dozens of chemicals, occurring simultaneously. Indeed, urban atmospheres have been referred to as giant chemical reactors. The most important reactions that occur in such air masses will be discussed in detail in Chapter 3.

The chief original reactants in an episode of photochemical smog are the **nitric oxide,** NO, and unburned hydrocarbons that are emitted into the air as pollutants from internal combustion engines, and also NO from electric power plants. The concentrations of these chemicals are orders of magnitude greater than those found in clean air. Recently it has been realized that gaseous hydrocarbons are also present in urban air as a result of the evaporation of solvents, liquid fuels, and other organic compounds. Collectively, the substances, including hydrocarbons and their derivatives, that readily vaporize into the air are called **volatile organic compounds,** or VOCs. For example, vapor is released into the air when a gasoline tank is filled unless the hose's nozzle is specially designed to minimize this loss. Evaporated, unburned gasoline is also emitted from the tailpipe of a vehicle before its catalytic converter has been warmed sufficiently to operate. Two-cycle engines such as those in outboard motor boats are particularly notorious for emitting significant proportions of their gasoline unburned into the air. Personal watercraft manufactured in the 1990s, before pollution controls came into effect, emitted more smog-producing emissions in a day's operation that an automobile of the same era driven for several years! Another vital ingredient in photochemical smog is sunshine, which serves to increase the concentration of free radicals that participate in the chemical processes of smog formation. Although the reactants—NO and VOCs—are relatively innocuous, the final products of the smog reaction—ozone, **nitric acid,** HNO_3, and partially oxidized (and in some cases nitrated) organic compounds—are much more toxic.

$$VOCs + NO + O_2 + sunlight \rightarrow \rightarrow mixture\ of\ O_3,\ HNO_3,\ and\ organics$$

Substances such as NO, hydrocarbons, and other VOCs that are emitted directly into air are called **primary pollutants;** the substances into which they are transformed, such as O_3 and HNO_3, are called **secondary pollutants.** A summary of the relative importance of various sectors in emissions of the primary pollutants sulfur dioxide, nitrogen oxides, and VOCs in the United States and Canada is given in Figure 2-1.

Other than those that absorb sunlight and subsequently decompose, most atmospheric molecules that are transformed in air begin by reacting with the hydroxyl free radical, OH, which is consequently the key reactive species in the troposphere. The most reactive VOCs in urban air are hydrocarbons that contain a carbon–carbon double bond, $C{=}C$, since their reaction with OH requires almost no activation energy and therefore is fast. Other hydrocarbons such as methane are also present in air, but due to the higher activation energy involved, their reaction with OH is sluggish. However, their reaction can become important in late stages of photochemical smog episodes.

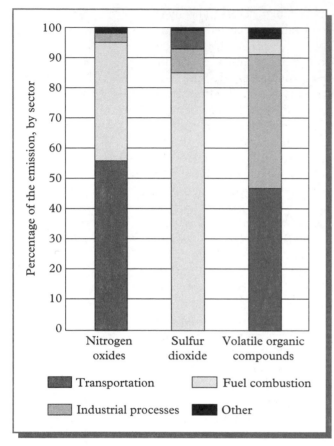

FIGURE 2-1 North American emissions of primary gaseous air pollutants from various sectors. [Source: U.S. EPA 1999 National Air Quality Trends Report.]

Nitrogen Oxides Production During Fuel Combustion

Nitrogen oxide gases are produced whenever a fuel is burned in air with a hot flame. At such high temperatures, some of the nitrogen and oxygen gases in the air passing through the flame combine to form nitric oxide, NO:

$$N_2 + O_2 \xrightarrow{\text{hot flame}} 2\,NO$$

The higher the flame temperature, the more NO is produced. Since this reaction is very endothermic, its equilibrium constant is very small at normal temperatures but increases rapidly as the temperature rises. One might expect that the relatively high concentrations of NO that are produced under combustion conditions would revert back to molecular nitrogen and oxygen as the exhaust gases cool, since the equilibrium constant for this reaction is so much smaller at lower temperatures. However, the activation energy for the reverse reaction is also quite high, so the process cannot occur to an appreciable extent except at high temperatures. Thus the relatively high concentrations of nitric oxide produced during combustion are maintained in the cooled exhaust gases; equilibrium cannot be quickly reestablished.

Nitric oxide produced by the reaction at high temperatures of atmospheric nitrogen is called **thermal NO.** Because the reaction between N_2 and O_2 has a high activation energy, it is negligibly slow except at very high temperatures, such as occur in the modern combustion engines of vehicles—particularly when they are traveling at high speeds—and in power plants.

Two distinct mechanisms are involved in the initiation of the reaction of molecular nitrogen and oxygen to produce thermal nitric oxide; in one it is atomic oxygen that attacks intact N_2 molecules, whereas in the other it is free radicals, such as CH, that are derived from the decomposition of the fuel. The initial reactions of the first mechanism are

$$O_2 \longrightarrow 2\,O$$
$$O + N_2 \longrightarrow NO + N$$

The rate of the second, slower step is proportional to $[O]$ $[N_2]$. However, since, from the equilibrium in the first step, $[O]$ is proportional to the square root of $[O_2]$, it follows that the rate of NO formation will be proportional to $[N_2]$ $[O_2]^{1/2}$.

Some additional nitric oxide is also produced from the oxidation of nitrogen atoms contained in the fuel itself; it is called **fuel NO.** About 30–60% of a fuel's nitrogen is converted to NO during combustion.

The nitric oxide in air is gradually oxidized to **nitrogen dioxide,** NO_2, over a period of minutes to hours, the rate depending upon the concentration of the pollutant gases. Collectively, NO and NO_2 in air are referred to as NO_X, pronounced "nox." The yellow color in the atmosphere of a smog-ridden city is due to the nitrogen dioxide present, since this gas absorbs visible light, especially near 400 nm (see its spectrum in Figure 2-2), removing sunlight's purple component while allowing most yellow light to be transmitted. The small levels of NO_X in clean air result in part from the operation of the above reaction in the very energetic environment of lightning flashes and in part from the release of NO_X and of **ammonia,** NH_3, from biological sources. Recently it has been discovered that NO_X is emitted from coniferous trees when sunlight shines on them and when the ambient concentrations of these gases are low.

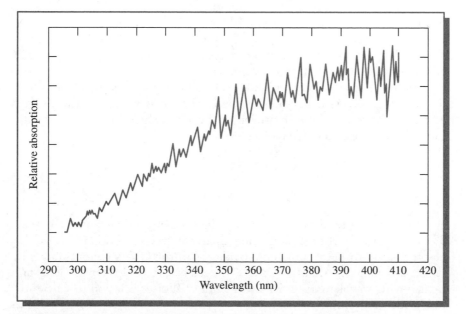

FIGURE 2-2 Absorption spectrum of NO_2 in the solar ultraviolet region. [Source: Reprinted from A.M. Bass et al., "Extinction Coefficients of NO_2 and N_2O_4," *J. Res. U.S. Natl. Bur. Stand.* A80 (1976): 143.]

PROBLEM 2-3

At combustion temperatures, the equilibrium constant for the reaction of N_2 with O_2 is about 10^{-14}. Calculate the concentration of nitric oxide that is in equilibrium with atmospheric levels of nitrogen and oxygen. Repeat the calculation for normal atmospheric temperatures, at which the equilibrium constant is about 10^{-30}. Given that the concentration of NO that exits from the combustion zone in a vehicle is much higher than the latter equilibrium value, what does that imply about equilibrium in the reaction mixture? [Hint: Use the stoichiometry of the reaction to reduce the number of unknowns in the expression for K.]

Ground-Level Ozone in Smog

In order for a city to generate photochemical smog, several conditions must be fulfilled. First there must be substantial vehicular traffic to emit sufficient NO, reactive hydrocarbons, and other VOCs into the air. Second, there must be warmth and ample sunlight for the crucial reactions, some of them photochemical, to proceed at a rapid rate. Finally, there must be relatively little movement of the air mass so that the reactants are not quickly diluted. For reasons of geography (e.g., the presence of mountains) and dense population, cities such as Los Angeles, Denver, Mexico City, Tokyo, Athens, Sao Paulo, and Rome all fit the bill splendidly and consequently are subject to frequent smog episodes. Indeed, the photochemical smog phenomenon was first observed in Los Angeles in the 1940s and has generally been associated with that city ever since, although pollution controls have partially alleviated the smog problem in recent decades.

As in the stratosphere, ground-level ozone is produced by the reaction of oxygen atoms with diatomic oxygen. The main source of the oxygen atoms in the troposphere, however, is the photochemical dissociation by sunlight of nitrogen dioxide molecules, NO_2:

$$NO_2 \xrightarrow{\text{UV}} NO + O$$

$$O + O_2 \longrightarrow O_3$$

According to the results of Problem 1-3, light having a wavelength less than 394 nm is capable of producing photochemical decomposition of NO_2. The absorption spectrum of nitrogen dioxide gas in the UV-A region, shown in Figure 2-2, indicates that the gas does indeed absorb in this region, and sunlight having wavelengths of about 394 nm or shorter is found to efficiently induce decomposition.

As discussed, it is predominantly NO rather than NO_2 that is emitted from vehicles and power plants into the air. In episodes of photochemical air pollution, NO is oxidized to the dioxide gradually over a period of

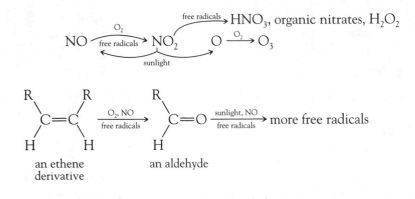

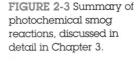

an ethene
derivative

an aldehyde

FIGURE 2-3 Summary of
photochemical smog
reactions, discussed in
detail in Chapter 3.

several hours in complex reaction sequences that involve free radicals as
catalysts (Figure 2-3). Indeed, one can see (Figure 2-4) the concentration
of NO first rising from emissions from early-morning vehicle traffic and
then falling during the morning as it is converted to NO_2 in urban atmospheres on smog days.

The concentration of ozone does not rise significantly in a city generating smog until late in the morning (see Figure 2-4), when the NO concentration has been greatly reduced. This occurs because residual nitric oxide
reacts with ozone to re-create nitrogen dioxide and oxygen, a reaction that
also occurs in the stratosphere:

$$NO + O_3 \longrightarrow NO_2 + O_2$$

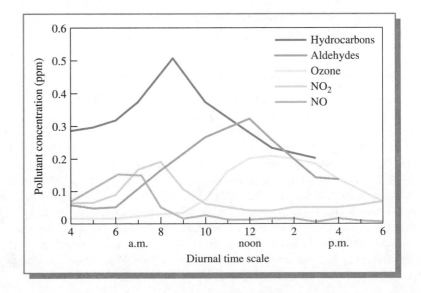

Diurnal time scale

FIGURE 2-4 Time-of-day
(diurnal) variation in the
concentration of gases
during days of marked
eye irritation in Los
Angeles in the 1960s.
[Source: Redrawn from
D.J. Speeding, *Air Pollution*
(Oxford: Oxford University
Press, 1974).]

The sum of the last three reactions constitutes a null cycle, whereby there is no net buildup of ozone or oxidation of NO to NO_2 by this mechanism.

PROBLEM 2-4

Perform a steady-state analysis on the three-step reaction mechanism discussed. Assume that both ozone and atomic oxygen are in a steady state, and derive an expression for $[NO_2]/[NO]$.

In fact, the oxidation of NO to NO_2 does occur rapidly, before the pollution has diffused away, partly because of weather conditions and partly because of the high concentration of catalytic free radicals that are generated during a smog episode. More free radicals are produced than are consumed in smog because of the reactions of VOCs, especially those containing highly reactive bonds such as C=C and C=O. For example, ethene and its derivatives react in a complex sequence of reactions with NO, free radicals, and atmospheric oxygen to produce nitrogen dioxide and aldehydes, the concentrations of which rise rapidly in the morning (see Figure 2-4). The aldehydes absorb sunlight with $\lambda < 350$ nm, i.e., UV-B and some UV-A light; some of them decompose photochemically in sunlight to produce additional free radicals, thereby increasing their concentration (see Figure 2-3). Once produced in significant amounts by decomposition of nitrogen dioxide, some of the ozone also reacts with VOCs to yield more hydroxyl radicals, further accelerating the smog reaction process.

Although our analysis has identified ozone as the main product of smog, the situation is actually more complicated, as a detailed study in Chapter 3 indicates. Some of the nitrogen dioxide reacts with hydroxyl radicals to generate nitric acid, HNO_3, and some reacts with organic free radicals to produce organic nitrates.

Governmental Goals for Reducing Ozone Concentrations

Many countries individually, as well as the *World Health Organization*, WHO, have established goals for maximum allowable ozone concentrations in air of about 100 ppb or less, averaged over a one-hour period. For example, the standard in Canada is 82 ppb, and that of WHO is 75–100 ppb. The United States has tentatively adopted a standard in which the ozone level over an eight-hour period is what is regulated, rather than the one-hour average; the average eight-hour limit was set at 80 ppb in 1997 for the United States, compared to the WHO eight-hour standard of 50–60 ppb. Generally speaking, the longer the period over which the concentration is averaged in a regulation, the lower the stated limit, since it is presumed that exposure to a higher level is acceptable only if it occurs for a short time. However, the new U.S. standard was challenged in court and has not yet been implemented.

The ozone level in clean air amounts to only about 30 ppb. By way of contrast, the levels of ozone in Los Angeles air used to reach 680 ppb, but peak levels have now declined to 300 ppb. Many major cities in North America, Europe, and Japan exceed ozone levels of 120 ppb typically for 5 to 10 days each summer.

Photochemical Smog Around the World

The air in Mexico City is so polluted by ozone, particulate matter, and other components of smog, and by airborne fecal matter, that it is estimated to be responsible for thousands of premature deaths annually; indeed, in the center of the city residents can purchase pure oxygen from booths to help them breathe more easily! In 1990 Mexico City exceeded the WHO air guidelines on 310 days; in 1992, ozone levels reached as high as 400 ppb. In contrast to temperate areas where photochemical smog attacks occur almost exclusively in the summer—when the air is sufficiently warm to sustain the chemical reactions—Mexico City suffers its worst pollution in the winter months, when temperature inversions prevent pollutants from escaping. Some of the smog in Mexico City originates from butenes that are a minor component of the liquefied gas that is used for cooking and heating in homes; some of it apparently leaks into the air.

Athens and Rome, as well as Mexico City, attempt to limit vehicular traffic during smog episodes. One strategy used by Athens and Rome is to allow only half the vehicles to be driven on alternate days, the allocation based upon license plate numbers (odd or even numbers).

Due to long-range transport of primary and secondary pollutants in air currents, many areas that themselves generate few emissions are subject to regular episodes of high ground-level ozone and other smog oxidants. Indeed, some rural areas, and even small cities, that lie in the path of such polluted air masses experience higher levels of ozone than do nearby larger urban areas. This occurs because in the larger cities some of the ozone transported from elsewhere is eliminated by reaction with nitric oxide released locally by cars into the air, as illustrated in the reaction of NO with O_3. Ozone concentrations of 90 ppb are common in polluted rural areas. When hot summertime weather conditions produce large amounts of ozone in urban areas but do not allow much vertical mixing of air masses as they travel to rural sites, elevated ozone levels are often observed in eastern North America and western Europe in zones that extend for 1000 km (600 miles) or more. Thus ozone control is a *regional* rather than a local air quality problem, in contrast to what was usually assumed in the past. Indeed, on occasion, polluted air from North America moves across the Atlantic to Europe, northern Africa, and the Middle East; that from Europe can move into Asia and the Arctic; and that from Asia can reach the west coast of North America.

A plot of ozone concentration contours for summer afternoon smog conditions in North America is shown in Figure 2-5a. The highest levels (100 ppb)

FIGURE 2-5a Ninetieth percentiles of summer afternoon ozone concentrations (ppb) measured in surface air over the United States. Ninetieth percentile means that concentrations are higher than this 10% of the time. [Adapted from A.M. Fiore, D.J. Jacob, J.A. Logan, and J.H. Yin, "Long-Term Trends in Ground-Level Ozone over the Contiguous United States, 1980–1995," *J. Geophys. Res.* 103 (1998): 1471–1480.]

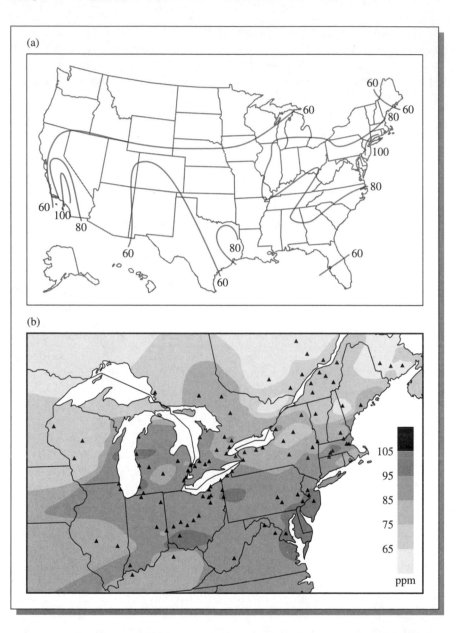

FIGURE 2-5b Maximum surface ozone levels, in ppm, for 1996–1998 in eastern North America. [Source: Environment Canada, "Interim Plan 2001 on Particulate Matter and Ozone," Government of Canada Publication (Ottawa: 2001).] Triangles represent monitoring stations.

occur in the Los Angeles and New York–Boston areas, but note the 80-ppb contour over a wide area south of the Great Lakes and into the Southeast, as well as one surrounding Houston. Ozone levels are particularly high over Houston—reaching 250 ppb on occasion—because of emissions of highly reactive VOCs containing C=C bonds from the region encompassing the petrochemical industry. Indeed, as of the late 1990s, Houston had overtaken Los Angeles in the number of days per year in which ozone standards were exceeded.

Considerable ozone is transported from its origin in the U.S. Midwest to surrounding states and Canadian provinces, especially around the Great Lakes (see Figure 2-5b). An example of a rural area subject to ozone is the farmland in southwestern Ontario, which often receives ozone-laden air from industrial regions in the United States that lie across Lake Erie. As a result of the damage caused by this pollution, crops such as white beans can no longer be grown successfully. Haze over China, produced by air pollution, so reduces sunlight intensity that it may be cutting food production by as much as 30% across a third of the country.

Elevated levels of ozone also affect materials: it hardens rubber, reducing the useful lifespan of consumer products such as automobile tires, and it bleaches color from materials such as fabrics.

The photochemical production of ozone also occurs during dry seasons in rural tropical areas where the burning of biomass for the clearing of forests or brush is widespread. Although most of the carbon is transformed immediately to CO_2, some methane and other hydrocarbons are released, as is some NO_X. Ozone is produced when these hydrocarbons react with the nitrogen oxides under the influence of sunlight.

Limiting VOC and NO Emissions to Reduce Ground-Level Ozone

In order to improve the air quality in urban environments that are subject to photochemical smog, the quantity of reactants, principally NO_X and hydrocarbons containing $C{=}C$ bonds plus other reactive VOCs, emitted into the air must be reduced. The control strategies that have been put in place in the United States have resulted in some reduction in ozone levels in the past few decades, notwithstanding the huge increase in total vehicle-miles driven—up to 100% more in the last 25 years—that has occurred.

For economic and technical reasons, the most common strategy has been to reduce hydrocarbon emissions. However, except in downtown Los Angeles, the percentage reduction in ozone and other oxidants that is achieved by this strategy usually has been much less than the percentage reduction in hydrocarbons. This happens because usually there is initially an overabundance of hydrocarbons relative to the amount of nitrogen oxides, and cutting back hydrocarbon emissions simply reduces the excess without slowing down the reactions significantly. In other words, it is usually the nitrogen oxides, rather than reactive hydrocarbons, that determine the overall rate of the reaction. This is especially true for rural areas that lie downwind of polluted urban centers.

Due to the large number of reactions that occur in polluted air, the functional dependence of smog production on reactant concentration is complicated, and the net consequence of making moderate decreases in primary pollutants is difficult to deduce without computer simulation. A 1997

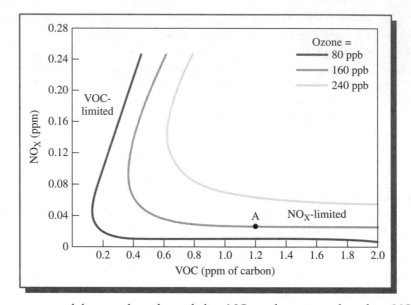

FIGURE 2-6 The relationship between NO_X and VOC concentrations in air and the resulting concentrations of ozone produced by their reaction. Point A represents a typical set of conditions under which ozone production is NO_X-limited. [Source: Redrawn from National Research Council, *Rethinking the Ozone Problem in Urban and Regional Pollution* (Washington, DC: National Academy Press, 1991).]

computer modeling study indicated that NO_X reduction, rather than VOC reduction, would be much more effective in reducing ozone in almost all of the eastern United States. An example of the predictions from the computer modeling studies is shown in Figure 2-6. The relationships between the NO_X and the VOC concentrations that produce three different values for the concentration of ozone are shown. Point A represents a typical set of conditions in which the ozone production is *NO_X-limited*. For example, reducing the concentration of VOCs from 1.2 ppm to 0.8 ppm has virtually no effect on the ozone concentration, which remains at 160 ppb since the curve in this region is almost linear and runs parallel to the horizontal axis. However, a reduction of the NO_X level, from about 0.03 ppm at point A to a little less than half this amount by dropping down to the curve directly below it, cuts the predicted ozone level in half, from 160 ppb to 80 ppb.

PROBLEM 2-5

Using Figure 2-6, and assuming an NO_X concentration of 0.20 ppm, estimate the effect on ozone levels of reducing the VOC concentration from 0.5 to 0.4 ppm. Do your results support the characterization of that zone of the graph as "VOC-limited"?

Some urban areas such as Atlanta, Georgia, and others located in the southern United States incorporate or border upon heavily wooded areas whose trees emit enough reactive hydrocarbons to sustain smog and ozone production, even when the concentration of **anthropogenic** hydrocarbons, i.e., those that result from human activities, is low. Deciduous trees and shrubs emit mainly the gas *isoprene*, whereas conifers emit *pinene* and *limonene*; all three hydrocarbons contain C=C bonds. The blue hazes that are observed over forested areas such as the Great Smoky Mountains in North Carolina and the Blue Mountains in Australia result from the reaction of such natural hydrocarbons in sunlight to produce carboxylic acids that condense to form suspended particles of the size that scatter sunlight and thereby produce a haze.

In urban atmospheres the concentration of these natural compounds normally is much less than that of the anthropogenic hydrocarbons, and it is not until the latter are reduced substantially that the influence of these natural substances becomes noticeable. In areas affected by the presence of vegetation, then, only the reduction of emissions of nitrogen oxides will reduce photochemical smog production substantially. As an air mass moves from an urban area to a rural one downwind, it often changes from being VOC-limited to being NO_X-limited, since there are few sources of nitrogen oxides but often substantial sources of reactive VOCs outside of cities.

Although hydrocarbons with C=C bonds are the most reactive type in photochemical smog processes, other VOCs play a significant role after the first few hours of a smog episode have passed and the concentration of free radicals has risen. For this reason, control of emissions of *all* VOCs is required in areas with serious photochemical smog problems. Gasoline, which is a complex mixture of hydrocarbons, is now formulated in order to reduce its evaporation, since gasoline vapor has been found to contribute significantly to atmospheric concentrations of hydrocarbons. The control of VOCs in air is discussed in more detail in Chapter 10. New regulations in California (with Los Angeles especially in mind) limit the use of hydrocarbon-containing products such as barbecue-grill starter fluid, household aerosol sprays, and oil-based paints that consist partially of a hydrocarbon solvent that evaporates into the air as the paint dries. The air quality in this region has improved because of current emission controls, but the increase in vehicle-miles driven and the hydrocarbon emissions from nontransportation sources such as solvents have thus far prevented a more complete solution. Recent research has also indicated that any substantial increase in the emissions of methane to the atmosphere could prolong and intensify the periods of high ozone levels in the United States, even though CH_4 is usually considered to be a rather unreactive VOC.

Catalytic Converters

The rate of creation of nitric oxide in a combustion system can be lessened by lowering the temperature of the flame. However, in recent decades, a more complete control of NO_X emissions from gasoline-powered cars and trucks has been attempted using **catalytic converters** placed just ahead of the mufflers in the vehicle's exhaust system. The original **two-way converters** controlled only carbon-containing gases, including carbon monoxide, CO, by completing their combustion to carbon dioxide. However, by use of a surface impregnated with a platinum–rhodium catalyst, the modern **three-way converter** changes nitrogen oxides back to elemental nitrogen and oxygen, using unburned hydrocarbons and the combustion intermediates CO and H_2 as reducing agents:

$$2\,NO \longrightarrow N_2 + O_2 \qquad \text{overall}$$

via, for example,

$$2\,NO + 2\,H_2 \longrightarrow N_2 + 2\,H_2O$$

The catalyst is dispersed as very tiny crystallites, initially less than 10 nm in size. The carbon-containing gases are oxidized almost completely to CO_2 and water:

$$2\,CO + O_2 \longrightarrow 2\,CO_2$$

$$C_nH_m + (n + m/4)\,O_2 \longrightarrow n\,CO_2 + m/2\,H_2O$$

An oxygen sensor in the vehicle's exhaust system is monitored by a computer chip that controls the intake air/fuel ratio of the engine to the stoichiometric amount required by the fuel to ensure a high level of conversion of the pollutants. The whole process is illustrated in Figure 2-7a. Since a fuel-rich mixture is fed to the engine in the first minute or so after the engine is

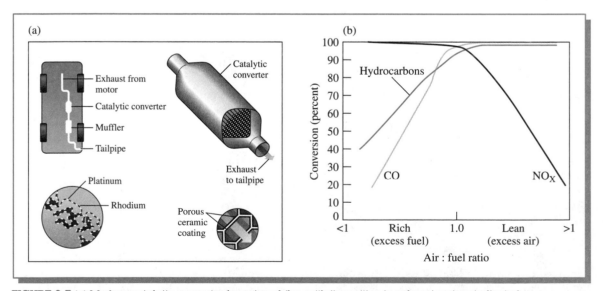

FIGURE 2-7 (a) Modern catalytic converter for automobiles, with its position in exhaust system indicated. [Source: L.A. Bloomfield, "Catalytic Converter," *Scientific American* (February 2000): 108.] (b) Efficiency in conversion of catalytic converter versus air/fuel ratio. [Source: B. Harrison, "Emission Control," *Education in Chemistry* 37 (2000): 127.]

started, and when high acceleration occurs, carbon monoxide and other pollutants are emitted directly into the air in these circumstances; in any event, the converter does not operate until exhaust gases have warmed it. If the air/fuel mix is not very close to the stoichiometric ratio, the warmed catalyst will not be effective for reduction (if there is too much air), causing nitrogen oxides to be emitted into the air, or for oxidation (if there is too little air), causing CO and hydrocarbons to be emitted, as illustrated in Figure 2-7b.

Some progress has been reported recently in the use of less valuable metals, such as copper and chromium, instead of the expensive platinum-group metals as catalysts in catalytic converters. Although the metals are recycled from old converters, a portion of them is inevitably lost in the process. Recently scientists have started to worry about the environmental problem of widely broadcasting the tiny particles of platinum, palladium, and rhodium that are lost from the converters themselves during their operation.

The small amounts of sulfur that remain in gasoline and diesel fuel after refining can partially deactivate catalytic converters. This occurs because the sulfate particles produced from the sulfur-containing molecules during gasoline combustion can become attached to the catalyst metal's sites, deactivating them. The maximum sulfur levels in gasoline, amounting to several hundred parts per million in the past, are due to be reduced to 30 ppm by the end of 2004 in both the United States and Canada, and in the European Union to 50 ppm by 2005. The sulfur is removed during refining, usually by **hydrodesulfurization,** itself a catalytic process, which reacts organic sulfur-containing molecules in the gasoline with hydrogen gas, H_2, to produce **hydrogen sulfide,** H_2S, which then is removed. Alternatively, the sulfur-containing molecules may be removed by absorbing them.

Once an engine has warmed up, properly working three-way catalytic converters eliminate 80–90% of the hydrocarbons, CO, and NO_X from the engine before the exhaust gases are released into the atmosphere. However, before the catalysts have warmed up to about 300 °C, and also during episodes of sudden acceleration or deceleration, the converters cannot operate effectively and there are bursts of emissions from the tailpipe. Approximately 80% of all the emissions from converter-equipped cars are produced in the first few minutes after starting. The catalyst that reduces nitric oxide to nitrogen also reduces sulfur dioxide, SO_2, to hydrogen sulfide. The emitted gases include reduced sulfur compounds such as H_2S, which often give vehicle emissions their characteristic odor of rotten eggs. Research and development is underway to develop catalytic converters that would convert start-up emissions so that these are not released into the air. Various approaches being investigated include

• devising a converter that will operate at lower temperatures or can be preheated so it begins to operate immediately,

- storing pollutants until the engine and converter are heated, and

- recirculating engine exhaust through the engine until the reactions are more complete.

Older cars (with no converters or just two-way converters) still on the road continue to pollute the atmosphere with nitrogen oxides even during normal operation.

The maximum amounts of emissions that can legally be released from light-duty motor vehicles such as cars have gradually been decreased in order to improve air quality. Some governments have recently instituted mandatory inspections of exhaust systems to ensure that they continue to operate properly. For example, in 2002 the automobile (a 1999 Honda Civic) owned by an author of this text was tested as part of the Ontario biennial license renewal process. The tailpipe emission of hydrocarbons was found to be 6 ppm (maximum allowed is 86 ppm for cars of this age), the carbon monoxide level was less than 100 ppm (480 allowed), and the nitric oxide concentration was 166 ppm (656 allowed); as expected, the NO level was the highest of the three since catalytic converters are less efficient for this gas than for carbon compounds.

Vehicles whose catalytic converters have been damaged or tampered with produce most of the emissions: typically, 50% of the hydrocarbons and carbon monoxide are released from 10% of the cars on the road. A 1996–1997 survey of carbon monoxide emitted by passing traffic in Denver, Colorado, produced the results shown in Figure 2-8. The great majority of Denver's vehicles of ages up to about 12 years were rated good in terms of their control of tailpipe emissions (Figure 2-8a). Most of the carbon monoxide came from cars 6–12 years old (Figure 2-8c), because they were so numerous (Figure 2-8b) and because of the presence in that fleet of cars with poor or fair emission levels (Figure 2-8a).

In the past, vehicle emission standards have been applied only to passenger vehicles. However, starting in 2004, new U.S. regulations will require for the first time that gasoline-powered sport-utility vehicles (SUVs)—which now account for about half of new vehicle sales—and light trucks also meet emission standards.

The catalytic converters used on vehicles with diesel engines are much less effective than those on gasoline-powered vehicles. They typically remove only about half the gaseous hydrocarbon emissions, compared to 80–90% achieved for gasoline engine emissions. This difference is due to the less active catalyst formulations that must be used with diesels because of the high sulfur content of diesel fuel; more active catalysts would oxidize the sulfur dioxide gas to sulfate particles, which would cover the catalyst surface and render it ineffective. However, diesel fuel intended for new on-road vehicles in the United States will have its maximum sulfur level drop from 500 ppm to 15 ppm in 2006; in the European Union the level will drop to 50 ppm. Furthermore, catalysts for diesel engines cannot remove NO_X since there must be excess oxygen in diesel

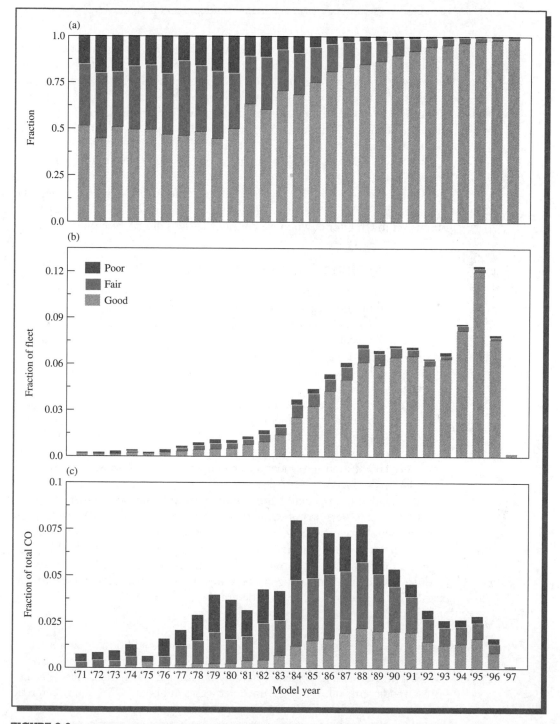

FIGURE 2-8 Fraction plots for CO emissions by vehicles by model year. The top plot (a) details the distribution fraction of Good/Fair/Poor readings by model year. The middle plot (b) shows the fleet distribution, and the bottom plot (c) shows the fleet-weighted CO contribution by model year. [Source: G.A. Bishop et al., "Drive-By Motor Vehicle Emissions: Immediate Feedback in Reducing Air Pollution," *Environ. Sci. Technol.* 34 (2000): 1110.]

exhaust; hence the required chemically reducing conditions cannot be attained. Even the hydrocarbons are not oxidized until the catalyst becomes quite hot, since sulfur dioxide rather than oxygen is absorbed strongly by the catalyst at low temperatures. However, *gaseous* emissions from diesel engines are inherently lower than from gasoline engines; it is the emissions of particles that are so high for diesels. Lowering the sulfur level in diesel fuel will also reduce particle emissions. New emission regulations that will take effect in 2005 will demand substantial reductions in NO_X emissions from diesel-powered vehicles.

In the future, vehicles using either gasoline or diesel fuel may well be replaced by emission-free ones powered by fuel cells. Such a prospect is discussed in Chapter 6, when we consider various alternative fuels.

Control of Nitric Oxide Emissions from Power Plants

In the United States approximately equal amounts of NO_X are emitted currently from vehicles and from electric power plants, and taken together they constitute the majority of the anthropogenic sources of these gases. To reduce their NO_X production, some power plants use special burners designed to lower the temperature of the flame. Alternatively, the recirculation of a small fraction of the exhaust gases through the combustion zone has the same effect.

Nitric oxide formation in power plants can also be greatly reduced by having the combustion of the fuel occur in stages. In the first, high-temperature stage, no excess oxygen is allowed to be present, thus limiting its ability to react with N_2. In the second stage, additional oxygen is supplied to complete the fuel's combustion but under lower temperature conditions, so that, again, little NO is produced.

Other plants, especially those in Japan and Europe, have been fitted with large-scale versions of catalytic converters to change NO_X back to N_2 before the release of stack gases into the air. The reduction of NO_X to N_2 in these **selective catalytic reduction** systems is accomplished to 80–95% completion by adding ammonia, NH_3, to the cooled gas stream. This highly reduced compound of nitrogen combines with the partially oxidized compound NO to produce N_2 gas in the presence of oxygen:

$$4\,NH_3 + 4\,NO + O_2 \longrightarrow 4\,N_2 + 6\,H_2O$$

However, tight control is needed in regulating the addition of ammonia to prevent its inadvertent oxidation to NO_X. The same reaction can be accomplished without expensive catalysts, though with much less efficient nitric oxide removal, using the uncooled gases at about 900 °C. The catalytic process occurs at 250–500°C, depending on the catalyst used.

The wet scrubbing of exhaust gases by an aqueous solution can also be used to prevent NO_X from being emitted into outside air. Since NO itself is rather insoluble in water and typical aqueous solutions, half or more of the NO_X must be in the form of the much more soluble NO_2 for such techniques to be effective. Solutions of *sodium hydroxide*, NaOH, react with equimolar amounts of NO and NO_2 to produce *sodium nitrite*, $NaNO_2$:

$$NO + NO_2 + 2\,NaOH \longrightarrow 2\,NaNO_2 + H_2O$$

PROBLEM 2-6

Deduce the balanced equation in which ammonia reacts with nitrogen dioxide to produce molecular nitrogen and water. Using the balanced equation, calculate the mass of ammonia that is required to react with 1000 L of air containing 10 ppm of NO_2. (Assume 27°C, 1 atm.)

PROBLEM 2-7

In a related technology, reduced nitrogen in the form of the compound urea, $CO(NH_2)_2$, is injected directly into the combustion flame to combine there, rather than later in the presence of a catalyst, with NO to produce N_2. Deduce the balanced equation that converts urea and nitric oxide and oxygen into N_2, CO_2, and water.

Future Reductions of Smog-Producing Emissions

Although direct emissions of five (CO, VOCs, SO_2, particulate matter, and lead) of the six major air pollutants in the United States fell significantly between 1970 and 2000, emissions of NO_X grew by 20%, with half that increase occurring during the 1990s. Since energy consumption grew 45% and vehicle distances traveled grew 143% in that period, restrictions on nitrogen oxide emissions have had some success in controlling some of the growth in this pollutant, but not enough to prevent an overall increase. NO emissions from electric power plants, their other major source, have recently been falling somewhat.

As a consequence of the increase in overall NO emissions, ground-level ozone concentrations increased in the southern (especially around Houston, Texas) and northeastern regions of the United States in the 1990s. The latter effect can be seen in Figure 2-5a, where high ozone levels center around New York–Boston and a few midwest sites, and somewhat lesser levels cover most of the East Coast and the Midwest, extending into southern Ontario. High ozone levels have been a problem in southern California for many decades.

To help reduce the incidence of summertime smog in south-central Canada and the northeastern United States, the two nations recently signed an Annex to their Air Quality Agreement. The United States has committed to reducing NO_X emissions originating in the northern and northeastern states by 35% by 2007, and also to reducing VOC emissions during summer months when most smog forms in this region. Canada has agreed to reduce its NO_X emissions from power plants in southern Ontario by 50% by the same date. One important mechanism by which both countries are attempting to reduce NO emissions from vehicles is by drastically lowering the allowed sulfur levels in gasoline, from an average in the hundreds of ppm

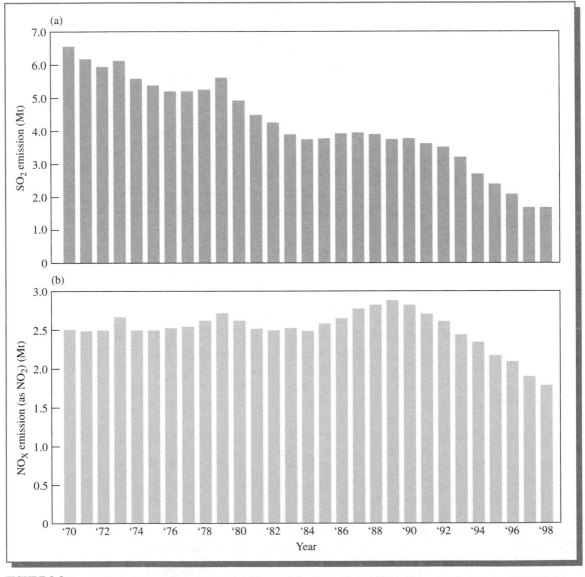

FIGURE 2-9 Annual UK emissions since 1970 of (a) SO_2; (b) NO_X (expressed as NO_2). [Source: M.R. Heal, "Acid Rain: Is the UK Coping?" *Education in Chemistry* (July 2002): 101.]

region to 15 ppm maximum in the future. Although this will not decrease sulfur emissions very much overall, it should significantly improve the efficiency of catalytic converters, many of whose reactive sites currently are blocked by sulfate particles produced in the engine when the fuel is burned. Emission standards for SUVs, trucks, and buses are also being tightened to bring them more in line with those for regular automobiles.

The *Gothenburg Protocol,* which controls the release of many pollutants in Europe, is expected to reduce NO_X emissions there by more than 40% by 2010, compared to 1990 levels. Great Britain—which saw emissions decline by the late 1990s by almost 40% compared to their peak in the late 1980s—will have to reduce its emissions by another third from 1998 levels by 2010 in order to meet these regulations (see Figure 2-9a). European VOC emissions are due to drop by 40% according to the protocol.

Green Chemistry: Replacement of Organic Solvents with Supercritical and Liquid Carbon Dioxide; Development of Surfactants for This Compound

In addition to their role in paints, organic solvents are used in many different products and processes in both commercial and household applications. It is estimated that over 15 billion kilograms of organic solvents are used worldwide each year in such areas as the electronics, cleaning, automotive, chemical, mining, food, and paper industries. These liquids include not only hydrocarbon solvents but also halogenated solvents. Both of these types of solvents contribute not only to air pollution as VOCs but also to water pollution (Chapter 10). Halogenated solvents may also contribute to the depletion of the ozone layer, as seen in Chapter 1. Discovering solvents with less environmental impact and even designing processes that use no solvents at all are the subjects of many green chemistry initiatives.

Carbon dioxide, CO_2, is one solvent that is receiving considerable attention as a replacement for traditional organic solvents. Although carbon dioxide is a gas at room temperature and pressure, it can be liquefied easily by the application of pressure. In addition to liquid carbon dioxide, there is considerable interest in supercritical carbon dioxide (a discussion of supercritical fluids can be found in Chapter 12) as a solvent in the electronics industry, as discussed in the green chemistry section in Chapter 4. The decaffeination of coffee and tea with carbon dioxide is a well-known application of this solvent.

Liquid carbon dioxide is attractive as a solvent because of its low viscosity and polarity and its wetting ability. Because of its low polarity, carbon dioxide is able to dissolve many small organic molecules. However, larger molecules including oils, polymers, waxes, greases, and proteins are generally

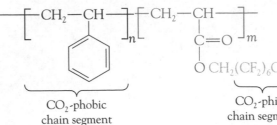

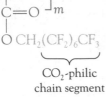

CO₂-phobic
chain segment

CO_2-philic
chain segment

FIGURE 2-10a A copolymer surfactant for carbon dioxide. [Source: M.C. Cann and M.E. Connelly, *Real-World Cases in Green Chemistry* (Washington, DC: American Chemical Society, 2000).]

insoluble in it. To increase the solubility of compounds in water, surfactants such as soaps and detergents have been developed that allow this very polar solvent to dissolve less polar materials such as oils and grease. In an analogous fashion, surfactants for carbon dioxide have been developed that increase the range of materials that will dissolve in it.

Joseph DeSimone, of the University of North Carolina and North Carolina State University, earned a Presidential Green Chemistry Challenge Award in 1997 for his preparation and development of polymeric surfactants for carbon dioxide. An example of such a surfactant is the block copolymer shown in Figure 2-10a. This molecule has nonpolar regions, which are CO_2-philic, and polar regions, which are CO_2-phobic. When dissolved in carbon dioxide, the CO_2-philic regions orient themselves to interact with the surrounding carbon dioxide solvent, while the CO_2-phobic regions aggregate with one another. The overall result is the formation of a structure known as a *micelle* (Figure 2-10b). Polar substances that normally do not dissolve in carbon dioxide will dissolve in the center polar region of the micelle.

DeSimone is one of the founders of a dry-cleaning chain know as Hangars that uses liquid carbon dioxide, along with surfactants that he developed, to clean clothes. Most dry cleaners presently use *perchloroethylene*, $Cl_2C{=}CCl_2$, known as PERC, as a solvent. PERC is a VOC, since it has a high vapor pressure and readily escapes into the troposphere if not carefully controlled. It is also is a groundwater contaminant (see Chapter 10) and is a suspected human carcinogen.

When carbon dioxide is used for dry cleaning, the spent liquid carbon dioxide is drained from the clothes after the wash cycle (in much the same way as the wash water in our washing machines at home is drained off after the wash cycle) and the carbon dioxide is allowed to evaporate by simply reducing the pressure. The carbon dioxide vapor is then captured, liquefied by increasing the pressure, and reused for another wash. Carbon dioxide is plentiful and inexpensive, since it can be recovered as a by-product

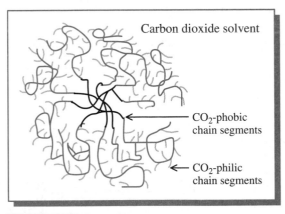

Carbon dioxide solvent

CO_2-phobic
chain segments

CO_2-philic
chain segments

FIGURE 2-10b A micelle in liquid carbon dioxide. [Source: M.C. Cann and M.E. Connelly, *Real-World Cases in Green Chemistry* (Washington, DC: American Chemical Society, 2000).]

from natural gas wells or ammonia production. Capture of carbon dioxide from these processes puts to good use this compound that would normally be released to the atmosphere and contribute to global warming (see Chapter 4). DeSimone is currently the director of the National Science Foundation Science and Technology Center for Environmentally Responsible

Solvents and Processes. This center focuses on discovering ways to replace conventional organic solvents and water with carbon dioxide in a multitude of processes.

Acid Rain

One of the most serious environmental problems facing many regions of the world is **acid rain.** This generic term covers a variety of phenomena, including acid fog and acid snow, all of which correspond to atmospheric precipitation of substantial amounts of acids. As will be discussed in a subsequent section of this chapter, acid rain has a variety of ecologically damaging consequences, and the presence of acid particles in air may also have direct effects on human health.

The phenomenon of acid rain was discovered by Angus Smith in Great Britain in the mid-1800s, but then it was essentially forgotten until the 1950s. It refers to precipitation that is significantly *more* acidic than "natural" (i.e., unpolluted) rain, which is mildly acidic due to the presence of dissolved atmospheric carbon dioxide, which forms **carbonic acid,** H_2CO_3.

$$CO_2(g) + H_2O(aq) \rightleftharpoons H_2CO_3(aq)$$

The weak acid H_2CO_3 then partially ionizes to release a **hydrogen ion,** H^+, with a resultant reduction in the pH of the system:

$$H_2CO_3(aq) \rightleftharpoons H^+ + HCO_3^-$$

Because of this source of acidity, the pH of unpolluted, natural rain is about 5.6. (The calculation of the pH values in such systems is discussed later in this chapter.) Only rain that is appreciably more acidic than this—i.e., with a pH of less than 5—is considered to be truly acid rain, since, because of natural trace amounts of strong acids, the acidity of rain in clean air is a little greater than that due to carbon dioxide alone. Strong acids such as HCl released by volcanic eruptions can produce natural acid rain temporarily in regions such as Alaska and New Zealand.

The two predominant acids in acid rain are **sulfuric acid,** H_2SO_4, and nitric acid, HNO_3, both of which are strong acids. Generally speaking, acid rain is precipitated far downwind from the source of the primary pollutants, namely sulfur dioxide, SO_2, and nitrogen oxides, NO_X. The strong acids are created during the transport of the air mass that contains the primary pollutants. Consequently, acid rain is a pollution problem that does not respect state or national boundaries because of the long-range transport that the atmospheric pollutants often undergo. For example, most acid rain that falls in Norway, Sweden, and the Netherlands originates as sulfur and nitrogen oxides emitted in other countries in Europe. Indeed, the modern recognition of acid rain as a problem stems from observations made in Sweden in the 1950s and 1960s, which arose from emissions from outside its borders. As the economist-philosopher John Kenneth Galbraith has noted, "Acid rain falls on the just and the unjust and also equally on the rich and poor."

In the next section, the sources of SO_2 air pollution and its potential abatement are discussed. Later in the chapter, the reaction conditions and mechanisms by which atmospheric sulfur dioxide is converted into sulfuric acid are analyzed in detail.

Sulfur Dioxide and Hydrogen Sulfide Sources and Abatement

On a global scale, most SO_2 is produced by volcanoes and by the oxidation of sulfur gases produced by the decomposition of plants. Because this natural sulfur dioxide is mainly emitted high into the atmosphere or far from populated centers, the background concentration of the gas in clean air is quite small (about 1 ppb). However, a sizable additional amount of sulfur dioxide is presently emitted into ground-level air, particularly over land masses in the Northern Hemisphere. The main anthropogenic source of SO_2 is the combustion of coal, a solid which, depending upon the geographic area from which it is mined, contains 1–6% sulfur. In many countries, including the United States, the major use of coal is to generate electricity. Usually half or more of the sulfur is trapped as inclusions in the mineral content of the coal. If the coal is pulverized before combustion, this type of sulfur can be mechanically removed, as discussed below. Any other sulfur, which usually amounts to about 1% of the coal's mass, is bonded in the complex carbon structure of the solid and cannot be removed without expensive processing.

Sulfur occurs to the extent of a few percent in crude oil (with higher levels in tar sands and shale oil), but it is reduced to the level of a few hundred ppm or less in products such as gasoline. Sulfur dioxide is emitted into the air directly as SO_2 or indirectly as H_2S by the petroleum industry when oil is refined and natural gas is cleaned before delivery. Indeed, the predominant component in natural gas wells is sometimes H_2S rather than CH_4! The substantial amounts of hydrogen sulfide obtained by its removal from oil and natural gas are often converted to solid, elemental sulfur, an environmentally benign substance, using the gas-phase process called the **Claus reaction:**

$$2\,H_2S + SO_2 \longrightarrow 3\,S + 2\,H_2O$$

One-third of the molar amount of hydrogen sulfide extracted from the fossil fuel is first combusted to sulfur dioxide to provide the other reactant for this process. Notice the analogy between the Claus reaction and the selective catalytic reduction process for nitric oxide control: they both involve the reaction together of oxidized and reduced forms of an element (N or S) to form the innocuous elemental form.

It is very important to remove hydrogen sulfide from gases before their dispersal in air because it is a highly poisonous substance, more so than sulfur dioxide. The concentration of H_2S sometimes becomes elevated in the area surrounding natural gas wells during *flaring*—the burning off of gas that cannot be immediately captured. Flaring burns only about 60% of the hydrogen sulfide content of the gas, so the remainder is dispersed into the surrounding air. Hydrogen sulfide is also a common pollutant in the air emissions from pulp and paper mills.

Several other smelly gases containing sulfur in a highly reduced state are emitted as air pollutants in petrochemical processes; these include CH_3SH, $(CH_3)_2S$, and CH_3SSCH_3. The term **total reduced sulfur** is used to refer to the total concentration of sulfur from H_2S and these three compounds. (The word *reduced* is not used here to denote a decrease in sulfur emissions but refers to the oxidation state of the sulfur.)

Large **point sources**—individual sites that emit large amounts of a pollutant—of SO_2 are also associated with the nonferrous smelting (i.e., the conversion of ores to free metals) industry. Many valuable and useful metals, such as copper and nickel, occur in nature as sulfide ores. In the first stage of their conversion to the free metals they were usually roasted in air to remove the sulfur, which is converted to SO_2, which was then traditionally released into the air. For example,

$$2\,NiS(s) + 3\,O_2(g) \longrightarrow 2\,NiO(s) + 2\,SO_2(g)$$

Ores such as copper sulfide can be smelted in a process that uses pure oxygen forced into the smelting chamber, and the very concentrated sulfur dioxide that is obtained from the reaction can be readily extracted, liquefied, and sold as a by-product. On the other hand, the SO_2 concentration in the waste gases from conventional roasting processes (such as that used for nickel) is high. Consequently it is feasible to pass the gas over an oxidation catalyst that converts much of the SO_2 to sulfur trioxide, to which water can be added to produce commercial concentrated sulfuric acid:

$$2\,SO_2(g) + O_2(g) \longrightarrow 2\,SO_3(g)$$
$$SO_3(g) + H_2O(aq) \longrightarrow H_2SO_4(aq)$$

The latter reaction (which represents only initial reactants and end-product) is, in fact, accomplished in two steps (not shown) to ensure that none of the substances escape into the environment: first the trioxide is combined with sulfuric acid, and then water is added to the resulting solution.

Clean Coal: Reducing Sulfur Dioxide Emissions from Power Plants

When the emitted sulfur dioxide is dilute, as in the case of power plant emissions, its extraction by oxidation is not feasible. Instead, the SO_2 gas is removed by an acid–base reaction with **calcium carbonate** (limestone), $CaCO_3$, or **calcium oxide** (lime), CaO, in the form of wet, crushed solid. The emitted gases are either passed through a slurry of the wet solid or bombarded by jets of the slurry. In some applications, fine grains of calcium oxide, rather than a slurry of calcium carbonate, are used to trap the sulfur dioxide from the emission gases. Up to 90% of the gas can be removed by such **scrubber** processes, more formally known as **flue-gas desulfurization.** In some operations, notably in Japan and Germany and to a minor extent in the United States, the product is fully oxidized by reaction with air, and the resulting **calcium sulfate,** $CaSO_4$, is dewatered and sold as *gypsum.* Otherwise, the product is a mixture of **calcium sulfite,** $CaSO_3$, and calcium sulfate in a slurry—or a dry solid if granular calcium oxide is used—and is usually buried in a landfill. The reactions with calcium carbonate are

$$CaCO_3 + SO_2 \longrightarrow CaSO_3 + CO_2$$

$$2\,CaSO_3 + O_2 \longrightarrow 2\,CaSO_4$$

Alternatively, the sulfur dioxide can be captured by the use of slurries of sodium sulfite or magnesium oxide or amine salts, and these compounds and concentrated SO_2 gas are later regenerated by thermally decomposing the product.

Recently, **clean coal technologies** have been developed to use coal in ways that are cleaner and often more energy efficient than those used in the past. In these technologies the cleaning can occur precombustion, during combustion, postcombustion, or by conversion of the coal to another fuel.

In **precombustion cleaning,** the coal has the sulfur associated with its mineral content—usually *pyritic sulfur,* FeS_2—removed so it cannot subsequently produce sulfur dioxide. The coal is first ground to a very small particle size, effectively into separate mineral particles and carbon particles. Since they have different densities, the two particle types can be separated by mixing the pulverized solid in a liquid of intermediate density and allowing the fuel portion to rise to the top, where it can be separated. As an alternative to such physical cleaning, biological or chemical methods can be employed. Thus a microorganism can be used to oxidize the insoluble Fe^{2+} pyrite in pulverized coal to the soluble Fe^{3+} form. Or bacteria cultured to consume the organic sulfur in coal can be utilized. The sulfur can be chemically leached with a hot sodium or potassium caustic solution.

In **combustion cleaning,** the combustion conditions can be modified to reduce the formation of pollutants, and/or pollutant-absorbing substances can be injected into the fuel to capture pollutants as they form. In **fluidized bed combustion,** pulverized coal and limestone are mixed and then suspended on jets of air (fluidized) in the combustion chamber. Virtually all the sulfur dioxide is thereby captured in solid form as calcium sulfite and sulfate before the gas can escape. This procedure allows for much-reduced combustion temperatures and therefore also greatly decreases the amount of nitrogen oxide that is formed and released.

Some of the advanced techniques used in **postcombustion cleaning**—such as the use of granular calcium oxide or sodium sulfite solutions—have already been described. In the *SNOX* process developed in Europe, cooled flue gases are mixed with ammonia gas to remove the nitric oxide by catalytic reduction to molecular nitrogen (by the reaction discussed previously). The resulting gas is reheated and the sulfur dioxide oxidized catalytically to sulfur trioxide, which is then hydrated by water to sulfuric acid, condensed, and removed.

In **coal conversion,** the fuel is first gasified by reaction with steam. The gas mixture is cleaned of pollutants, and the purified gas is then burned in a gas turbine that generates electricity. The waste heat of the combustion gases is used to produce steam for a conventional turbine and thus to generate more electricity. Alternatively, as we shall discuss in Chapter 6, the gasified coal can be converted into liquid fuels suitable for vehicular use.

Sulfur dioxide emissions from power plants can also be reduced by substituting oil, natural gas, or low-sulfur coal for the coal, although these fuels are sometimes more expensive than high-sulfur coal.

PROBLEM 2-8

What mass of calcium carbonate is required to react with the sulfur dioxide that is produced by burning 1 tonne of coal that contains 5.0% sulfur by mass?

PROBLEM 2-9

Write the balanced equations whereby sodium hydroxide can be used to scrub sulfur dioxide from exhaust gases by a reaction that produces water and sodium sulfite, Na_2SO_3. What substance would you have to react with this solution to produce calcium sulfite and regenerate the sodium hydroxide? Write the balanced equation for the latter process, and deduce the net reaction for the cycle.

The alternative to sulfur dioxide control—to simply allow the pollutant gas to be emitted into the air—can cause devastation by SO_2 to the plant life in the surrounding area unless extremely high smokestacks are used. The tallest such stacks in the world are located at Sudbury, Ontario, and reach 400 m high. However, using tall stacks simply solves a local SO_2 problem at the expense of creating a problem downwind. For example, emissions from mainland North America can sometimes be detected in Greenland.

Because of federal regulations, the amount of sulfur dioxide emitted into the air in North America has fallen substantially—by about 35% from 1973 levels by 1997 in the United States and 45% in Canada by 2000. The 1991 Air Quality Accord between the United States and Canada required both countries to reduce substantially their sulfur dioxide emissions beyond those of previous laws and agreements. Such emissions in the United States are restricted in accordance with the Clean Air Act, especially the 1990 amendments to it; by the year 2000 there was a substantial reduction in SO_2 emissions compared to values from the 1970s and 1980s. The average concentration of sulfur dioxide in air in the United States fell by 43%, from 13.2 ppb to 7.5 ppb, from 1975 to 1991. Whereas Phase I (1995 deadline) of the Clean Air Act imposed controls on only the largest coal-fired plants, Phase II, which began in 2000, imposes more stringent requirements and applies to almost all plants. There is an SO_2 tonnage limit for each power plant that emits this gas, based upon the power it produces, and a cap on overall national emissions as well. By 2010, the SO_2 emissions from these power plants should have been reduced by 50% compared to 1980 levels. Indeed, emissions from the regulated plants had dropped by 38% in 1995–1997, compared to their 1993–1994 levels.

The reductions in sulfur dioxide emissions by power plants in the U.S. Midwest has been achieved at lower-than-expected cost, due in part to the availability of cheap, low-sulfur coal (1% S, versus more than 3% S in the high-sulfur coal used previously) and inexpensive scrubbers and in part to the implementation of a system of tradable emission permits. This permit system, in operation since the early 1980s, allows industries to buy emission allowances if they need to exceed their allowed levels or to sell excess allowances on the open market (through the Chicago Board of Trade) if they do not need their whole allowance. A similar program has been started for nitrogen oxide emissions.

In Europe, the EU issued a directive in 1988 specifying reductions of SO_2 emissions from large power plants of 40–60% from 1980 levels by 1998, and of 50–70% by 2003. The decline in SO_2 emissions in Great Britain from 1970 to 1998 amounted to 75%, as illustrated in Figure 2-9b. This cutback was achieved mainly by switching from coal to natural gas in power stations and the use of low-sulfur coal and scrubbing of emissions in facilities where coal is still burned. According to the 1999 Gothenburg Protocol, Europe's sulfur dioxide emissions are to be cut beyond those of 1990 by another 63% by 2020.

Global SO_2 emissions are predicted to keep rising until about 2020, due mainly to increased releases from Asia. China has become a world leader in emitting sulfur dioxide; that country now has serious acid rain problems, and some acid is carried by the wind to Japan and on occasion all the way to North America. In contrast, Japan initiated tight controls on SO_2 emissions in the 1970s, and by 1980 its power plants had almost eliminated such emissions by the widespread installation of scrubbers. The high rate of SO_2 emissions from the former Soviet bloc countries in Europe has declined in recent decades, due more to economic problems than to intentional control, although in places emissions continue to make acidification a problem.

The Ecological Effects of Acid Rain and of Photochemical Smog

Nitric oxide is not especially soluble in water, and the acid (sulfurous) that sulfur dioxide produces upon dissolving in water is a weak one. Consequently, the primary pollutants NO and SO_2 themselves do not make rainwater particularly acidic. However, some of the mass of these primary pollutants is converted over a period of hours or days into the secondary pollutants sulfuric acid and nitric acid, both of which are very soluble in water and are strong acids. Indeed, virtually all the acidity in acid rain is due to the presence of these two acids. In eastern North America, sulfuric acid greatly predominates because much electrical power is generated from power plants that use high-sulfur coal. In western North America, nitric acid attributable to vehicle emissions is predominant, since the coal mined and burned there is low in sulfur.

PROBLEM 2-10

The pH of a sample of rain is found to be 4.0. Calculate the percentage of HSO_4^- that is ionized in this sample, given that the acid dissociation constant for the second stage of ionization of H_2SO_4 is 1.2×10^{-2} mol/L. Repeat the calculation for a pH of 3.0. Is the trend shown by these calculations consistent with qualitative predictions made according to Le Châtelier's principle (which states that the position of equilibrium shifts so as to minimize the effect of any stress)? [Hint: Write the expression for the acid dissociation constant for the weak acid in terms of the concentrations of the reactants and products and use the stoichiometry of the balanced equation to reduce the number of unknowns to one.]

Figure 2-11 shows a contour map of the average pH in precipitation in different regions of the world. At each point along any solid line, the annual average pH has the same value; hence contours connect regions having equally acidic rain. The lowest pH ever recorded, 2.4, occurred for a rainfall in April 1974 in Scotland. Indeed, central-west Europe, including the United Kingdom, has a serious acid rain problem, as can be seen from the pH = 4.0 and 4.5 contours surrounding the area in Figure 2-11. In North America the greatest acidity occurs in the eastern United States and in southern Ontario, since both regions lie in the path of air originally polluted

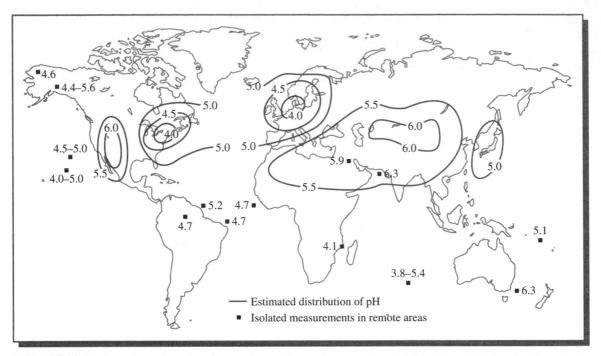

FIGURE 2-11 Global pattern of acidity of precipitation. [Source: Redrawn from J.H. Seinfeld and S.N. Pandis, *Atmospheric Chemistry and Physics* (Chichester: John Wiley, 1998).]

by emissions from power plants in the Ohio Valley. On the other hand, much of the acidity that falls in upper New York State stems from emissions in southern Ontario.

In addition to the acids delivered to ground level during precipitation, a comparable amount is deposited on the Earth's surface by means of **dry deposition,** the process by which nonaqueous chemicals are deposited as pollutants on solid and liquid surfaces at ground level when air containing them passes over the surfaces. Much of the original SO_2 gas is never oxidized in the air but rather is removed by dry deposition from air before reaction can occur: oxidation and conversion to sulfuric acid occur after deposition. **Wet deposition** processes include the transfer of pollutants to the Earth's surface by rain, snow, or fog—i.e., by aqueous solutions.

Neutralization of Acid Rain by Soil

The extent to which acid precipitation affects biological life in a given area depends strongly on the composition of the soil and bedrock in that area. Strongly affected areas are those that have granite or quartz bedrock, since

FIGURE 2-12 Regions of North America with low soil alkalinity for neutralizing acid rain. [Source: D.J. Jacob, *Introduction to Atmospheric Chemistry* (Princeton, NJ: Princeton University Press, 1999), p. 233.]

the soil there has little capacity to neutralize the acid. Figure 2-12 shows areas of North America that have low soil alkalinity. Large areas susceptible to acidity are the Precambrian Shield regions of Canada and Scandinavia. In contrast, if the bedrock is limestone or chalk, the acid can be efficiently neutralized (buffered), since these rocks are composed of calcium carbonate, which acts as a base and reacts with acid:

$$CaCO_3(s) + H^+(aq) \longrightarrow Ca^{2+}(aq) + HCO_3^-(aq)$$

$$HCO_3^-(aq) + H^+(aq) \longrightarrow H_2CO_3(aq) \longrightarrow CO_2(g) + H_2O(aq)$$

The reactions here proceed almost to completion due to the excess of H^+ that is present. Thus the rock dissolves, producing carbon dioxide and calcium ion to replace the hydrogen ion. These same reactions are responsible for the deterioration of limestone and marble statues; fine detail, such as ears, noses, and other facial features, are gradually lost as a result of reaction with acid and with sulfur dioxide itself.

Acidity from precipitation leads to the deterioration of soil. When the pH of soil is lowered, plant nutrients such as the cations potassium, calcium, and magnesium are exchanged with H^+ and thereupon leached from it.

Although sulfur dioxide emission levels fell significantly in recent decades in both Europe and North America, there has not been as large a corresponding change in the pH of the precipitation, especially in northeastern North America. The lack of a corresponding reduction in acidity is attributed to a decline over the same period of fly ash emissions from smokestacks and of other solid particles, all of which were alkaline and neutralized a fraction of the sulfur dioxide and sulfuric acid in the same way that calcium carbonate does in soil. Thus the decline in acidity in precipitation in the northeastern United States from 1983 to 1994 amounted to only 11%, although the sulfate ion molar concentrations in precipitation fell, not by 5.5% (half that of H^+) but by 15%. The much smaller nitrate levels remained essentially unchanged in this period for this region. The change in sulfate deposition in the northeastern United States and south-central Canada from the early 1980s to the late 1990s is shown in Figure 2-13. In Great Britain, rainfall acidity declined by about 40% in the 1986–1997 period because of emission controls there.

Because of acid rainwater falling and draining into them, tens of thousands of lakes in the Shield regions of both Canada and Sweden have become strongly acidified, as have lesser numbers in the United States, Great Britain, and Finland. Lakes in Ontario are particularly hard hit, since they lie directly in the path of polluted air and since the soil there contains little limestone. In a few cases, attempts have been made to neutralize the acidity by adding limestone or **calcium hydroxide,** $Ca(OH)_2$, to the lakes; however, this process must be repeated every few years to sustain an acceptable pH. Adding phosphate to lakes can also control acidity, since it stimulates plant growth, by which **nitrate ion,** NO_3^-, is converted to reduced nitrogen with the consumption of hydrogen ions (see below).

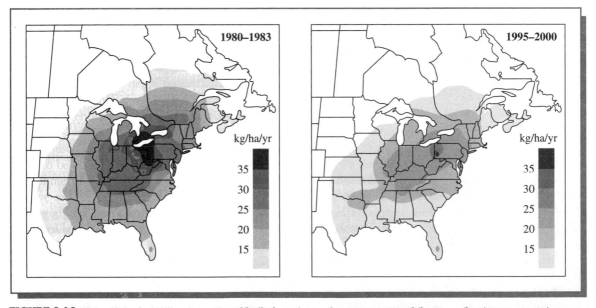

FIGURE 2-13 Wet sulfate deposition in eastern North America as four-year mean (kilograms/hectare per year). [Source: Canadian National Atmospheric Chemistry Database, Meteorological Service of Canada, Environment Canada.]

In recent years a new source of sulfuric acids in lakes has appeared—the oxidation of sulfur in shallow wetlands dried up by global warming and thereby exposed to the air.

In Australia, soil acidity has a completely different origin. Acidification is associated there with the removal by harvest of plant and animal crops and by soil leaching of nitrate ion. Nitrate ion, when it undergoes the natural process of denitrification to N_2, consumes large quantities of hydrogen ion, as shown in the reduction half-reaction

$$2\,NO_3^- + 12\,H^+ + 10\,e^- \longrightarrow N_2 + 6\,H_2O$$

Presumably the loss of nitrate prevents this natural buffering of acidity. As in Canadian lakes, the effects of the acidification have been partially reversed in Australia by the addition of lime to the soil.

As Problem 2-11 shows, the oxidation of ammonium ion to nitrate produces hydrogen ions. Indeed, the large emissions of ammonia into the air from manure in areas of livestock and poultry farming result in the atmospheric deposition of **ammonium ion,** NH_4^+, which then is oxidized by soil microbes. The resulting H^+ contributes to the acidification of soil.

PROBLEM 2-11

Deduce the balanced half-reaction for conversion of ammonium ion, NH_4^+, to nitrate ion, NO_3^-, and thereby show that H^+ is also produced in this process. [Hint: Consult your general chemistry text if you have forgotten how to balance redox half-reactions.]

Until recently, acid rain in the United States was considered to be a problem for its northeastern region. Indeed, one of the hardest-hit regions is the Catskill Mountains in New York State, whose surface rocks consist of calcium-poor sandstone and from which most of the nutrients have not been leached. At the Hubbard Brook Experimental Forest in New Hampshire, half the calcium and magnesium in soil was leached by 1996 and, as a result, vegetative growth almost stopped. However, acid rain now is also a concern in the southeastern United States. Here soils are generally thicker and were able to neutralize more acid. Much of that leaching ability now has been exhausted and acid levels in many waterways have increased substantially. It has been discovered that the recovery of such soils, and of soils in Germany, is slowed once acid precipitation has declined, because previously stored sulfate ion is then released, causing more cation leaching and penetration of acidity deeper into the ground.

The regulatory scheme originated in the United States of requiring reductions in sulfur dioxide emissions in certain geographical regions has been extended by European scientists and regulators into the concept of **critical load.** This concept recognizes that different levels of risk from acid rain are faced in different regions. Geographic areas having buffering capacity can withstand a much greater load of acid rain before damage occurs than those without the capacity. Thus higher sulfur dioxide emissions from a particular region can be allowed if the area in which the resulting sulfuric acid is usually deposited has a high critical load. To determine the critical loads, scientists use computer models that incorporate soil chemistry, rainfall, topography, etc. Use of the concept has had great success in Sweden, for example.

In using critical loads, pollution control becomes effects-based rather than source-based. The critical loads concept has been implemented in regulations in Europe and Canada and is favored by many scientists and politicians in the United States, but it has not been implemented there. Although significant progress has been made in reducing SO_2 emissions, and more reductions are scheduled both for it and for NO, scientists predict that these efforts will be insufficient to allow a full recovery to lakes and forests in the northeastern United States and south-central Canada.

Acidification reduces the ability of some plants to grow, including those in fresh-water systems. Because of the decrease in this productivity in lakes and in the streams that feed them, the amount of **dissolved organic carbon** (DOC) in the surface water has declined. The DOC contains molecules that absorb UV from sunlight; thus a decline in DOC levels has allowed more penetration of UV light into the lower layers of the lake. In addition, global warming (see Chapters 4 and 5) has resulted in the drying up of some streams that supplied DOC to lakes. Furthermore, stratospheric ozone depletion has also allowed more UV to reach the Earth's surface, including lakes, in the first place. Thus fresh-water lakes have suffered a "triple whammy" from global environmental problems.

Release of Aluminum into Soil and Water Bodies by Acid Rain

Acidified lakes characteristically have elevated concentrations of dissolved **aluminum,** Al^{3+}, which is leached from rocks by hydrogen ions (H^+); under normal, near-neutral pH conditions, the aluminum is immobilized in the rock by its insolubility. Plots of dissolved aluminum concentration versus water acidity for lakes in the Adirondack Mountains of New York State and for lakes in Sweden are illustrated in Figure 2-14. (The chemistry underlying these processes and the reasons why natural waters have pH values of 7 or 8 rather than the 5.6 of rain are discussed in Chapter 9.)

Scientists believe that the acidity itself and the high concentrations of aluminum together are responsible for the devastating decreases in fish populations that have been observed in many acidified water systems. Different types of fish and aquatic plants vary in their tolerances for aluminum and acid, so the biological composition of a lake varies as it gradually becomes increasingly acidic. Generally speaking, fish reproduction is severely diminished even

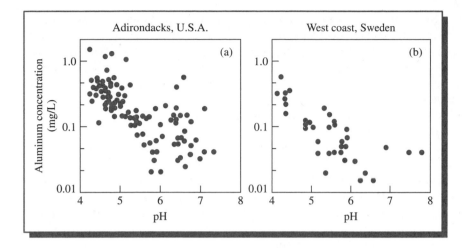

FIGURE 2-14 Aluminum concentrations versus pH of the water in different fresh-water lakes in (a) the Adirondacks and (b) western Sweden. Notice the logarithmic vertical axis. [Source: M. Havas and J.F. Jaworski, *Aluminum in the Canadian Environment,* (Ottawa: National Research Council of Canada Report 24759, 1986).]

at low levels of acidity that can, however, be tolerated by adult fish. Very young fish, hatched in early spring, also are subject to the shock of very acidic water that occurs when the acidic winter snow all melts in a short time and enters the water systems.

Healthy lakes have a pH of about 7 or a little higher; few fish species survive and reproduce when the pH drops much below 5. As a result, many lakes and rivers in affected areas are now devoid of their valuable fish; e.g., some rivers in Nova Scotia are too acidic for Atlantic salmon. The water in many acidified lakes is crystal clear due to the death of most of the flora and fauna. Research reported recently found that aluminum levels draining from soils at medium-to-high elevations in the United States declined significantly over the period from 1982 to 1998 and, if the trend continues, will not be a threat to fish by about 2012.

Effect on Trees and Crops of Air Pollution

In recent years it has become clear that air pollution can also have a severe effect on trees. The phenomenon of forest decline was first observed on a large scale in western Germany and occurs mainly at high altitudes. However, the cause-and-effect relationship behind this forest decline has been very difficult for scientists to untangle. As discussed previously, acidification of the soil can leach nutrients from it and, as occurs in lakes, solubilize aluminum. This element may interfere with the uptake of nutrients by trees and other plants. Apparently both the acidity of the rain falling on affected forests and the tropospheric ozone and other oxidants in the air to which they are exposed pose significant stresses to the trees. These two stresses alone will not kill them, but when they are combined with drought, temperature extremes, disease, or insect attack, the trees become much more vulnerable.

Forests at high altitudes are most affected by acid precipitation, possibly because they are exposed to the base of low-level clouds, where the acidity is most concentrated. Fogs and mists are even more acidic than precipitation, since there is much less total water to dilute the acid. For example, white birch trees along the shores of Lake Superior experience dieback in regions where acid fog occurs, as it frequently does there. Deciduous trees (i.e., those that lose their leaves annually) affected by acid rain gradually die from their tops downward; the outermost leaves dry and fall prematurely and are not replenished the following spring. The trees become weakened as a result of these changes and become more susceptible to other stressors. In some regions of Europe and North America, forest soils are limed to combat the effects of acidity on trees.

Ground-level ozone itself has an effect on some agricultural crops due to its ability to attack plants. Apparently the ozone reacts with the ethene (ethylene) gas that the plants emit, generating free radicals that then damage plant tissue. The rate of photosynthesis is slowed, and hence the total amount of plant material is reduced, by the action of ozone. As in the case of trees, air pollution acts as a stressor to plants. The collective damage to North American crops, e.g., alfalfa in the United States and white beans in Canada, is estimated to be $3 billion a year. Other crops whose yields are adversely affected by current levels of ozone include wheat, corn, barley, soybeans, cotton, and tomatoes. The fraction of the world's cereal crops that are grown in regions of high ozone, and therefore subject to damage, is predicted to more than triple by 2025.

Particulates in Air Pollution

The black smoke released into the air by a diesel truck is often the most obvious form of pollution that we routinely encounter. The smoke is composed largely of particulate matter. **Particulates** are tiny solid or liquid particles—other than those of pure water—that are suspended in air and that are usually individually invisible to the naked eye. Collectively, however, such particles often form a haze that restricts visibility. Indeed, on many summer days the sky over North American and European cities is milky white rather than blue, owing to the scattering of light by suspended particulates in the air.

The particles that are suspended in a given mass of air are neither all of the same size or shape nor do they all have the same chemical composition. The smallest suspended particles are about 0.002 μm (i.e., 2 nm) in size; by contrast, the size of typical gaseous molecules is 0.0001 to 0.001 μm (0.1–1 nm). The upper limit for suspended particles corresponds to dimensions of about 100 μm (i.e., 0.1 mm). When atmospheric water droplets coalesce to form particles bigger than this, they are raindrops and fall out of

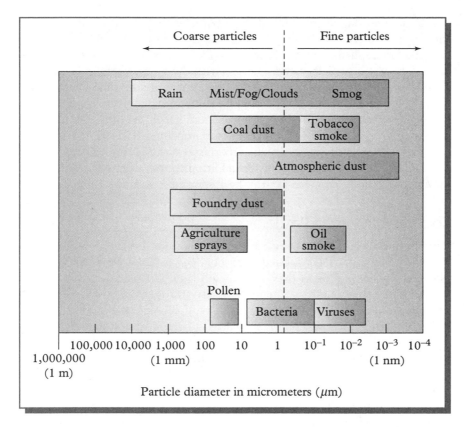

FIGURE 2-15 Sizes of common airborne fine and coarse particulates. [Source: Adapted from J.G. Henry and G.W. Heinke, *Environmental Science and Engineering* (Upper Saddle River, NJ: Prentice-Hall, 1989).]

the air so quickly they are not considered to be suspended. The range of particle sizes for common types of suspended particulates is illustrated in Figure 2-15.

Although few of the particles suspended in air are exactly spherical in shape, it is convenient and conventional to speak of all particles as if they were. Indeed, the **diameter** of particulates is their most important property. Qualitatively, individual particles are classified as **coarse** or as **fine** depending upon whether their diameters are greater or less than 2.5 μm. (About 100 million particles of diameter 2.5 μm would be required to cover the surface of a small coin.)

There are many common names for atmospheric particles: dust and soot refer to solids, whereas mist and fog refer to liquids, the latter denoting a high concentration of water droplets. An **aerosol** is a collection of particulates, whether solid particles or liquid droplets, dispersed in air. A true aerosol (as opposed, say, to the fairly large droplets from a hair-spray dispenser) has very small particles: their diameters are less than 100 μm.

Intuitively, one might think that all particles should settle out under the influence of gravity and be deposited on the Earth's surface rapidly, but this is not true for the smaller ones. According to *Stoke's law*, the rate, in distance per second, at which particles settle increases with the square of their diameter. In other words, a particle half the diameter of another falls four times more slowly. The small ones fall so slowly that they are suspended almost indefinitely in air (unless they stick to some object they encounter). As we shall see later, the very small ones aggregate to form larger ones, usually still in the fine size category. Fine particulates usually remain airborne for days or weeks, whereas coarse particulates settle out fairly rapidly. In addition to this sedimentation process, particles also can be removed from air by absorption into falling raindrops.

Sources of Coarse Particles

The primary-versus-secondary distinction made between atmospheric gaseous pollutants is also applied to suspended particles. Most coarse particles are primary, although they often begin their existence as even coarser matter, since they originate chiefly from the disintegration of larger pieces of matter. Minerals constitute one type of the coarse particulates in air. Because many of the large particles in atmospheric dust, particularly in rural areas, originate as soil or rock, their elemental composition is similar to that of the Earth's crust, namely high concentrations of Al, Ca, Si, and O in the form of aluminum silicates, some of which also contain calcium ions. Wind storms in deserts sweep large amounts of fine sand into the air. Near and above oceans, the concentration of solid NaCl is very high, since sea spray leaves sodium chloride particles airborne when the water evaporates. Indeed, sea salt aerosols are by far the largest mass of primary particles in air, followed by soil dusts and debris from natural fires. The wind generates coarse particles by the mechanical disintegration of leaf litter. Pollen released from plants also consists of coarse, primary particles. Wildfires and volcanic eruptions generate both fine and coarse particulate matter.

Although most coarse particulates originate from natural sources, human activities such as stone crushing in quarries and land cultivation result in particles of rock and topsoil being picked up by the wind. Coarse particles in many areas are basic, reflecting the calcium carbonate and other such salts in soils. Acidity from acid rain is often neutralized by the carbonate ion in these particles:

$$H^+ + CO_3^{2-} \longrightarrow HCO_3^-$$

Sources of Fine Particles

Primary fine particles of anthropogenic origin include those from the wearing of tires and vehicle brakes, and dust from metal smelting. The incomplete combustion of carbon-based fuels such as coal, oil, gasoline, and diesel fuel produces many fine soot particles, which are mainly crystallites (miniature crystals) of carbon. Consequently, one of the main sources of carbon-based primary atmospheric particulates, both fine and coarse, is the exhaust from vehicles, especially those having diesel engines. About half the organic content from heavy-duty diesel vehicles is elemental carbon; one can easily observe this soot as the black smoke that emanates from such equipment. Most carbon-containing emissions from gasoline-powered engines are composed of organic compounds rather than elemental carbon.

Whereas coarse particles result mainly from the breakup of larger ones, fine particles are formed mainly by chemical reactions between gases and by the coagulation of even smaller species, including molecules in the vapor state; they are therefore classified as secondary particles. Although most of the atmospheric mass of fine particles arises from natural sources, that over urban areas often has chiefly an anthropogenic origin. Most ultrafine particles are anthropogenic.

The average organic content of fine particles is generally greater than that of coarse ones. In areas such as Los Angeles, up to half the organic compounds in the particulate phase are formed from the reaction of VOCs and nitrogen oxides in the photochemical smog reaction and correspond to partially oxidized hydrocarbons that have incorporated oxygen to form carboxylic acids, etc. and nitrogen to form nitro groups, etc. Aromatic hydrocarbons with at least seven carbon atoms (e.g., toluene) that enter the air of such cities from the evaporation of gasoline also form aerosols. Hydrocarbons having fewer than seven carbons give oxidation products with substantial vapor pressures and therefore remain in the gas phase.

The other important fine particles suspended in the atmosphere consist predominantly of inorganic compounds of sulfur and of nitrogen. Much of the natural sulfur in air originates as *dimethyl sulfide*, $(CH_3)_2S$, emitted from the oceans. A by-product of its oxidation in air is *carbonyl sulfide*, COS, a long-lived trace atmospheric component that also results from the atmospheric oxidation of *carbon disulfide*, CS_2, and from direct emissions from oceans and biomass. Some of the COS makes its way into the stratosphere, where it is oxidized and produces the natural sulfate aerosol found at those altitudes. Both dimethyl sulfide and hydrogen sulfide are oxidized in air, mainly to sulfur dioxide, SO_2. Sulfur dioxide gas also is emitted directly in large quantities both by natural sources such as volcanoes and as pollution from power plants and smelters. It becomes oxidized over a period of hours or days to sulfuric acid and sulfates in air. Sulfuric acid, H_2SO_4, itself travels in air, not as a gas but as an aerosol of fine droplets, since it has such a great affinity for water molecules.

Another natural source of atmospheric particles was recently discovered. Alkyl iodine compounds such as CH_2I_2 are emitted by seaweed into the air above coastal regions. Absorption of the ultraviolet component of light is sufficient to detach iodine atoms from such gaseous molecules.

In subsequent reactions analogous to those of chlorine in the stratosphere, the iodine atoms react with ozone to form *iodine monoxide*, IO, which in turn dimerizes to form I_2O_2. The dimer and other iodine–oxygen compounds condense to form fine particles.

PROBLEM 2-12

By analogy with the reactions of atomic chlorine, write balanced equations for the reaction of atomic iodine with ozone and for the dimerization of IO.

Fine particles in many areas are acidic, due to their content of sulfuric and nitric acids. The nitric acid is the end-product of the oxidation of nitrogen-containing atmospheric gases such as NH_3, NO, and NO_2. Because HNO_3 has a much higher vapor pressure than does H_2SO_4, there is less condensation of nitric acid onto preexisting particles than occurs with H_2SO_4.

Both sulfuric and nitric acids in tropospheric air often eventually encounter ammonia gas that is released as a result of biological decay processes occurring at ground level. The acids undergo an acid–base reaction with the ammonia, which transforms them into **ammonium sulfate** and **ammonium nitrate** salts. For example,

$$H_2SO_4(aq) + 2\,NH_3(g) \longrightarrow (NH_4)_2SO_4(aq)$$

The neutralization of acidity by ammonia gas released into the air from livestock and from the use of fertilizers, and by carbonate ion suspended in air from the dust raised by farming activities, are the main reasons why precipitation over the central United States is not acidic (see Figure 2-11). The ammonia that is produced from animal urine originates in the liquid as **urea,** $CO(NH_2)_2$, which then hydrolyzes:

$$CO(NH_2)_2 + H_2O \longrightarrow 2\,NH_3 + CO_2$$

Although the nitrate and sulfate salts initially are formed from acids in aqueous particles, evaporation of the water can result in the production of solid particles. The predominant ions in fine particles are the anions **sulfate,** SO_4^{2-}, **bisulfate,** HSO_4^-, and nitrate, NO_3^-; and the cations ammonium, NH_4^+, and hydrogen ion, H^+. Aerosols dominated by oxidized sulfur compounds are called **sulfate aerosols.** On the west coast of North America, nitrate rather than sulfate is the predominant anion because more pollution results initially from nitrogen oxides than from sulfur dioxide, since coal mined in the western United States tends be low in sulfur. In Great Britain, most of the fine particles in the winter months originate as soot from car exhaust and pollution from industry, whereas in the summer they arise from the oxidation of sulfur and nitrogen oxides.

If there is substantial ammonia gas in the air, nitric acid will react with it to form the salt ammonium nitrate, a solid that will occur in the particulate phase. Recent simulations of smog formation in southern California indicate that, although reductions in VOC concentrations without any change in NO_X would reduce ozone formation, the production of nitrate-based particulates would actually increase because more nitrogen dioxide would react to produce nitric acid and then nitrate ion. The simultaneous control of ozone and particulates presents regulators with a formidable challenge!

In summary, coarse particles are usually either soot or inorganic (soil-like) in nature, whereas fine ones are mainly either soot, or sulfate or nitrate aerosols. Fine particles are usually acidic due to the presence of unneutralized acids, whereas coarse ones are usually basic because of their soil content.

Air Quality Indices and Size Characteristics for Particulate Matter

The PM Indices

When air quality is monitored, the most common measure of the concentration of suspended particles is the **PM** index, which is the amount of *particulate matter* that is present in a given volume. Since the matter involved usually is not homogeneous, no molar mass for it can be quoted and thus concentrations are given in terms of the mass, rather than the number of moles, of particles. The usual units are *micrograms* of particulate matter *per cubic meter* of air, i.e., $\mu g/m^3$. Because smaller particles have a greater detrimental effect on human health than do larger ones, as we shall see later in this chapter, usually only those having a lower than specified diameter are collected and reported; this cut-off diameter, in μm, is listed as the subscript to PM.

In recent years government agencies in many countries, including the United States and Canada, have monitored PM_{10}, i.e., the total concentration of all particles having diameters less than 10 μm, which corresponds to all of the fine-particle range plus the smallest members of the coarse range. These are called **inhalable** particles, since they can be breathed into the lungs. A typical value for PM_{10} in an urban setting is 20–30 $\mu g/m^3$. Increasingly, regulators are using the $PM_{2.5}$ index, i.e., one which includes all and only fine particles, which are also called **respirable** particles. The respirable range includes only particles that can penetrate deep into the lungs, where there are no natural mechanisms such as the cilia that line the walls of bronchial tubes to catch particles and move them up and out. Urban $PM_{2.5}$ values are usually in the 10- to 20-$\mu g/m^3$ range in North America, although background concentrations are only 1–5 $\mu g/m^3$. The new term **ultrafine** is applied to particles with very small diameters, usually taken to be less than 0.1 μm, although different scientists use different cutoff values. In the past, the **total suspended particulates,** abbreviated TSP, which is the concentration of all particulates suspended in air, was often reported instead of a PM index.

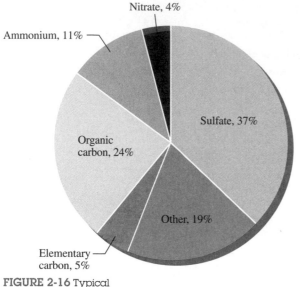

FIGURE 2-16 Typical composition of fine continental aerosol. [Adapted from J. Heintzenberg, *Tellus* 41B (1989): 149–160.]

PROBLEM 2-13

What would be the correct PM symbol for an index that included only ultrafine particles? What would be the PM symbol for the TSP index? Numerically, would the value for the ultrafine component of a given air mass be larger or smaller than its TSP?

Haze

Particles whose diameter is about that of the wavelength of visible light, i.e., 0.4–0.8 μm, can interfere with the transmission of light in air, reducing visual clarity, long-distance visibility, and the amount of light reaching the ground. A high concentration in air of particles of diameters between 0.1 and 1 μm produces a haze. Indeed, one conventional technique of measuring the extent of particulate pollution in an air mass is to determine its haziness. The widespread haze in the Arctic atmosphere in winter is due to sulfate aerosols that originate from the burning of coal, especially in Russia and Europe. The enhanced haziness in summertime over much of North America is due mainly to sulfate aerosols arising from industrialized areas in the United States and Canada. Fine particles also are largely responsible for the haze associated in Los Angeles and other locations with episodes of photochemical smog. The smog aerosols contain nitric acid that has been neutralized to salts such as ammonium nitrate. Also present in these aerosols are carbon-containing products that are intermediates in the photochemical smog reactions; however, intermediates formed from fuel molecules having short carbon chains usually have vapor pressures so high that they exist as gases rather than condense onto particles. The typical composition of the fine component of an aerosol suspended over continental areas is illustrated in Figure 2-16.

Since most fine particles in urban air are secondary, their number can only be controlled by reducing emissions of the primary pollutant gases—NO, VOCs, SO_2—from which they are created. Thus governments have successively required more and more stringent emission controls on vehicles, power plants, etc. As previously mentioned, the switch to low-sulfur gasoline and diesel fuels should make catalytic converters more efficient in reducing emissions.

Only recently have emissions from vehicles with diesel engines begun to be controlled. The use of catalytic converters to reduce gaseous emissions has already been discussed. However, typical catalytic converters do not work well with light-duty diesel vehicles because the engines don't operate at high temperatures; hence the exhaust gases entering the converter are relatively cool.

Some progress in the reduction of particulate emissions has been accomplished by a combination of engine modifications and the use of **particle traps** (filters). The traps, which can remove up to 90% of PM_{10} emissions, prevent organic particulate emissions from escaping from the exhaust system of light-duty diesel vehicles. In order to prevent a build-up of soot, which would restrict engine exhaust flow or melt the trap, the system is designed so that the soot will ignite and burn away once a temperature of at least 500 °C is attained. In many instances, however, this regeneration of the filter by combustion happens too infrequently for smooth operation, since the exhaust from diesel engines is typically relatively cool. One promising solution to this problem is to add tiny amounts of a metal catalyst compound containing iron or copper to the fuel; the addition of such metals to the soot lowers the temperature at which ignition will occur and so assures more frequent regeneration of the filter.

The Distribution of Particle Sizes in an Air Sample

Because particles suspended in the atmosphere have different origins and compositions, and were often formed and interacted with each other over a period of time in haphazard ways, there is a wide distribution of particle sizes present in any air mass.

One way of looking at the distribution of sizes is to plot the *number* of particles having a given diameter against the diameter; this is done in Figure 2-17 for a typical urban air sample. Notice that logarithmic scales are used on both axes, in order that the details of the distribution for particles of many sizes can be seen clearly. The peak in the distribution occurs at about 0.01 μm, and it has "shoulders" at about 0.1 μm and 1.0 μm. As indicated by the dashed lines, the net distribution appears to be the sum of three symmetrical (bell-shaped) distributions, with peaks at diameters of 0.01μm, 0.1μm, and 1 μm. The particles with the smallest diameter (0.01-μm peak) are formed by the condensation of vapors of pollutants formed by chemical reactions, such as the sulfuric acid formed by the oxidation of gaseous sulfur dioxide and the soot particles formed by combustion. Particles that form in this way constitute the **nuclei mode.** The coagulation of such particles into larger ones (which can occur in minutes) and the deposition of gas molecules on them give rise to the intermediate, **accumulation mode,** distribution (peak at 0.1 μm). Growth beyond this size is slow because the larger the particle, the more slowly it moves and the less likely it is to encounter and coagulate with particles of comparable size.

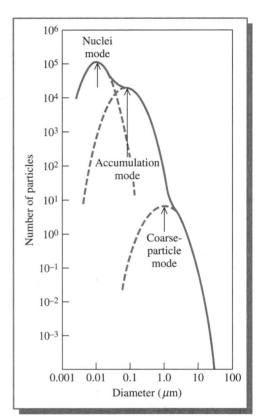

FIGURE 2-17 Distribution of numbers of aerosol particles versus size in a typical urban environment. [Source: Redrawn from K.T. Whitby, "The Physical Characteristics of Sulfate Aerosols," *Atmospheric Environment* 12 (1978): 135–159.]

Growth by condensation of gases is also slow for larger particles since their surface-to-mass ratio is smaller than that of small particles. Accumulation mode particles are also created when water in aqueous droplets containing dissolved solids evaporates.

The particles associated with the third distribution, the **coarse-particle mode,** notwithstanding that it includes some fine particles and peaks at 1 μm, are mainly soot or consist of material produced by mechanical disintegration of soil particles, etc. There are few particles *larger* than a few microns in diameter, because such particles quickly settle out of the air, although large particles that have settled on roadways often become resuspended temporarily by the action of vehicular traffic.

In summary, the distribution of particles suspended in air peaks in the micrometer region because

• smaller particles coagulate to form ones of this size and further growth is slow, and

• much larger ones rapidly settle.

The average time particles of various sizes remain suspended in the air is illustrated in Figure 2-18.

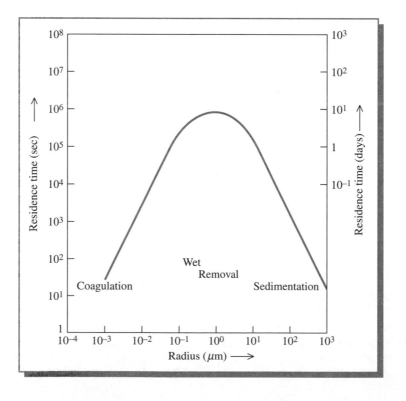

FIGURE 2-18 The average residence time for aerosol particles as a function of their size. The dominant sink for each size range is also shown. [Source: Redrawn from R. Jaenicke, "Aerosols: Anthropogenic and Natural Sources and Transport," *Annals of the New York Academy of Sciences* 338 (1980): 317–329.]

The plots of particle *numbers* can be misleading for some purposes because tiny particles of very small mass and surface area dominate the samples and thus the distributions. One alternative way to represent the data in a more meaningful way is to plot the total *mass* of all particles of a given size in an air sample against the diameter to see how mass is distributed among the different sizes. This type of plot is shown in Figure 2-19 for an urban air mass; for technical reasons, it is actually the volume rather than the mass that is plotted, but for particles of equal density the distributions are identical.

The distribution function for mass is displaced to larger diameters compared to that for particle numbers for the following reason: the mass (or volume) of a particle is proportional to the cube of its diameter d (since for a sphere, volume is proportional to the cube of the radius), so the height of the distribution curve at any diameter in Figure 2-19 corresponds to the value for the number distribution for this air mass times d^3. Consequently, in the mass distributions, the peak heights for larger particles are emphasized more than those for smaller ones, and the whole distribution appears to shift to larger diameters. Two symmetrical distribution curves, one centered in the fine region at about 0.3 μm and the other in the coarse region at about 7 μm, appear to be superimposed to produce the final bimodal distribution; in fact, they correspond to the second and third peaks of the number distribution. The nuclei mode peak does not show up at all in this plot since the total mass (and volume) of the tiny particles is so small. Notice that the total mass of the coarse-particle range (i.e., the sum of the area under the solid curve for $d > 2.5$ μm) in Figure 2-19 is greater than that for the fine region; this ratio is even larger for clean, rural air masses.

Substances that dissolve into the body of a particle are said to be **absorbed** by it; those that simply stick to the surface of the particle are said

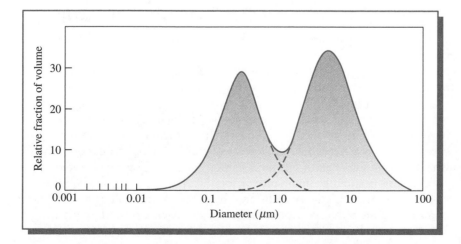

FIGURE 2-19 Distribution of particle volumes (and masses) for a typical urban model aerosol. [Source: Redrawn from K.T. Whitby and G.M. Svendrup, "California Aerosols: Their Physical and Chemical Characteristics," *Advances in Environmental Science and Technology* 10 (1980): 447.]

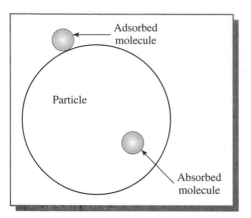

Adsorbed
molecule

Particle

Absorbed
molecule

FIGURE 2-20 Contrast between adsorption and absorption of molecules on/in an airborne particle (schematic).

to be **adsorbed** (see Figure 2-20). An important example of the latter is the adsorption of large organic molecules onto the surfaces of carbon (soot) particles. Many insoluble airborne particles are surrounded by a film of water, which can itself dissolve other substances. The large increase in surface area that occurs when a large particle is split into smaller ones is explored in Problem 2-15.

PROBLEM 2-14

The total surface area of the particles in an air mass is an important property since the transmission of contaminants by adsorption onto them is crucial in determining the effects of inhaled particles on human health. Since the surface area of a spherical particle is proportional to the *square* of its diameter, a plot of d^2 times the number of particles of a given size yields the distribution of surface areas. Predict qualitatively the location and relative sizes of the peaks for the three modes in the distribution plot of surface area versus diameter, given the curve positions in Figures 2-17 and 2-19.

PROBLEM 2-15

Let k be a given measure of length; then suppose a cubic particle of dimensions $3k \times 3k \times 3k$ is split up into 27 particles of size $k \times k \times k$. Calculate the relative increase in surface area when this occurs. From your answer, deduce whether the total surface area of a given mass of atmospheric particles is larger or smaller when it occurs as a large number of small particles rather than a small number of large ones.

The Health Effects of Outdoor Air Pollutants

The effect that pollutants have on human health cannot be deduced from general laws of biology or physiology but must be established by experimentation. One can imagine experiments involving animals or human volunteers in which the health effects of exposure to brief periods of artificially produced high-level pollution are studied. However, the extrapolation of information gained from short-term studies of high-level pollution to long-term exposures at low levels is difficult. In particular, for some pollutants, there may exist a **threshold** pollutant concentration, or an exposure below which a particular health effects does not occur; in these cases predictions obtained by assuming direct proportionality between exposure and effect would be unwarranted. In addition, there could be deleterious effects of chronic exposure that do not come into play when exposure, even intense exposure, to pollutants occurs only for brief periods of time.

For these reasons, the best information regarding the effects of pollutants on health comes from the large-scale experiment in which we are all enrolled as test animals—namely, living in a society in which we are routinely exposed to these pollutants for our whole lives. Because the level of exposure to any given pollutant varies considerably from place to place, scientists can collect information on health and on pollution levels in different locations and correlate them using statistics to establish the effect of one upon the other.

As would be expected, the major effects on human health from air pollution occur in the lungs. For example, asthmatics suffer worse episodes of their disease when the sulfur dioxide or the ozone or the particulate concentration rises in the air that they breathe. In one U.S. study it was established that asthma attacks increased by 3% for each increase of 10 $\mu g/m^3$ in the PM_{10} index. A recent study in California found that asthma can be *caused* by air pollution, specifically by ozone and especially among highly active children, who naturally inhale more air into their lungs.

Another gaseous pollutant of some concern is **1,3-butadiene,** which has the structure $CH_2\!=\!CH\!-\!CH\!=\!CH_2$. This hydrocarbon is known as an **air toxic** since there is evidence that it causes cancer—leukemia and *non-Hodgkin's lymphoma* especially—and may also negatively affect human reproduction. It is produced as a by-product of the incomplete combustion of fuels and in forest fires and wildfires; it is also a component of cigarette smoke.

The Human Health Effects of Smogs

In the middle decades of the twentieth century, several Western industrialized cities experienced wintertime episodes of smog from soot and sulfur pollution that were so serious that the death rate increased noticeably. For example, in London, England, in December 1952 about 4000 people died within a few days—plus 8000 more in the next few months—as a result of the high concentrations of these pollutants that had built up in a stagnant, foggy air mass trapped by a temperature inversion close to the ground. Those at most risk were elderly persons already suffering from bronchial problems and young children. A ban on household coal burning, from which most of the pollutants originated, has now largely eliminated such problems. Scientists are still unsure whether the main sulfur-containing agent that caused such serious problems in London was the SO_2, the sulfuric acid droplets, or the sulfate particulates.

Today, due to pollution controls, *soot-and-sulfur smogs* are no longer a major problem in Western countries. For example, deaths from bronchitis have fallen by over half in the United Kingdom, the result of changes in air quality (and smoking habits). However, the quality of winter air in some areas of what was the Soviet bloc of countries, such as southern Poland, the Czech Republic, and eastern Germany, until very recently at least was very poor on account of the burning of large amounts of high-sulfur (up to 15% S) "brown" coal for both industrial and home-heating purposes. For example, although the acceptable limit for the concentration of SO_2 in air is 80 $\mu g/m^3$ in many countries, the level of this gas in Prague has surpassed 3000 $\mu g/m^3$ on occasion. Indeed, four out of five children admitted to the hospital in some areas of Czechoslovakia in the early 1990s were there for treatment of respiratory problems. However, the average SO_2 level in Prague decreased by about 50% from the early 1980s to the early 1990s. The tremendous improvement of air quality in eastern Germany since 1990, where mean SO_2 levels have dropped from 113 to 6 $\mu g/m^3$, has produced a decrease in childhood respiratory infections and an increase in lung function.

The effects of sulfur dioxide are also evident in cities such as Athens, where the death rate is found to increase by 12% when the concentration of the gas exceeds 100 $\mu g/m^3$. Detail on the ancient statues and monuments of Athens is also being seriously eroded by sulfur dioxide and its secondary pollutants. A 1996 study indicated that high levels of sulfur dioxide and of fine particulates, both mainly from diesel-fueled vehicles, caused about 350 premature deaths in Paris annually in the late 1980s. And the air in London is not so improved that it does not affect human health; a recent study concluded that one in every 50 heart attacks was triggered by outdoor air pollution, from a combination of smoke, CO, SO_2, and NO_2.

European cities are not the only ones affected by air pollution. Both sulfur dioxide and particulate matter levels regularly exceed *World Health Organization* guidelines (see below) in Beijing, Seoul, and Mexico City. Particulate-level guidelines are also exceeded regularly in Bangkok, Bombay (now Mumbai), Cairo, Calcutta, Delhi, Jakarta, Karachi, Manila, Seoul, and Shanghai. In many of these megacities, coal is still the predominant fuel and in some cases diesel-powered vehicles substantially worsen the problem. In Beijing, high SO_2 emissions from burning coal to heat buildings plus smoke from smelters on the edges of the city and windblown dust and sand from the Gobi Desert combine to produce poor air quality. Indeed, there are a number of cities in China in which the air quality is among the poorest in the world. According to recent projections, if no attempts are made to reduce SO_2 emissions as industrialization increases, by 2020 the concentrations of the gas in Bombay and the Chinese cities of Shanghai and Chongqing will be about four times the WHO maximum safe limit.

Although acute smog episodes from soot and sulfur-based chemicals have been eliminated in the West, many residents in these countries still are chronically exposed to measurable levels of suspended particles containing sulfuric acid and sulfates due to the long-range transport of these substances from industrialized regions that still emit SO_2 into the air. For example, research has shown a positive correlation between atmospheric concentrations of oxidized sulfur and ozone and hospital admissions for respiratory problems in southern Ontario. There is some evidence that the acidity of the pollution is the main active agent in causing lung dysfunction, including wheezing and bronchitis in children. Asthmatic individuals appear to be adversely affected by acidic sulfate aerosols, even at very low concentrations.

Photochemical smog, which arises from nitrogen oxides, is now more important than sulfur-based smog in most cities, particularly those with high population and vehicle densities. It consists of gases such as ozone and an aqueous phase containing water-soluble organic and inorganic compounds in the form of suspended particles. In contrast to London smogs, which chemically were reducing in nature due to sulfur dioxide, photochemical smogs are oxidizing.

Ozone itself is a harmful air pollutant. In contrast to sulfur-based chemicals, its effect on the robust and healthy is as serious as on those with pre-existing respiratory problems. Experiments with human volunteers have shown that ozone produces transient irritation in the respiratory system, giving rise to coughing, nose and throat irritation, shortness of breath, and chest pains upon deep breathing. People with respiratory problems can often tell from symptoms—such as the tightening of the chest or the beginning of a cough—when the air quality is poor. Even healthy young people often experience such symptoms while exercising outdoors by cycling or jogging during smog episodes. Indeed, there is evidence that the daily race times of cross-country runners increase with increasing ozone concentration in the air that they inhale. One recent study indicated that a few percent of the day-to-day fluctuations in mortality rate in Los Angeles is explained by variations in the concentrations of air pollutants. It is not yet clear what, if any, long-term lung dysfunction results from exposure to ozone, and indeed this is a controversial subject among scientists.

One anticipated effect of ozone is a decreased resistance to infectious disease because of the destruction of lung tissue. Many scientists believe that chronic exposure to high levels of urban ozone leads to the premature aging of lung tissue. At the molecular level, ozone readily attacks substances containing components with $C{=}C$ bonds, such as those that occur in the tissues of the lung. As discussed later, the fine particulates produced in the photochemical smog process can have a deleterious health effect on humans.

Most industrialized nations have enacted standards (see Table 2-2) that regulate the maximum concentrations in air of sulfur dioxide, nitrogen dioxide, and carbon monoxide, as well as ozone and in some cases total reduced sulfur, since all these gases cause health effects at sufficiently high concentrations. For example, several recent North American studies have statistically linked the rate of hospitalization for congestive heart failure among elderly people to the daily carbon monoxide concentration in outside air. The average outdoor concentration of CO in the United States fell from 10 ppm to 6 ppm from 1975 to 1991. Mexico City currently has the highest levels of carbon monoxide among the world's most polluted cities. Both CO and NO_2 are usually more of a problem in indoor air and will be discussed in detail in a later section.

TABLE 2-2	Air Quality Standards for Gaseous Pollutants		
Pollutant	Time Span to Average	U.S. EPA Standard	EU Standard[a]
SO_2	1 day	140 ppb	47 ppb
	1 year	30 ppb	8 ppb
NO_2	1 hour		105 ppb
	1 year	53 ppb	21 ppb
CO	1 hour	35 ppm	
	8 hour	9 ppm	
O_3	1 hour	120 ppb	90–180 ppb
	8 hour	80 ppb (proposed)	55 ppb

[a]Some European Union standards allow exceedance of these values several times per year. E.U. values are for 20°C.

It has been speculated that pollution due to SO_2 and sulfates causes a decrease in resistance to colon and breast cancer in people living in northern latitudes. The suggested mechanism of this action is a reduction in the amount of available UV-B, which is necessary to form vitamin D, a protective agent for both types of cancers. Since sulfur dioxide absorbs UV-B and sulfate particles scatter it, significant concentrations of either substance in air will reduce the amount of UV-B reaching ground level. Thus too little UV-B can have detrimental health effects, just as too much of it can, as was outlined in Chapter 1.

Finally, we note that there are some positive effects of air pollution on human health! For example, the rate of skin cancers in areas heavily polluted by ozone is probably reduced because of the ability of the gas to filter UV-B from sunlight.

Particulates as Health Risks

Particulate matter in the form of smoke from coal burning has been an air pollution problem for many hundreds of years, especially in the United Kingdom. John Evelyn wrote in his January 1684 diary that "London by reason of the excessive coldness of the air, hindering the ascent of the smoke, was so filled with the fuliginous [sooty] steam of sea-coal, that hardly could one see across the street, and this filling the lungs with its gross particles exceedingly obstructed the breast, so as one would scarce breathe." Indeed, unsuccessful attempts to control coal burning and punish offenders had begun in the thirteenth century in Britain. Perhaps Shakespeare was referring to this type of air pollution in the quotation from *Hamlet* that opened this chapter.

Although serious episodes of such soot-and-sulfur smogs have been largely eliminated in Western industrialized countries, the air pollution parameter that correlates most strongly with increases in the rate of disease or mortality in these regions is the concentration of respirable (fine) particulates, $PM_{2.5}$. It appears that particulate-based air pollution has a greater effect on human health than that produced directly by pollutant gases.

Large particles are of less concern to human health than are small ones for several general reasons:

• Since coarse particles settle out quickly, human exposure to them via inhalation is reduced.

• When inhaled, coarse particles are efficiently filtered by the nose (including its hairs) and throat and generally do not travel as far as the lungs. In contrast, inhaled fine particles usually travel through to the lungs (which is why they are called respirable), can be adsorbed on cell surfaces there, and can consequently affect health.

• The ratio of surface area to mass of large particles is smaller than that of small ones; thus, gram for gram, their ability to transport adsorbed gas molecules to any parts of the respiratory system and there to catalyze chemical and biochemical reactions is correspondingly smaller.

• Devices such as electrostatic precipitators, spray towers, and cyclone collectors that are used to remove particulates from air are efficient only for coarse particles. Although a device may remove 95% of the total particulate mass, the reduction of surface area and of respirable particles is a much lower fraction; see Problem 2-16. Baghouse filters, which are finely woven fabric bags through which air is forced, are highly efficient in removing fine particles in the $1\text{-}\mu\text{m}$ size range and all the larger ones as well.

The exhaust from diesel engines has been classified as "likely to be carcinogenic to humans" by the U.S. Environmental Protection Agency (EPA). Studies in California and in Seattle conclude that 70% or more of the risk to health from air toxics arises from diesel exhaust. Following court decisions, the United States will institute by 2010 a series of new regulations limiting emissions from on-road diesel vehicles.

PROBLEM 2-16

An air-filtering device is tested and is found to remove all particles larger than $1 \mu\text{m}$ in diameter but almost none of the smaller ones. Calculate the percentage of the surface area removed by the device for a sample of particulates, 95% of the mass of which is particles of diameter $10 \mu\text{m}$ and 5% of which is particles of diameter $0.1 \mu\text{m}$. Assume all particles are spherical and of equal density.

A number of studies have correlated day-to-day urban morbidity (sickness) rates, as measured by hospital admission rates, for respiratory problems with the pollution levels during the same short time period. For example, there have been several reports concerning the immediate effects on the population of southern Ontario of the pollutants—ozone gas and sulfate particulates—to which they are most exposed. In one study, the average number of hospital admissions for respiratory problems correlated best with the ozone level of the previous day, and to a slightly lesser degree with the sulfate level of the previous day, for the summers of 1983–1988. Air pollution was found to account for about 6% of summertime hospital respiratory admissions, a magnitude close to that found in previous investigations in Ontario and New York State. Another recent study found that respiratory admissions correlated significantly with both PM_{10} and ozone concentrations in Spokane, Washington—an area where atmospheric sulfur dioxide is essentially nonexistent and therefore can be ruled out as a culprit in causing the illnesses.

The strongest evidence linking the deterioration of human health to airborne particulates comes from statistical studies correlating death rates in different cities with their particulate air pollution levels. In such studies, the rates of death—either total rates or rates from diseases such as lung cancer or cardiovascular disease—are plotted against the average concentration of particulates in order to determine whether they are related.

The question of whether particulate air pollution actually does shorten human lifespans by increasing mortality rates has been intensively researched in the past few decades, particularly in the United States. These epidemiological studies have shown consistent and significant associations between daily and long-term mortality rates and particulate concentrations in air, especially the fine component ($PM_{2.5}$). These correlations are consistently much stronger than those with concentrations of any gaseous pollutant.

One of the most influential studies of the effect of air pollution on human health was carried out for six medium-sized American cities using data extending from the late 1970s to the late 1980s. The cities were Portage, Wisconsin; Topeka, Kansas; Watertown, Massachusetts; Harriman, Tennessee; St. Louis, Missouri; and Steubenville, Ohio. The correlation between the *daily* death rate and $PM_{2.5}$ concentration that resulted from this study is shown in Figure 2-21. The shaded area in the figure represents the uncertainty in the various data points. The best straight line through the data goes close to the origin, indicating that there is no threshold: any amount of fine particles in the air increased the mortality rates. The plot means that typical U.S. cities in this period with high $PM_{2.5}$ (~20 $\mu g/m^3$) had about a 1.4% higher daily death rate than typical low-$PM_{2.5}$ (~10 $\mu g/m^3$) cities due to short-term air pollution. A reduction of 10 $\mu g/m^3$ across the board would result in about 36,000 fewer early deaths per year in the United States, about the same number that die in automobile accidents.

Recently the authors of the six-cities study have analyzed their data further and established that the greatest increase in mortality arises from particles that originated with emissions from vehicles, followed by those—mainly sulfates—resulting from coal combustion. Fine particles originating from

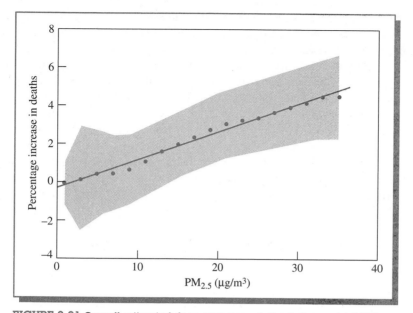

FIGURE 2-21 Overall estimated does–response relation between total $PM_{2.5}$ and daily deaths in six U.S. cities. The estimate is obtained by combining the estimated smoothed curves in each of the cities, after controlling for weather, season, and day of the week. The shaded area indicates the pointwise 95% confidence intervals at each point. The line shown is a least-squares regression line through the estimated points. [Source: J. Schwartz et al., "The Concentration–Response Relation Between $PM_{2.5}$ and Daily Deaths," *Environmental Health Perspectives* 110 (2002): 1025.]

dust and soil had no effect on mortality. It should be noted that, at the time the results were gathered, leaded gasoline was still in use, so the composition and/or effects of particles from automobiles may have changed since then. There are hints in the six-cities data and in other studies that it may be the ultrafine particulates that are especially dangerous to health. Some scientists have warned that the drive to decrease $PM_{2.5}$ levels will be counterproductive if by doing so the number of ultrafine particles is greatly increased; e.g., by converting from diesel to natural gas vehicles. Indeed, even gasoline engines produce more ultrafine particulates than do diesel engines.

A recent Canadian study found that short-term increases in fine-particle concentrations affected mainly people with acute lower respiratory diseases, chronic coronary artery diseases (especially the elderly), and congestive heart failure but not a number of other conditions, including chronic upper respiratory diseases and acute coronary artery disease. Several other studies in North America have established that daily mortality rates, including that of very young children who succumb to sudden infant death syndrome, correlate with PM_{10} values. Particulates and gaseous pollutants have also been linked to an increase in fatal strokes.

Studies have also been done to determine the effects of *chronic* particulate air pollution on overall mortality rates. These studies have involved a large number of American cities and towns. Particulate levels correlated with long-term mortality rates from lung cancer and cardiopulmonary causes but, as expected, not from death from all other causes combined in the most detailed of the analyses. In the most recent many-city study, a 4% increase in a city's overall mortality rate, due to 6% and 8% increases in death rates from cardiopulmonary causes (respiratory and cardiovascular) and lung cancer, respectively, were observed for every 10 μg/m^3 increase in the PM$_{2.5}$ index. The health risk from breathing polluted air is comparable to that of a nonsmoker living with a cigarette smoker.

The overall results obtained on the basis of a large number of studies correlating health with particulate levels are summarized by the graph in Figure 2-22 for the effects of both acute and chronic exposure. A number of analyses have also been performed to determine what subsets of the

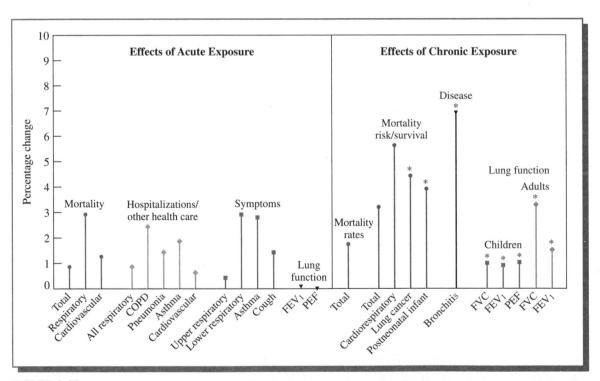

FIGURE 2-22 Stylized summary of observed health effects, presented as approximate percentage changes in health end points per 5 μg/m^3 increase in PM$_{2.5}$. Abbreviations: COPD, chronic obstructive pulmonary disease; FVC, forced vital capacity; PEF, peak expiratory flow. Asterisk (*) indicates estimate based on very limited or inconsistent evidence. [Source: C.A. Pope III, "Epidemiology of Fine Particle Air Pollution and Human Health: Biologic Mechanisms and Who's at Risk?" *Environmental Health Perspectives* 108, Supplement 4 (2000): 713.]

population, if any, are most susceptible to particulates in the air, and the results of these studies are shown in Table 2-3. The elderly, infants, and those with certain preexisting diseases are found to be the most susceptible to acute exposures to high pollution levels, though many of the deaths in these cases may only be advanced by a few days by the temporarily high pollution level. There is no evidence that deleterious effects of long-term pollution are restricted to these groups. Lifetime exposure to high particulate levels appears to reduce average life expectancy by one to three years; by contrast, smoking decreases it by about nine years.

Of course, the effects of particulate air pollution are not restricted to North America. Fine particles may be responsible for up to 10,000 premature deaths in Great Britain, for example. A report by the United Nations Environmental Program estimates that deaths worldwide from all forms of air pollution amounted to 2.7–3.0 million in 2001, a figure which may rise to 8 million by 2020.

Notwithstanding the substantial amount of circumstantial evidence from correlations of the type discussed, some scientists do not believe that the case has yet been proven concerning the causal link between particulate air pollution and human mortality. They point out that most people spend most of their time indoors and that consequently their personal exposure to particulates is not tightly linked to outdoor pollution levels. In addition, no biological mechanism has as yet been firmly established to account for the effect of the particles upon health.

The 1987 U.S. Air Quality Standards called for a maximum 24-hour PM_{10} level of 150 $\mu g/m^3$ and a maximum annual average of 50 $\mu g/m^3$. The United Kingdom has instituted a 24-hour PM_{10} standard of 50 $\mu g/m^3$ that cannot legally be exceeded on more than four days each year. In 1997 the U.S. EPA decided to regulate $PM_{2.5}$ levels—to an average of no more than 15 $\mu g/m^3$ annually and 65 $\mu g/m^3$ daily. The EPA estimated that the new particulate standards could prevent 15,000 premature deaths, as well as 250,000 person-days of aggravated asthma, annually.

Smoke

The burning of wood in domestic fireplaces produces large quantities of particulates, which are emitted from the chimneys into outdoor air unless catalytic converters are fitted to the smokestack. Indeed, in residential neighborhoods where wood is the predominant fuel used for heating, wood stoves contribute up to 80% of the fine particles in the air during the winter months. Some newer wood stoves have catalytic converters or secondary combustion chambers in which particulates and unburned gases are more fully oxidized, thus reducing their emissions to outside air.

Serious episodes of smoky haze pollution over large areas of land have occurred in recent years in Southeast Asia, especially in Malaysia and Indonesia. The smoke originates mainly from forest fires that are intentionally

TABLE 2-3	Summary of Who's Susceptible to Adverse Health Effects from PM Exposure and Overall Health Relevance	
Heath Effects	Who's Susceptible?	Overall Health Relevance
Acute exposure Mortality	Elderly, infants, persons with chronic cardiopulmonary disease, influenza, or asthma.	Obviously relevant. How much life shortening is involved and how much is due to short-term mortality displacement (harvesting) is uncertain.
Hospitalization/other health care visits	Elderly, infants, persons with chronic cardiopulmonary disease, pneumonia, influenza, or asthma.	Reflects substantive health impacts in terms of illness, discomfort, treatment costs, work or school time lost, etc.
Increased respiratory symptoms	Most consistently observed in people with asthma and children.	Mostly transient effects with minimal overall health consequences, although for a few there may be short-term absence from work or school due to illness.
Decreased lung function	Observed in both children and adults.	For most, effects seem to be small and transient. For a few, lung function losses may be clinically relevant.
Plasma viscosity, heart rate, heart rate variability, pulmonary inflammation	Observed in both healthy and unhealthy adults. No studies of children.	Effects seem to be small and transient. Overall health relevance is unclear, but may be part of pathophysiologic pathway linking PM with cardiopulmonary mortality.
Chronic exposure Increased mortality rates, reduced survival times, chronic cardiopulmonary disease, reduced lung function	Observed in broad-based cohorts or samples of adults and children (including infants). All chronically exposed potentially are affected.	Long-term, repeated exposure appears to increase the risk of cardiopulmonary disease and mortality. May result in lower lung function. Population average loss of life expectancy in highly polluted cities may be as much as a few years.

Source: C.A. Pope III, "Epidemiology of Fine Particle Air Pollution and Human Health: Biologic Mechanisms and Who's at Risk?" *Environmental Health Perspectives* 108, Supplement 4 (2000): 713.

started in order to clear land that can be subsequently used for agriculture and to grow trees for their rubber, palm oil, or pulp content. A secondary source of the smoke is the smoldering underground fires that slowly burn in underground coal and peat deposits. Indeed, there are estimated to be a quarter million individual coal fires currently burning in Indonesia, as well as many in China and India, and there are also many peat fires in Malaysia. The fires are initiated when an outcropping of coal, or of a peat deposit that has dried after draining, is ignited, typically during one of the fires set to clear the land. Fires can also be ignited in coal, once exposed to the air, by lightning strikes, and even by spontaneous combustion when the surface pyrite is oxidized and the exothermicity of this reaction sets the carbon ablaze. These underground fires can continue to burn for decades after the original forest fires have stopped.

Overall, a so-called *Asian brown cloud* of particles and gases from forest fires, vehicle exhausts, and domestic cookers—especially in rural areas—that burn wood, dung, and agricultural waste hangs over most of eastern and southeastern Asia annually from December to May, the main season for home heating. This haze lowers sunlight levels by 10%, with a corresponding decline in the yield of crops such as rice. In contrast to the pollution aerosol over North America and Europe, to which it is comparable in magnitude, the "black carbon" content of the Asian cloud is significant. The absorption of sunlight by this elemental carbon alters the local hydrological cycle and hence the weather over the northern Indian Ocean. The lack of nitric oxide produced in the low-temperature flames of burning biomass currently limits ozone production over the area, but that will likely be reversed in the future with increased use of fossil fuels for vehicles.

Large forest fires in northern Canada produce huge quantities of carbon monoxide and VOCs, which have been found to travel as far as the U.S. Southeast and which may well increase ozone and particulate concentrations in the air of this region.

Indoor Air Pollution

The levels of some common air pollutants often are greater indoors than outdoors, although pollutant concentrations do vary significantly from one building to another. Since most people spend more time indoors than outdoors, exposure to indoor air pollutants is an important environmental problem. Indeed, the inadequate ventilation practices encountered in developing countries that burn coal, wood, crop residues, and other unprocessed biomass fuels create smoke and carbon monoxide pollution that produces respiratory problems and ill health among huge numbers of people in these countries. Women and young children are particularly affected since they spend more time indoors. Cooking smoke from biomass fuels increases asthma rates among elderly men and women. The particulate emissions from traditional cookstoves used indoors in developing countries can be reduced by 90% by switching from wood to charcoal.

Pollution of indoor air by radon is discussed in Chapter 13, that by pesticides in Chapter 7, and that by polycyclic aromatic hydrocarbons (PAHs) in Chapter 8. Chloroform in indoor air is considered when water purification is discussed, along with indoor air contamination by chlorinated organic solvents, in Chapter 10.

Formaldehyde

The most controversial indoor organic air pollutant gas is **formaldehyde,** $H_2C{=}O$. It is a widespread trace constituent of the atmosphere since it occurs as a stable intermediate in the oxidation of methane and of other VOCs. While its concentration in clean outdoor air is too small to be important—about 10 ppb in urban areas, except during episodes of photochemical smog—the level of formaldehyde gas *indoors* is often orders of magnitude greater, in certain cases exceeding 1000 ppb (1 ppm). A survey of U.S. homes in the late 1990s found that the indoor formaldehyde concentration usually was in the 5- to 20-ppb range, with outdoor levels lower still.

The chief sources of indoor exposure to this gas are emissions from cigarette smoke and from synthetic materials that contain formaldehyde resins (a type of plastic) used in urea formaldehyde foam insulation and in the adhesive employed in manufacturing plywood and particleboard (chipboard). Many useful resins (which are rigid polymeric materials) are prepared by combining formaldehyde with another organic substance. Formaldehyde itself is used in the dyeing and glueing of carpets, carpet pads, and fabrics. In the first few months and years after their manufacture, however, such materials release small amounts of free formaldehyde gas into the surrounding air. Consequently, new prefabricated structures such as mobile homes that contain chipboard generally have much higher levels of formaldehyde in their air than do older, conventional homes. Many manufacturers of pressed-wood products have now modified their production processes in order to reduce the rate at which formaldehyde is released.

The rate of formaldehyde emission from synthetic materials increases with temperature and relative humidity and declines as the materials age. Initially, formaldehyde temporarily trapped as a gas or simply adsorbed onto the materials is released into the surrounding air. There is also release of formaldehyde due to the rearrangement and dissociation of amide end-groups on resin polymers, from R—NH—CH$_2$OH to R—NH$_2$ + H$_2$CO. Later, slow but continuous reactions of water vapor in humid air with the methylene bridges joining amide groups within the polymer backbone provide a constant emission of formaldehyde:

$$R{-}NH{-}CH_2{-}NH{-}R + H_2O \longrightarrow 2\,R{-}NH_2 + H_2CO$$

Formaldehyde has a pungent odor, with a detection threshold in humans of about 100 ppb; its odor is often noticeable in stores that sell carpets and synthetic fabrics. At somewhat higher levels, many people report problems of irritation to their eyes, especially if they wear contact lenses, and to their noses, throats, and skin. The formaldehyde in cigarette smoke can cause eye irritation. Chronic exposure to low levels of formaldehyde produces similar effects and respiratory symptoms. Formaldehyde in air may cause children to develop asthma and to have more respiratory infections and allergies and asthma, although evidence for these effects is controversial. Dampness in homes, allowing the proliferation of dust mites, fungi, and bacteria, also plays a large role in increasing lower respiratory tract illnesses, especially in children.

Formaldehyde is thought to be the most important VOC in producing what is known as **sick building syndrome.** This term is used to describe situations in which the occupants of a building experience acute health and comfort effects that seem to be linked to the time they spend in a particular building, though no specific illness or cause is apparent. Complaints commonly include

- headaches;

- irritation of eyes, nose, or throat; dry cough;

- dizziness and nausea; fatigue;

- difficulty in concentrating; and

- dry or itchy skin.

In addition to VOCs emitted from indoor sources, other causes include inadequate ventilation, pollutants entering from outside the building, and biological contamination of the air from bacteria, molds, pollen, and viruses that have bred in stagnant water that has accumulated in air vents, etc.

Formaldehyde is established as a carcinogen (a cancer-causing agent) in test animals and may also be carcinogenic to humans; it was classified as a probable human carcinogen by the U.S. EPA in 1987. The expected cancer sites are in the respiratory system, including the nose, and cancers at these sites have been found for some people who are exposed to the gas in occupational settings. However, studies of human populations exposed to formaldehyde have led to no clear-cut conclusions concerning an increase in cancer frequency arising from nonoccupational exposure to it. From animal studies, an upper limit to the possible effect in humans can be estimated: it corresponds to an increase in the cancer rate of one or two cases per 10,000 people after 10 years of living in a high-formaldehyde house or trailer. However, the lower limit to the effect could well be a zero increase in cancer rate. In summary, no scientific consensus has yet been reached on the dangers to human health of low-level exposure to formaldehyde.

Benzene

Like formaldehyde, benzene is classified as a **hazardous air pollutant,** HAP. **Benzene,** C_6H_6, is a stable, volatile liquid hydrocarbon that through the modern age has found a variety of uses. It is a minor constituent of gasoline and was commonly used as a solvent for many organic products, including paints and inks. The public is exposed to benzene vapor indoors from the use of solvents and gasoline, through smoking (mainly for the smoker but to a lesser extent for those inhaling second-hand smoke), and from the importing of benzene from outdoor air into the house. The levels of benzene generally are smaller outdoors and in large buildings than in individual homes, especially those with smokers living in them. A significant fraction of benzene vapor exposure occurs while riding in motor vehicles and refueling them at gas stations.

Benzene is classified by the EPA as a *known human carcinogen*. Chronic exposure at high occupational levels increases the rate of leukemia in individuals. Indeed, there were many deaths among workers in the first half of the twentieth century from exposure to benzene from petroleum-based solvents, such as those used in the rubber and glue industries; in the making of paints, adhesives, and coatings; and in dry cleaning. It also causes *aplastic anemia*, a condition in which an individual is chronically tired and is especially subject to infections because the bone marrow produces insufficient red blood cells. There continues to be some uncertainty, however, about whether occupational or domestic exposure to low levels of benzene vapor does indeed increase the risk for leukemia and multiple myeloma.

Because of the serious health problems it causes, the use of benzene as a solvent has largely been phased out. In addition, its maximum allowable level in gasoline has been reduced. Benzene can be replaced in many applications by *toluene*, which is benzene with one hydrogen atom replaced by a methyl group. The —CH_3 group in toluene provides liver enzymes with a site that is much easier to attack and thereby initiate metabolism than the very strong bonds in benzene itself.

Nitrogen Dioxide

Indoor concentrations of NO_2 often exceed outdoor values in homes that contain stoves, space heaters, and water heaters that are fueled by gas. The flame temperature in these appliances is sufficiently high that some nitrogen and oxygen in the air combine to form NO, which eventually is oxidized to nitrogen dioxide. In one study, it was established that NO_2 levels in homes that use gas for cooking or that have a kerosene stove average 24 ppb, compared to 9 ppb for homes that have neither. Peak concentrations near gas cooking stoves can exceed 300 ppb.

Some nitric oxide is also released from the burning of wood and other biomass fuels, since these natural materials contain nitrogen. However, the flame temperature in burning such fuels is much lower than in burning gas, so little thermal NO is produced from nitrogen in the air.

Nitrogen dioxide is soluble in biological tissue and is an oxidant, so its effects on health, if any, are expected to occur in the respiratory system. There have been many studies of the effects on respiratory illness in children due to exposure to low levels of NO_2 emitted by gas appliances, but the results of different studies are not mutually consistent and are inadequate for establishing a cause-and-effect relationship. One study, conducted by researchers at Harvard University, found that a 15-ppb increase in the mean NO_2 concentration in a home leads to about a 40% increase in lower respiratory system symptoms among children aged 7 to 11 years. Nitrogen dioxide is the only oxide of nitrogen that is detrimental to health at concentrations likely to be encountered in residences.

Nitrogen dioxide is probably responsible for the finding that indoor concentrations of *nitrous acid*, HNO_2, exceed those found outdoors, since the gas reacts with water to form nitrous and nitric acids:

$$2\,NO_2 + H_2O \longrightarrow HNO_2 + HNO_3$$

Indoor nitrous acid concentrations were found to correlate inversely with ozone gas concentrations, presumably because the acid is oxidized to nitric acid by the gas.

Carbon Monoxide

Carbon monoxide is a colorless, odorless gas whose concentration indoors can be greatly increased by the incomplete combustion of carbon-containing fuels such as wood, gasoline, kerosene, or gas. High indoor concentrations usually are the result of a malfunctioning combustion appliance, such as a kerosene heater. Even properly functioning kerosene or gas heaters in poorly ventilated rooms can result in CO levels in the 50- to 90-ppm range. Average indoor and outdoor CO concentrations usually amount to a few parts per million, though elevated values in the 10- to 20-ppm range are common in parking garages, due to the carbon monoxide emitted by motor vehicles. Exhaust fumes containing CO and other pollutants can enter homes having attached garages. In developing countries, carbon monoxide poisoning is a serious hazard when biomass fuels are used to heat poorly ventilated rooms in which people sleep.

People such as traffic police who work outdoors in areas of high vehicular traffic can be exposed to elevated CO levels for long periods. Between 1986 and 1995, the average outdoor CO levels in the United States fell by 37%. Among major cities, Mexico City is considered to have the worst outdoor CO problem. The introduction of **oxygenated** substances, which are hydrocarbons in which some of the atoms have been replaced by oxygen, into American gasoline was expected to reduce CO emissions from vehicles (see Chapter 6).

The major danger from carbon monoxide arises from its ability, when inhaled, to complex strongly with the hemoglobin in blood and thus to impair its ability to transport oxygen to cells. Hemoglobin's affinity for CO is 234 times that for oxygen, and once one CO is bound, the rate of release of oxygen molecules to cells is reduced. Recent research has found that mental functioning is reduced during short-term exposure to high levels of CO and perhaps also as a result of long-term exposure to low concentrations.

On average, nonsmokers have about 1% of their hemoglobin tied up as the complex with CO (called carboxyhemoglobin); the figure for smokers is many times this value because of the carbon monoxide that they inhale during smoking and that arises from the incomplete combustion of the cigarettes. Studies have shown that increased mortality from heart disease can result even if only several percent of hemoglobin is chronically tied up as the CO complex. Exposure to very high concentrations of CO results in headache, fatigue, unconsciousness, and eventually death (if the exposure is sustained for long periods).

Low-priced, easily installed carbon monoxide detectors suitable to warn residents in homes and offices when high CO levels occur are now on the market. However, scientists have begun to worry about the poorly known health effects of chronic exposure to low levels of CO and the fact that such exposure may be quite common.

Environmental Tobacco Smoke

It is well established that smoking tobacco is the leading cause of lung cancer and is one of the main contributors to heart disease. Nonsmokers are often exposed to cigarette smoke, although in lower concentrations than smokers, since it is diluted by air. This **environmental tobacco smoke,** ETS (second-hand smoke), has been the subject of many investigations in order to determine whether or not it is harmful to people who are exposed to it.

ETS consists of both gases and particles. The concentration of some toxic products of partial combustion is actually *higher* in sidestream smoke than in mainstream, since combustion occurs at a lower temperature—and so is less complete—in the smoldering cigarette than in one through which air is being inhaled. Since the sidestream smoke is usually diluted by air before being inhaled, however, the concentrations of pollutants reaching the lungs of nonsmokers are much lower than those reaching the lungs of smokers themselves.

The chemical constitution of tobacco smoke is complex: it contains thousands of components, several dozen of which are carcinogens. The gases in smoke include

- carbon monoxide, carbon dioxide;

- formaldehyde and several other aldehydes, ketones, and carboxylic acids;

- nitrogen oxides, hydrogen cyanide, ammonia, and a number of organic nitrogen compounds;

- methyl chloride;

- 1,3-butadiene;

- toluene, benzene, and several hundred different PAHs, discussed in Chapters 6 and 8; and

- cadmium and radioactive elements such as polonium.

Included in the nitrogen compounds are several *nitrosamines*, organic nitrogen compounds of formula $R_2N-N=O$, which, together with the PAHs, are probably the most important respiratory carcinogens in the smoke.

The particulate phase of cigarette smoke is called the **tar,** and much of it is respirable in size. The zone in a cigarette that actively burns when a smoker inhales a puff is quite hot ($700-950°C$) and produces CO and H_2 as well as the expected CO_2 and water vapor. Immediately downstream of this area is a cooler zone ($200-600°C$) where smoke constituents such as nicotine distill out of the tobacco. When this vapor cools farther along the cigarette path toward the smoker, much of it condenses to aerosol particles that constitute the particulate phase of the smoke.

Many people experience irritation of their eyes and airways from exposure to ETS. The gaseous components of ETS, especially formaldehyde, hydrogen cyanide, *acetone*, toluene, and ammonia, cause most of the odor and irritation. Exposure to ETS aggravates the symptoms of many people who suffer from asthma or from **angina pectoris,** chest pains brought on by exertion. ETS, particularly when it originates with maternal smoking, is known to induce new cases of asthma in children, especially those of preschool age. Some recent studies have established correlations between the rate of acute respiratory illness and the level of indoor $PM_{2.5}$ (which would include the total amount of respirable particulates from all sources, including tobacco smoke). **Passive smoking**—which involves inhalation of sidestream as well as already exhaled smoke—is believed by scientists to cause bronchitis, pneumonia, and other infections such as those of the ear in up to 300,000 infants and several thousand instances of sudden infant death in the United States each year. Second-hand smoke may even reduce the cognitive abilities of children, whether they are exposed prenatally or when young. One recent study in Finland established that being exposed to second-hand smoke, whether on the job or by living with a smoker, approximately doubles a nonsmoker's chance of developing asthma.

In 1993, the U.S. EPA classified ETS as a *known human carcinogen* and estimated that it causes about 3000 lung cancer deaths annually. ETS is also considered to be responsible for killing as many as 60,000 Americans annually from heart disease; this effect is surprisingly large and occurs because of the nonlinear relationship between the disease and exposure to smoke. In a study of American nurses it was found that nonsmoking women regularly exposed to ETS had a 91% greater rate of heart attacks than women who had no exposure. Apparently the smoke leads to hardening of the arteries, a main cause of heart attacks. An analysis of all recent studies on passive smoking led to the conclusion that the risks of developing lung cancer and heart disease each are increased by about one-quarter for nonsmoking spouses of smokers. Longtime workers in bars and restaurants in which smoking is permitted also have an increased rate of lung cancer, even if they themselves do not smoke. A recent British study estimated that ETS kills 140,000 Europeans annually through cancer and heart disease.

Asbestos

The term **asbestos** refers to a family of six naturally occurring silicate minerals that are fibrous. Structurally, they are composed of long double-stranded networks of silicon atoms connected through intervening oxygen atoms; the net negative charge of this silicate structure is neutralized by the presence of cations such as magnesium.

The most commonly used form of asbestos, **chrysotile,** has the formula $Mg_3Si_2O_5(OH)_4$. It is a white solid whose individual fibers are curly. Chrysotile, mined mainly in Quebec and Russia, is the principal type of asbestos used in North America. It has been employed in huge quantities because of its resistance to heat, its strength, and its relatively low cost. Common applications of asbestos include its use as insulation and spray-on fireproofing material in public buildings, in automobile brake-pad lining, as an additive to strengthen cement used for roofing and pipes, and as a woven fiber in fireproof cloth.

The use of asbestos has been sharply reduced since the 1970s in developed countries because it is now recognized from studies on the health of asbestos miners and other asbestos workers to be a human carcinogen. It causes *mesothelioma,* a normally rare, incurable cancer of the lining of the chest or abdomen. In addition, airborne asbestos fibers and cigarette smoke act **synergistically:** their combined effect is greater than the sum of their individual effects (in this case equal to the product of the two) in causing lung cancer.

There is much controversy concerning whether chrysotile should be banned outright from further use and whether or not existing asbestos insulation in buildings should be removed. Many experts feel that existing asbestos should be left in place unless it becomes sufficiently damaged that there is a chance that its fibers will become airborne. Indeed, its removal can increase dramatically the levels of airborne asbestos in a building unless extraordinary precautions are taken. One scientist stated, "Removing asbestos is like waking up a pit bull terrier by poking a stick in its ear. We should let sleeping dogs lie." Some environmentalists, however, feel that existing asbestos is a ticking time bomb—that it should be removed as soon as possible, as one can never predict when building insulation will be damaged.

Most of the initial concern about asbestos was related to **crocidolite,** *blue asbestos*, and **amosite,** *brown asbestos*. Evidence implicating crocidolite in the causation of cancer in humans was already well established several decades ago. It is a material with thin, straight, and relatively short fibers that more readily penetrate lung passages, and it is a more potent carcinogen than the white form. Crocidolite and amosite are mined in South Africa and Australia and were not used much in North America but were used in many areas of Europe, including the United Kingdom.

Review Questions

1. In the micrograms per cubic meter concentration scale, to what substances do micrograms and cubic meters refer?

2. In general terms, what is meant by *photochemical smog*? What are the initial reactants in the process? Why is sunlight required?

3. What is meant by a *primary pollutant* and by a *secondary pollutant*? Give examples.

4. What is the chemical reaction by which *thermal NO* is produced? From which two sources does most urban NO arise? What is meant by the term NO_X?

5. Describe the strategies by which reduction of urban ozone levels have been attempted. What difficulties have been encountered in these efforts?

6. Describe the operation of the *three-way catalyst* in transforming emissions released by an automobile engine. Does the catalyst operate when the engine is cold? Why is it important in converters that the level of sulfur in gasoline be minimized?

7. Describe the reaction used in the *selective catalytic reduction* of nitrogen oxides.

8. What is *acid rain*? What two acids predominate in it?

9. What are the main anthropogenic sources of sulfur dioxide? Describe the strategies by which these emissions can be reduced. What is the *Claus reaction*?

10. What species are included in the air pollution index called *total reduced sulfur*?

11. Explain why the predominant acid in acid rain differs in eastern and western North America.

12. Describe the various strategies used in *clean coal*.

13. What is the difference between *dry* and *wet deposition*?

14. Using chemical equations, explain how acid rain is neutralized by limestone that is present in soil.

15. Describe the effects of acid precipitation on (a) dissolved levels of aluminum, (b) fish populations, and (c) trees.

16. Define the term *aerosol*, and differentiate between *coarse* and *fine particulates*. What are the usual origins of these two types of atmospheric particles?

17. What are the usual chemical components of a *sulfate aerosol*?

18. Write a balanced equation illustrating the reactions that occur between one molecule of ammonia and (a) one molecule of nitric acid, and (b) one molecule of sulfuric acid.

19. What are the usual concentration units for suspended particulates? What would the designation PM_{40} mean? What do the terms *respirable* and *ultrafine* mean?

20. Discuss the relationship between atmospheric particulates and haze.

21. What are the three modes of particles in plots of particle numbers versus diameter? What is the origin of each mode?

22. What is the difference in meaning between *absorbed* and *adsorbed* when they refer to particulates?

23. Describe the major health effects of outdoor air pollutants.

24. List four important reasons why coarse particles usually are of less danger to human health than are fine particles.

25. What are the names and formulas for six of the gases that are released into our atmosphere from biological or volcanic sources? What chemical species initiates their oxidation?

26. What is the two-step mechanism by which the hydroxyl free radical is produced in clean air?

27. What are the main sources of formaldehyde in indoor air? What are its effects?

28. What are the main sources of nitrogen dioxide and of carbon monoxide in indoor air? of benzene?

29. What are the three forms of asbestos called? Why is asbestos of environmental concern?

Green Chemistry Questions

1. *PERC* replaced gasoline and kerosene in the dry-cleaning process.

(a) Describe any environmental problems or worker hazards that would be associated with these solvents.

(b) Would the same environmental problems or worker hazards be eliminated by the use of PERC?

(c) By the use of carbon dioxide?

2. The development of surfactants for carbon dioxide by Joseph DeSimone won a Presidential Green Chemistry Challenge Award.

(a) Which of the three focus areas (see the Introduction to Green Chemistry) for these awards does this one best fit into?

(b) List two of the twelve principles of green chemistry (see the Introduction to Green Chemistry) that are addressed by the green chemistry developed by DeSimone.

Additional Problems

1. The rate constant for the oxidation of nitric oxide by ozone is 2×10^{-14} cm^3/molecule sec, whereas that for the competing reaction in which it is oxidized by oxygen, that is

$$2\,NO + O_2 \longrightarrow 2\,NO_2$$

is 2×10^{-38} cm^6/molecule^2 s. For typical concentrations encountered in morning smog episodes, namely, 40 ppb for ozone and 80 ppb for nitric oxide, deduce the rates of these two reactions and decide which one is the dominant process.

2. In the overall reaction that produces nitric oxide from N_2 and O_2, the slow step in the mechanism is the reaction between atomic oxygen and molecular nitrogen to produce nitric oxide and atomic nitrogen. (a) Write out the chemical equation and the rate equation for the slow step. (b) Given that its rate constant at 800°C is 9.7×10^{10} L/mol sec and its activation energy is 315 kJ/mol, calculate the amount by which the rate constant increases if the temperature is raised to 1100°C.

3. The concentration of ozone in ground-level air can be determined by allowing the gas to react with an aqueous solution of potassium iodide, KI, in a redox reaction that produces molecular iodine, molecular oxygen, and potassium hydroxide. (a) Deduce the balanced equation for the overall process. (b) Determine the ozone concentration, in ppb, in a 10.0-L sample of outdoor air if it required 17.0 μg of KI to react with it.

4. The pH in a lake of size 3.0 km $\times$ 8.0 km and an average depth of 100 m is found to be 4.5. Calculate the mass of calcium carbonate that must be added to the lake water in order to raise its pH to 6.0.

5. An experimenter isolates a sample of an air mass from further inputs of particulate matter. What will happen to the **relative** heights of the three peaks of Figure 2-17 as time goes on?

6. Calculate the mass of fine particles inhaled by an adult each year, assuming he/she inhales about 350 L of air per hour and that the average PM$_{2.5}$ index of this air is 10 μg/m^3. Assuming that each such particle has a diameter of about 1 μm and that the density of the particles is about 0.5 g/mL, calculate the total surface area of this annual load of particles. [Hint: The surface area of a spherical particle is equal to $4\pi r^2$, where r is its radius.]

7. The percentage of sulfur in coal can be determined by burning a sample of the solid and passing the resulting sulfur dioxide gas into a solution of hydrogen peroxide, which oxidizes it to sulfuric acid, and then titrating the acid. Calculate the mass percent of sulfur in a sample if the gas from a 8.05-g sample required 44.1 mL of 0.114 M NaOH in the titration of the diprotic acid.

8. A sample of acidic precipitation is found to have a pH of 4.2. Upon analysis it is found to have a total sulfur concentration of 0.000010 M. Calculate the concentration of nitric acid in the sample, and, from the ratio of nitric to total acid, decide whether the air sample probably originated in eastern or in western North America.

9. If the pH of rainfall in upstate New York is found to be 4.0, and if half the acidity is due to nitric acid and half to the two hydrogen ions released by sulfuric acid, calculate the masses of the primary pollutants nitric oxide and sulfur dioxide that are required to acidify 1 L of such rain.

10. What mass of formaldehyde gas must be released from building materials, carpets, etc. in order to produce a concentration of 0.50 ppm of the gas in a room having dimensions of 4 m $\times$ 5 m $\times$ 2 m?

11. Calculate the volume (at 20°C and 1.00 atm) of SO_2 produced by the conventional roasting of 1.00 tonnes (1000 kg) of nickel sulfide ore, NiS. What mass of pure sulfuric acid could be produced from this amount of SO_2 if it were oxidized to SO_3 and then reacted with water?

12. According to Stokes' law, the settling rate of particulates in the atmosphere is directly proportional to the difference in density ($\Delta\rho$) of the particulars compared to the atmosphere and proportional to the square of the diameter d: rate $\propto \Delta\rho \, d^2$. Given that particulates of a particular form of fly ash with a diameter of 5.2 μm are found to settle out after two days, how long would it take for particulates of the same material that are half as large (diameter of 2.6 μm) to settle out?

13. The detection threshold of formaldehyde by humans is about 100 ppb. Would a typical human be able to detect formaldehyde at a concentration of 250 μg/m^3? (Assume 23°C and 1.00 atm.)

Further Reading

1. O. Klemm, "Local and Regional Ozone: A Student Study Project," *Journal of Chemical Education* 78 (2001): 1641–1646.

2. R. J. Chironna and B. Altshuler, "Chemical Aspects of NO$_X$ Scrubbing," *Pollution Engineering* (April 1999): 32–36; R. K. Agrawal and S. C. Wood, "Cost-Effective NO$_X$ Reduction," *Chemical Engineering* (February 2001): 78–82.

3. A. Sheth and T. Giel, "Understanding the PM-2.5 Problem," *Pollution Engineering* (March 2000): 32–35.

4. C. T. Driscoll et al., "Acidic Deposition in the Northeastern United States: Sources and Inputs, Ecosystem Effects, and Management Strategies," *Bioscience 51* (2001): 180–198; J. A. Lynch et al., "Acid Rain Reduced in Eastern United States," *Environmental Science and Technology* 34 (2000): 940–949; M. Heal, "Acid Rain: Is the UK Coping?" *Education in Chemistry* (July 2002): 101–104.

5. J. G. Calvert et al., "Chemical Mechanisms of Acid Generation in the Troposphere," *Nature* 317 (1985): 27–35.

6. D. Mage et al., "Urban Air Pollution in Megacities of the World," *Atmospheric Environment* 30 (1996): 681–686.

7. M. Lippmann, "Health Effects of Tropospheric Ozone," *Environmental Science and Technology* 25 (1991): 1954–1961; M. Raizenne et al.,

"Air Pollution Exposures and Children's Health," *Canadian Journal of Public Health* 89, supplement 1 (1998): S43–S48.

8. F. Laden et al., "Association of Fine Particulate Matter from Different Sources with Daily Mortality in Six U.S. Cities," *Environmental Health Perspectives* 108 (2000): 941–947; D. W. Dockery et al., "An Association Between Air Pollution and Mortality in Six U.S. Cities," *New England Journal of Medicine* 329 (1993): 1753–1759; J. Schwartz et al., "The Concentration–Response Relation Between PM$_{2.5}$ and Daily Deaths," *Environmental Health Perspectives* 110 (2002): 1025–1029; C. A. Pope III et al., "Review of Epidemiological Evidence of Health Effects of Particulate Air Pollution," *Inhalation Toxicology* 7 (1995): 1–18.

9. W. H. Smith, "Air Pollution and Forest Damage," *Chemical and Engineering News* (11 November 1991): 30.

10. J. Kaiser, "Showdown over Clean Air Science," *Science* 277 (1997): 466–469.

11. "Fires from Hell," *New Scientist* 31 (August 2002): 34–37.

12. D. R. Gold, "Environmental Tobacco Smoke, Indoor Allergens, and Childhood Asthma," *Environmental Health Perspectives* 108, supplement 4 (2000): 643–646; "Secondhand Smoke—Is It a Hazard?," *Consumer Reports* (January 1995): 27–33.

Websites of Interest

Log on to www.whfreeman.com/envchem3e/ and click on Chapter 2.

The Detailed Chemistry of the Atmosphere

In Chapters 1 and 2 we have discussed, in broad terms, chemical processes in the stratosphere and troposphere, respectively, emphasizing the environmental concerns that have arisen. In this chapter the reactions that occur in clean tropospheric air and the processes encountered in the polluted air of modern cities are analyzed in more detail. In addition, the processes that deplete ozone in the stratosphere are systematized. It is only by understanding the science underlying such complicated environmental problems that we can hope to solve them.

This Mexico City newspaper salesman wears a mask to help protect himself from air pollution during a photochemical smog episode. (Guillermo Gutierrez/AP)

One of the important characteristics that determines the reactivity of a species is whether or not it has unpaired electrons. To emphasize that an atomic or molecular species is a free radical, in this chapter we shall place a superscript dot at the end of its molecular formula, signaling the presence of the unpaired electron. For example, the notation OH˙ is used for the **hydroxyl free radical.** The position of the unpaired electron in the Lewis structure is often important in determining the reaction of a free radical. Box 3-1 discusses the deduction of the Lewis structure for free radicals.

BOX 3-1 | Lewis Structures of Simple Free Radicals

Most of the free radicals that are important in atmospheric chemistry have their unpaired electron located on a carbon, oxygen, hydrogen, or halogen atom. In a formula showing the location and position of bonds, the specific atomic location can be denoted by placing a dot above the relevant atom symbol to represent the unpaired electron, for example, in the notation Ḟ. Characteristically, such an atom forms one less bond than usual—its unpaired electron is not in actual use as a bonding electron. Thus a carbon atom on which an unpaired electron is located forms three rather than four bonds, an oxygen forms one rather than two bonds, and a halogen or hydrogen forms no bonds if it is the radical site. Usually, the unpaired electron exists as a nonbonding electron localized on one atom, not as a bonding electron shared between atoms.

For many polyatomic free radicals, the choice of atom to which the unpaired electron is to be assigned in deducing the Lewis structure is obvious from the atom–atom connections. Thus in the hydroperoxy radical HOO, the hydrogen atom cannot be the radical site since to be part of the molecule it must form a bond to the adjacent oxygen, and neither can the central oxygen be the site since it must form two bonds, one to each neighbor (or one of them would not be part of the molecule). This leaves the terminal oxygen as the radical site, and we can show the bonding network as H—O—Ȯ. If desired, the unbonded electron pairs can also be shown:

$$H-\ddot{O}-\dot{O}:$$

The procedure is more complicated in molecules that contain multiple bonds. Thus in HOCO it is not initially obvious whether it is the carbon or the terminal oxygen that carries the unpaired electron. After a little manipulation with various bonding schemes, it becomes clear that the unpaired electron could not be located on an oxygen, since to fulfill its valence requirement of four the carbon would have to form three bonds to the other oxygen. Thus the only reasonable structure is H—O—Ċ=O.

If only a simple formula rather than a partial or complete Lewis structure is drawn for a radical, the superscript dot is placed following the formula and does not indicate which atom carries the unpaired electron. An example is HCO˙, in which the actual location of the unpaired electron is at carbon, not oxygen.

For a few free radicals involving unusual bonding, such as $NO_2^{\cdot}$, these rules generate a Lewis structure that is not the dominant one; further discussion of such systems is beyond the scope of this book.

PROBLEM 1

Draw simple Lewis structures, showing the locations of the bonds and of the unpaired electron, for the following free radicals.
(a) OH˙   (b) $CH_3^{\cdot}$ (c) $CF_2Cl^{\cdot}$
(d) $H_3COO^{\cdot}$ (e) $H_3CO^{\cdot}$ (f) ClOO˙
(g) ClO˙   (h) HCO˙   (i) NO˙

Tropospheric Chemistry

The Principles of Reactivity in the Troposphere

Most gases in the troposphere are gradually oxidized by a sequence of reactions involving free radicals. For a given gas, the sequence can be predicted from the principles discussed below, which are also systematized in Figure 3-1. These tropospheric reactions are similar in many ways to those encountered in the stratosphere, which are discussed later in the chapter.

As mentioned in Chapter 2, the usual initial step in the oxidation of an atmospheric gas is reaction with the hydroxyl free radical rather than with **molecular oxygen,** O_2. With molecules that contain a multiple bond, the hydroxyl radical usually reacts by *adding* itself to the molecule at the position of the multiple bond. Recall the general principle that radical reactions that are spontaneous are those which produce stable products, i.e., products containing strong bonds. Thus it is understandable that $OH^{\cdot}$ addition does *not* occur to an oxygen atom since the O—O bonds that would result are weak. Similarly, $OH^{\cdot}$ addition does *not* occur to CO *double* bonds since they are very strong relative to the single O—O or C—O bond that would be produced. For example, the $OH^{\cdot}$ radical adds to the sulfur atom, forming a strong bond, but not to an oxygen atom, in **sulfur dioxide,** SO_2:

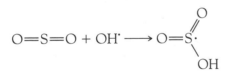

(Here and elsewhere in this book we write Lewis structures that assume that *d* orbitals in atoms such as sulfur and phosphorus allow these elements to form double bonds.) Hydroxyl radical does not add to **carbon dioxide,** O=C=O, since the molecule contains only very strong C=O bonds. However, $OH^{\cdot}$ addition does occur to the carbon atom in **carbon monoxide,** CO, since the triple bond is thereby converted to the very stable double bond and a new single bond is also formed:

$$^{\ominus}C{\equiv}O^{\oplus} \;\; + OH^{\cdot} \longrightarrow HO{-}\overset{\cdot}{C}{=}O$$

This process is exothermic because the third C—O bond in carbon monoxide is weak relative to the other two.

Generally, $OH^{\cdot}$ does not add to multiple bonds in any fully oxidized species such as CO_2, SO_3, and N_2O_5, since such processes are endothermic and therefore are very slow to occur at atmospheric temperatures. Similarly, N_2 does not react with $OH^{\cdot}$ because the component of the nitrogen-to-nitrogen bond that would be destroyed is stronger than the N—O bond that would be formed. It doesn't react with O_2 because a high activation energy is required for this reaction to occur.

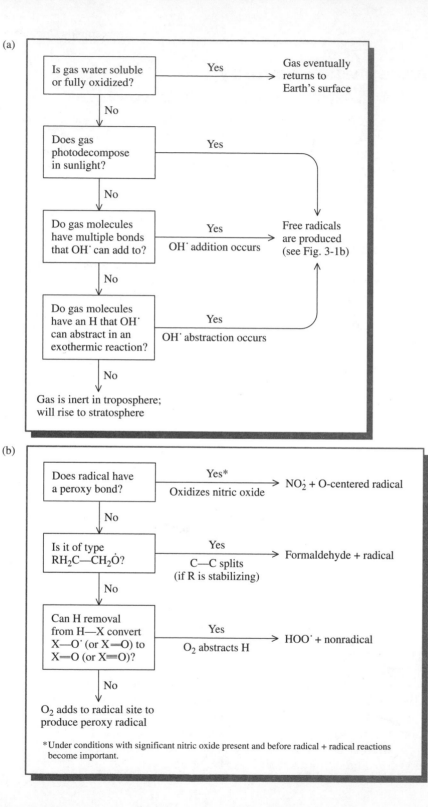

FIGURE 3-1 (a) Decision tree illustrating the fate of gases emitted into the air. (b) Decision tree illustrating the fate of airborne free radicals.

For molecules that do *not* have a reactive multiple bond but do contain hydrogen, OH$^{\cdot}$ reacts with them by the **abstraction** of a hydrogen atom to form a water molecule and a new reactive free radical. For CH_4, NH_3, H_2S, and CH_3Cl, for instance, the reactions are

$$CH_4 + OH^{\cdot} \longrightarrow CH_3^{\cdot} + H_2O$$

$$NH_3 + OH^{\cdot} \longrightarrow NH_2^{\cdot} + H_2O$$

$$H_2S + OH^{\cdot} \longrightarrow SH^{\cdot} + H_2O$$

$$CH_3Cl + OH^{\cdot} \longrightarrow CH_2Cl^{\cdot} + H_2O$$

Because the H—OH bond formed in these reactions is very strong, the processes are all exothermic; thus only small activation energy barriers exist to impede these reactions (see Box 1-2).

PROBLEM 3-1

Why aren't gases such as CF_2Cl_2 (a CFC) readily oxidized in the troposphere? Would the same be true for CH_2Cl_2?

PROBLEM 3-2

The abstraction of the hydrogen atom in HF by OH$^{\cdot}$ is endothermic. Comment briefly on the expected rate of this reaction: would it be (at least potentially) fast, or necessarily very slow in the troposphere?

PROBLEM 3-3

The hydroxyl radical does not react with gaseous nitrous oxide, N_2O, even though the molecule contains multiple bonds. What can you deduce about the probable energetics (endothermic or exothermic character) of this reaction from the observed lack of reactivity?

A few gases emitted into air can absorb some of either the UV-A or the visible component of sunlight, and this input of energy is sufficient to break one of the bonds in the molecule, thereby producing two free radicals. For example, most molecules of atmospheric **formaldehyde** gas, H_2CO, react by photochemical decomposition after absorption of UV-A from sunlight:

$$H_2CO \xrightarrow{\text{UV-A} (\lambda < 338\,\text{nm})} H^{\cdot} + HCO^{\cdot}$$

In all the cases discussed, the initial reaction of a gas emitted into air produces free radicals, almost all of which are extremely reactive. The predominant fate in tropospheric air for most simple radicals is reaction with diatomic oxygen, often by an addition process: one of the oxygen atoms attaches, or "adds on," to the other reactant, usually at the site of the unpaired electron. For instance, O_2 reacts by addition with the **methyl radical,** $CH_3^{\cdot}$:

$$CH_3^{\cdot} + O_2 \longrightarrow CH_3OO^{\cdot}$$

Notice that $CH_3OO^{\cdot}$ itself is a free radical; the terminal oxygen forms only one bond and carries the unpaired electron:

$$H_3C - \overset{..}{\underset{..}{O}} - \overset{\cdot}{\underset{..}{O}} : \qquad \text{or just} \qquad H_3C - O - \overset{\cdot}{O}$$

Species such as $HOO^{\cdot}$ and $CH_3OO^{\cdot}$ are called **peroxy** radicals since they contain a peroxide-like O—O bond; recall that $HOO^{\cdot}$ is the **hydroperoxy radical.**

As radicals go, peroxy radicals are less reactive than most. They do *not* readily abstract hydrogen since the resulting peroxides would not be very stable energetically. Since the transfer of H to the peroxy radical would be endothermic and thus would possess a large activation energy, abstraction reactions for peroxy radicals are usually so slow that they are of negligible importance (in contrast to those for $OH^{\cdot}$). Peroxy radicals in the troposphere do not react with atomic oxygen because of the extremely low concentrations of the free atom in this region of the atmosphere. The most common fate of peroxy radicals in tropospheric air, except for the very cleanest type of air, such as that over oceans, is reaction with **nitric oxide,** $NO^{\cdot}$, by the transfer of the "loose" oxygen atom (see the later section on stratospheric chemistry), thereby forming **nitrogen dioxide,** $NO_2^{\cdot}$, and a radical that has one fewer oxygen atoms:

$$HOO^{\cdot} + NO^{\cdot} \longrightarrow OH^{\cdot} + NO_2^{\cdot}$$
$$CH_3OO^{\cdot} + NO^{\cdot} \longrightarrow CH_3O^{\cdot} + NO_2^{\cdot}$$

It is by this type of reaction that most atmospheric $NO^{\cdot}$ *is oxidized to* $NO_2^{\cdot}$, at least in polluted air. Recall that this reaction also is typical of the types encountered in stratospheric chemistry (see Chapter 1) and that $NO^{\cdot}$ oxidation by ozone in sunlit conditions yields a null reaction.

For free radicals that contain nonperoxy oxygen atoms, the reaction with molecular oxygen frequently involves the abstraction of an H atom by O_2. This process occurs *provided that*, as a result, a new bond within the system is formed: a single bond involving oxygen is converted to a double one, or a double bond involving oxygen is converted to a triple one. As examples, consider the three reactions below in which a C—O single (or double) bond is converted to a double (or triple) one as a consequence of the loss of a hydrogen atom:

$$CH_3\text{—}\dot{O} + O_2 \longrightarrow H_2C\text{=}O + HOO^{\bullet}$$

$$HO\text{—}\dot{C}\text{=}O + O_2 \longrightarrow O\text{=}C\text{=}O + HOO^{\bullet}$$

$$H\text{—}\dot{C}\text{=}O + O_2 \longrightarrow {}^{\ominus}C\text{≡}O^{\oplus} + HOO^{\bullet}$$

Such processes do not occur unless a new bond is created in the product free radical, since the strength of the newly created H—OO bond alone is not sufficient to compensate for the breaking of the original bond to hydrogen.

If there is no suitable hydrogen atom for O_2 to abstract, then when it collides with a radical, it instead *adds* to it at the site of the unpaired electron, as was previously discsussed for simple radicals. For example, radicals of the type R—$\dot{C}$=O, where R is a chain of carbon atoms, add O_2 to form a peroxy radical:

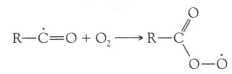

The only exception to the generalization that oxygen-containing radicals react with O_2 occurs when the radical can decompose spontaneously in a thermoneutral or exothermic fashion. An example of this rare phenomenon is discussed later in the section on photochemical smog.

These generalizations are summarized in the form of the *decision trees* diagrammed in Figure 3-1. By using these diagrams you can deduce the sequence of reactions by which most atmospheric gases in the troposphere are oxidized.

The Tropospheric Oxidation of Methane

Gaseous **methane,** CH_4, is released into the atmosphere in large quantities as a result of **anaerobic** (i.e., O_2-free) biological decay processes and of the use of coal, oil, and, especially, natural gas. It is the predominant hydrocarbon in the atmosphere. Details concerning its production, and the effects on climate of atmospheric methane, are discussed in Chapter 4. Here, however, we shall be concerned with its conversion to carbon dioxide. A similar series of reactions is followed by other alkanes and other VOCs lacking multiple bonds.

The sequence of reactions by which methane is slowly oxidized in the atmosphere can be deduced by applying the principles outlined above and summarized in Figure 3-1, as discussed below.

Since CH_4 is not very soluble in water, does not absorb sunlight, and contains no multiple bonds, the sequence is initiated by a hydroxyl radical abstracting a hydrogen atom from a methane molecule, giving the *methyl radical,* $CH_3^{\bullet}$:

$$CH_4 + OH^\cdot \longrightarrow CH_3^\cdot + H_2O \qquad (1)$$

Since the $CH_3^\cdot$ radical contains no oxygen, we deduce that it adds O_2, producing a peroxy radical:

$$CH_3^\cdot + O_2 \longrightarrow CH_3OO^\cdot \qquad (2)$$

Further, since $CH_3OO^\cdot$ is a peroxy radical, we deduce that except in very clean air, it reacts with $NO^\cdot$ molecules in air to oxidize them by transfer of an oxygen atom:

$$CH_3OO^\cdot + NO^\cdot \longrightarrow CH_3O^\cdot + NO_2^\cdot \qquad (3)$$

The radical $CH_3O^\cdot$ contains a C—O bond that can become C=O upon loss of one hydrogen, so we conclude from our principles that in the next step O_2 abstracts an H atom, producing the nonradical product formaldehyde, H_2CO:

$$CH_3O^\cdot + O_2 \longrightarrow H_2CO + HOO^\cdot \qquad (4)$$

Thus methane is converted to formaldehyde as the first stable intermediate in its oxidation. Since formaldehyde is reactive as a gas in the atmosphere, the mechanism is not complete at this point. After several hours or days in the sunlight, most formaldehyde molecules decompose photochemically by the absorption of UV-A from sunlight, resulting in the cleavage of a C—H bond and the consequent formation of two radicals:

$$H_2CO \xrightarrow{\text{UV-A } (\lambda < 338 \text{ nm})} H^\cdot + HCO^\cdot \qquad (5)$$

A minority of formaldehyde molecules react with $OH^\cdot$ by H atom abstraction, yielding the same $HCO^\cdot$ radical; see Problem 3-7 for the implications of this alternative route.

The hydrogen atom from formaldehyde photolysis is itself a simple radical, and therefore it reacts by addition to O_2 to yield $HOO^\cdot$:

$$H^\cdot + O_2 \longrightarrow HOO^\cdot \qquad (6)$$

Meanwhile, the $H—\overset{\cdot}{C}{=}O$ radical reacts by yielding an $H^\cdot$ atom to O_2, to produce carbon monoxide and $HOO^\cdot$, since by this route a double C—O bond is converted to a triple one:

$$HCO^\cdot + O_2 \longrightarrow CO + HOO^\cdot \qquad (7)$$

Thus carbon monoxide also is an intermediate in the oxidation of methane. Indeed, most of the CO in a clean atmosphere is derived from this source. Since CO is not a radical and does not absorb visible or UV-A light, we deduce that it reacts ultimately by hydroxyl radical addition to its triple bond:

$$^{\ominus}C{\equiv}O^{\oplus} + OH^\cdot \longrightarrow H—O—\overset{\cdot}{C}{=}O \qquad (8)$$

This radical can convert its O—C bond to O=C by loss of H, so we deduce that O_2 readily abstracts the hydrogen:

$$H\text{—}O\text{—}\dot{C}{=}O + O_2 \longrightarrow O{=}C{=}O + HOO^{\bullet} \quad (9)$$

Carbon in its fully oxidized form of carbon dioxide is ultimately produced from methane by this sequence of steps, which are summarized in Figure 3-2. If we add up the nine steps involved and cancel common terms, the overall reaction is seen to be

$$CH_4 + 5\,O_2 + NO^{\bullet} + 2\,OH^{\bullet} \xrightarrow{\text{UV-A}}$$
$$CO_2 + H_2O + NO_2^{\bullet} + 4\,HOO^{\bullet}$$

If to this result is added the conversion of the four $HOO^{\bullet}$ radicals back to $OH^{\bullet}$ by reaction with four $NO^{\bullet}$ molecules, the revised overall reaction is

$$CH_4 + 5\,O_2 + 5\,NO^{\bullet} \xrightarrow{\text{UV-A}} CO_2 + H_2O + 5\,NO_2^{\bullet} + 2\,OH^{\bullet}$$

We conclude that $NO^{\bullet}$ is oxidized to $NO_2^{\bullet}$ synergistically, i.e., in a mutually cooperative process, when methane is oxidized to carbon dioxide. Note also that the number of $OH^{\bullet}$ free radicals is increased as a result of the process, due to the photochemical decomposition of formaldehyde. Thus hydroxyl radical is not only a catalyst in the overall reaction but also a product of it.

The initial step of the mechanism—the abstraction by $OH^{\bullet}$ of a hydrogen atom from methane—is a relatively slow process, requiring about a decade to occur on average. Once this has happened, however, the subsequent steps leading to formaldehyde occur very rapidly. The slowness of the initial step in methane oxidation, and the increasing amounts of the gas released from the surface of the Earth, have led to an increase in the atmospheric concentration of CH_4 in recent times, as discussed further in Chapter 4.

Under conditions of low nitrogen oxide concentration, such as occur over oceans, the mechanism of methane oxidation differs in some of the steps. In particular, instead of oxidizing $NO^{\bullet}$, the peroxy radicals often react with each other, combining to produce a (nonradical) peroxide:

$$2\,HO_2^{\bullet} \longrightarrow H_2O_2 + O_2$$

Under these conditions, then, the oxidation of methane in clean air decreases, rather than increases, the concentration of free radicals.

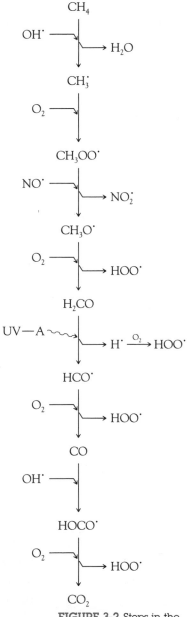

FIGURE 3-2 Steps in the atmospheric oxidation of methane to carbon dioxide.

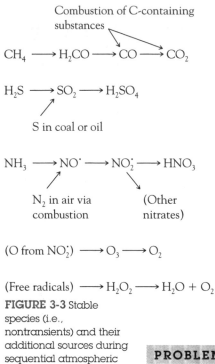

FIGURE 3-3 Stable species (i.e., nontransients) and their additional sources during sequential atmospheric oxidation processes.

In general, during the atmospheric oxidation of any of the hydrides (simple hydrogen-containing molecules such as CH_4, H_2S, and NH_3), one or more stable species are encountered along the reaction sequence before the totally oxidized product is formed. These intermediates are also formed independently by various pollution processes. Figure 3-3 summarizes the sequences for hydrides and partially oxidized materials from the viewpoint of the stable species; close reflection will persuade you that the net result is the $OH^{\cdot}$-induced oxidation of the reduced and partially oxidized gases emitted into the air from both natural and pollution sources. In a few cases, for instance, for methane and methyl chloride, the initiation reaction is sufficiently slow that a few percent of these gases survive long enough to penetrate to the stratosphere by the upward diffusion of tropospheric air. Most hydrocarbons react much more quickly than methane (since their C–H bonds are weaker or fast reactions with $OH^{\cdot}$ other than by hydrogen abstraction are possible) and are classified as **nonmethane hydrocarbons** (NMHC) to emphasize this distinction.

PROBLEM 3-4

Using the reaction principles developed above (the decision trees in Figure 3-1 will help here), predict the sequence of reaction steps by which atmospheric H_2 gas will be oxidized in the troposphere. What is the overall reaction?

PROBLEM 3-5

Deduce two short series of steps by which molecules of methanol, CH_3OH, are converted to formaldehyde, H_2CO, in air. The mechanisms should differ according to which hydrogen atom you decide will react first, that of CH_3 or that of OH.

PROBLEM 3-6

Write equations showing the reactions by which atmospheric carbon monoxide is oxidized to carbon dioxide. Then, by adding the process by which $HOO^{\cdot}$ is returned to $OH^{\cdot}$, deduce the overall reaction.

PROBLEM 3-7

Deduce the series of steps, and the overall reaction as well, for the oxidation of a formaldehyde molecule to CO_2, assuming that for the particular H_2CO

molecule involved, the initial reaction is abstraction of H by OH˙ rather than photochemical decomposition. Overall, is there any increase in the number of free radicals as a result of the oxidation if it proceeds in this manner?

Photochemical Smog: The Oxidation of Reactive Hydrocarbons

Notwithstanding the great complexity of the process, the most important features of the photochemical smog phenomenon can be understood by considering only its few main categories of reactions; these differ in speed, but not much in type, from those occurring in clean air.

We shall restrict our attention to the most reactive VOCs, namely hydrocarbons that contain a C═C bond. (Readers unfamiliar with the basics of organic chemistry are advised to consult the Appendix, where the nature of such molecules is explored.) The simplest example is **ethene** (ethylene), C_2H_4; its structure can be written in condensed form as $H_2C═CH_2$; its full structural formula is

In similar hydrocarbons, one or more of the four hydrogens are replaced by other atoms or groups, often an alkyl group containing a short chain of carbon atoms such as CH_3— or CH_3CH_2—, which will be designated simply as R, since it is generally not the chain but rather the C═C part of the molecule that is the reactive site in atmospheric reactions.

Consider a general hydrocarbon RHC═CHR. In air it reacts with hydroxyl radical by *addition* to the C═C bond:

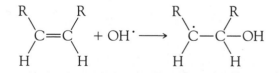

This addition reaction is a faster process, due to its lower activation energy, than the alternative of abstraction of hydrogen, so we can neglect the abstraction process in molecules containing a C═C link. Because the reaction of addition of OH˙ to a multiple bond is much faster than H abstraction from methane and other alkane hydrocarbons, RHC═CHR molecules in general are much faster to react than are alkanes.

As anticipated from the reaction principles, the carbon-based radical produced from the reaction of hydroxyl radical with the hydrocarbon adds O_2 to yield a peroxy radical, which in turn oxidizes $NO^{\cdot}$ to $NO_2^{\cdot}$:

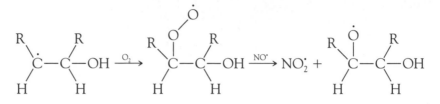

Once much of the $NO^{\cdot}$ has been oxidized to $NO_2^{\cdot}$, photochemical decomposition by sunlight of the latter gives $NO^{\cdot}$ plus O, which then quickly combines with molecular oxygen to give ozone, as discussed in Chapter 2. It is a characteristic of air pollution driven by photochemical processes that ozone from $NO_2^{\cdot}$ photodecomposition builds up to much higher levels than are found in clean air. Nitrogen dioxide is the only significant tropospheric source of the atomic oxygen from which ozone can form.

As mentioned previously, the ozone concentration does not build up substantially as a result of this sequence until most of the $NO^{\cdot}$ has been converted to $NO_2^{\cdot}$, since $NO^{\cdot}$ and O_3 mutually self-destruct if both are present in significant concentrations. It is only after most $NO^{\cdot}$ has been oxidized to $NO_2^{\cdot}$ as a result of reactions with peroxy free radicals that the characteristic buildup of **urban ozone** occurs, as can be seen in Figure 2-4; the transition occurred at about 9 a.m. on the particular smoggy day in the 1960s in Los Angeles (when smog levels were higher than in more recent years) that is illustrated.

One might anticipate from our reactivity principles (Figure 3-1) that the two-carbon radical mentioned above ($RCH\dot{O}CH(R)OH$) would lose H by abstraction by O_2, but instead it decomposes spontaneously by cleavage of the C—C bond to give a nonradical molecule containing a C=C bond and another radical, $RH\dot{C}OH$:

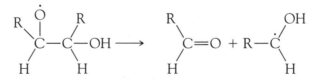

It happens that the reaction requires no energy input, i.e., ΔH is close to zero, because in this case the formation of a C=O bond from C—O compensates energetically for loss of the C—C bond. Since the decomposition of this radical is not endothermic, its activation energy is small and thus the process occurs spontaneously in air.

Molecules of the RHCO type are called **aldehydes.** The simplest example is formaldehyde, H_2CO, which was encountered previously in the process of methane oxidation. The carbon-based radical $R\overset{\cdot}{H}COH$ produced in the preceding reaction subsequently reacts with an O_2 molecule. Since loss of the hydroxyl hydrogen from this radical allows the C—O bond to become C=O, the oxygen molecule abstracts the H atom:

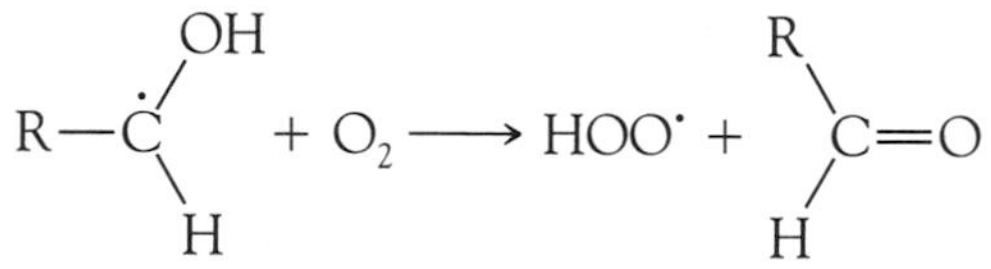

$$R-\overset{\overset{\textstyle OH}{\diagup}}{\underset{\underset{\textstyle H}{\diagdown}}{\overset{\cdot}{C}}} + O_2 \longrightarrow HOO^{\cdot} + \overset{\overset{\textstyle R}{\diagdown}}{\underset{\underset{\textstyle H}{\diagup}}{C}}{=}O$$

If we add all the above reactions, the net reaction thus far is

$$RHC{=}CHR + OH^{\cdot} + 2\,O_2 + NO^{\cdot} \longrightarrow 2\,RHC{=}O + HOO^{\cdot} + NO_2^{\cdot}$$

Thus the original RHC=CHR pollutant molecule is converted into two aldehyde molecules, each possessing half the number of carbon atoms. Indeed, as shown in Figure 2-4, by about noon on the very smoggy day in Los Angeles, most of the reactive hydrocarbons emitted into the air by morning rush-hour traffic had been converted to aldehydes. By midafternoon, most of the aldehydes had disappeared, since they had largely been photochemically decomposed into $HCO^{\cdot}$ and $R^{\cdot}$ (alkyl) free radicals.

$$RHCO \xrightarrow{\text{sunlight}} R^{\cdot} + HCO^{\cdot}$$

The sunlight-induced decomposition of aldehydes and of ozone leads to a huge increase in the number of free radicals in the air of a city undergoing photochemical smog, although in absolute terms the concentration of radicals is still very small.

The steps in the conversion of the original RHC=CHR molecule into aldehydes, and then of the latter to carbon dioxide (see Problem 3-8), are summarized in Figure 3-4. As indicated by the results of Problem 3-8, the net effect of the synergistic oxidation of nitric oxide and RHC=CHR is the production of carbon dioxide, nitrogen dioxide, and more hydroxyl radicals. Thus the reaction is **autocatalytic**—its net speed will increase with time since one of its products, here $OH^{\cdot}$, catalyzes the reaction for other reactant molecules.

PROBLEM 3-8

Rewrite the net reaction shown above for RHC=CHR, assuming that R is H. Deduce the series of steps by which the formaldehyde molecules will subsequently undergo photochemical decomposition and by a further series of steps be oxidized to carbon dioxide. Add these steps to the net reaction. Also add the reactions by which $HOO^{\cdot}$ oxidizes $NO^{\cdot}$. What is the final net reaction obtained by adding all these processes together?

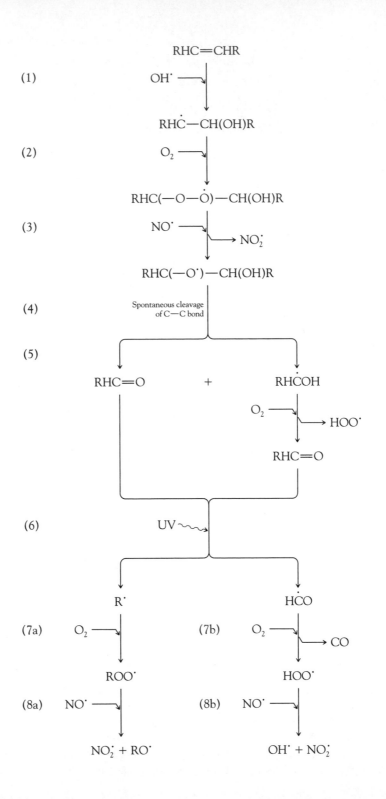

FIGURE 3-4 Mechanism of the RHC=CHR oxidation process in photochemical smog.

PROBLEM 3-9

Repeat Problem 3-8 but this time assume that the alkyl group R in the aldehyde RHCO produced by photochemical smog is a simple methyl group, CH_3, and that, when the aldehyde undergoes photochemical decomposition by sunlight, the radicals $CH_3^{\cdot}$ and $HCO^{\cdot}$ are produced. Using the air reactivity principles, deduce the sequence of reactions by which these radicals are oxidized to carbon dioxide, and determine the overall reaction of conversion of RHCO to CO_2. Assume formaldehyde photolyzes.

Photochemical Smog: The Fate of the Free Radicals

In later stages of photochemical smog formation, reactions that occur between two radicals are no longer insignificant, since their concentrations have become so high. Because their rates are proportional to the *product* of two radical concentrations, these processes are important when the radical concentrations are high; i.e., they occur quickly under such conditions. Generally, the reaction of two free radicals yields a stable, nonradical product:

$$\text{radical} + \text{radical} \longrightarrow \text{nonradical molecule}$$

One important example of a radical–radical reaction is the combination of hydroxyl and nitrogen dioxide to yield **nitric acid,** HNO_3, a process that, as we saw in Chapter 1, also occurs in the stratosphere:

$$OH^{\cdot} + NO_2^{\cdot} \longrightarrow HNO_3$$

This reaction is the main tropospheric sink for hydroxyl radicals. The average lifetime for an HNO_3 molecule is several days. By then it either has dissolved in water and been rained out or has been photochemically decomposed back into its components.

Similarly, combination of $OH^{\cdot}$ with $NO^{\cdot}$ gives **nitrous acid,** HONO, also written HNO_2. In sunlight the nitrous acid is almost immediately photochemically decomposed back to $OH^{\cdot}$ and $NO^{\cdot}$, but at night it is stable and therefore its concentration climbs. The observed gigantic increase by dawn in the concentration of $OH^{\cdot}$ radicals in the air of smog-ridden cities, which serves to start the oxidation of hydrocarbons, is due largely to the decomposition of the HONO that had been created the previous evening:

$$OH^{\cdot} + NO^{\cdot} \longrightarrow HONO \xrightarrow{\text{sunlight}} OH^{\cdot} + NO^{\cdot}$$

It is a characteristic of the later stages in the day of a smog episode that oxidizing agents such as nitric acid are formed in substantial quantities.

The reaction of two OH$^\cdot$ radicals, or of two hydroperoxy radicals, HOO$^\cdot$, produces another atmospheric oxidizing agent, **hydrogen peroxide, H_2O_2,** which, as we have already seen, is also produced in this way in clean atmospheres devoid of nitrogen oxides:

$$2\,OH^\cdot \longrightarrow H_2O_2$$

$$2\,HOO^\cdot \longrightarrow H_2O_2 + O_2$$

The latter reaction occurs also in clean air when the concentration of NO_X is especially low and was encountered in stratospheric chemistry in Chapter 1.

The fate of the $R\!-\!\overset{\cdot}{C}\!=\!O$ radicals, produced by H atom abstraction by OH$^\cdot$ from aldehydes in the ways discussed above, is to combine with O_2 and so produce the free radical

When NO$^\cdot$ is plentiful, this complex species, as expected, behaves as a peroxy radical and oxidizes nitric oxide. In the afternoon, when the concentration of NO$^\cdot$ is very low, the radical reacts instead in a radical–radical process by *adding* to $NO_2^\cdot$ to yield a nitrate. For the common case for which $R = CH_3$, the nitrate product formed is **peroxyacetylnitrate,** or PAN, which is a potent eye irritant in humans and is also toxic to plants.

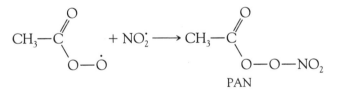

Overall, then, the afternoon stage of a photochemical smog episode is characterized by a build-up of oxidizing agents such as hydrogen peroxide, nitric acid, and PAN, as well as ozone.

Another important species that is present in the later stages of smog episodes is the **nitrate radical,** $NO_3^\cdot$, produced when high concentrations of $NO_2^\cdot$ and ozone occur simultaneously:

$$NO_2^\cdot + O_3 \longrightarrow NO_3^\cdot + O_2$$

Although $NO_3^\cdot$ is photochemically dissociated to $NO_2^\cdot$ and O rapidly during the daytime, it is stable at night and plays a role similar to OH$^\cdot$ in attacking hydrocarbons in the hours following sundown:

$$NO_3^\cdot + RH \longrightarrow HNO_3 + R^\cdot$$

Thus at night, when the concentration of the short-lived hydroxyl radicals goes almost to zero since no new ones are being produced due to the absence of O^*, NO_3^- rather than $OH^{\cdot}$ initiates the oxidation of reduced gases in the troposphere. The similarity between $OH^{\cdot}$ and NO_3^- is not surprising since both react as $-\dot{O}$ radicals and form very stable $O-H$ bonds when they abstract hydrogens.

In summary, an episode of photochemical smog in a city such as Los Angeles begins at dawn, when sunlight initiates the production of hydroxyl radicals from the nitrous acid and from the ozone left over from the previous day. The initial input of nitric oxide and reactive hydrocarbons from morning rush-hour vehicle traffic reacts first to produce aldehydes (see Figure 2-4), the photolysis of which increases the concentration of free-radical reactions and thereby speeds up the overall reaction. The increase in free radicals in the morning serves to oxidize the nitric oxide to nitrogen dioxide; photolysis of the latter causes the characteristic rise in ozone concentrations about midday. Oxidants such as PAN and hydrogen peroxide are also produced, especially in the afternoons due to the high free-radical concentration present at that time. The late afternoon rush-hour traffic produces more nitric oxide and hydrocarbons, which presumably react quickly under the conditions of high free-radical concentration. The smog reactions largely cease at dusk due to the lack of sunlight, but some oxidation of hydrocarbons continues due to the presence of the nitrate radical. The nitrous acid that forms after dark is stable until dawn, when its decomposition helps initiate the process for another day.

Finally, we return to the matter of ozone control by the reduction of NO_X emissions. When the ratio of VOCs to NO_X is lower than normal, reducing NO_X can actually increase ozone production! This occurs because under these conditions, VOCs and NO_X molecules compete with each other for the hydroxyl radicals. Increasing the NO_X concentration then reduces the $OH^{\cdot}$ concentration, due to the increased rate of radical–radical reactions that produce nitrous and nitric acid and remove $OH^{\cdot}$ from air. Then, since there are fewer $OH^{\cdot}$ radicals, the rate of attack on the VOCs, and hence the rate of production of ozone, declines.

PROBLEM 3-10

Annotate (in pencil) the top of Figure 2-4 to show the *dominant* reaction occurring in the polluted air in the following time segments: (a) 5 a.m.– 8 a.m.; (b) 8 a.m.– 12 noon.

PROBLEM 3-11

Some formaldehyde molecules photochemically decompose to the molecular products H_2 and CO rather than to free radicals. Deduce the mechanism and overall reaction for the oxidation to CO_2 for formaldehyde molecules that initially produce the molecular products.

PROBLEM 3-12

Deduce the series of steps by which ethylene gas, $H_2C{=}CH_2$, is oxidized to CO_2 when it is released into an atmosphere undergoing a photochemical smog process. (Assume in this case that aldehydes react completely by photochemical decomposition rather than by $OH^{\cdot}$ attack.)

PROBLEM 3-13

Radical–radical reactions can also occur in *clean* air, particularly when the nitrogen oxide concentration is very low. Predict the product that will be formed when the $CH_3OO^{\cdot}$ radical intermediate of methane oxidation combines with the $HOO^{\cdot}$ free radical. Note that long oxygen chains are unstable with respect to O_2 expulsion.

Oxidation of Atmospheric SO$_2$: The Homogeneous Gas-Phase Mechanism

When the sky is clear or when clouds occupy only a few percent of the tropospheric volume, the predominant mechanism for the conversion of SO_2 to H_2SO_4 is a homogeneous gas-phase reaction that occurs by several sequential steps. As usual for atmospheric trace gases, the hydroxyl radical initiates the process. Since SO_2 molecules contain multiple bonds but no hydrogen, it is expected (see Figure 3-1) that the $OH^{\cdot}$ will *add* to the molecule at the sulfur atom:

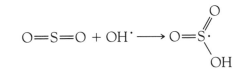

Since a stable molecule, namely **sulfur trioxide,** SO_3, can be produced from this radical by the removal of the hydrogen atom, the reaction principles predict that the next reaction in the sequence is that between the radical and an O_2 molecule to abstract H:

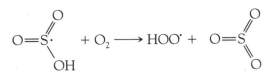

The sulfur trioxide molecule rapidly combines with a gaseous water molecule to form **sulfuric acid.** Finally, the H_2SO_4 molecules react with water, whether in the form of water vapor or as a mist, to form an aerosol of droplets, each of which is an aqueous solution of sulfuric acid. The sequence of steps from gaseous SO_2 to aqueous H_2SO_4 is

$$SO_2 + OH^\bullet \longrightarrow HSO_3^-$$

$$HSO_3^- + O_2 \longrightarrow SO_3 + HOO^\bullet$$

$$SO_3 + H_2O \longrightarrow H_2SO_4(g)$$

$$H_2SO_4(g) + many\ H_2O \longrightarrow H_2SO_4(aq)$$

The sum of these reaction steps is

$$SO_2 + OH^\bullet + O_2 + many\ H_2O \longrightarrow HOO^\bullet + H_2SO_4(aq)$$

When we include the return of $HOO^\bullet$ to $OH^\bullet$ via reaction with $NO^\bullet$, the overall reaction is seen to be $OH^\bullet$-catalyzed co-oxidation of SO_2 and $NO^\bullet$:

$$SO_2 + NO^\bullet + O_2 + many\ H_2O \xrightarrow{OH^\bullet\ catalysis} NO_2^- + H_2SO_4(aq)$$

For representative concentrations of the $OH^\bullet$ radical in relatively clean air, a few percent of the atmospheric SO_2 is oxidized per hour by this mechanism. The rate is much faster for air masses undergoing photochemical smog reactions since the concentration of $OH^\bullet$ there is much higher. However, generally only a small amount of sulfur dioxide is oxidized in cloudless air; the rest is removed by dry deposition before the reaction has time to occur.

Oxidation of Atmospheric SO_2: The Aqueous-Phase Mechanism

Since sulfur dioxide is somewhat soluble in water, a fraction of the atmospheric SO_2 exists in dissolved aqueous form if there is a significant cloud, fog, or mist content in the air. Under these circumstances, most of its oxidation to sulfuric acid occurs in the liquid phase (an aqueous solution) rather than in the gas phase, since the process is inherently faster in aqueous solutions. Nevertheless, the production of the acid does not occur within raindrops themselves, since their lifetime of only a few minutes is insufficient for much oxidation to occur.

Some of the sulfur dioxide dissolved in the aqueous phase occurs as **sulfurous acid,** H_2SO_3; the process by which the gas dissolves to yield this weak acid is

$$SO_2(g) + H_2O(aq) \rightleftharpoons H_2SO_3(aq)$$

The relative concentrations of SO_2 and of H_2SO_3 are related by the equilibrium constant for this reaction. When gases are dissolved in liquids, the equilibrium constant is expressed as the **Henry's law** constant, K_H, which equals the *equilibrium* molar concentration of the substance in the liquid phase divided by its partial pressure in the gas phase. For this reaction, K_H is equal to

$$K_H = [H_2SO_3(aq)]/P$$

where P is the atmospheric partial pressure of gaseous SO_2. Typically the concentration of gaseous SO_2 is about 0.1 ppm, which at 1 atm total air pressure corresponds to a partial pressure of 0.1×10^{-6} atm, i.e., $P = 1 \times 10^{-7}$ atm. Since at 25°C, $K_H = 1$ M/atm, after rearrangement and substitution we obtain

$$[H_2SO_3(aq)] = PK_H = 1 \times 10^{-7} M$$

This value of about 10^{-7} M for the equilibrium concentration of $H_2SO_3(aq)$ is deceptive since it by no means represents *all* the dissolved sulfur dioxide. Although it is a weak acid, H_2SO_3 has an **acid dissociation constant,** K_a, which is sufficiently large (1.7×10^{-2}) that in dilute solutions—as are encountered in atmospheric droplets—most of the sulfurous acid that forms initially subsequently ionizes to **bisulfite** (or hydrogen sulfite) ion, HSO_3^-,

$$H_2SO_3 \rightleftharpoons HSO_3^- + H^+$$

Thus $[H_2SO_3]$ represents the concentration only of the H_2SO_3 that does *not* ionize; its value is held fixed at 10^{-7} M owing to the equilibrium with gaseous SO_2.

If we assume that the above reaction is the only significant source of acidity, then from its stoichiometry it follows that $[HSO_3^-] = [H^+]$, and thus the K_a expression

$$1.7 \times 10^{-2} = [HSO_3^-] [H^+]/[H_2SO_3]$$

becomes, after substituting for $[H_2SO_3]$,

$$1.7 \times 10^{-2} = [HSO_3^-]^2/1 \times 10^{-7}$$

Solving this equation, we find that

$$[HSO_3^-] = 4 \times 10^{-5} M$$

Thus the equilibrium ratio of bisulfite ion to sulfurous acid is 400:1. Consequently, the total dissolved sulfur concentration is about 4×10^{-5} M, rather than just the 1×10^{-7} M that represents only the contribution from the un-ionized acid. Since the concentration of hydrogen ion produced by the reaction is also 4×10^{-5}, the pH of the raindrops is 4.4. Rain does not become highly acidic if only weak acids are dissolved in it.

PROBLEM 3-14

Confirm by calculation that the pH of CO_2-saturated water at 25°C is 5.6, given that the CO_2 concentration in air is 365 ppm, i.e., 0.00037 atm, and that for carbon dioxide the Henry's law constant $K_H = 3.4 \times 10^{-2}$ mol/L atm at 25°C. Furthermore, the ionization constant, K_a, for H_2CO_3 has a value of 4.5×10^{-7} at this temperature. [Hint: Use the same techniques employed to calculate the pH of sulfur dioxide dissolved in water.] Recalculate the pH for a carbon dioxide concentration of 560 ppm, i.e., double the amount in the preindustrial age. Given that at 15°C, $K_H = 0.047$ mol/L atm and $K_a = 3.8 \times 10^{-7}$, calculate the pH of natural rainwater at this temperature for a CO_2 concentration of 365 ppm.

PROBLEM 3-15

Calculate the pH of rainwater in equilibrium with SO_2 in a polluted air mass for which the sulfur dioxide concentration is 1.0 ppm.

PROBLEM 3-16

Calculate the concentration of SO_2 that must be reached in polluted air if the dissolved gas is to produce a pH of 4.0 in raindrops without any oxidation of the gas.

If strong acids are already present in the water droplet, they, rather than sulfurous acid, control its pH. Under these conditions, the bisulfite ion concentration can be calculated through rearrangement of the K_a expression to give

$$[HSO_3^-] = 1.7 \times 10^{-2} \times 10^{-7}/[H^+]$$
$$= 1.7 \times 10^{-9}/[H^+]$$

Here the H^+ concentration is overwhelmingly due to its release from the strong acids, since the additional amount contributed from the H_2SO_3 ionization will be negligible in comparison. Thus, under these conditions, the bisulfite ion concentration, $[HSO_3^-]$, is inversely proportional to $[H^+]$.

Dissolved sulfur dioxide, SO_2, is oxidized to **sulfate ion**, SO_4^{2-}, by trace amounts of the well-known oxidizing agents hydrogen peroxide, H_2O_2, and ozone, O_3, that are present in the airborne droplets (see Figure 3-5). Indeed, these reactions currently are thought to constitute the main oxidation pathways for atmospheric SO_2, except under clear sky conditions when the gas-phase, homogeneous mechanism predominates. The ozone and hydrogen peroxide result mainly from sunlight-induced reactions in photochemical smog. Consequently, oxidation of SO_2 occurs most rapidly in air that has also

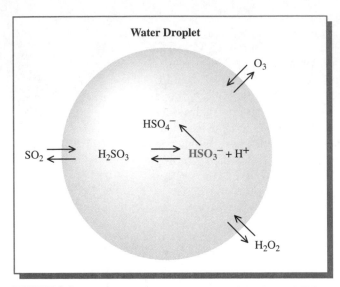

FIGURE 3-5 Dissolution of atmospheric gases SO_2, O_3, and H_2O_2 into a water droplet and their subsequent reactions.

been polluted by reactive hydrocarbons and nitrogen oxides. Since the smog reactions occur predominantly in summer, rapid oxidation of SO_2 to sulfate also is characteristic of the summer season.

The very small concentration of hydrogen peroxide present in acidic atmospheric droplets arises from the gaseous form of the compound present in the air. Hydrogen peroxide is highly soluble in water, since it readily forms hydrogen bonds with water molecules. Its Henry's law constant K_H is 7.4×10^4 M/atm. If its atmospheric gas-phase concentration is 1 ppb, then its partial pressure P' is 1×10^{-9} atm. Further, since

$$K_H = [H_2O_2(aq)]/P'$$

then

$$[H_2O_2(aq)] = P' K_H$$
$$= 1.0 \times 10^{-9} \text{ atm} \times 7.4 \times 10^4 \text{ M/atm}$$
$$= 7 \times 10^{-5} \text{ M}$$

Although this value of 7×10^{-5} M, which is 70 μM when expressed in micromolar units, seems very small by comparison with the concentration values usually encountered in laboratory solutions, it is sufficient to oxidize dissolved sulfur dioxide at an appreciable rate. Measured concentrations of H_2O_2 in clouds and rainwater range from about 0.01 μM to more than 200 μM. Because gas-phase concentrations are highest in the afternoons and in summer, and lowest at night and in the winter, sunlight is a dominant factor in determining the H_2O_2 concentration.

Although both sulfurous acid and bisulfite ion are present in atmospheric droplets when sulfur dioxide dissolves in them, the ion is much more easily oxidized than is the neutral species. Consequently, the rate of the oxidation by ozone slows substantially as the acidity in the droplet increases, since the equilibrium shifts and the bisulfite ion is converted back to relatively unreactive sulfurous acid. The rate at which sulfur dioxide is oxidized by hydrogen peroxide is essentially independent of pH, however, since the rate-determining step of this process is acid-catalyzed. Thus the competing effects of H^+ on bisulfite concentration and on catalysis rate cancel each other. As a consequence of the different effects of acidity on the rate of oxidation, hydrogen peroxide dominates the oxidation process when the pH is below 5. Above this pH, oxidation by ozone or catalysis by transition metals becomes dominant.

The uncatalyzed oxidation of dissolved SO_2 by dissolved molecular oxygen, O_2, is thought to proceed at too slow a rate to make a significant contribution to the overall speed of the reaction. However, if Fe^{3+} or the ions of other transition metals are present in the droplets (due to their release into air from pollution sources or even just from dust), this generalization may not be valid. If neutral or basic conditions occur within the droplet, oxidation of dissolved sulfur dioxide by Fe^{3+} and ozone usually dominates that by hydrogen peroxide.

Systematics of Stratospheric Chemistry

There are many similarities between the chemical reactions discussed in Chapter 1 for the stratosphere and those outlined above for the troposphere. For example, a characteristic process in both regions of the atmosphere is hydrogen atom abstraction. The stratosphere and troposphere differ, however, in which reactions are dominant. In the stratosphere $OH^{\bullet}$, O^{*}, $Cl^{\bullet}$, and $Br^{\bullet}$ are all important in abstracting a hydrogen atom from stable molecules such as methane, whereas only the hydroxyl and nitrate radicals are important in this respect in the troposphere. In the following material we systematize the Chapter 1 chemistry that is important in the stratosphere, especially in regard to processes of ozone depletion.

Processes Involving Loosely Bound Oxygen Atoms

Many of the species in the stratosphere have a **loosely bound oxygen atom,** denoted Y, which is readily detached from the rest of the molecule in several characteristic ways. In Table 3-1 we list the molecules Y—O that contain a loose oxygen. In every case, dissociation of this oxygen atom requires much less energy than is required to break any of the remaining bonds, so the resulting Y units remain intact. Notice that the Y species, except for O_2, are the free radicals that in Chapter 1 we called X when we discussed ozone

TABLE 3-1	Molecules Containing Loose Oxygen Atoms		
Molecule Y—O	Structure of Y—O	Y—O Bond Energy in kJ/mol	Comment
O_3	O_2—O	107	The most loose oxygen
$BrO^{\bullet}$	Br—O	235	
$HOO^{\bullet}$	HO—O	266	
$ClO^{\bullet}$	Cl—O	272	
$NO_2^{\bullet}$	ON—O	305	The least loose oxygen

destruction catalysts. In terms of electronic structure, all "loose" oxygens are joined by a single bond to another electronegative atom that possesses one or more nonbonding electron pairs. The interaction between the nonbonded electron pairs on this atom and those on the oxygen weakens the single bond.

The characteristic reactions involving loose oxygen are collected below:

- *Reaction with Atomic Oxygen*   Here the oxygen atom detaches the loose oxygen atom by combining with it:

$$Y{-}O + O \longrightarrow Y + O_2$$

These reactions are all exothermic since the $O{=}O$ bond in O_2 is much stronger than the $Y{-}O$ bond.

- *Photochemical Decomposition*   The $Y{-}O$ species absorbs UV-B, and in some cases even longer wavelength light, from sunlight and subsequently releases the loose oxygen atom:

$$Y{-}O + \text{ sunlight} \longrightarrow Y + O$$

- *Reaction with* $NO^{\cdot}$ Nitric oxide abstracts the loose oxygen atom:

$$Y{-}O + NO^{\cdot} \longrightarrow Y + NO_2^{\cdot}$$

This reaction is exothermic since the $ON{-}O$ bond strength (see Table 3-1) is the greatest of those involving a loose oxygen. (Recall the general principle that exothermic free-radical reactions are relatively fast.)

- *Abstraction of Oxygen from Ozone*   Abstraction of the loose oxygen atom from ozone (only) to form the $Y{-}O$ species is characteristic of $OH^{\cdot}$, $Cl^{\cdot}$, $Br^{\cdot}$, and $NO^{\cdot}$. Thus all these radicals act as catalytic ozone destroyers, X:

$$O_2{-}O + X \longrightarrow O_2 + XO$$

The reaction involving ozone is exothermic since ozone contains the weakest of the bonds involving a loose oxygen. The other YO species do not undergo this reaction to an important extent either because it is endothermic and therefore negligibly slow or because the atmospheric X species react more quickly with other chemicals.

- *Combination of Two YO Molecules*   If the concentration of YO species becomes high, they may react by the collision of two of them (identical or different species). If at least one is O_3 or $HOO^{\cdot}$, an unstable chain of three or more oxygen atoms is created when they collide and join; in these circumstances, the loose oxygens combine to form one or more molecules of O_2, which are expelled:

$$2\,O_2{-}O \longrightarrow 3\,O_2$$

$$2\,HO{-}O^{\cdot} \longrightarrow HOOH + O_2$$

$$HO{-}O^{\bullet} + O{-}O_2 \longrightarrow OH^{\bullet} + 2\,O_2$$

$$HO{-}O^{\bullet} + {}^{\bullet}O{-}Cl \longrightarrow HOCl + O_2$$

When neither is O_3 or $HOO^{\bullet}$, the two Y—O molecules combine to form a larger molecule, which subsequently often decomposes photochemically:

$$2\,NO_2^{\bullet} \longrightarrow N_2O_4$$

$$2\,ClO^{\bullet} \longrightarrow ClOOCl \xrightarrow{\text{sunlight}} \longrightarrow 2\,Cl^{\bullet} + O_2$$

$$ClO^{\bullet} + NO_2^{\bullet} \underset{\text{sunlight}}{\rightleftharpoons} ClONO_2$$

$$ClO^{\bullet} + BrO^{\bullet} \longrightarrow Cl^{\bullet} + Br^{\bullet} + O_2$$

The Y—O—O—Y molecules have little thermal stability, and even at moderate temperatures may dissociate back to their Y—O components before light absorption and photolysis has time to occur.

PROBLEM 3-17

Which of the following species do(es) *not* contain a loose oxygen?
(a) $HOO^{\bullet}$ (b) $OH^{\bullet}$ (c) $NO^{\bullet}$ (d) O_2 (e) $ClO^{\bullet}$

PROBLEM 3-18

From which Y—O species
(a) does $NO^{\bullet}$ abstract an oxygen atom?
(b) does atomic oxygen abstract an oxygen atom?
(c) does sunlight detach an oxygen atom?
(d) do the Y—O—O—Y species (with identical Y groups) form in the stratosphere?
(e) is O_2 produced when two identical Y—O species react?

PROBLEM 3-19

Using the principles above, predict what would be the likely fate of $BrO^{\bullet}$ molecules in a region of the stratosphere that was particularly (a) high in atomic oxygen, (b) high in $ClO^{\bullet}$, (c) high in $BrO^{\bullet}$ itself, and (d) high in sunlight intensity.

PROBLEM 3-20

Using the principles above, deduce what reaction(s) could be sources of atmospheric (a) $ClONO_2$ (b) $ClOOCl$ (c) $Cl^{\bullet}$ atoms

PROBLEM 3-21

Draw the Lewis structure for the free radical $FO^\bullet$. On the basis of this structure, could you predict whether it contains a loose oxygen?

PROBLEM 3-22

What is the expected product when $ClO^\bullet$ reacts with $NO^\bullet$? What are the possible fates of the product(s) of this reaction? Devise a mechanism incorporating (a) this reaction, (b) the reaction of $Cl^\bullet$ with ozone, and (c) the photochemical decomposition of $NO_2^\bullet$ to $NO^\bullet$ and atomic oxygen. Is the net result of this cycle, which operates in the lower stratosphere, the destruction of ozone?

Review Questions

1. Explain why $OH^\bullet$ reacts more quickly than $HOO^\bullet$ to abstract hydrogen from other molecules.

2. How does $OH^\bullet$ react with molecules that contain hydrogen but not multiple bonds?

3. What are the two different initial steps by which atmospheric formaldehyde, H_2CO, is decomposed in air?

4. What is the reaction by which most nitric oxide molecules in the troposphere are oxidized to nitrogen dioxide?

5. What are the two common reactions by which diatomic oxygen reacts with free radicals?

6. Explain why photochemical smog is an autocatalytic process.

7. What is the fate of $OH^\bullet$ radicals that react with $NO^\bullet$? with $NO_2^\bullet$? with other $OH^\bullet$?

8. What is the fate of $NO_2^\bullet$ molecules that photodissociate? that react with ozone? that react with $RCOO^\bullet$ radicals?

9. Why does the production of high concentrations of $NO_2^\bullet$ lead to an increase in ozone levels in air? Why does this not occur if much $NO^\bullet$ is present?

10. What is the formula of nitrate radicals? Explain how they are similar in reactivity to hydroxyl radicals.

11. Explain how atmospheric sulfur dioxide is oxidized by (a) homogeneous and (b) heterogeneous reactions in the atmosphere. What is the role of hydrogen peroxide and ozone in the latter process?

12. Explain what is meant by the term *loosely bound oxygen*. What are its four characteristic reactions in stratospheric chemistry?

Additional Problems

1. Write the two-step mechanism by which CO is oxidized to CO_2. Also include the sequence of reactions by which the hydroperoxy radical so produced oxidizes $NO^\bullet$ to $NO_2^\bullet$, the nitrogen dioxide is photolyzed to $NO^\bullet$ and atomic oxygen, and oxygen atoms produce ozone. By adding the steps, show that the atmospheric oxidation of carbon monoxide can increase the ozone concentration by a catalytic process.

2. Using the reactivity principles developed in this chapter, deduce the series of steps and the overall reaction by which ethane, H_3C-CH_3, is oxidized in the atmosphere. Assume that the aldehydes produced in the mechanism undergo photochemical decomposition to $R\cdot$ and $HCO\cdot$.

3. When the concentration of nitrogen oxides in a region of the air is very low, peroxy radicals combine with other species rather than oxidizing nitric oxide. Deduce the mechanism, including the overall equation, for the process by which carbon monoxide is oxidized to carbon dioxide under these conditions, assuming that the hydroperoxy radicals react with ozone. From your result, would you predict that ozone levels would be abnormally high or low in air masses having low nitrogen oxide concentration? [Hint: See the generalities in the section on the systematics of stratosphere chemistry concerning reactions that produce long oxygen chains.]

4. Predict the most likely reaction (if any) that would occur between a hydroxyl radical and each of the following atmospheric gases:

(a) $CH_3CH_2CH_3$ (b) $H_2C=CHCH_3$

(c) $H_2C=CHCl$ (d) HCl

(e) H_2O (f) CH_3OH

5. Draw complete Lewis structures for NO_2, HONO, and HNO_3, illustrating the radical nature of the first species and the nonradical nature of the last two species. (Resonance structures and formal charges are not required.)

6. In Problem 1-2, the longest wavelength of light that could dissociate an O atom from O_3 was calculated and determined to occur in the IR region of the spectrum. Using the information in Table 3-1, calculate the longest wavelength of light that could photolytically cleave the loose hydrogen atom in the case of each of the remaining molecules listed in that table. What region of the electromagnetic spectrum does each correspond to?

Further Reading

1. R. Atkinson, "Atmospheric Chemistry of VOCs and NO_X," *Atmospheric Environment* 34 (2000): 2063–3101.

2. B. J. Finlayson-Pitts and J. N. Pitts, *Atmospheric Chemistry* (New York: Wiley, 1986). [A comprehensive guide to the detailed chemistry of the atmosphere.]

3. B. J. Finlayson-Pitts and J. N. Pitts, Jr., "Tropospheric Air Pollution: Ozone, Airborne Toxics, Polycyclic Aromatic Hydrocarbons, and Particles," *Science* 276 (1997): 1045–1051.

4. J. H. Seinfeld and S. N. Pandis, *Atmospheric Chemistry and Physics* (New York: Wiley, 1998). [Another comprehensive guide to the detailed chemistry of the atmosphere.]

Websites of Interest

Log on to www.whfreeman.com/envchem3e/and click on Chapter 3.

$HCOR \rightarrow HCO\cdot + R$

CHAPTER 4

The Greenhouse Effect and Global Warming

Everyone has heard the prediction that the "greenhouse effect" will significantly affect climates around the world in the future. The terms *greenhouse warming* and *global warming* in ordinary usage simply mean that average global air temperatures are expected to increase by several degrees as a result of the buildup of carbon dioxide and other "greenhouse" gases in the atmosphere. Indeed, most atmospheric scientists believe that such **global warming** has been under way already for some time and is largely responsible for the air temperature increase of about two-thirds of a Celsius degree that has occurred since 1860.

The phenomenon of rapid global warming—with its demands for large-scale adjustments—is generally considered to be our most crucial worldwide environmental problem, although both positive and negative effects would be associated with any significant increase in the average global temperature. Unlike stratospheric ozone depletion, which has manifested itself in spectacular fashion in the form of the ozone hole, the phenomenon of global warming due to the greenhouse effect has yet to be observed in a fashion that convinces everyone of its existence. No one currently is sure of the extent or timing of future temperature increases, nor is it likely that reliable predictions for individual regions will ever be

Global warming may have led to the dramatic breakup of the Larsen Ice Shelf off Antarctica in 2002. [GSFC/LaRC/JPL/MISR Team/NASA.]

available much in advance of the events in question. If current models of the atmosphere are correct, however, significant warming will occur in coming decades. It is important that we understand the factors that are driving this increase so that we can, if we wish, take steps to avoid potential catastrophes caused by rapid climate change in the future.

In this chapter the mechanism by which global warming could arise is explained, and the nature and sources of the chemicals that are responsible for the effect are analyzed. The extent of the atmospheric warming to date and other indications that change is under way are also discussed. The predictions concerning global warming in the future, and an analysis of steps that could be taken to minimize it, are presented in Chapters 5 and 6.

The Mechanism of the Greenhouse Effect

The Earth's Energy Balance

The Earth's surface and atmosphere are kept warm primarily by energy from the Sun. The wavelength λ_{peak}, in micrometers, of maximum energy emission by a radiating **blackbody** is determined by its Kelvin temperature, T, according to the equation

$$\lambda_{peak} = 2897/T$$

(A blackbody is 100% efficient in absorbing energy that encounters it and 100% efficient in radiating energy.) Since for the surface of the Sun $T \sim 5800$ K, then, from the equation $\lambda_{peak} = 0.50$ μm, which is in the visible region. Indeed, the maximum observed solar output (see the dashed portion of the curve in Figure 4-1) lies in the range of visible light—that of wavelengths between 0.40 and 0.75 μm (400–750 nm). Beyond the

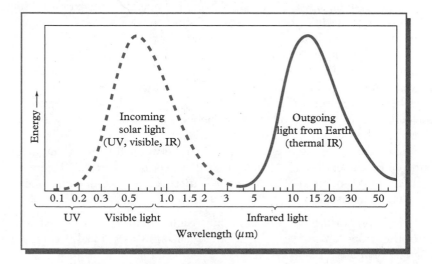

FIGURE 4-1 Wavelength distributions (using different scales) for light emitted by the Sun (dashed curve) and by Earth's surface and troposphere (solid curve). [Source: Redrawn from J. Gribbin, "Inside Science: The Greenhouse Effect," *New Scientist*, supplement (22 October 1988).]

red limit (i.e., the maximum wavelength for visible light), we receive **infrared** (IR) light in the 0.8–3 μm region from the Sun. Of the energy received above the Earth's atmosphere from the Sun, slightly over half is infrared and most of the remainder is visible light.

As explained in Chapter 1, much of the ultraviolet light (wavelengths < 0.4 μm) from the Sun is filtered out in the stratosphere and warms the air there rather than at the surface of the Earth. Of the total incoming light of all wavelengths that impinges upon the Earth, about 50% reaches the surface and is absorbed by it. A further 20% of the incoming light is absorbed by gases — UV by stratospheric ozone and diatomic oxygen, IR by CO_2 and H_2O — and by water droplets in air. The remaining 30% of incoming light is reflected back into space by clouds, ice, snow, sand, and other reflecting bodies, without being absorbed.

Historical Temperature Trends

The trends in average surface temperature in the Northern Hemisphere for the past 2000 years, as reconstructed for most of that period from indirect evidence such as tree ring growth, is shown in Figure 4-2a. (The Medieval Warm Period early in the previous millennium was apparently restricted to the North Atlantic region, so it is not very evident on the global plot.) Notice the overall downward trend in temperature until the beginnings of the Industrial Revolution.

The warming of the climate during the twentieth century stands in stark contrast to the gradual cooling trend in the previous 900 years of the millennium, producing a "hockey stick" shape in the overall temperature plot. Air temperature did not increase *continuously* throughout the twentieth century. A plot of the variations in the surface air temperature over the last century and a half is shown in Figure 4-2b. Periods of overall temperature increases and some of temperature declines become evident when the variations are smoothed over a few decades. The main trends during four periods in the 140 years are illustrated in Figure 4-2b. The average rate of change during the last quarter-century amounted to 2.0°C (3.5°F) *per century*, so the total change within the 25-year period was $0.25 \times 2.0 = 0.5$°C. The average air temperature *declined* slightly in the preceding period (1940–1975), as it also did at the beginning of the century. There was also a warming period from 1910 to 1940, with a rate about half that of the recent rise.

Scientists have discovered correlations between the Sun's output and Earth's air temperature for some of the time periods covered by Figure 4-2a without fully understanding how they arise. Three possible mechanisms have been suggested:

• Variations in the Sun's total output result in changes in its heat input into the troposphere. This could only account for global temperature changes of about 0.1°, however.

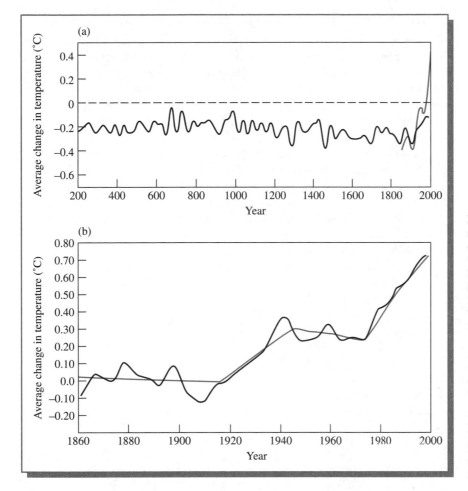

FIGURE 4-2 (a) Reconstruction of changes in average global surface temperatures over the last two millennia. [Source: M.E. Mann and P.D. Jones, "Global Surface Temperatures over the Past Two Millennia," *Geophysical Research Letters* 30 (2003): 1820.] (b) Trends in average global surface temperature (black line) 1860–1999. The green line shows linear trends within four characteristic periods. [Source: Adapted from World Wildlife Fund report.]

• The much larger fluctuations in the UV component of the Sun's output produce changes in the concentration of stratospheric ozone, which in turn change the extent of heating of the stratosphere by ozone, an effect that is transmitted to the air of the troposphere.

• Variations in the strength of the solar wind result in changes to the magnetic shielding that it provides to the Earth. When less magnetic shielding is in place, more cosmic rays (mostly protons) from outer space penetrate the atmosphere and produce increases in the amount of cloud cover, thereby increasing the greenhouse effect (see below) and further warming the Earth.

Earth's Energy Emissions and the Greenhouse Effect

Like any warm body, the Earth emits energy; indeed, the amount of energy that the planet absorbs and the amount that it releases must be equal if its temperature is to remain constant. The emitted energy (see the solid portion of the curve in Figure 4-1) is neither visible nor UV light. Since the temperature of the Earth's surface and lower atmosphere are approximately 300 K, then according to the equation above, the wavelength of maximum emission is expected to be about 10 μm. Indeed, the Earth's emission does peak near 10 μm and consists of infrared light of wavelengths from 4 to 50 μm; this is called the **thermal infrared** region, since the energy is a form of heat, the same kind of energy a heated iron pot would radiate.

Some gases in the air can absorb thermal infrared light of specific wavelengths, so not all the IR emitted from the Earth's surface and atmosphere escapes directly to space. Shortly after its absorption by airborne molecules such as CO_2, this infrared light is re-emitted in all directions—completely randomly. Consequently, some of this thermal IR is redirected back toward the Earth's surface, is reabsorbed, and further heats both the surface and the air. This phenomenon, the redirection of thermal IR toward the Earth, as shown in Figure 4-3, is called the **greenhouse effect** and is responsible for the average temperature at the Earth's surface being about +15°C rather than about −15°C, the temperature it would be if there were no IR-absorbing gases in the atmosphere. The very fact that our planet is not entirely covered by a thick sheet of ice is due to the natural operation of the greenhouse effect. The surface is warmed as much by this mechanism as it is by the solar energy it receives directly!

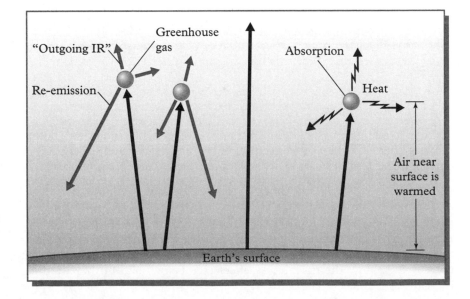

FIGURE 4-3 The greenhouse effect: Outgoing IR absorbed by greenhouse gases is either re-emitted (left side of diagram) or converted to heat (right side).

The atmosphere operates in the same way as a blanket, retaining within the immediate region some of the heat released by a body and thereby increasing the local temperature. The phenomenon that worries environmental scientists is that increasing the concentration of the trace gases in air that absorb thermal infrared light (piling on more blankets, so to speak) would result in the redirection of even more of the outgoing thermal infrared energy and thereby increase the average surface temperature beyond 15°C. This phenomenon is referred to as the **enhanced greenhouse effect** (or *artificial global warming*) to distinguish it from the warming that has been operating naturally for millennia.

The principal constituents of the atmosphere, N_2, O_2, and Ar, are incapable of absorbing infrared light; the reasons for this will be discussed in the following section. The atmospheric gases that in the past have produced most of the greenhouse warming are water (responsible for about two-thirds of the effect) and carbon dioxide (responsible for about one-quarter). Indeed, the absence of water in the dry air of desert areas leads to low nighttime temperatures there, even though the daytime temperatures are quite high on account of direct absorption of solar energy. More familiar to people living in temperate climates is the crisp chill in winter air on cloudless days and nights.

Molecular Vibrations: Energy Absorption by Greenhouse Gases

Light is most likely to be absorbed by a molecule when its frequency almost exactly matches the frequency of an internal motion within the molecule. For frequencies in the infrared region, the relevant internal motions are the **vibrations** of the molecule's atoms relative to each other.

The simplest vibrational motion in a molecule is the oscillatory motion of two bonded atoms X and Y relative to each other. In this motion, called a **bond-stretching vibration,** the X to Y distance increases beyond its average value R, then returns to R, then contracts to a lesser value, and finally returns to R, as illustrated in Figure 4-4a. Such oscillatory motion occurs in all bonds of all molecules under all temperature conditions, even at absolute zero. A huge number (about 10^{13}) of such vibrational cycles occur each second. The exact frequency of the oscillatory motion depends primarily on the type of bond—i.e., whether it is single or double or triple—and on the identity of the two atoms involved. For many bond types, e.g., the C—H bond in methane and the O—H bond in water, the stretching frequency does not fall within the thermal infrared region. The stretching frequency of carbon–fluorine bonds does, however, occur within the thermal infrared range (4–50 μm); thus any molecules in the atmosphere with C—F bonds will absorb outgoing thermal IR light and enhance the greenhouse effect.

(a) Bond-stretching vibration

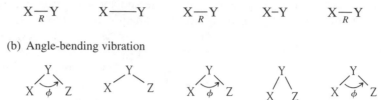

(b) Angle-bending vibration

FIGURE 4-4 The two kinds of vibrations within molecules. Bond stretching (a) is illustrated for a diatomic molecule XY. The variable R represents the average value of the X–Y distance. In (b), the angle-bending vibration is shown for a triatomic molecule XYZ. The average XYZ angle is indicated by ϕ.

The other relevant type of vibration is an oscillation in the distance between two atoms X and Z bonded to a common atom Y but not bonded to each other. Such motion alters the XYZ bond angle from its average value ϕ and is called a **bending vibration.** All molecules containing three or more atoms possess bending vibrations. The oscillatory cycle of bond angle increase, followed by a decrease, and then another increase, etc., is illustrated in Figure 4-4b. The frequencies of many types of bending vibrations in most organic molecules occur within the thermal infrared region.

If infrared light is to be absorbed by a vibrating molecule, there must be a difference in the relative positions of the molecule's center of positive charge (its nuclei) and its center of negative charge (its electron "cloud") at some point during the motion. More compactly stated, in order to absorb IR light, the molecule must have a dipole moment during some stage of the vibration. Technically, there must be a *change* in the magnitude of the dipole moment during the vibration, but this is more or less guaranteed to be the case if there is a nonzero dipole moment at any point in the vibration. The positive and negative centers of charge coincide in free atoms and (by definition) in homonuclear diatomic molecules like O_2 and N_2, and the molecules have dipole moments of zero at all times in their stretching vibration. Thus argon gas, Ar, diatomic nitrogen gas, N_2, and diatomic oxygen, O_2, do not absorb IR light.

For carbon dioxide, during the vibratory motion in which both C—O bonds lengthen and shorten simultaneously, i.e., synchronously, there is at no time any difference in position between the centers of positive and negative charges, since both lie precisely at the central nucleus. Consequently, during this vibration, called the **symmetric stretch,** the molecule cannot absorb IR light. However, in the **antisymmetric stretch** vibration in CO_2, the contraction of one C—O bond occurs when the other is lengthening, or vice versa, so that during the motion the centers of charge no longer necessarily coincide. Therefore, IR light at this vibration's frequency *can* be absorbed since, at some points in the vibration, the molecule does have a dipole moment.

$$\overleftarrow{O}=C=\overrightarrow{O}$$
symmetric
stretch

$$\overrightarrow{O}=C=\overrightarrow{O}$$
antisymmetric
stretch

Similarly, the bending vibration in a CO_2 molecule, in which the three atoms depart from a colinear geometry, is a vibration that can absorb IR light since the centers of positive and negative charge do not coincide when the molecule is nonlinear.

PROBLEM 4-1

Deduce whether the following molecules will absorb infrared light due to internal vibrational motions:

(a) H_2 (b) CO (c) Cl_2 (d) O_3 (e) CCl_4 (f) NO

PROBLEM 4-2

None of the four diatomic molecules listed in Problem 4-1 actually absorb much, if any, of the Earth's outbound light in the *thermal* infrared region. What does this imply about the frequencies of the bond-stretching vibrational motion of those molecules that can, in principle, absorb IR light?

The Major Greenhouse Gases
Carbon Dioxide: Absorption of Infrared Light

As stated previously, the absorption of light by a molecule occurs most efficiently when the frequencies of the light and of one of the molecule's vibrations match almost exactly. However, light of somewhat lower or higher frequency than that of the vibration is absorbed by a collection of molecules. This ability of molecules to absorb infrared light over a short range of frequencies rather than at just a single frequency occurs because it is not only the energy associated with vibration that changes when an infrared photon is absorbed; there is also a change in the energy associated with the rotation (tumbling) of the molecule about its internal axes. This **rotational energy** of a molecule can be either slightly increased or slightly decreased when IR light is absorbed to increase its **vibrational energy.** Consequently, photon absorption occurs at a slightly higher or lower frequency than that corresponding to the frequency of the vibration. Generally, the absorption tendency of a gas falls off for light frequency that lies farther and farther in either direction from the vibrational frequency.

The absorption spectrum for **carbon dioxide** in a portion of the infrared range is shown in Figure 4-5. For CO_2, the maximum absorption of light in the thermal infrared range occurs at a wavelength of 15.0 μm, which corresponds

FIGURE 4-5 The infrared absorption spectrum for carbon dioxide. The scale for wavelength is linear when expressed in wavenumbers, which have units of cm^{-1}; wavenumber = 10,000/ wavelength in nm. [Source: Redrawn from A.T. Schwartz et al., *Chemistry in Context: Applying Chemistry to Society*, American Chemical Society (Dubuque, IA: Wm. C. Brown, 1994).]

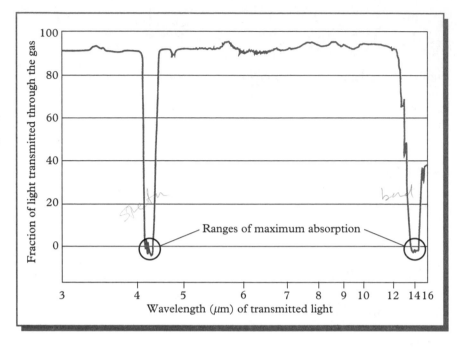

to a frequency of 2×10^{13} cycles per second (hertz). The absorption occurs at this particular frequency because it matches that of one of the vibrations in a CO_2 molecule, namely the OCO angle-bending vibration. Carbon dioxide also strongly absorbs IR light having a wavelength of 4.26 μm, which corresponds to the 7×10^{13} cycles per second (hertz) frequency of the antisymmetric OCO stretching vibration.

PROBLEM 4-3

Calculate the energy absorbed per mole and per molecule of carbon dioxide when it absorbs infrared light (a) at 15.0 μm and (b) at 4.26 μm. Express the per mole energies as fractions of that required to dissociate CO_2 into CO and atomic oxygen, given that the enthalpies of formation of the three gaseous species are -393.5, -110.5, and $+249.2$ kJ/mol, respectively. [Hint: Recall the relationship between wavelength and energy in Chapter 2. Avogadro's constant = 6.02×10^{23}.]

The carbon dioxide molecules that are now present in air collectively absorb about half of the outgoing thermal infrared light with wavelengths in the 14–16 μm region, together with a sizable portion of that in the 12–14 and 16–18 μm regions. It is because of CO_2's absorption that the solid curve in Figure 4-6, representing the amount of IR light that actually escapes from our

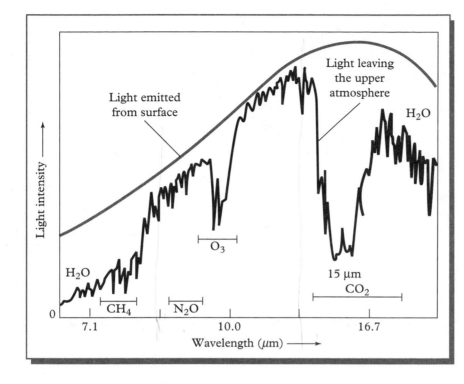

FIGURE 4-6 Experimentally measured intensity (black curve) of thermal IR light leaving the Earth's surface and lower atmosphere (above the Sahara desert) compared with the theoretical intensity (green curve) that would be expected without absorption by atmospheric greenhouse gases. The regions in which the various gases have their greatest absorption are indicated. [Source: E. S. Nesbit, *Leaving Eden* (Cambridge: Cambridge University Press, 1991).]

atmosphere, falls so steeply around 15 μm; the vertical separation between the curves is proportional to the amount of IR of a given wavelength that is being absorbed rather than escaping. Further increases in the CO_2 concentration in the atmosphere will prevent more of the remaining IR from escaping, especially in the "shoulder" regions around 15 μm, and will further warm the air. (Although carbon dioxide also absorbs IR light at 4.3 μm due to the antisymmetric stretching vibration, there is little energy emitted from the Earth at this wavelength—see Figure 4-1—so this potential absorption is not very important.)

Carbon Dioxide: Past Concentration and Emission Trends

Measurements on air trapped in ice-core samples from Antarctica and Greenland indicate that the atmospheric concentration of carbon dioxide in preindustrial times (i.e., before about 1750) was about 280 ppm. The concentration had increased by one-third, to 373 ppm, by 2002. A plot of the increase in the annual atmospheric CO_2 concentration over time is shown in Figure 4-7. The insert in the figure shows a detail of the increase in recent times. In the 1990s the concentration grew at an average annual rate of about 0.4%, or 1.6 ppm, almost double that of the 1960s, although there were considerable year-to-year fluctuations of up to ±1 ppm in the rate of increase.

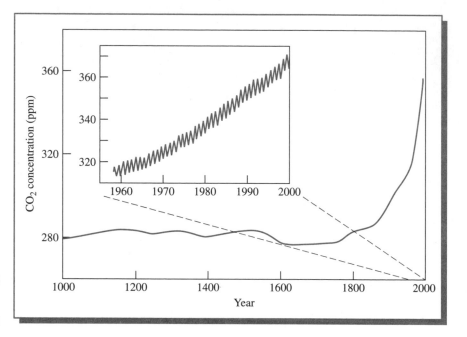

FIGURE 4-7 The historical variation in the atmospheric concentration of carbon dioxide. The insert shows the trend, with seasonal fluctuations, in recent times. [Source: J. L. Sarmiento and N. Gruber, "Sinks for Anthropogenic Carbon," *Physics Today* 55 (2002): 30.]

These seasonal fluctuations in the CO_2 concentrations are due to the spurt in the growth of vegetation in the spring and summer, which removes CO_2 from air, and the vegetation decay cycle in fall and winter that increases it. In particular, huge quantities of CO_2 are extracted from the air each spring and summer by the process of plant photosynthesis:

$$CO_2 + H_2O \xrightarrow{\text{sunlight}} O_2 + \text{ polymeric } CH_2O$$

The term *polymeric CH$_2$O* used for the product in this equation is an umbrella word for plant fiber, typically the cellulose that gives wood its mass and bulk. The CO_2 "captured" by the photosynthetic process is no longer free to function as a greenhouse gas—or as any gas—while it is packed away in this polymeric form. The carbon that is trapped in this way is called **fixed carbon.** However, the biological decay of this plant material, the reverse of the reaction, which occurs mainly in the fall and winter, frees the withdrawn carbon dioxide. Notice that the global carbon dioxide fluctuations follow the seasons of the Northern Hemisphere, since there is so much more land mass—and hence much more vegetation—there compared to the Southern Hemisphere.

Much of the considerable increase in anthropogenic contributions to the increase in carbon dioxide concentration in air is due to the combustion of **fossil fuels**—chiefly coal, oil, and natural gas—that were formed eons ago when plant and animal matter was covered by geological deposits before it could be broken down by air oxidation.

On average, each person in the industrial countries is responsible for the release of about 5 metric tons (a metric ton is 1000 kg, i.e., 2200 lb, whereas a conventional ton is 2000 lb) of CO_2 from carbon-containing fuels each year! There is considerable variation in the per capita releases among different industrialized countries; this is discussed in Chapter 5. Some of the per capita carbon dioxide output is direct, e.g., that released as gases when vehicles are driven and homes are warmed by burning a fossil fuel. The remainder is indirect, arising when energy is used to produce and transport goods; heat and cool factories, classrooms, and offices; produce and refine oil—in fact, to accomplish virtually any constructive economic purpose in an industrialized society. This topic is also discussed in greater depth in Chapter 5.

A significant amount of carbon dioxide is added to the atmosphere when forests are cleared and the wood burned to provide land for agricultural use. This sort of activity occurred on a massive scale in temperate climate zones in past centuries (consider the immense deforestation that accompanied the settlement of the United States and southern Canada) but has now shifted largely to the tropics. The greatest single amount of current deforestation occurs in Brazil and involves both rain forest and moist deciduous forest, but the annual rate of deforestation on a percentage basis is actually greater in Southeast Asia and Central America than in South America. Overall, deforestation accounts for about one-quarter of the annual anthropogenic release of CO_2, the other three-quarters originating mainly in the combustion of fossil fuels. Notwithstanding forestry harvesting operations, the total amount of carbon contained in the forests of the Northern Hemisphere (including their soils) is increasing, and in the 1980s the annual increment approximately equaled the decreases in stored carbon cited above in Asia and South and Central America.

PROBLEM 4-4

Carbon dioxide is also released into the atmosphere when calcium carbonate rock (limestone) is heated to produce the quicklime, i.e., calcium oxide, used in the manufacture of cement:

$$CaCO_3(s) \longrightarrow CaO(s) + CO_2(g)$$

Calculate the mass, in metric tons, of CO_2 released per metric ton of limestone used in this process. What is the mass of carbon that the air gains for each gram of carbon dioxide that enters the atmosphere? Note that at least as much carbon dioxide is released from combustion of the fossil fuel needed to heat the limestone as is released from the limestone itself.

The growth in recent times of the total annual *emissions*, in terms of the mass of carbon, of carbon dioxide from fossil-fuel combustion is shown in Figure 4-8. The current rate of growth is about five times that for the period

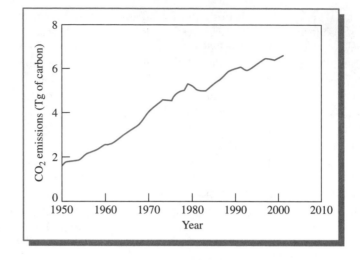

FIGURE 4-8 Annual global emissions of carbon dioxide from fossil-fuel combustion, in Tg (10^{12} g) of carbon. [Source: L.R. Brown et al., *Vital Signs 2002* (New York: Norton, 2002).]

before World War II. The emission rate growth pattern showed a temporary decline in the early 1980s, and again in the early 1990s, the latter due in part to the rapid decay of the economies of the former Soviet bloc nations. The current growth rate in emissions, about 1% per year, is considerably less than that of the 1960s and 1970s.

Carbon Dioxide: Atmospheric Lifetime and Fate of Its Emissions

The lifetime of a carbon dioxide molecule emitted into the atmosphere is a complicated measurement since, in contrast to most gases, it is not decomposed chemically or photochemically. On average, within a few years of its release into the air, a CO_2 molecule will likely dissolve in surface seawater or be absorbed by a growing plant. However, many such carbon dioxide molecules are released back into the air a few years later on average, so this disposal is only a *temporary* sink for the gas. The only *permanent* sink for it is deposition in the deep waters of the ocean and/or precipitation there as insoluble calcium carbonate. However, the top few hundred meters of seawater mix slowly with deeper waters; thus carbon dioxide that is newly dissolved in surface water requires hundreds of years to penetrate to the ocean depths. Consequently, although the oceans will ultimately dissolve much of the increased CO_2 now in the air, the time scale associated with this permanent sink is very long, hundreds of years.

Because the processes involving the interchange of carbon dioxide among the air and the biomass and shallow ocean waters, and between shallow and deep seawater, are complicated, it is not possible to cite a meaningful average lifetime for the gas in air alone. Rather we should think of new

CO_2 fossil-fuel emissions as being rather quickly allocated among air, the shallow ocean waters, and biomass, with interchange among these three compartments occurring continuously. Then slowly, over a period of many decades and even centuries, almost all this new carbon dioxide will eventually enter its final sink, the deep ocean. In effect, the atmosphere rids itself of almost half of any new carbon dioxide within a decade or two but requires a much longer period of time to dispose of the rest. It is commonly quoted as taking 50 to 200 years for the carbon dioxide level to adjust to its new equilibrium concentration if a source of it increases. In summary, the effective lifetime of additional CO_2 in the atmosphere should be considered to be long, on the order of many decades or centuries, rather than the few years required for its initial dissolution in seawater or absorption by biomass.

The annual inputs and outputs of carbon dioxide to and from our atmosphere, as of the mid-1980s, are summarized in Figure 4-9. Fossil-fuel combustion and cement production released 5.4 gigatonnes (Gt—i.e., billions of tonnes, equivalent to petagrams, 10^{15} g) of the carbon component (only) of CO_2 per year into the air, of which 3.3 Gt, or 60%, did not find a sink. The upper layers of the oceans absorbed about 92 Gt of carbon

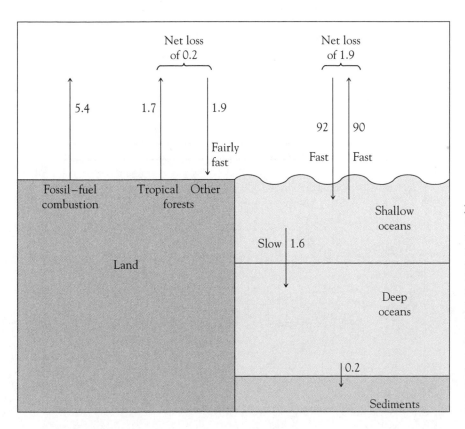

FIGURE 4-9 Annual fluxes of CO_2 to and from the atmosphere as of the mid-1980s, in units of gigatonnes of carbon. Note that the air/ocean flux is the total of both natural and anthropogenic contributions. [Source: Data is from J.T. Houghton et al., *Climate Change 2001: The Scientific Basis* (Intergovernmental Panel on Climate Change). (Cambridge: Cambridge University Press, 2001).]

but released 90 Gt, and the net absorption of this principal sink was 1.9 Gt yearly. Only 1.6 Gt was removed from the upper layers into the intermediate and lower depths, and only 0.2 Gt found its way into the bottom sediments. Although tropical deforestation contributed 1.7 Gt of carbon per year into the air, this was slightly more than offset by the withdrawal of about 1.9 Gt by temperate zone forests.

In the 1990s the average anthropogenic carbon dioxide emissions increased (from 5.4 to 6.3 Gt of C), but this change was slightly more than offset by an accelerated net withdrawal by the biosphere (from about 0.2 to about 1.4 Gt/year) and a withdrawal by the oceans only slightly less than in the 1980s. Because only about half of the anthropogenic CO_2 emissions are quickly removed, over the short and medium term, the gas continues to accumulate in the atmosphere.

There are significant variations in the effectiveness of these sinks from year to year—more variation than the fluctuations in emissions from fossil fuels—with the result that the yearly increase in carbon dioxide in the atmosphere varies greatly from one year to the next, as shown by the dark green curve in Figure 4-10. The straight light green line shows that on average the net input to air is increasing with time. Oceanic uptake of carbon dioxide varies by 1 or 2 Gt/year, depending partially on surface water temperature: the colder the water, the more CO_2 dissolves. Higher net additions to air seem to happen during El Niño periods, though this effect was canceled in the early 1990s by the atmospheric effects of the eruption of Mt. Pinatubo.

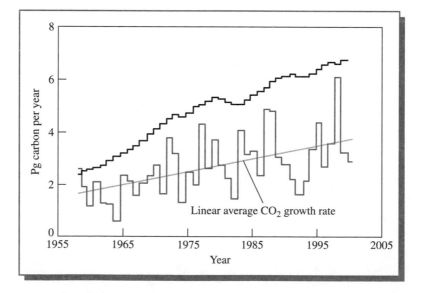

FIGURE 4-10 The annual atmospheric growth rate for CO_2 (green lines) and the emissions of the gas from fossil-fuel combustion (black line) for the same year. [Source: P. Quay, "Ups and Downs of CO_2 Uptake," *Science* 298 (2002): 2344.]

The increase in growth rate of certain types of trees due to the increased concentration of carbon dioxide in the air is called **CO_2 fertilization**. Some scientists suspect that the rate of photosynthesis is speeding up as the level of CO_2 and the air temperature increase and that the formation of greater amounts of fixed carbon represents an important sink for the gas. Indeed, an increase in the biomass of northern temperate forests is the most likely sink to account for the annual atmospheric CO_2 loss for which scientists had previously been unable to assign a cause. This increased activity in photosynthesis has been confirmed by satellite data for the region between 45°N and 70°N. European forests accumulated about 100 million tonnes of carbon annually in the 1970s and 1980s. Much of the increase in the biomass of temperate forests at high latitudes occurs in the soil, especially as peat. The anthropogenic releases of CO_2 amount only to about 4% of the enormous amounts produced by nature, so a very small variation in the rate at which carbon is absorbed into biomass could have a large effect on the residual amount of CO_2 that accumulates in the atmosphere. Unfortunately, scientists still do not completely understand the global carbon cycle. As Figures 4-7 and 4-10 indicate, however, there is no doubt that the atmospheric CO_2 concentration is increasing.

PROBLEM 4-5

Given that the atmospheric burden of carbon increases by about 3.2 Gt annually, calculate the annual increase in the concentration of carbon dioxide on the ppm scale. Given that its total concentration was 365 ppm, calculate the total mass of CO_2 that was present in the atmosphere in 1998. Note that the atmospheric mass = 5.1×10^{21} g; air's average molar mass = 29.0 g/mol.

Green Chemistry: Supercritical Carbon Dioxide in the Production of Computer Chips

In this example of green chemistry, we see how waste CO_2 (which would normally be vented to the atmosphere) can be put to good use as a solvent. We will also see how using CO_2 as a solvent pays additional environmental dividends in terms both of energy and resource conservation and of reduction of wastes.

As technology relentlessly pervades our planet, the demand for integrated circuits (ICs) and computer chips increases dramatically each year. Computer chips are used in almost any electronic device that one can imagine, including telephones, televisions, radios, automobiles, trucks, computers, airplanes, rockets, smart bombs, calculators, and cameras. It is estimated that the combination of the average personal computer, keyboard, monitor, and printer has a mass of about 25 kg and contains about 9 g of silicon and metal in the ICs that are the heart of each computer.

The manufacture of computers, other electronic devices, and chips involves high-tech, high-paying, highly skilled, and highly sought-after jobs. Facilities involved in these activities are considered by most as "clean" industries, especially when compared to the automobile and chemical industries. It is a little-known fact, except to those who work in the field or study the chip-manufacturing process, that chip making creates more waste than any other process involved in the manufacture of computers and is very energy intensive. By some measures, chip manufacturing is orders of magnitude more wasteful and polluting than the production of automobiles. It is estimated that the fabrication of the chips in your computer generated about 196 kg of waste (4500 times the weight of the average chip) and used about 10,600 L of water. The ratio of the weight of the materials (chemicals and fossil fuels) needed to produce a chip to the weight of the chip is estimated at 630:1, while the analogous ratio for the production of an automobile is approximately 2:1. Consequently there are ongoing efforts to find less resource-intensive and less wasteful methods of chip production.

The process of producing a 2-g computer chip entails many steps and requires 72 g of chemicals, 32 L of water (mostly for rinsing), and 700 g of process gases. For production and use over a four-year lifetime, a 2-g chip requires 1.6 kg of fossil fuels. A typical chip-manufacturing facility uses millions of gallons of highly purified water per month. A few of these steps are outlined in Figure 4-11. The process begins with the mechanical or chemical cleaning of the surface of highly purified silicon, followed by deposition of silicon dioxide, and then a process known as photolithography. Photolithography defines the shape and pattern of individual components on an IC.

Photolithography begins with the deposition of a photoresist polymer, followed by baking and exposure of selected areas of the polymer to light. The light causes the polymer to cross-link (i.e., form bonds that link the polymer chains to one another at many positions along each chain — Figure 4-12). The chip is then developed (a process that removes the photoresist polymer from the unexposed areas) and hard baked, the SiO_2 is etched, and the remaining photoresist polymer is removed, creating a pattern on the surface of the chip. The removal of the photoresist is accomplished with large amounts of aqueous solutions of strong acid (sulfuric or hydrochloric) or base, or by use of organic solvents (halogenated or polycyclic aromatics). The chip is then rinsed several times with copious amounts of highly purified water and dried with alcohol. The removal of the photoresist is very resource- and energy-intensive and creates large amounts of waste. The layering, developing, and etching process is repeated several times for each chip.

Scientists at Los Alamos National Laboratories in New Mexico and SC Liquids in Nashua, New Hampshire, were awarded a Presidential Green Chemistry Challenge Award in 2002 for their development of a new process for removing photoresist in chip manufacturing, known as *SCORR* (supercritical

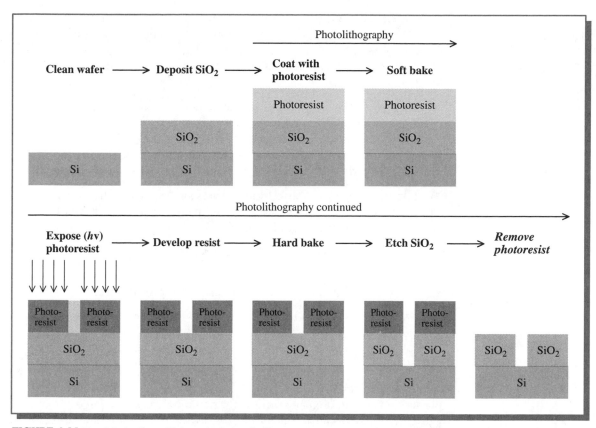

FIGURE 4-11 The fabrication of integrated circuits. [Source: L. Rothman, G. Jacobson, and C. Taylor, "Supercritical CO_2 Resist Remover–SCORR," a proposal submitted to the Presidential Green Chemistry Challenge Awards Program, 2002.]

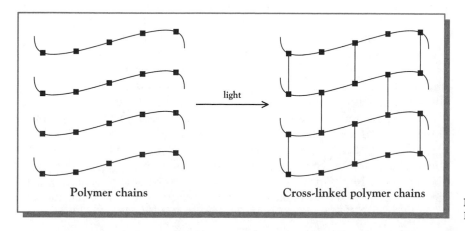

FIGURE 4-12 The cross-linking of polymer chains.

carbon dioxide resist remover). The process employs supercritical carbon dioxide (a discussion of which can be found in Chapter 12) as the solvent for removal of the photoresist (the last step of Figure 4-11). The use of SCORR offers several environmental benefits over traditional methods, including the following:

•   The rinse step is no longer necessary, thus eliminating the need for millions of liters of highly purified water, the energy required to produce this water, and the associated wastewater. This also reduces the amount of fossil fuel required to produce the highly purified water and the accompanying formation of carbon dioxide.

•   The need for (and waste from) hazardous and toxic chemicals such as strong acids or bases or organic solvents in the photoresist removal step is eliminated or reduced. This also enhances worker safety.

•   The need for alcohols used for drying after the aqueous rinse step is eliminated.

•   The carbon dioxide is recovered after each use and reused.

•   The only waste left after evaporation (and recovery) of the carbon dioxide is the spent photoresist, which is unregulated.

The carbon dioxide used in this process can be obtained as waste by-product of other processes (as indicated in Chapter 1 during the discussion of using carbon dioxide as a blowing agent). Processes such as the production of ammonia and drilling for natural gas produce large amounts of this gas, which would normally be released to the atmosphere and add to the concentration of carbon dioxide. If we can capture this "unwanted" by-product and find constructive (and environmentally sound) uses for carbon dioxide, then we have not only prevented its release into the atmosphere, but we may have also reduced our reliance on valuable resources and prevented the formation of other pollutants. SCORR, the use of carbon dioxide as a blowing agent (green chemistry section, Chapter 1), and the use of carbon dioxide as a solvent for various cleaning purposes (green chemistry section, Chapter 2) all are examples of this. In general, chemists are seeking ways to find beneficial uses for by-products of other processes and reactions that would normally be considered waste and would add to the environmental burden of the planet. The use of carbon dioxide as a blowing agent and as a solvent for cleaning and photoresist removal offers three significant examples of this paradigm.

Employing the SCORR process reaps several additional advantages. As the architecture of computer chips continually becomes smaller, water (because of its high surface tension) will no longer be able to penetrate these small spaces. Supercritical fluids have low viscosity, low surface tension, and high diffusivity. Because of these properties, they are ideal for cleaning rough, irregular surfaces with small openings. Supercritical carbon dioxide offers the answer to the cleaning problems associated with the smaller features of the new generations of chips. Other advantages of the SCORR process include the facts that (1) cleaning times are cut in half, (2) eliminating the rinse step allows for greater throughput (more chips in less time), (3) carbon dioxide is less costly than conventional solvents.

Water Vapor: Its Infrared Absorption and Role in Feedback

Water molecules, always abundant in air, absorb thermal IR light through their H—O—H bending vibration; the peak in the spectrum for this absorption occurs at about 6.3 μm. As a consequence, almost all the relatively small amount of outgoing IR in the 5.5- to 7.5-μm region is intercepted by water vapor (see Figure 4-6). (The symmetric and antisymmetric stretching vibrations for water occur near 2.7 μm, outside the thermal IR region. The symmetric stretch in a symmetric but nonlinear molecule like H_2O does absorb IR.) Absorption of light leading to increases in the rotational energy of water molecules, without any change in vibrational energy, removes thermal infrared light of 18-μm and longer wavelengths. In fact, water is the most important greenhouse gas in the Earth's atmosphere, in the sense that it produces more greenhouse warming than does any other gas, although on a per molecule basis it is a less efficient IR absorber than is CO_2.

Although human activities, such as the burning of fossil fuels, produce water as a product, the concentration of water vapor in air is determined primarily by temperature and by other aspects of the weather. Virtually all the H_2O in the troposphere arises from the evaporation of liquid and solid water on the Earth's surface and in clouds. The rate at which water evaporates and the maximum amount of water vapor that an air mass can hold both increase sharply with increasing temperature. Indeed, the equilibrium vapor pressure of liquid water increases exponentially with temperature. The rise in air temperature that is caused by increases in the concentration of the other greenhouse gases, and by other global warming factors, heats the surface water and ice, thereby causing more evaporation to occur.

The consequent increase in the water vapor concentration from global warming due to increased CO_2, etc. produces an *additional* amount of global warming due to $H_2O(g)$ that is comparable in magnitude to the original amount due to the other greenhouse gases, because water vapor is a greenhouse gas! This behavior of water is an example of the general phenomenon called **positive feedback:** *the operation of a phenomenon produces a result that itself further amplifies the result.* Feedback is a reaction to change; with positive feedback, the reaction accelerates the pace of future change. On the other hand, a system whose output reduces the subsequent level of output displays **negative feedback.** An example of negative feedback from daily life is the attempt by a business to raise its profits by increasing its prices. However, the rise in price often results in a reduction in demand for the item of concern, and the rise in profits is less than anticipated. (No value judgment as to the desirability of the effect is implied by the terms *positive* and *negative*; only the increase or decrease in the pace of change is meant.)

Since it comes about as an indirect effect of increasing the levels of other gases, and since it is not within our control, the warming increment due to water is usually apportioned without further comment into the direct warming effects of the other gases. Consequently, water is not usually listed explicitly among gases whose increasing concentrations are enhancing the greenhouse effect.

Water in the form of liquid droplets in clouds also absorbs thermal IR. However, clouds also reflect some incoming sunlight, both UV and visible, back into space. It is not yet clear whether the additional cloud cover produced by increasing the atmospheric water content will have a net positive or a net negative contribution to global warming. Clouds over tropical regions are known to have a zero net effect on temperature, but those in northern latitudes produce a net cooling effect since their ability to reflect sunlight outweighs their ability to absorb IR. Thus, if increased air temperatures produce more of the latter type of cloud, the enhancement by the greenhouse effect of global warming would be damped. However, no one is sure that additional northern cloud cover would occur at the same altitudes and act in the same manner as do the current clouds. Overall, the net effect of clouds on global warming is still subject to some uncertainty.

The Atmospheric Window

As a result of absorption of IR light of other wavelengths, mainly by carbon dioxide, methane, and water, it is essentially only infrared light from 8 to 13 μm that escapes the atmosphere efficiently (see Figure 4-6). Since light of these wavelengths passes unimpeded, this portion of the spectrum is called the **atmospheric window.**

The injection into the atmosphere, even in trace amounts, of gases that can absorb thermal IR light will lead to additional global warming, i.e., to enhancing the greenhouse effect. Particularly serious are pollutant gases that absorb thermal IR in the atmospheric window region, since the absorption by H_2O and CO_2 in other regions is already so great that there remains little such light for trace gases to absorb.

In considering which potential pollutants might contribute to global warming, recall that we can dismiss free atoms and homonuclear diatomic molecules, as they cannot absorb IR light. Heteronuclear diatomic molecules such as CO and NO are also not of direct concern since their only vibration—bond stretching—has a frequency that lies outside the thermal IR region. In general, however, most long-lived gases consisting of

TABLE 4-1	Summary of Information About Some Greenhouse Gases		
Gas	Current concentration	Residence time, in years	Relative global warming efficiency, 100-year horizon
CO_2	373 ppm	50–200	1
CH_4	1.77 ppm	12	23
N_2O	316 ppb	120	296
CFC-11	0.26 ppb	45	4600
HCFC-22	0.15 ppb	12	1700
HFC-134a	0.01 ppb	14	1300
Halon-1301	0.003 ppb	65	6900

molecules with three or more atoms are of concern since they possess many vibrations that absorb IR, one or more of which usually fall in the thermal infrared region. The important trace greenhouse gases, i.e., those whose concentration is small in absolute terms but whose ability even at these levels to warm the air is substantial, are detailed below, following a discussion of their average lifetimes in the atmosphere. The nature, abundance, atmospheric lifetime, and relative effectiveness in promoting global warming of some important greenhouse gases are summarized in Table 4-1.

Atmospheric Residence Time

The extent to which a substance accumulates in some compartment of the environment, such as the atmosphere, depends upon the rate, R, at which it is received from the source and the mechanism by which it is eliminated, i.e., its sink. Commonly, the rate of elimination via the sink is directly proportional to the concentration, C, of the substance in the organism or environmental compartment. In chemistry, this is known as a *first-order* relationship, since the power to which the independent variable is raised is unity. If the rate constant for the elimination process is defined as k, the rate of elimination is kC:

$$\text{rate of intake} = R$$

$$\text{rate of elimination} = kC$$

In some cases, such as the atmospheric sink for methane, reaction involving a second substance is involved, and k incorporates the steady-state concentration of this other substance.

If none of the substance is initially present, that is, if $C_0 = 0$, then initially the rate of elimination must be zero. The concentration then builds up solely due to its input or ingestion, as illustrated near the origin in Figure 4-13. However, as C rises, the rate of elimination rises and increases since it is proportional to C; eventually it matches the rate of intake if R is a constant. Once this equality is achieved, C does not vary thereafter; it is in a steady state, which as we saw in Chapter 1 is defined as the state for which $dC/dt = 0$. Since under these steady-state conditions

$$\text{rate of elimination or loss} = \text{rate of intake}$$

$$kC = R$$

It follows that the steady-state value for the concentration, C_{SS}, is

$$C_{SS} = R/k$$

Often the speed of elimination or loss of a substance is discussed in terms of the **half-life period,** $t_{0.5}$, the length of time required for half of it to decay under the condition that all new input has now ceased. Under the latter condition, we know

$$dC/dt = -kC$$

Bringing to the left side of the equation all the terms involving C and putting to the right side the time dependence, we have

$$\int dC/C = -k \int dt$$

If we integrate both sides of the equation, we can find how C changes with time:

$$\int dC/C = -k \int dt$$

Performing the integration, we obtain

$$\ln C = -kt + \text{constant}$$

Thus we see that the logarithm of the concentration will decline linearly with time for the substance. At time $t = 0$, we have $\ln C = $ a constant, so we see that the (integration) constant equals the logarithm of the original concentration, C_0, the concentration at $t = 0$. Thus

$$\ln C - \ln C_0 = -kt$$

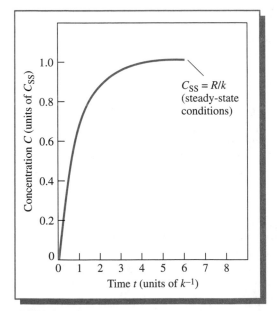

$C_{SS} = R/k$ (steady-state conditions)

FIGURE 4-13 Increase in concentration with time to eventually reach the steady-state value, C_{SS}

From the property of logarithms that $\log x - \log y = \log(x/y)$, we obtain the simpler equation

$$\ln(C/C_0) = -kt$$

or, in exponential form,

$$(C/C_0) = e^{-kt}$$

It is convenient to discuss the rate of decline of a substance in terms of its half-life. Substituting $C = 0.5C_0$ into the logarithmic equation above yields

$$\ln(0.5C_0/C_0) = -kt_{1/2}$$

But $\ln(0.5) = -0.693$, so we obtain

$$t_{1/2} = 0.693/k$$

We can use this final result to advantage to substitute for k in our equation for the steady-state concentration a chemical achieves when it is being both created and destroyed:

$$C_{SS} = R/k$$

We obtain

$$C_{SS} = R\, t_{0.5}/0.693 \quad \text{or} \quad C_{SS} = 1.44R\, t_{0.5}$$

Clearly, *the longer the half-life of a substance in its elimination process, the higher the steady-state accumulation level* it will achieve. The variations with time of concentrations and rates for systems of this type illustrated in Figure 4-13 are for the specific case where R and k (and hence C) are expressed in units of C_{SS}, and time t is expressed in units of k.

Every atmospheric gas that is present at or near a steady state has its own characteristic **residence time,** t_{avg}, which equals the average amount of time one of its molecules exists in air before it is removed by one means or another. The *average lifetime* or residence time, t_{avg}, of a substance is equal mathematically to the time required for its overall concentration to fall to $1/e$ times its initial value, where e is the base of natural logarithms. Since at that time $C = C_0/e$, then

$$\ln\left(\frac{C_0/e}{C_0}\right) = -kt_{avg}$$

But $\ln(1/e) = -1$, so

$$t_{avg} = 1/k$$

Substituting this expression for k into C_{SS} gives us $C_{SS} = R/t_{avg}$. Sometimes this expression is more useful in the form

$$t_{avg} = C_{SS}/R$$

since it tells us the lifetime of a substance if we know its steady-state concentration and rate of input into the environment.

In order to assess the impact of any substance on the enhanced greenhouse effect, it is necessary to know how long the substance is expected to remain in the atmosphere, since the longer its atmospheric lifetime, the greater will be its total effect. Thus, e.g., if the steady-state atmospheric concentration of a gas is 6.0 ppm, and if its global rate of input, as determined by dividing the yearly amount of input by the volume of the atmosphere, is 2.0 ppm/year, then according to the preceding equation, its average residence time is 6.0 ppm/2.0 ppm/year, or 3.0 years.

The effective residence times of the greenhouse gases carbon dioxide, nitrous oxide, and the CFCs are all many decades, so the influence of gases now being emitted into the atmosphere will extend over long periods of time. In contrast, methane has a residence time of only about a decade (see Table 4-1).

PROBLEM 4-6

If the average steady-state residence time of a trace atmospheric gas is 50 years and its input rate is 2.0×10^6 kg/year, what is the total amount of it in the atmosphere?

PROBLEM 4-7

The steady-state concentration of an atmospheric gas of molar mass 42 g/mol is 7.0 μg/g of air and its residence time is 14 years. What is the annual total release of the gas into the atmosphere as a whole? See Problem 4-5 for additional data.

PROBLEM 4-8

Explain qualitatively why the average lifetime of a substance that decays by a first-order process is greater than its half-life period. If you have difficulty in understanding this, do a calculation for the average lifetime in which you assume that the half of the substance that decays in the first half-life exists for about $0.5 t_{0.5}$, the quarter that decays in the second half-life exists for about $1.5 t_{0.5}$, etc.

Other Greenhouse Gases

Methane: Absorption and Sinks

After carbon dioxide and water, **methane,** CH_4, is the next most important greenhouse gas. A methane molecule contains four C—H bonds. Although C—H bond-stretching vibrations occur well outside the thermal IR region, HCH bond-angle-bending vibrations absorb at 7.7 μm, near the edge of the thermal IR window; consequently atmospheric methane absorbs IR in this region. Per molecule, increasing the amount of methane in air causes 23 times the warming effect as does adding more carbon dioxide, since CH_4 molecules absorb a greater fraction of the thermal IR photons that pass through them than do CO_2 molecules. However, because the CO_2 concentration has increased 80 times more than the CH_4 concentration, methane has been less important in producing global warming. To date, methane is estimated to have produced about one-third as much global warming as has carbon dioxide.

In contrast to the century-long lifetime of carbon dioxide emissions, molecules of methane in air have an average lifetime of only about a decade. As discussed in Chapter 2, the dominant sink for atmospheric methane, accounting for about 90% of its loss from air, is its reaction with molecules of **hydroxyl,** OH, the very reactive gas present in air in very low concentration:

$$CH_4 + OH \longrightarrow CH_3 + H_2O$$

This reaction is the first step of a sequence that transforms methane ultimately to CO and then to CO_2 (see Chapter 3 for details).

$$CH_4 \longrightarrow \longrightarrow CH_2O \longrightarrow \longrightarrow CO \longrightarrow \longrightarrow CO_2$$

The yearly loss of methane by this reaction is about 480 Tg (where 1 Tg, or teragram, is 10^{12} — 1 million metric tons) and the net sink from all sources amounts to about 530 Tg/year.

The other two sinks for methane gas are its reaction with soil and its loss to the stratosphere. A minor sink for methane exists in the stratosphere, to which a few percent of emissions eventually rise. It reacts there with OH, or atomic chlorine or bromine, or excited atomic oxygen; reaction with the latter produces hydroxyl radicals and eventually water molecules:

$$O^* + CH_4 \longrightarrow OH + CH_3$$
$$OH + CH_4 \longrightarrow H_2O + CH_3$$

Stratospheric water vapor acts as a significant greenhouse gas. About one-quarter of the total global warming caused by increased methane emissions is not brought about directly. Rather, it is due to this effect in the stratosphere, by which the region's water content is increased. In fact, due to decreased levels of ozone and increased levels of carbon dioxide, the stratosphere has undergone cooling in recent decades; the increase in water vapor has reduced the amount of this cooling, and so has contributed to the warming of the atmosphere overall.

Methane: Emission Sources

About 70% of current methane emissions are anthropogenic in origin. The manner in which total methane *emissions* rose over the last century is illustrated by the black line in Figure 4-14. As in the case of carbon dioxide, post–World War II rates increased annually much more quickly than had been the case before. In the last 20 years, however, the emission rate for methane has leveled off and may even have decreased slightly (Figure 4-14).

Most of the methane produced from plant decay results from the process of **anaerobic decomposition,** which is the decomposition of formerly living matter in the absence of air, i.e., under oxygen-starved conditions. This process converts cellulose into methane and carbon dioxide:

$$2 \,(\text{polymeric } CH_2O) \;\rightarrow\rightarrow\; CH_4 + CO_2$$

Anaerobic decomposition occurs on a huge scale where plant decay occurs under water-logged conditions, e.g., in natural wetlands such as swamps and bogs, and in rice paddies. Indeed, the original names for methane were *swamp gas* and *marsh gas*. Wetlands are the largest *natural* source of methane emissions.

The expansion of wetlands that occurs by the deliberate flooding of land to produce more hydroelectric power adds to the total natural emissions of the gas. Deep, small reservoirs produce and emit much less methane than do shallow ones that contain large volumes of flooded biomass, such as those in the Brazilian Amazon, especially if the trees are not first removed. Indeed, the combined global warming effects of the methane and carbon

FIGURE 4-14 Changes over the twentieth century in the annual emission (black line) and annual growth in atmospheric concentration (green line) of methane. [Source: "Global Methane Rise Slows," *Atmosphere* (CSIRO Newsletter) (April 1999).]

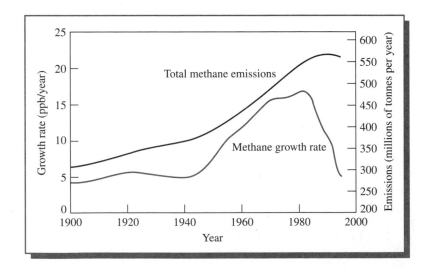

dioxide produced by a large, shallow reservoir created to generate hydroelectric power can, for many years, exceed the effect of the carbon dioxide that would have been emitted by a coal-fired power plant used to generate the same amount of electrical power! Hydroelectric power is not a zero-emission form of energy production if land must be flooded to create it.

Ruminant animals—including cattle, sheep, and certain wild animals—produce huge amounts of methane as a by-product in their stomachs when they digest the cellulose in their food. The animals subsequently emit the methane into the air by belching or flatulence. The decrease in the population of some methane-emitting wild animals (e.g., buffalo) in recent centuries has been far exceeded by the huge increase in the population of cattle and sheep. The net result has been a large increase in emissions of methane from animal sources.

The anaerobic decomposition of the organic matter in garbage in landfills is another important source of methane in the air. Food waste in the landfill produces the greatest amount of methane. In some communities, methane from landfills is collected and burned to generate heat, rather than being allowed to escape into the air. Although the combustion of methane produces an equal number of molecules of carbon dioxide, because the *per molecule* effect of CO_2 molecules is so much lower than that of CH_4 molecules, the net greenhouse enhancement effect of the emission is thereby greatly reduced.

The burning of biomass, such as forests and grasslands in tropical and semitropical areas, releases methane to the extent of about 1% of the carbon consumed, along with larger amounts of carbon monoxide (both compounds are products of incomplete—poorly ventilated—combustion).

Methane is released into the air when natural gas pipelines leak, when coal is mined and the CH_4 trapped within it is released into the air, and when the gases dissolved in crude oil are released or incompletely flared into the air when the oil is collected or refined. Emissions from these sources have likely leveled off in the last decade. The technique by which scientists determine the component of methane in ancient fossil-fuel sources is discussed in Box 4-1.

In summary, there are six different significant sources of atmospheric methane: natural wetlands, fossil fuels, landfills, ruminant animals, rice paddies, and biomass burning. The magnitude of each contribution is subject to enough uncertainty that even their relative ordering is uncertain. There is general agreement as to the current relative importance of some of the anthropogenic sources of atmospheric methane:

energy production/distribution and livestock > landfills > biomass burning

Research studies differ as to whether rice production gives methane in amounts comparable to energy and livestock production or is comparable to landfills and biomass burning.

BOX 4-1	Determining the Emissions of "Old Carbon" Sources of Methane

The relative abundances of carbon isotopes in atmospheric carbon dioxide can be used to help deduce its origin by the following logic. The carbon in all living matter contains a small, constant fraction of a radioactive isotope, carbon-14 (^{14}C), taken in via the carbon cycle when photosynthesis captures atmospheric CO_2 and when animals in turn feed off plant matter. This fact underlies the radiocarbon dating methods used by archaeologists and anthropologists: when an organism dies, its ^{14}C decays at a known first-order rate that makes the date of its death calculable. (The assumptions justifying these methods are that biotic carbon and atmospheric carbon in CO_2 are balanced—in equilibrium with one another—and that the level of atmospheric ^{14}C is constant. The principles underlying radioactive decay are discussed in Chapter 13.)

However, in the case of atmospheric methane, the average fraction of ^{14}C is less than the value found in living tissue. This indicates that a significant fraction of the CH_4 escaping into air must be "old carbon" that has been trapped in the ground for so long that its ^{14}C content has diminished to almost zero as a result of radioactive decay through the ages. Most methane containing old carbon is released into the air as a by-product of the mining, processing, and distribution of fossil fuels. Methane trapped in coal is released into the atmosphere when this material is mined, as is methane in oil when it is pumped from the ground. The transmission of natural gas, which is almost entirely methane, involves losses into the air due to leakage from pipelines and is the largest of the atmospheric sources of old carbon. Measurements of the methane levels in the air of various cities indicated that much of the loss from pipelines in the past occurred in eastern Europe. Finally, there is probably a small contribution to the old carbon source from methane trapped in permafrost in far northern latitudes; the methane was formed by the decay of plant matter that lived there many thousands of years ago when the polar climate was much warmer than it is today.

Methane: Concentration Trend and Possible Future Increases

Historically (i.e., before 1750), the methane concentration in air was approximately constant at about 0.75 ppm, i.e., 750 ppb (Figure 4-15a). It has more than doubled since preindustrial times, to 1.77 ppm (in 2002); almost all of this increase occurred in the twentieth century (Figures 4-15a and b) and was quite fast in the 1950–1980 period (green line in Figure 4-14). By the early 1990s, however, the rate of concentration increase had declined to half that of the 1980s, and since that time it has fallen to zero in some years (see Figure 4-15c). The rise in the atmospheric CH_4 level that has occurred since preindustrial times is presumed to be the consequence of such human activities as increased food production, fossil-fuel use, and forest clearing.

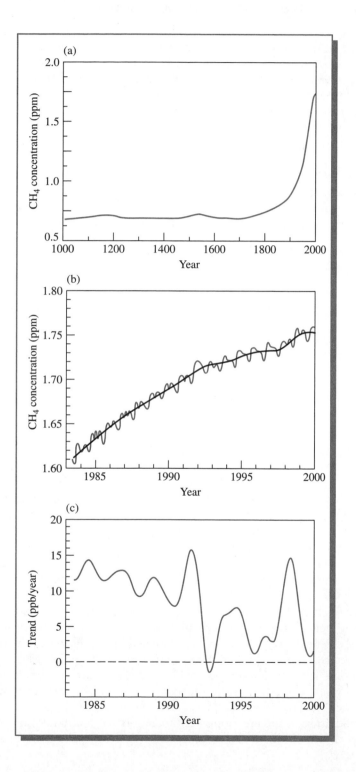

FIGURE 4-15 Variation with time of the atmospheric concentration of methane (a) in the last millennium, (b) in recent decades, and (c) fluctuations in the growth rate in recent decades. [Source: J. T Houghton et al., *Climate Change 2001: The Scientific Basis* (Intergovernmental Panel on Climate Change) (Cambridge: Cambridge University Press, 2001).]

It is not known with certainty why the growth rate of methane concentration decreased recently. Since the rate of change in concentration is proportional to the difference between the rate of change in the emission rate and the rate of change in the destruction rate, change in either one or both of the two rates could be responsible; neither can be directly measured very accurately. Stabilized emissions from rice paddies in China and decreasing emissions from gas pipelines, especially in the former Soviet Union, contributed to a possible decline in the emission rate. Some scientists speculated that the declines in the early 1990s were related to the air temperature decreases associated with the eruption of Mt. Pinatubo. The rate of oxidation of CH_4 by OH would also have increased if the concentration of hydroxyl radical increased overall, though there is controversy concerning how [OH] has changed.

PROBLEM 4-9

The 2002 concentration of atmospheric methane was 1.77 ppm, and the rate constant for the reaction between CH_4 and OH is 3.6×10^{-15} cm^3/molecule s. Calculate the rate, in Tg per year, of methane destruction by reaction with hydroxyl radical, the concentration of which is 8.7×10^5 molecules/cm^3. See Problem 4-5 for additional data.

Some scientists have speculated that the rate of release of methane into air could greatly increase in the future as a *consequence* of rising temperature from the enhanced greenhouse effect. For instance, higher temperatures would accelerate the anaerobic biomass decay of plant-based matter, as occurs in a common landfill. In turn, the additional release of methane would itself cause a further rise in temperature. This is another example of positive feedback.

Methane release from biomass decay among the extensive bogs and tundra in Canada, Russia, and Scandinavia could also increase with increasing air temperature and would also constitute positive feedback. However, the rate of biomass decay, and thus of CH_4 production, also depends on soil moisture and therefore on rainfall, which probably would be affected by climate change in an as yet uncertain direction, so the net feedback from this source could be positive or negative.

There is much methane currently immobilized in the permafrost of far northern regions; it was produced from the decay of plant materials during warm spells in the region but became trapped due to glaciation as temperatures became lower and lower at the start of the last ice age. Melting of the permafrost due to global warming could release large amounts of this methane. Melting would also allow the decomposition of organic matter currently present in the permafrost, with the consequent release of more methane.

In addition, there are monumental amounts of methane trapped at the bottom of the oceans and on continental shelves in the form of *methane hydrate*. This substance has the approximate formula $CH_4 \cdot 6\,H_2O$ and is an example of a **clathrate compound,** i.e., a rather remarkable structure that forms when small molecules occupy vacant spaces (holes) in a cage-like polyhedral structure formed by other molecules. In this case, methane is *caged* in a 3-D ice-lattice structure formed by the water molecules. The melting point of the structure is +18 °C, somewhat higher than that of pure ice. Clathrates form under conditions of high pressure and low temperature, such as are found in cold waters and under ocean sediments. The methane was produced over thousands of years by bacteria that facilitated the anaerobic decomposition of organic matter in the sediments.

If seawater warmed by the enhanced greenhouse effect penetrates to the bottom of the oceans, the clathrate compounds could decompose and release their own methane, as well as reservoirs of pure methane currently trapped below them, to the air above. Methane trapped far below the permafrost in northern areas and in offshore areas in the Arctic also exists in the form of clathrates; it would be released eventually if the Arctic warmed sufficiently. Measurements made thus far do not indicate any significant emissions from these sources. It has even been suggested by some scientists that CH_4 released from clathrates may be oxidized to CO_2 before it reaches the air.

Although the uncertainties concerning methane feedback are large, the stakes are higher than with any other gas. A few scientists believe that several positive climate feedback mechanisms, including those involving methane, could possibly combine to trigger an unstoppable warming of the globe. This worst-case scenario is called the *runaway greenhouse effect*. Such a climate change would threaten all life on Earth, as the temperature would rise markedly, ocean currents would probably shift, and rainfall patterns would be very different from those we know. The possibility that the North Atlantic ocean current that brings warm water from the south and thereby warms Europe could become nonoperational because of rapid global warming—induced by rapid increases in methane or carbon dioxide—is one of the most dramatic predictions about the possible consequences of the enhanced greenhouse effect.

PROBLEM 4-10

Calculate the mass of methane gas trapped within each kilogram of methane hydrate.

Nitrous Oxide

Another significant greenhouse trace gas is **nitrous oxide,** N_2O, "laughing gas," the molecular structure of which is NNO rather than the more symmetrical NON. Its bending vibration absorbs IR light in a band at 8.6 μm, i.e., within the window region, and in addition one of its bond-stretching vibrations is centered at 7.8 μm, on the shoulder of the window and at the same wavelength as one of the absorptions for methane. Per molecule, N_2O is 296 times as effective as CO_2 in causing an immediate increase in global warming. Like that of methane, the atmospheric concentration of nitrous oxide was constant until about 300 years ago, at which time it began to increase, although the level has increased from 275 ppb (preindustrial) by only 13% to 316 ppb. The yearly growth rate in the 1980s was about 0.25% but fell significantly in the early 1990s for reasons that are uncertain. The increased amounts of nitrous oxide that have accumulated in the air since preindustrial times have produced about one-third of the amount of the additional warming that methane has induced.

Less than 40% of nitrous oxide emissions currently arise from anthropogenic sources. In 1990 it was discovered that the traditional procedure, using *nitric acid*, HNO_3, of synthesizing *adipic acid* (a raw material in the preparation of nylon) resulted in the formation and release of large amounts of nitrous oxide. Since that time, nylon producers have instituted a plan to phase out N_2O emissions.

The greater part of the natural supply of nitrous oxide gas comes from release from the oceans, and most of the remainder is contributed by processes occurring in the soils of tropical regions. The gas is a by-product of the biological denitrification process in aerobic (oxygen-rich) environments and in the biological nitrification process in anaerobic (oxygen-poor) environments; the chemistry of both processes is illustrated in Figure 4-16. In **denitrification,** fully oxidized nitrogen in the form of the **nitrate ion,** NO_3^-, is reduced mostly to molecular nitrogen, N_2. In **nitrification,** reduced nitrogen in the form of ammonia or the ammonium ion is oxidized mostly to **nitrite,** NO_2^-, and nitrate ions. Chemically, the existence of the nitrous oxide by-product in both processes is simple to rationalize: nitrification (oxidation) under oxygen-limited conditions yields some N_2O, which has less oxygen than the intended nitrite ion, and denitrification (reduction) under oxygen-rich conditions yields some N_2O, which has more oxygen than the intended nitrogen molecule. Nitrification is more important than denitrification as a global source of N_2O. Normally, about 0.001 mol of nitrous oxide is emitted per mol of nitrogen oxidized, but this value increases substantially when the ammonia or ammonium concentration is high and relatively little oxygen is present. Overall, the increased use of fertilizers for agricultural purposes probably accounts for the majority of anthropogenic emissions of nitrous oxide. Landfills may also be a significant source of nitrous oxide from denitrification processes.

Apparently nitrous oxide released from new grasslands is particularly significant in the years following the burning of a forest. Some portion of the nitrate and ammonium fertilizers used agriculturally, particularly in tropical areas, is similarly converted (an unintended effect, to be sure) to nitrous oxide and released into the air. Tropical forests in wet areas are probably a huge natural source of the gas.

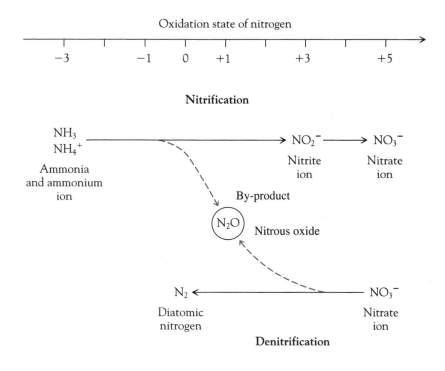

FIGURE 4-16 Nitrous oxide production as a by-product during the biological cycling of nitrogen.

At one time it was believed that fossil-fuel combustion released nitrous oxide as a by-product of the chemical combination of the N_2 and O_2 in air, but this belief was based on faulty experiments. Only when the fuel itself contains nitrogen, as do coal and biomass (but not gasoline or natural gas), does N_2O form; apparently N_2 from air does not enter into this process at all. However, some of the NO produced from atmospheric N_2 during fuel combustion in automobiles is unavoidably converted to N_2O rather than to N_2 in the three-way catalytic converters currently in use and is subsequently released into air. Some of the newer catalysts developed for use in automobiles do not suffer from this flaw of producing and releasing nitrous oxide during their operation.

As mentioned in Chapter 1, there are no sinks for nitrous oxide in the troposphere. Instead, all of it rises eventually to the stratosphere, where each molecule absorbs UV light and decomposes, usually to N_2 and atomic oxygen, or reacts with atomic oxygen.

CFCs and Their Replacements

Gaseous compounds consisting of molecules with carbon atoms bonded exclusively to fluorine and/or chlorine atoms have perhaps the greatest potential among trace gases to induce global warming, since they are both very persistent and absorb strongly in the 8- to 13-μm atmospheric window region. Absorption due to the C—F bond stretch is centered at 9 μm. The C—Cl bond stretch and various bond-angle-bending vibrations involving carbon atoms bonded to halogens also occur at frequencies that lie within the window region.

As discussed in Chapter 1, the **chlorofluorocarbons** $CFCl_3$ and CF_2Cl_2 have already been released into the atmosphere in large quantities and have long residence times. Due to this persistence, and to their high efficiency in absorbing thermal IR in the window region, each CFC molecule has the potential to cause the same amount of global warming as do tens of thousands of CO_2 molecules. The *net* effect of CFCs on global temperature is small, however. The heating that the CFCs produce by the redirection of thermal infrared is partially canceled by a separate effect, the cooling that they induce in the stratosphere due to their destruction of ozone. (Recall from Chapter 1 that the stratosphere is heated when oxygen atoms, recently detached photochemically from ozone molecules, collide with O_2 molecules to produce an exothermic reaction.) However, the decrease in stratospheric ozone allows more UV light to reach the lower atmosphere and the surface and to be absorbed there. The cooling and heating effects produced by CFCs occur at very different altitudes, so that their net effect on the Earth's weather may be substantial.

Ironically, the use of CFCs in insulating freezers, refrigerators, and air conditioners has reduced the energy requirements of this equipment and so has reduced CO_2 emissions resulting from electricity production.

The influence of CFCs on climate in the future will be reduced as a result of the requirements of the Montreal Protocol, which banned further production in developed countries after 1995, as discussed in Chapter 1. Most of the **HCFC** and **HFC** replacements for CFCs (with the notable exception of HFC-143a) have shorter atmospheric lifetimes and absorb less efficiently in the center of the atmospheric window region, and thus on a molecule-for-molecule basis they pose less of a greenhouse threat (see Table 4-1). However, if their levels of production and release become high in future decades because of expanding world population and increasing affluence, they will make significant contributions to global warming if they are released into the air. For this reason, many people feel that these substances must be used only in closed systems from which leakage to the atmosphere does not occur and that they must be recovered from equipment before its eventual disposal. Indeed, prevention of the chronic release of long-lived gases of all types to the atmosphere is a principle now agreed to by many scientific, business, and governmental groups. Measurements reported in 2002 indicate that the loss of the refrigerant HFC-134a (see Chapter 1) from the air-conditioning units of modern cars has a global warming impact that is about 4–5% of that from the carbon dioxide emitted from the cars.

PROBLEM 4-11

Fully fluorinated compounds such as tetrafluoromethane and hexafluoroethane are released as by-product wastes into the air in the production of aluminum. They were also briefly considered as CFC replacements. Will such molecules have a sink in the troposphere? Will they act as greenhouse gases? Would your answers be the same for monofluoromethane and -ethane?

Sulfur Hexafluoride

Sulfur hexafluoride, SF_6, is a little-known greenhouse gas. It has some importance, however, since it is such a good absorber of thermal IR— 23,900 times greater than CO_2 in global warming potential—and because, like other fully fluorinated compounds, it is so long-lived in the atmosphere (3200 years). It is used by electric utilities and in the semiconductor industry as an insulating gas. Formerly it was vented to the air during routine maintenance of equipment but now is mainly recycled instead.

Tropospheric Ozone

Like methane and nitrous oxide, tropospheric **ozone,** O_3, is a "natural" greenhouse gas, but one which has a short tropospheric residence time. The symmetric stretching vibration of ozone molecules occurs between 9 and 10 μm, i.e., in the atmospheric window region. The dip near 9 μm in the window region of the outgoing thermal IR distribution (Figure 4-6) is due to absorption by this vibration in atmospheric ozone molecules. Ozone's bending vibration occurs at 14.2 μm, near that for CO_2, and thus does not contribute much to the enhancement of the greenhouse effect, since atmospheric carbon dioxide already removes much of the outgoing light at this frequency. The antisymmetric stretching vibration of O_3 occurs at 5.7 μm, where there is very little outgoing IR.

As explained in Chapter 2, ozone is formed in the troposphere as a result of pollution from power plants and motor vehicles, from forest fires and grass fires, as well as from natural processes. As a result of these anthropogenic activities, the levels of ozone in the troposphere probably have increased since preindustrial times. The best guess is that approximately 10% of the increased global warming potential of the atmosphere results from increases in tropospheric ozone, although this value is very uncertain. The amount of thermal IR absorbed by stratospheric ozone has probably dropped slightly due to the recent decline in ozone levels.

The Climate-Modifying Effects of Aerosols

In Chapter 1 we saw that the initial neglect by scientists of the effects of atmospheric aerosol particles, specifically ice crystals in the stratosphere, led to a large underestimation of the amount of ozone that would be destroyed by chlorine. A similar neglect of the effect of aerosols initially led scientists to overestimate the amount of short-term global warming to be expected from enhancement of the greenhouse effect. Only in the last few years has the vital role of aerosols in affecting the potential for global warming from greenhouse gas accumulation been recognized and included explicitly in computer simulations of the atmosphere. The types of particulate matter of most importance in this context are particles expelled by powerful volcanic eruptions into the upper atmosphere *and* those produced by industrial processes and expelled into the lower troposphere. In order to understand how aerosols can affect global warming, it is necessary to understand how they interact with light.

The Interaction of Light with Particles

All solids and liquids—including atmospheric particles—have some ability to **reflect** light. Atmospheric particles can reflect incoming sunlight, with the consequence that some of it is directed back into space and so is unavailable later for absorption at the surface (see Figure 4-17). The particles can also reflect outgoing infrared light, with the consequence that some of it is redirected back toward the surface rather than escaping from the atmosphere. The redirection of light by a particle is sometimes called **scattering;** reflection backward is *backscattering*.

FIGURE 4-17 Interaction of sunlight with suspended atmospheric particles. (a) Modes of interaction. (b) Illustration of the indirect effect of producing increased reflection by small water droplets as compared to larger ones having the same total volume.

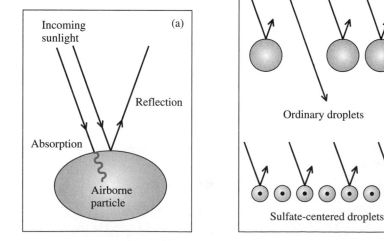

In general, the fraction of sunlight reflected by a surface is called its **albedo.** For example, the albedos of fresh snow and of the top surfaces of many clouds are close to unity, whereas the albedo for bare ground is only about 0.1. Certain types of suspended particulates in air reflect some of the sunlight that shines on them back into space and so have a significant albedo value; this reflection of sunlight by the aerosol cools the air mass and the surface below it, since none of the reflected light is subsequently absorbed and then converted to heat.

Some types of aerosol particles can **absorb** certain wavelengths of light (Figure 4-17a). Once absorbed, the energy that was associated with the light is rapidly converted into heat, which then is shared with the surrounding air molecules as a result of their collisions with the heated particle. Thus absorption of light by a particle leads to warming of the air immediately surrounding it. The absorption of sunlight, with consequent warming, is significant only for dark-colored particles such as those composed primarily of soot, often called *carbon black,* and of ash particles from volcanoes. The contribution of carbon black to global warming has only recently been fully appreciated. The emission into the atmosphere of carbon black is greatest in developing countries, where incomplete combustion of coal and biomass is widespread. Its effect globally is to increase air temperatures by its absorption of sunlight, with the subsequent export of this tropospheric air to other areas. However, its *local* effect may be cooling, since it blocks sunlight from reaching the surface. Carbon black's effects on local climate may be substantial, increasing drought in some areas and flooding in others.

Recall from Chapter 2 that the *sulfur dioxide* gas predominantly released as a pollutant from the burning of fossil fuels—especially coal—and from the smelting of nonferrous metals creates a **sulfate aerosol.** Pure sulfate aerosols do not absorb sunlight since none of their constituents—water, sulfuric acid, and the ammonium salts thereof—absorb light in the visible or the UV-A regions. The sulfate aerosols are not particularly effective in trapping outgoing thermal IR emissions. Only if tropospheric sulfate aerosols incorporate some soot will absorption of sunlight by these particles be significant.

The ability of anthropogenically derived aerosols to cool the atmosphere by the reflection of sunlight has been recognized for some time; indeed, in the early 1970s some scientists worried that over the long term this cooling might prematurely induce another ice age. But it is only since 1991 that aerosol effects arising from pollution have been approximately quantified and incorporated into the computer models that are used to forecast long-term climate changes. Specifically, the sulfate-rich tropospheric aerosols that are daily produced from air pollutants, especially over urban areas in the Northern Hemisphere, reflect sunlight back into space more effectively than they absorb it and thereby increase the Earth's average albedo. Consequently, their existence means that less sunlight is available to be absorbed by the surface and in the lower troposphere and converted to heat. Thus the net effect of the sulfate aerosols is to cool the air near ground level and thereby to offset some of the effects of global warming induced by greenhouse gases. Some scientists argue that in addition to the **direct effect** of sulfate aerosols in reflecting sunlight, there is a substantial **indirect effect** that arises because the sulfate particles act as nuclei for the formation of additional small water droplets. Such small droplets are more effective in backscattering light than are larger ones, and their formation thereby reflects additional incoming sunlight and cools the Earth's surface (see Figure 4-17b).

A short-term, dramatic example of the effects of atmospheric aerosols on climate occurred as a consequence of the massive eruption of substances into the troposphere and stratosphere by the Mount Pinatubo volcano in the Phillipines in 1991. Initially, the lower stratosphere was warmed due to the dominant effect of the large volcanic ash particles, which absorbed some of the incoming sunlight and subsequently converted it to heat, and by their interception of outgoing infrared from the surface. Due to their relatively large size, the ash particles were not long-lived in the stratosphere. The longer-term effect of the Pinatubo eruption on air temperatures at ground level was a significant decrease. The stratospheric aerosol that remained suspended after a few months was formed by the oxidation of the 30 million tonnes of SO_2 that the volcano had blasted directly into the lower parts of this region. The sulfate aerosol remained there for several years, during which time it efficiently reflected sunlight back into space. Inspection of Figure 4-2b indicates a slowing in the rise in the globally averaged surface temperature in 1992 and 1993, due apparently to this volcanic cooling. You may recall that many regions, including North America, experienced several cool summers in the early 1990s as a result. Due to the gradual sedimentation of the aerosol, we returned to 1990–1991 temperatures by 1995.

Aerosols and Global Warming

The cooling effect of the sulfate aerosol is concentrated almost entirely in the Northern Hemisphere because most industrial activity takes place in that half of the globe, so it is there that most emissions occur. The relatively short lifetime of such sulfate aerosols preclude their spreading to the Southern Hemisphere, and consequently the concentration of sulfate particulates is much higher over the Northern Hemisphere. The short lifetime of the sulfate particles can be understood by considering the processes of their removal from air. The average diameter of the tropospheric sulfate aerosol particles is about 0.4 μm and their average altitude is about 0.5 km. For particles of this size and altitude the expected atmospheric lifetime before gravitational settling to the surface is several years. However, since the sulfate aerosol droplets are also removed efficiently by rain, their actual lifetime in the lower troposphere is of the order of days rather than years.

The increase of global SO_2 emissions from fossil-fuel combustion over the last century and a half is shown in Figure 4-18. Up to 20% more sulfur dioxide is emitted by smelting, etc. Presumably the trend in anthropogenic sulfate aerosol production has approximately followed the pattern of SO_2 emissions in Figure 4-18. The initial approximately linear time increase of the sulfur dioxide emission rate changed to one with a much steeper slope after World War II, a behavior we saw previously for CO_2 and CH_4. However, the ratio of global SO_2 to CO_2 emission rates, expressed as a percentage of sulfur in the carbon of the fuel, fell from about 2.2% in the 1930s and 1940s to a constant 1.1% in recent decades, due presumably to the gradual replacement of coal by oil and natural gas.

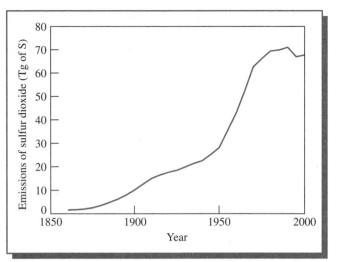

FIGURE 4-18 Estimated historical emissions of sulfur dioxide from anthropogenic sources. [Source: Adapted from S. J. Smith, H. Pitcher, and T. M. L. Wigley, "Global and Regional Anthropogenic Sulfur Dioxide Emissions," *Global and Planetary Change* 29 (2001): 99.]

As illustrated by the contour diagrams in Figure 4-19, the bulk of the anthropogenically produced aerosols in North America is centered above the Ohio Valley and directly reflects sunlight mostly above that area. Equal or even larger effects are observed over southern Europe and the Middle East and over portions of China. Indeed, according to some calculations, the cooling effect from aerosols outweighs the heating effect due to

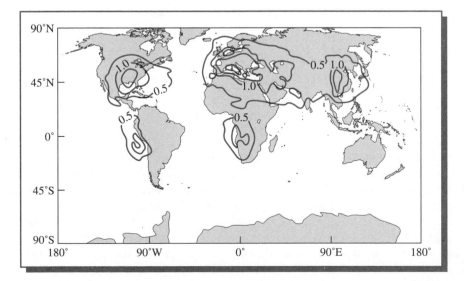

FIGURE 4-19 The amount of sunlight reflected into space by anthropogenic aerosols by the direct mechanism, in units of watts per square meter of the Earth's surface. [Source: J. T. Houghton et al., *Climate Change 1994— Radiative Forcing of Climate Change* (Intergovernmental Panel on Climate Change) (Cambridge: Cambridge University Press, 1995).]

greenhouse gases for some regions in the eastern United States, south-central Europe, and eastern China.

It is not clear how the amount of tropospheric sulfate aerosol will change in the future. Emissions of sulfur dioxide from power production in North America and western Europe are now more tightly controlled in order to combat acid rain, so the SO_2/CO_2 ratio in emissions from these areas should decline. However, the anthropogenic sulfate aerosol concentrations over southern Europe, the Middle East, and parts of Russia and China are considerably higher than the current maximum values in North America and will not be affected by these legislative controls. The only substantial domestic energy source currently available to China for its rapid industrialization is coal, so the SO_2 emissions from this source and from India may well continue to rise.

Aerosols also result from the oxidation of the gas **dimethyl sulfide** (DMS), $(CH_3)_2S$, which is produced by marine phytoplankton and subsequently released into the air over oceans. Once in the troposphere, DMS undergoes oxidation, some of it to SO_2 which then can oxidize to sulfuric acid, and some to *methanesulfonic acid*, CH_3SO_3H. Both of these acids form aerosol particles, which in turn lead to the formation of water droplets and hence of clouds over the oceans. The particles and droplets both deflect incoming light from the Sun. Some scientists believe that increased emissions of dimethyl sulfide from the oceans will occur when seawater warms as a result of the enhancement of the greenhouse effect and that this negative feedback will temper global warming.

Although the sulfate aerosol has a short lifetime, new supplies of it are constantly being formed from the sulfur dioxide pollution that pours into the atmosphere on a daily basis. Consequently there is a steady-state amount of the aerosol in the troposphere; sulfur dioxide emissions keep postponing the full effects of global warming induced by the rise in greenhouse gas concentrations.

Global Warming to Date

Earth's Energy Balance

The current energy inputs and outputs from the Earth, in watts (i.e., joules per second) per square meter of surface and averaged over day and night, all latitudes and longitudes, and all seasons, are summarized in Figure 4-20. Of the 342 W/m^2 present in sunlight outside the Earth's atmosphere, 235 are absorbed by the atmosphere and the surface and must be re-emitted into space if the planet is to maintain a steady temperature. To achieve this balance, 390 W/m^2 are emitted from the surface, 155 of which are deflected back by greenhouse gases and do not escape. (In fact, much more than 155 W/m^2 of the IR emitted by the surface is temporarily absorbed by the greenhouse gases, but subsequently re-emitted from various levels of the atmosphere into space.)

Infrared light is emitted both at the Earth's surface and by its atmosphere, though in different amounts since the emission rate is very temperature sensitive. The rate of release of energy by a blackbody in the form of light increases in proportion to the fourth power of its Kelvin temperature:

$$\text{rate of energy release} = kT^4$$

where k is a proportionality constant. It follows that, for average surface conditions (~300 K), a one degree rise in temperature increases the rate of energy release by 1.3%. Consequently, the amount of IR released by the Earth's surface and lower atmosphere will increase as they heat up, which will compensate for the additional IR absorbed by the higher levels of greenhouse gases present. Overall, the same amount of IR as at present, 235 W/m^2, will escape from the top of the atmosphere into space.

Ironically, an increase in greenhouse gas concentrations is predicted to cause a *cooling* of the stratosphere. This phenomenon occurs for two reasons. In the first place, more outgoing thermal IR is absorbed at low altitudes (the troposphere), so less is left to be absorbed by and to warm the gases in the stratosphere. Second, at stratospheric temperatures CO_2 emits more thermal IR to space than it absorbs as photons—most of the absorption at these altitudes is due to water vapor and ozone—so increasing its concentration cools the stratosphere. Indeed, the observed cooling of the stratosphere has been taken to be a signal that the greenhouse effect is indeed undergoing enhancement.

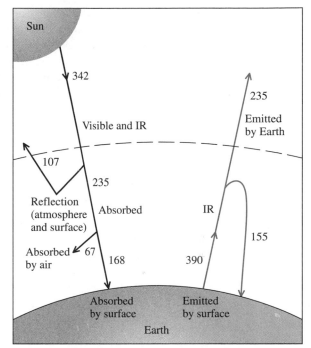

FIGURE 4-20 Globally and seasonally averaged energy fluxes to and from the Earth, in watts per square meter of surface. [Source: Data from Chapter 1 of J. T. Houghton et al., *Climate Change 1995— The Science of Climate Change* (Intergovernmental Panel on Climate Change) (Cambridge: Cambridge University Press, 1996).]

Effective CO_2 Concentration

The combined greenhouse effect enhancement from the *increases* in concentrations of the trace gases methane, nitrous oxide, ozone, and the CFCs is now almost as large as that from the *increase* in carbon dioxide. To conveniently summarize the temperature-enhancing effects of all greenhouse gases by a single number, the concept of an **effective** (or equivalent) **carbon dioxide concentration** has been devised. For this scale, one considers the increases in the concentration of greenhouse gases (other than carbon dioxide) that have occurred since preindustrial times, calculates the global temperature change that should result from these increases, and deduces the additional increase in CO_2 concentration that *instead* would have produced the same effect. The sum of the actual carbon dioxide concentration plus that equivalent to the increases in the other gases gives the effective CO_2 concentration. The value of this parameter is now more than 400 ppm, since the other gases have added the equivalent of about 50 ppm of CO_2 to the actual concentration of 373 ppm. If current emission trends continue, the effective CO_2 concentration will be double the preindustrial CO_2 concentration of 280 ppm by the year 2025, although the real CO_2 concentration itself will not have doubled until about 2100. It should be realized that the global warming induced by greenhouse gases is not necessarily linearly proportional to their concentrations; the reasons for this are explored in Additional Problem 6.

Allocation of Warming to Natural and Anthropogenic Factors

The best estimates of global warming or cooling since 1850 arising from each of the factors discussed is summarized by the bar graphs in Figure 4-21. Because no precise estimate of the indirect effect of aerosols is available, only an estimate of the total aerosol effect is shown. According to one calculation published in the early 1990s, the cooling due to the reflection of incoming sunlight by the tropospheric sulfate haze produced by anthropogenic SO_2 emissions is, on a globally averaged basis, approximately one-third of the energy increase since preindustrial times due to anthropogenic greenhouse gas emissions that remain in the air. Several recent estimates predict a direct cooling effect by the

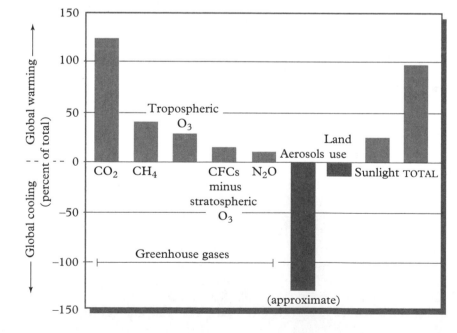

FIGURE 4-21 Contributions to global warming and cooling produced by various factors discussed in this chapter. Note that some factors individually exceed 100% but are partially canceled by others. [Source: Adapted from J. T. Houghton et al., *Climate Change 2001: The Scientific Basis* (Intergovernmental Panel on Climate Change) (Cambridge: Cambridge University Press, 2001).]

sulfate aerosol that is quite large, but at least one other recent study proposes that aerosols have very little direct cooling effect and that cooling is due to stratospheric ozone depletion that was underestimated in the past.

PROBLEM 4-12

Using a ruler, measure the vertical heights, positive or negative relative to the zero axis, for the individual contributions—including those from individual greenhouse gases—to global warming shown in Figure 4-21, and re-express each as a percentage of the net warming and of the contribution of carbon dioxide. Does the total contribution from sources other than greenhouse gases (not including ozone) amount to a significant fraction of the total?

Global Warming: Chronology and Geography

The year-to-year variations in the average worldwide air temperature for the period 1860–1998 (the latter being the hottest year on record, 2002 being the second hottest) are illustrated by the black line in Figure 4-22. As previously noted, periods of warming occurred from about 1910 to 1940, and from 1970 to the present, with a period of slight cooling between them. Since 1975 the global surface air temperature has increased by about 0.5°C.

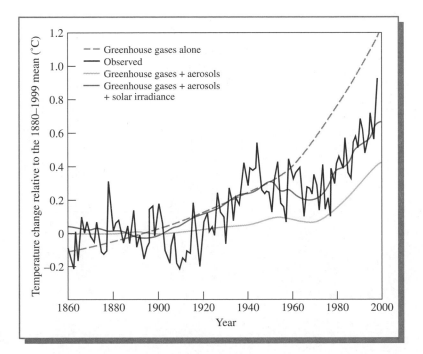

FIGURE 4-22 Global surface air temperatures observed (black curve) and simulated (green curves) by global circulation models of varying levels of sophistication. [Source: Tom M. L. Wigley in E. Claussen, ed., "Climate Change," *The Science of Climate Change* (Arlington, VA: Pew Center on Global Climate Change, 2001).]

The changes that occurred over the twentieth century in the surface air temperature—averaged over night and day, over all seasons, at various places on Earth—are illustrated in Figure 4-23. Although most of the United States and Canada, and indeed most of the world, became warmer—as indicated by the green dots in Figure 4-23—the Arctic region warmed most of all. Portions of the southeastern and south-central United States experienced a slight cooling (gray dots), as did regions of China (see Box 4-2). Air temperatures over land generally experienced a greater increase than those over the seas. Over the last few decades less warming of the air has been observed in the Northern Hemisphere than in the relatively unpolluted Southern Hemisphere, as anticipated from the geographic distribution of sulfate aerosols (Figure 4-19). The aerosols resulting from SO_2 released over the United States keep the temperatures there about one Celsius degree cooler than they would be otherwise, thus explaining why average summer temperatures have fallen rather than risen in eastern North America.

Most of the changes in air temperature documented in Figure 4-23 corresponded to an increase in the daily *minimum*—the low experienced just before dawn—rather than in the daily high temperature, usually achieved in the afternoon. This occurs presumably because aerosols exert their effects mainly by blocking incoming sunlight; consequently their existence is much more important in countering warming during the daytime than at night.

The increased lowering of air temperatures due to higher sulfate aerosol levels does not *permanently* cancel out all the warming due to greenhouse gases because of the very different atmospheric lifetimes of the particles as compared to the gases. The tropospheric aerosols last only a few days, so they do not accumulate with time and their effect is short-term. In contrast, today's emissions of carbon dioxide into the atmosphere will exert effects for decades or centuries to come: CO_2 emissions are cumulative over the

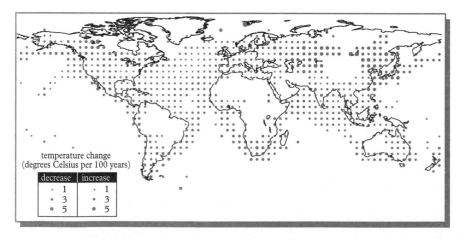

FIGURE 4-23 Trends in average annual temperature from 1901 to 1998. [Source: P. Martens, "How Climate Change Affects Human Health," *American Scientist* 87 (1999): 534–542.]

temperature change
(degrees Celsius per 100 years)

decrease	increase
1	1
3	3
5	5

BOX 4-2 | Cooling over China from Haze

Measurements of the amount of sunshine reaching the surface in China indicate a significant decrease over the last half-century (Figure 1). The blockage of solar intensity results from the aerosols in the air above the region produced mainly from the sulfur dioxide emitted by coal burning. As a consequence of the increasing presence of the aerosols, maximum summer temperatures in heavily polluted eastern China have fallen by about 0.6°C per decade. Similar effects are observed in the Brazilian Amazon region due to soot and ash emitted into the local air from forest and grass fires lit to clear land, and in the African country of Zambia from grass fires. Indeed, Figure 4-23 shows no warming over the last century in all three areas.

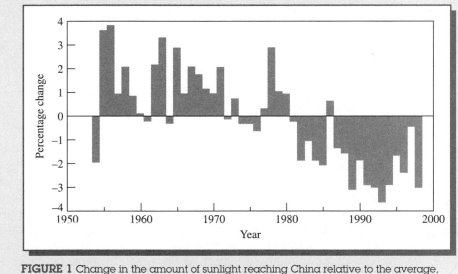

FIGURE 1 Change in the amount of sunlight reaching China relative to the average, over 50 years. [Source: F. Pearce, "Pollution Is Plunging Us into Darkness," *New Scientist* (14 December 2002):6.]

medium term. This is also true for CFCs and nitrous oxide, which are also important greenhouse gases, but it is less so for methane since its half-life is only about one decade. Thus, although sulfate-producing SO_2 emissions can temporarily cancel the effects of CO_2 emissions, eventually the cumulative effects of the carbon dioxide and the other greenhouse gases win out.

In summary, global warming has been experienced by most areas in the last half-century. Most of the warming is due to emissions of carbon dioxide into the atmosphere, with lesser amounts of warming from increased levels of methane, tropospheric ozone, nitrous oxide, and the introduction of

CFCs. The increased water vapor in the atmosphere that resulted from the warming by these gases has itself produced at least as much additional warming. The increased emissions of sulfur dioxide that accompanied fossil-fuel combustion have produced aerosols that canceled out some but not all of the warming produced by the greenhouse gases.

Global Circulation Models

In continuing research that began in the 1980s, scientists have attempted by computer modeling to predict the consequences of the increases in atmospheric gases and particles on the past, present, and future climate of the planet. There are some uncertainties in such an endeavor, including the fact that we don't yet fully understand all the sources and sinks of the green-house gases or the net effect of aerosols. Probably the most important prob-lem remaining with these **global circulation models** is in their treatment of clouds. In particular, is the net feedback from the increased cloudiness expected for a warmed atmosphere negative, positive, or zero? Clouds oper-ate both to cool and to heat the atmosphere. They cool it by reflecting incoming light back to space; we experience this dramatically when the Sun goes behind the clouds on a warm, sunny day and we immediately feel the air cooling. We also know that the water droplets in clouds absorb infrared light emitted from below them.

- *Low-lying* clouds are warm, so they re-emit almost all the energy absorbed in random directions, rather than converting it to heat and thereby warming the air in their immediate surroundings. Since some of the IR emitted by these low clouds is directed downward, the surface is warmed—think of the general phenomenon that cloudy nights are generally warmer than cloudless ones. The fact that the IR is all re-emitted, however, means that these clouds don't warm the atmosphere much by this effect. Thus the *net* effect over a full day of low-lying clouds is to cool the Earth, because their reflection of incom-ing sunlight is dominant.

- In contrast, *high-lying* clouds absorb outgoing IR but don't re-emit much of it since they are cold; all the absorbed IR is converted to heat, warming the nearby air. Thus the net effect of high-lying clouds is to warm the Earth, since their conversion of outgoing IR into heat is more important than their reflection of incoming sunlight.

Because we don't accurately know whether global warming will produce more additional low-lying or high-lying clouds, it is uncertain whether the feedback from this factor will be positive or negative.

Notwithstanding these problems, current global circulation models reproduce the climate of the last century and a half quite well. The solid light green curve in Figure 4-22 represents the reconstruction, i.e., the "retrospective prediction," of computer simulations of the temperature trends that should have occurred due to the increase of greenhouse gases *and* atmospheric aerosols in the 1860–1998 period. The analogous simulation *ignoring the effects of aerosols* is shown by the dashed green curve. Comparison of these two curves leads to the conclusion that during the 1940–1970 temperature flattening, the warming effects of increasing greenhouse gas concentrations was masked by the shielding effects of increased concentrations of the aerosols.

The agreement between the observed temperature trends and the calculations including both greenhouse gases and aerosols is not quantitative, owing perhaps to the influence of the natural fluctuations. Indeed, if estimates of changes in solar irradiance are included in the calculations along with greenhouse gas and aerosol effects, the agreement between observed and calculated temperatures (black line and solid green line of Figure 4-22, respectively) is quite good. Although transient year-to-year oscillations of observed temperatures in Figure 4-22 are not included in the calculations, the size of the overall observed temperature increase, the temporary flattening (or decline) in temperature from the early 1940s to the mid-1970s, and the rapid rise in the last two decades are well accounted for. The dips around 1860 and around 1910 are still unexplained.

It is clear from the difference between the black line and the light green line of Figure 4-22 that natural effects dominated global warming until the mid-1970s but have changed little since then. The lack of volcanic activity, which normally puts cooling aerosols into the air, in the 1920–1960 period probably also contributed to the warming; indeed, volcanic effects may well have been the dominant source of temperature changes over the last millennium.

It is not absolutely certain that all or indeed that any part of the observed increases in the last 100 years are attributable to the anthropogenic-induced enhancement of the greenhouse effect. However, natural, century-long global warming or cooling trends that are as large as the half-a-Celsius-degree we have recently experienced appear only once or twice a millennium. There is an 80–90% likelihood that the increase in the twentieth century was not a wholly natural climatic fluctuation. Furthermore, on the basis of computer simulations such as that discussed above, the UN-sponsored **Intergovernmental Panel on Climate Change** (IPCC) group concluded in their 2001 report that "most of the observed warming over the last 50 years is likely to have been due to the increase in greenhouse gas concentrations."

Some skeptics have pointed out that the evidence cited in favor of the argument that global warming has begun refers to air temperatures at the Earth's surface only, and that satellite data indicate that the lower troposphere as a whole has cooled rather than warmed. However, this cooling has been claimed by other scientists to be spurious, and apparently arises from problems in merging data from two different satellites with slightly different calibrations and other artifacts of uncertainties in the data.

Other Signs of Global Warming

In addition to a rise in average global surface air temperature, there have been a number of other changes that indicate that the climate indeed is changing:

- *Winters have become shorter, by about 11 days.* In the Northern Hemisphere, spring has been arriving sooner and autumn has been starting later. Over the last three decades, the advent of spring—as observed by the appearance of buds, the unfolding of leaves, and the flowering of plants—has advanced by an average of six days in Europe, while the start of autumn—as defined by the date at which leaves change color and begin to fall—has been delayed by about five days. In Alaska and northwestern Canada average temperatures have risen one degree per decade recently, resulting in an earlier date for the last frost and significant thawing of the permafrost. Consequently, in many regions there are fewer "frost days" now than there used to be. The response of plants is driven by the increase in average daily air temperatures. Indeed, the range boundaries of some plants have increased toward the poles, and phenologies (season-dependent behavior) have shown an earlier start for spring for the great majority of plants and animals.

- *The Earth's ice cover is shrinking fast.* Glaciers, polar icecaps, and polar sea ice are melting and disappearing at unprecedented rates, as a consequence of global warming. For example, the remaining glaciers in Glacier National Park in the U.S. Rocky Mountains could disappear in 30 years if current melting rates continue. About 10% of the world's winter snow cover has disappeared since the late 1960s. Sea ice in the Arctic has not only decreased in area, by about 9% per decade recently, but it has also thinned dramatically. Warmer weather has also delayed the seasonal formation of sea ice. All these changes have caused a sharp decline in some populations of Antarctic penguins and Arctic caribou.

- *Warming water is killing much of the coral in ocean reefs and threatening sea life.* Coral reefs in tropical waters nurture and protect fish and attract scuba divers. As water warms, corals "bleach" themselves by expelling the algae that give them color and provide nutrition. Over 95% of the coral is already dead in some parts of the Seychelles Islands. Thus far, reefs in the central Pacific Ocean have escaped bleaching, and it is just beginning in the Caribbean. The death of the coral reefs not only affects the tourist trade but also threatens fishing for species that depend on the reefs for food. Beaches will erode if the reefs break up and no longer provide protection.

- *Mosquito-borne diseases have reached higher altitudes.* Because of warmer temperatures, mosquitoes are now able to survive in regions where they formerly were not viable. As a result, mosquitoes have carried malaria to higher mountain regions in parts of Africa and *dengue fever* to new regions in Central America. Outbreaks of malaria have occurred in Texas, Florida, Michigan, New York, New Jersey, and even southern Ontario in the last decade. Warmer weather and changing precipitation patterns allowed the West Nile virus, another mosquito-borne disease, to become established in the New York City area in the late 1990s and in southern Canada and virtually all the contiguous United States by the early 2000s.

- *Rising sea levels are threatening to engulf Pacific islands.* Warming the air eventually leads to a rise in sea levels. The average level of the sea has risen by about 10 cm since 1940, a sharp increase in rate for this natural process.

- *Precipitation has increased in most areas.* An important aspect of climate is the amount of precipitation—rain and snow—that falls at various locations on Earth. In the twentieth century, the total annual precipitation in the Northern Hemisphere increased, as you can see from Figure 4-24, since the green-dot areas (increased precipitation) greatly outnumber the gray-dot areas (decreased precipitation). Many areas just north and just south of the Equator, especially in Africa, became much drier, with disastrous consequences for food production, though there has been more rain at or near the Equator. Most mid- and high-latitude regions of North America and Europe became somewhat wetter. An overall increase in global precipitation is expected, since warming of the air warms the surface waters of lakes and oceans, and warmer water evaporates faster and thereby increases the water content of the atmosphere.

- *Extreme weather is becoming more common.* The frequency of extreme and violent weather events has increased in many areas of the world. Such events include stronger blizzards and storms with heavy snow and freezing rain in northern areas but record heat waves, hurricanes, and drought in others. For example, the number of heat waves lasting three days or longer almost doubled in the United States between 1949 and 1995. Moreover, the frequency of storms with heavy or extreme precipitation increased in the United States and in many other countries in the latter half of the twentieth century. The economic damage caused by storms in the 1990s greatly exceeded that in previous decades.

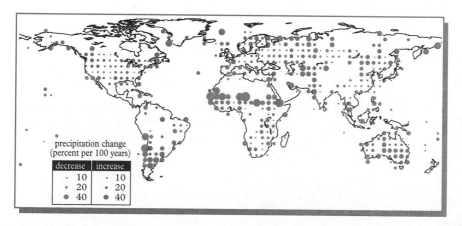

FIGURE 4-24 Trends in average annual global precipitation from 1901 to 1998. [Source: P. Martens, "How Climate Change Affects Human Health," *American Scientist* 87 (1999): 534–542.]

Review Questions

1. Sketch a plot showing how global air temperature has changed over the last two millennia.

2. What is the wavelength range, in μm, for infrared light? In what portion of this range does the Earth receive IR from the Sun? What are the wavelength limits for the *thermal IR* range?

3. Explain in terms of the mechanism involved what is meant by the *greenhouse effect*. Explain what is meant by the enhancement of the greenhouse effect.

4. Explain what is meant by the terms *symmetric* and *antisymmetric bond-stretching* vibrations and by *bending* vibrations.

5. Explain the relationship between the frequency of vibrations in a molecule and the frequencies of light it will absorb.

6. Why don't N_2 and O_2 absorb thermal IR? Why don't we consider CO and NO to be trace gases that could contribute to enhancing the greenhouse effect?

7. What are the two main anthropogenic sources of carbon dioxide in the atmosphere? What is its main sink? What is *fixed carbon*?

8. Is water vapor a greenhouse gas? If so, why is it not usually present on lists of such substances?

9. Explain what is meant by *positive* and *negative feedback*. Give an example of each as it affects global warming.

10. What is meant by the term *atmospheric window* as applied to the emission of IR from the Earth's surface? What is the range of wavelengths of this window?

11. What is meant by the *residence time* of a gas in air? How is it related to the gas's rate R of input/output and to its total concentration C?

12. What are four important trace gases that contribute to the greenhouse effect?

13. What are the six most important sources of methane?

14. What are the three most important sinks for methane in the atmosphere? Which one of them is dominant? What is meant by the term *clathrate compound*?

15. Is the enhancement of the greenhouse effect by release of methane from clathrates due to increased temperature an example of feedback? If so, is it positive or negative feedback? Would an increase in the rate and amount of photosynthesis with increasing temperatures and CO_2 levels be a case of positive or negative feedback?

16. Explain in chemical terms what is meant by *nitrification* and *denitrification*. What are the conditions under which nitrous oxide production is enhanced as a by-product of these two processes?

17. What are the main sources and sinks for N_2O in the atmosphere?

18. Are the proposed CFC replacements themselves greenhouse gases? Why is their emission considered to be less of a problem in enhancing the greenhouse effect than was that of the CFCs themselves?

19. By which two mechanisms does light interact with atmospheric particles?

20. Explain how sulfate aerosols in the troposphere affect the air temperature at the Earth's surface, by both the direct and indirect mechanisms.

21. Explain how the *effective CO_2 concentration* differs conceptually and numerically from the real CO_2 concentration in air.

22. List four important signs, other than increases in average air temperature, that global warming is occurring.

Green Chemistry Questions

1. The development of supercritical carbon dioxide for removal of photoresist (SCORR) won a Presidential Green Chemistry Challenge Award.

(a) Which of the three focus areas (see the Introduction to Green Chemistry) for these awards does this award best fit into?

(b) List four of the twelve principles of green chemistry (see the Introduction to Green Chemistry) that are addressed by the green chemistry developed by Los Alamos and SC fluids.

2. What environmental advantages does the SCORR process offer versus conventional methods for photoresist removal?

Additional Problems

1. The common tropospheric pollutant gases SO_2 and NO_2 have molecular structures which, like that of CO_2, have the central atom connected to two oxygen atoms but, unlike CO_2, they are nonlinear. The wavelengths for their vibrations are given in the table below. **(a)** Which of the vibrations are capable of absorbing infrared energy? **(b)** Based on the wavelengths for the IR-absorbing vibrations and the spectrum in Figure 4-6, decide which, if any, vibrations could contribute much to global warming. **(c)** What lifetime characteristic of these gases would limit their role in global warming?

Gas	Symmetric stretch	Antisymmetric stretch	Bending
SO_2	8.7 μm	7.3 μm	19.3 μm
NO_2	7.6 μm	6.2 μm	13.3 μm

2. **(a)** How can the fact that nitrous oxide has three vibrations that absorb infrared light be used to prove that its linear structure is NNO rather than NON? **(b)** Would methane molecules absorb IR during the vibration in which all four C—H bonds stretched or contracted in phase?

3. Anthropogenic carbon dioxide emissions into the atmosphere amounted to 178 Gt from January 1990 to December 1997. Calculate the fraction of this emitted carbon dioxide that remained in the air, given that in that same eight-year period, the carbon dioxide concentration in air rose by 11.1 ppm. Note that the molar masses of C, O, and air, respectively, are 12.0, 16.0, and 29.0 g, that the mass of the atmosphere is 5.1×10^{21} g, and that 1 Gt is 10^{15} g.

4. The total amount of methane in the atmosphere in 1992 was about 5000 Tg and was then increasing by about 0.6% annually due to the fact that the annual input rate exceeded the annual output rate of 530 Tg/yr. Calculate the percentage by which anthropogenic releases of methane, which account for two-thirds of the total, had to be reduced if the atmospheric concentration of this gas was to be stabilized in 1992.

5. What is the ratio of the rates of energy emission by two otherwise identical blackbodies, one of which is at 0°C and the other at 17°C? [Hint: Use the T^4 dependence of the rates of emission.]

6. The fraction F of light that is absorbed by any gas in air is logarithmically related to the concentration c of the gas and the distance d through which the light travels; this relationship is called the Beer–Lambert law:

$$\log_e(1 - F) = -Kcd$$

where K is a proportionality constant. Show that for concentrations near zero (e.g., where $Kcd = 0.001$), F is related almost linearly to c, whereas for larger Kcd values (e.g., near 2), doubling the concentration does not nearly double the light absorption. (The former situation is analogous to that for trace molecules that absorb in the window region, whereas the latter situation is that relevant to absorption by carbon dioxide.)

7.   The vapor pressure P of a liquid rises exponentially when it is heated according to the equation

$$\ln(P_2/P_1) = -\Delta H/R \, (1/T_1 - 1/T_2)$$

Here P_2 and P_1 are the vapor pressures of the liquid at the Kelvin temperatures T_2 and T_1 after and before the temperature increase, R is the gas constant 8.3 J/K mol, and ΔH is the liquid's enthalpy of vaporization, which for water is 44 kJ/mol. Calculate the percentage increase in the vapor pressure of water that occurs if the temperature is raised from 15°C to 18°C. Give several reasons why the amount of outgoing thermal infrared in water's absorption bands may not be increased by exactly the percentage you •calculate if the average surface air temperature is increased to 18°C.

8.   Suppose that some climatic crisis inspired the Earth's population to switch to energy systems that did not emit carbon dioxide and that the transition occurred within a decade. What would be the predicted immediate effect on the Earth's average air temperature of this change, given that both carbon dioxide and sulfur dioxide emissions from fossil fuels would have rapidly declined?

9.   Given the information in the text and in Table 4-1, calculate the contributions that methane and nitrous oxide have made to date in increasing the value of the effective carbon dioxide concentration. Explain why the per molecule relative warming efficiencies in Table 4-1 for short-lived gases such as methane and HCFC-22 are considerably less than their relative absorption values.

10.   Explain why CHF_2Cl (HCFC-22) has more IR absorption bands with a greater range of IR wavelengths absorbed, and absorbs IR much more efficiently on a per molecule basis, than CH_4, the hydrocarbon from which it is derived.

11.   Calculate the volume of CO_2 produced at 1 atm and 20.0°C from the complete combustion of 1.00 L of gasoline. Although gasoline is a mixture of C_7 and C_8 hydrocarbons (as described in Chapter 6), for the purposes of this calculation consider gasoline to have the chemical formula C_8H_{18} and the same density as n-octane: 0.702 g/mL. Calculate the volume of CO_2 produced by driving 100 miles on the highway in a mid-size sedan compared to an SUV, given their highway fuel efficiencies of 33 and 19 mpg, respectively. Note that 1 gal = 3.785 L.

12.   Given that the 2002 concentration of CH_4 in the atmosphere was 1.77 ppm, calculate the total mass of CH_4 in the atmosphere in 2002. Note that the total mass of the atmosphere is 5.1×10^{18} kg and the average molar mass of the atmosphere is 29.0 g/mol.

Further Reading

1.   J. T. Houghton et al., *Climate Change 2001: The Scientific Basis* (Cambridge: Cambridge University Press, 2001). (Published for the Intergovernmental Panel on Climate Change.)

2.   E. Claussen, ed., *Climate Change: Science, Strategies, and Solutions* (Arlington, VA: Pew Center on Global Climate Change, 2001).

3. D. Rind, "The Sun's Role in Climate Variations," *Science* 296 (2002): 673–677.

4. V. Ramanathan et al., "Aerosols, Climate, and the Hydrological Cycle," *Science* 294 (2001): 2119–2124.

5. F. W. Zwiers and A. J. Weaver, "The Causes of 20th Century Warming," *Science* 290 (2000): 2081–2137.

6. W. M. Post et al., "The Global Carbon Cycle," *American Scientist* 78 (1990): 310–326.

7. A. S. Moffat, "Resurgent Forests Can Be Greenhouse Gas Sponges," *Science* 277 (1997): 315–316.

8. D. S. Schimel et al., "Recent Patterns and Mechanisms of Carbon Exchange by Terrestrial Ecosystems," *Nature* 414 (2001): 169–172.

9. C. Li et al., "Reduced Methane Emissions from Large-Scale Changes in Water Management of China's Rice Paddies During 1980–2000," *Geophysical Research Letters* 29 (2002): 33.

10. E. J. Dlugokencky et al., "Continuing Decline in the Growth Rate of the Atmospheric Methane Burden," *Nature* 393 (1998): 447–450.

11. C. Parmesan and G. Yohe, "A Globally Coherent Fingerprint of Climate Change Impacts Across Natural Systems," *Nature* 421 (2003): 37–42.

12. S. H. Schneider, "Global Warming" (San Francisco: Sierra Club Books, 1989).

13. B. Lomberg, "The Skeptical Environmentalist" (Cambridge: Cambridge University Press, 2001), Chapter 24.

Websites of Interest

Log on to www.whfreeman.com/envchem3e/and click on Chapter 4.

CHAPTER 5

Climate Change in the Future: Predictions, Consequences, and Controls

As we saw in Chapter 4, the Earth's weather has probably already been affected by the enhancement of the greenhouse effect due to increasing atmospheric concentrations of carbon dioxide and other gases. The continuing build-up of CO_2 in the air leads to the conclusion that we are in store for further increases in global air temperatures and other changes to our climate.

In this chapter we shall inquire into predictions of the likely trends in energy usage, and consequently of carbon dioxide emissions, that will occur for the rest of the twenty-first century and of the additional global warming that can be expected as a result. Before beginning our analysis of energy trends, however, we shall summarize what projections tell us qualitatively about the climate changes to expect in the coming decades and some of their consequences for human health.

Waste methane is burned in these stacks rather than being emitted into the air. (Photo Disc)

The Potential Consequences of Global Warming

Those of us who currently suffer through severe winters each year may look forward to the warmer climate associated with the enhanced greenhouse effect. After all, in the eleventh and twelfth centuries, an increase of a few tenths of a degree in the northern temperate zone was sufficient to allow farming on the coast of Greenland, for vineyards to flourish extensively in England, and for the Vikings to travel the North Atlantic and settle in Newfoundland.

However, the climate changes predicted for the twenty-first century and beyond do not present a uniformly pleasant prospect. The *rate* of change in our climate, which to date has been modest, will be dramatic by the middle of the century. Indeed, the rapid rate of global change will probably be the greatest problem with which humanity will have to contend. A more gradual transition, even to the same end result, would be much easier to handle, not only for humans but for all living organisms on the planet.

It is very difficult for scientists to model the climate—even with the assistance of the fastest computers in the world—in order to make definitive statements about what changes will occur in particular regions in the future. We know that there will be substantial changes in the climate, but we are unable to specify exactly what they will be.

Predictions for Climate Change by 2100

The changes in the Earth's climate that have occurred in the past half-century and are predicted to continue into the twenty-first century, as judged by the U.N.'s Intergovernmental Panel on Climate Change (IPCC) in its 2001 report, are summarized in Table 5-1.

According to sophisticated computer simulations of the future climate reported by the IPCC in 2001, the rise in the average global air temperature by 2100 (compared to 1990) could be as small as 1.4°C or as high as a staggering 5.8°C. As in the twentieth century, more of the warming will occur at night. The magnitude of the temperature increase will depend greatly on whether emissions (including those of sulfur dioxide) are controlled or not. At a minimum, however, the world will warm more than twice as fast in this century as it did in the last. Part of the rather wide range of the predicted values is due to uncertainty about exactly how sensitive the climate is to carbon dioxide. Indeed, research reported in 2003 indicates that aerosols have counterbalanced more greenhouse warming in the past than was previously thought; consequently scientists may have significantly underestimated the sensitivity of temperature to CO_2, in which case the predicted increases will have to be revised upward.

An increase of a few degrees may seem small, but our current average air temperature is less than 6°C warmer than that in the coldest periods of the ice ages! Snow cover and sea-ice area will continue to decline. There may well be enough melting of ice in the Arctic region for the Northwest Passage to be used for commercial transport, since the warming in winter of all Arctic regions is projected to be much greater than the global average.

TABLE 5-1 Estimates of Confidence in Observed and Projected Changes in Extreme Weather and Climate Events

Confidence in observed changes (latter half of the twentieth century)	Changes in phenomenon	Confidence in projected changes (during the twenty-first century)
Likely	Higher maximum temperatures and more hot days[a] over nearly all land areas	Very likely
Very likely	Higher minimum temperatures, fewer cold days and frost days over nearly all land areas	Very likely
Very likely	Reduced diurnal temperature range over most land areas	Very likely
Likely, over many areas	Increase of heat index[b] over land areas	Very likely, over most areas
Likely, over many Northern Hemisphere mid- to high-latitude land areas	More intense precipitation events[c]	Very likely, over many areas
Likely, in a few areas	Increased summer continental drying and associated risk of drought	Likely, over most mid-latitude continental interiors (lack of consistent projections in other areas)
Not observed in the few analyses available	Increase in tropical cyclone peak wind intensities[d]	Likely, over some areas
Insufficient data for assessment	Increase in tropical cyclone mean and peak precipitation intensities[d]	Likely, over some areas

[a] Hot days refers to a day whose maximum temperature reaches or exceeds some temperature that is considered a critical threshold for impacts on human and natural systems. Actual thresholds vary regionally, but typical values include 32°C, 35°C, or 40°C.
[b] Heat index refers to a combination of temperature and humidity that measures effects on human comfort.
[c] For other areas, there are either insufficient data or conflicting analyses.
[d] Past and future changes in tropical cyclone location and frequency are uncertain.
Source: J. T. Houghton et al., *Climate Change 2001: The Scientific Basis* (Cambridge: Cambridge University Press, 2001).

The total amount of global rainfall is projected to increase, since more water will evaporate at the higher surface temperatures. The global average precipitation increases by about 2% for every Centigrade degree rise in temperature. Although the world overall will become more humid, some areas will become drier. To make matters worse, most areas of the world that currently suffer from drought are predicted to become even drier. Continental interiors at mid-latitudes will have a continuing risk of drought in summer due to continued drying of the soil, the increased rate of evaporation from higher air temperatures being greater than the increase in the rate of precipitation. Subtropical areas will experience less precipitation, and equatorial regions and high-latitude regions will experience more, continuing the twentieth-century trends.

An increase in the average atmospheric temperature means that the air and water at the Earth's surface contain more energy and that more violent weather disturbances could result. Indeed, more extreme weather may result from global warming, and it is the phenomenon that will affect many of us the most. The number of days per year having intense rain showers or very high temperatures will increase. Storm wind intensities and heavy downpours will increase in some tropical areas.

Predictions About Sea Levels

Although air and land surfaces warm quickly with an increase in average global temperature, the same is not true of seawater. It takes many centuries for an increase in air temperature to gradually make its way down deeper and deeper into the ocean. For this reason, the rise in sea levels resulting from any particular amount of global warming is largely delayed for many years. Consequently, even if atmospheric carbon dioxide levels did not increase at all beyond today's values and no further global warming were to occur, sea levels would *continue* to rise for centuries to come, as deeper and deeper layers of the oceans became warmer—and expanded—by absorbing heat from air that has already been warmed.

Sea levels are predicted to rise by about half a meter by 2100—in addition to the 10–25 cm rise experienced over the last 100 years—although there is a large uncertainty in this value. Some of the predicted rise in sea level is due to the melting of glaciers, but most arises from the **thermal expansion** of seawater. The expansion occurs because the density of water *decreases* gradually as water warms beyond 4°C, the temperature at which it reaches its maximum density, as illustrated in Figure 5-1. Since density is mass divided by volume, and since the mass of a given sample of water cannot change, the volume it occupies must increase if its density decreases. As seawater warms (beyond 4°C), the volume occupied by a gram or kilogram of it increases; when this occurs, its depth increases and therefore the sea level rises.

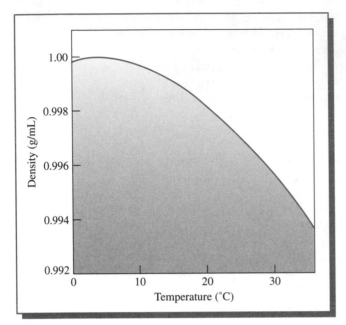

FIGURE 5-1 Density
of liquid water versus
temperature.

Although an increase of half a meter in sea level does not seem very large, there are countries, like Bangladesh and some small island nations such as Tuvalu in the South Pacific, in which much of the population currently lives on land that would be flooded by a rise in sea level of this amount. Damage from tropical storms would increase because of the higher sea levels. However, in contrast to some fears expressed in the 1980s, it now seems unlikely that temperature increases in the twenty-first century will be large enough to cause rapid melting of the Antarctic and Greenland ice sheets. Parts of both of these sit on land above sea level and their transfer, by the breaking off of icebergs, into the oceans would cause a major increase in sea levels.

In the long term, the most dramatic—though unlikely—effect of substantial global warming would be a change in the circulation patterns of water in the Atlantic Ocean. Currently, warm surface waters flow northward from the tropics into the North Atlantic, bringing heat to Europe and to a lesser extent to eastern North America. Some scientists have speculated that a rapid rise in temperature and rainfall levels could weaken or even eliminate this circulation pattern, as geological records indicate has happened in the past.

Climate Predictions for Specific Regions

It is much harder to make specific, reliable predictions for individual regions than for the globe as a whole. However, the higher latitudes of the Northern Hemisphere are expected to experience temperature increases substantially greater than the global average. Warming over some land areas, including the United States and Canada, should be noticeably faster than the average rate for the globe.

In the midwest regions of the United States and the areas just north of them in Canada, as well as in southern Europe, the soil probably will become much less moist because of increased rates of evaporation in the warmer air. This could affect the continued suitability of the land in North America and southern Europe for the growing of grain. High-latitude regions in Canada and northern Europe, however, could experience increased productivity, at least where the soil is suitable for agriculture. Already, yields of crops such as corn and sugar beets have fallen in Italy, whereas they have increased in Ireland and the Scandinavian countries. In areas that become drier, the positive CO_2 fertilization effect on plants will cancel some of the negative effects of decreased rainfall. There will be longer frost-free growing seasons at northern latitudes but increased chances that heat stress will affect crops grown there. Food production in temperate areas will probably also be affected by the attack of insects that in the past have been largely killed off during the winters but could survive and flourish under warmer conditions.

Temperature and moisture changes will occur quickly compared to those that have taken place in the past, and consequently some ecosystems will be destabilized. Coastal ecosystems such as coral reefs are also particularly at risk. The species compositions of forests are likely to change, especially in regions far removed from the Equator. For example, the hardwood forests in eastern North America may be at risk of extinction if climate zones shift more quickly than their migration can keep up with. The boreal forest of central Canada could be eliminated by fire by 2050; indeed, the frequency of fires in these woodlands is already climbing.

Predicted Effects on Human Health

Some scientists have speculated that human health will be affected adversely by global warming. There probably will be more extreme heat waves in summers but fewer prolonged cold snaps in winters. The expected doubling in the annual number of very hot days in temperate zones will affect people who are especially vulnerable to extreme heat, such as the very young, the very old, and those with chronic respiratory diseases, heart disease, or high blood pressure. This will be particularly acute for poor people, who have less access to air conditioning. The heat wave in the summer of 2003 in Europe was a factor in the death of at least 10,000 people in France alone.

Domestic violence and civil disturbances could also increase, as they tend to occur more frequently in hot weather. On the other hand, there would probably be a decrease in cold-related illnesses because of the milder winters. Air quality in summer will probably degrade further, as the background concentration of ground-level ozone will probably increase substantially. Higher CO_2 concentrations and warmer weather will increase the production and release of plant pollen such as ragweed, thus exacerbating allergic responses.

Less directly, global warming may extend the range of insects carrying diseases such as malaria into regions where people have developed no immunity and may intensify transmission in regions where such diseases are already prevalent. It has been predicted that malaria could claim an additional million victims annually if the temperature rise is sufficient to allow parasite-bearing mosquitoes to spread into areas not now affected. In North America, suitable habitat for the mosquitoes is being lost, so fortunately this should not be as much of a problem for Americans and Canadians. A different type of mosquito carries the dengue and yellow fever viruses, and its range could increase with warming, thereby spreading these diseases. There is also evidence that cholera rates increase with warming of ocean surface waters, because coastal blooms of algae are breeding grounds for the disease, and they increase with water temperature. Some experts in disease control discount these predictions, arguing that other effects such as increased rainfall could well negate or even reverse the effects of increases in air temperature on disease rates.

Animal health could also be affected by the spreading of disease by parasites. In addition, species such as polar bears and caribou that live in very cold regions could be at risk of extinction from habitat changes that threaten their hunting practices.

Energy Reserves and Usage

Ever since the Industrial Revolution, the worldwide use of **commercial energy**—that sold to users and usually derived on a large scale from fossil-fuel combustion, hydroelectricity, and nuclear power, as opposed to the biomass collected and used by individual families—has risen almost every year; the current annual global growth rate is about 2%. As discussed for fossil fuels in Chapter 4, the period of the most rapid increase began after the Second World War, at a time when annual global commercial energy consumption was only about one-tenth the current level.

Because the amounts are so large, it is useful to discuss global quantities of energy in terms of the large energy unit the **exajoule,** EJ, which is 10^{18} joules. The total amount of commercial energy consumed by humans amounts currently to about 400 EJ annually, with the United States consuming about 100 EJ of that total.

Determinants of Growth in Energy Use

Although increases in energy use are sometimes thought to be directly tied to population growth, this is a dominant factor only for less developed countries, for which energy use per capita is small anyway. The usage of commercial energy by a country depends on many factors, including not only its population, geography, and climate, but also the cost of energy. However, the most important factor in total energy usage appears to be the **gross national** (or **domestic**) **product** (GNP or GDP) of the country. In industrialized societies, about 20 megajoules (MJ; 20 million joules) of energy, equivalent to a little over 1 kg of carbon dioxide if the energy is produced by burning fossil fuels, are needed on average to produce one (U.S.) dollar's worth of goods. The ratio of energy to GDP usually rises when a country begins to industrialize but then drops gradually as the infrastructure becomes more substantial and efficient. The drop in the energy/GDP ratio for the United States over the second half of the twentieth century is illustrated in Figure 5-2a.

The ratio of CO_2 equivalent emissions per unit of GDP is known as the **carbon intensity** (or *greenhouse gas intensity*). The decline in the global carbon intensity over the last half-century is illustrated in Figure 5-2b. As an alternative to cutting greenhouse gas emissions in line with the Kyoto treaty, the U.S. government proposed in 2003 to reduce the carbon intensity of the U.S. economy by 18% by 2012. Based on recent trends, a 14% decline would have been expected without any new initiatives.

The fantastic rise in global energy usage in the second half of the twentieth century was due mainly to industrial expansion and to increases in the standard of living in the now-developed countries. The energy consumption in these countries continues to expand, though now only slowly, as the graph in Figure 5-3 indicates. However, economic growth in the developing countries—which contain three-quarters of the world's population—is rising more quickly, and with it, their total energy consumption. The annual per capita release of carbon dioxide in developing countries is about one-tenth as large as it is in developed countries, but it is increasing. Although the developing countries used only 30% of the world's commercial energy in 1993, they are expected to consume more than half of it starting sometime in the first two decades of the present century, the exact year depending on their rate of economic growth (see Figure 5-3). According to the *International Energy Agency*, between 1994 and 2010 the rate of increase for the developing countries collectively is expected to be 4% annually, which, if compounded, would produce a doubling over that period. For developed countries, the annual rise over the same period is expected to be 1.5%, amounting to 28% if compounded over that period. The overall global increase in energy use is expected to be about 1.7% annually over the next three decades, with fossil-fuel combustion meeting more than 90% of the increased demand.

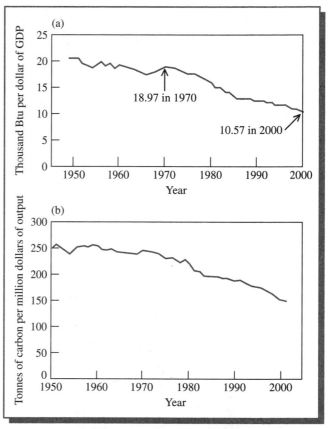

FIGURE 5-2 (a) The energy intensity of the U.S. economy in the second half of the twentieth century. [Source: R. A. Kerr, "More Science and a Carrot, Not a Stick," *Science* 295 (2002): 1439.] (b) Carbon intensity of the global economy. [Source: L. R. Brown et al., *Vital Signs 2003* (New York: W.W. Norton, 2003).]

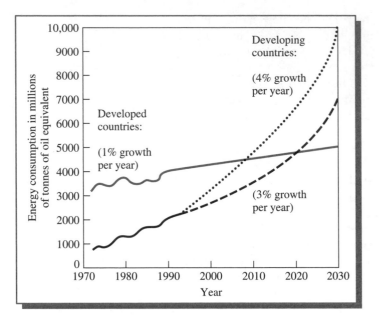

FIGURE 5-3 Historical and projected energy consumption. [Source: Adapted from J. Goldemberg, *Science* 269 (1995): 1058–1059.]

PROBLEM 5-1

Any quantity V whose value increases in a time period t by a percentage of its previous value exhibits exponential growth according to the equation

$$V = V_0 \, e^{kt}$$

where V_0 is the initial value and k is the fractional increase in each time period. Given that this equation will apply to the growth in energy usage if it increases by the same percentage each year, derive a general formula relating the number of years required for energy usage to double as a function of the annual fractional increase k. What is the doubling time when 4% annual growth (i.e., $k = 0.04$) is in operation? What about 3%, 1.5%, and 1.0% annual growth rates? If $k = 0.02$, how many years does it take for the energy use to increase tenfold? If energy usage grew by a factor of ten over 50 years, what was the annual compounded rate of growth in this period?

Energy Reserves: Coal

The main problem with fossil-fuel usage in this century may well be the CO_2 emissions that result from its combustion rather than a shortage in supply.

The main fossil-fuel reserve is coal, which is available in abundance in many regions of the world, including developing countries, and which is cheap to mine and to transport. At today's rate of consumption, coal reserves are estimated to last another 200 years, much longer than oil or gas (see below). Indeed, some observers believe the world will return to a greater reliance on coal as the major fossil fuel later in this century. Currently, coal produces about half the electric power in the United States.

Although it is a mixture, to a first approximation, coal is graphitic carbon, C. Like the other fossil fuels, it was formed from the very small proportion of ancient plant matter that was covered over by water and could not be recycled back to CO_2 at that time. This also accounts for the build-up of O_2 in the atmosphere. Coal was formed from the highly aromatic, polymeric component of land-based woody plants called *lignin*; an approximate empirical formula for lignin is C_3H_3O. Over long periods of time during which the material was subjected to high pressures and temperatures, both water and carbon dioxide were lost. The material polymerized further in the process to yield the very carbon-rich, hard material known as coal. Unfortunately, the coal also incorporated measurable quantities of virtually every naturally occuring element during its formation, so that when it is burned, it emits not only CO_2 and H_2O but also substantial quantities of many air pollutants, notably sulfur dioxide, fluorides, uranium and other radioactive metals, and heavy metals. Thus coal has a reputation for being a "dirty fuel." The removal of some of these imputities, especially sulfur, by various modern technologies was discussed in Chapter 2.

The burning of coal domestically in stoves and furnaces produces a great deal of soot, and it therefore has been largely discontinued in developed countries. However, coal is still used in most developed and developing countries for electric power production. When it is burned in power plants, the soot problem is readily solved, but emissions of sulfur and nitrogen dioxides and of mercury require more sophisticated and expensive equipment and it is not done, even in most plants in developed countries.

The heat that combustion of the fossil fuel produces is used to generate steam, which in turn is used to turn turbines and thereby produce electricity. As discussed later in this chapter, however, the ratio of CO_2 to energy produced from coal is substantially greater than for the other fossil fuels. Coal can also be used to produce alternative fuels—also as discussed later in this chapter—but unfortunately, the conversion processes are not very energy efficient. Although the emission of carbon dioxide is not reduced by such conversions, they do allow the removal of sulfur dioxide and other pollutants and so are "clean" ways to use coal.

Energy Reserves: Petroleum and Natural Gas

Petroleum and natural gas are essentially hydrocarbons. They also originated from the small fraction of photosynthetically produced plant matter that was buried rather than oxidized. Sulfur is also an important impurity in these fossil fuels; as previously mentioned, some natural gas deposits contain more H_2S than CH_4! The sulfur compounds can be removed relatively easily from gas and oil, making these fuels inherently cleaner than coal. Petroleum and fuels made from it, such as gasoline, have the great advantage that they are energy-dense liquids that are convenient, relatively safe to use, and relatively cheap to produce. Virtually all transportation systems in both the developed and developing worlds are based on cheap petroleum fuels. The possibility of switching to alternative fuels for transportation is discussed in detail in Chapter 6. It will be much more difficult to switch from oil to other chemical feedstocks for the production of pharmaceuticals and polymers once oil runs out.

Although it is commonly said that we are running out of oil and gas, this will probably not occur globally in the short-to-medium term. Proven reserves of natural gas continue to climb, notwithstanding the fact that its use is growing more than that of either coal or oil. Improvements in the technology of petroleum extraction allow greater and greater proportions of the oil in a given deposit to be used. At today's rate of usage, there is estimated to be about 30 to 40 years' reserves of petroleum still available, and about 60 years for natural gas. There remains the problem of extracting the significant fraction of oil left in any reserve once the initial amounts have been withdrawn. The remaining oil occurs mostly in pores as droplets that are larger than the connecting necks of the porous formation and will not flow because of surface tension. Using surfactants or pressurized carbon dioxide can lower the surface tension, but the process is expensive and challenging to accomplish.

Some analysts believe that global oil production will peak sometime between 2005 and 2015. A recent estimation of the manner in which oil production is expected to vary with time is shown as a modified bell curve in Figure 5-4. The solid parts of each curve show the actual trends. Notice that

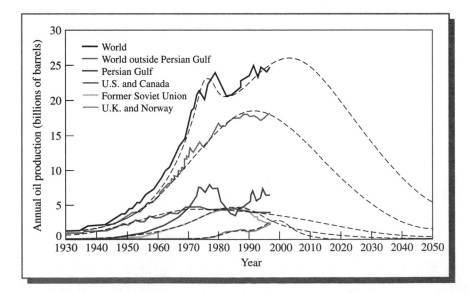

FIGURE 5-4 The global production of oil—real and predicted. [Source: C. C. Campbell and J. H. Laherrere, "The End of Cheap Oil," *Scientific American* (March 1998): 81.]

production outside the Persian Gulf area (second curve down, in light gray) has probably already peaked. Indeed, U.S. and Canadian production combined (dark green curve) reached a maximum in 1972.

The enormous quantity of natural gas held in methane hydrates (clathrates) in ocean sediments and permafrost, as mentioned in Chapter 4, would double the fossil-fuel reserves if they could be tapped. The technology to extract the clathrates, many of which are in dilute form and mixed with sediments that lie far below the seabed, does not yet exist. More oil is available in tar sands and oil shale. However, there are at present high economic, energy, and environmental costs for exploiting these potential reserves in a large-scale manner. For example, about two tonnes of the asphalt-like mixture of sand and tar in northern Alberta's heavy oil deposits are required to eventually produce one barrel of oil. Energy must be expended to separate the oil from the sands and to split the long-chain hydrocarbon molecules of the tar into shorter ones for use in gasoline.

Green Chemistry: Polylactic Acid—Biodegradable Polymers from Renewable Resources: Reducing the Need for Petroleum and the Impact on the Environment

Our everyday lives are permeated by the chemicals in products such as pharmaceuticals, plastics, pesticides, personal hygiene products, cleaners, fibers, dyes, paints, clothes, building materials, computer chips, packaging, and food. The vast majority of these chemicals are ultimately made from oil, consuming approximately 2.7% of the production of this natural resource. The compounds that are isolated from oil and used to produce these chemicals are known as *chemical feedstocks*. Approximately 290 billion kilograms of feedstocks are employed to create 130 billion kilograms of polymers (many are loosely referred to as plastics) each year. Some of the more familiar polymers (as will be discussed later, in Chapter 12) that are produced from crude oil include *polyethylene terephthalate* (PET), which is used to make plastic beverage bottles and fibers for clothes; polyethylene, which is used to produce plastic grocery and trash bags; and polystyrene, which we discussed in the green chemistry section in Chapter 1. Trade names, such as *Dacron, Teflon, Styrofoam,* and *Kevlar,* represent polymers that are part of our everyday lexicon.

Over 9 billion kilograms of PET are produced each year. PET is one of the main targets for recycling of plastics, yet less than a quarter of this total is recycled in the United States, with the rest being landfilled or incinerated. Even when PET is recycled, it generally can't be reused as beverage bottles; it is downward recycled into polyester fiber products such as carpets, T-shirts, fleece jackets, sleeping bags, and car trunk linings, or into thermoformed sheet products such as laundry scoops, nonfood containers, and containers for fruits.

When we use oil to produce items that we dispose of or incinerate (including the use of oil as a fuel), we are consuming a resource that it has taken nature millions of years to produce. Petroleum is a finite, nonrenewable resource. Although there are still considerable oil reserves, at the rate of our current use we will deplete the supply of cheap, readily accessible oil within the next 30 to 40 years. We must learn to use renewable resources such as biomass rather than petroleum to produce chemical feedstocks.

Scientists at Cargill Dow LLC in Minnetonka, Minnesota, have developed a method for producing a polymer called **polylactic acid** (PLA) from renewable resources—such as corn (called maize in the United Kingdom and elsewhere) and sugar beets—for which they won a Presidential Green Chemistry Challenge Award in 2002. Cargill Dow produces PLA at a plant in Blair, Nebraska. Ultimately, the goal is to utilize waste biomass as the source of this polymer. As in the steps shown in Figure 5-5, the corn is milled into starches, which are then reacted with water to yield glucose, which is then converted to lactic acid by natural fermentation. This naturally occurring compound is then converted to its dimer, followed by polymerization to PLA.

The environmental advantages of PLA over petroleum-based polymers include the following:

- It is made from annually renewable resources (corn, sugar beets, and eventually waste biomass).

- Production of PLA consumes 20–50% less fossil-fuel resources than petroleum-based polymers.

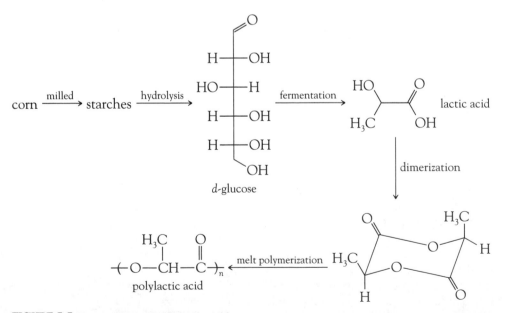

FIGURE 5-5 The synthesis of polylactic acid.

- It uses natural fermentation to produce lactic acid; no organic solvents or other hazardous substances are used.

- It uses catalysts, resulting in reduced energy and resource consumption.

- High yields of > 95% are obtained.

- The use of recycle streams helps to reduce waste.

- PLA can be recycled: converted back to its monomer via hydrolysis, then repolymerized to produce virgin polymer (i.e., closed-loop recycling).

- PLA can be composted (it is biodegradable); complete degradation occurs in a few weeks under normal composting conditions.

Another environmental consideration is that the plants, such as corn, used to produce this polymer consume atmospheric carbon dioxide, thus reducing the concentrations of this greenhouse gas. When PLA biodegrades, it releases this carbon dioxide back into the atmosphere in amounts equal to the carbon dioxide absorbed by the plants used to produce it. However, the advantage of PLA is that calculations indicate the greenhouse gas emission rate of PLA is 1600 kg CO_2/metric ton, while the petroleum-based polymers—polypropylene, polystyrene, PET, and nylon—have significantly higher values of 1850, 2740, 4140, and 7150.

PLA can used to produce products that are currently made from petroleum-based polymers such as cups, rigid food containers, food wrappers/bags, bags for refuse, furnishings for homes and offices (carpet tile, upholstery, awnings, and industrial wall panels), and fibers for clothing, pillows, and diapers.

Biodegradable polymers produced from renewable resources help to reduce our consumption of oil and have the potential to offer significant environmental and economic advantages over petroleum-based polymers. However, we must remember that even production of chemicals from annually renewable resources such as biomass does not offer complete solutions to energy and environmental problems. Growing crops, whether they are used to produce food or chemicals, requires fertilizers and pesticides. Energy is needed to plant, cultivate, and harvest; to produce, transport, and apply fertilizers and pesticides; to make and run tractors; to transport seeds, biomass, monomers, and polymers. Use of land to produce crops for chemicals also removes land that could be used to produce food and feed.

Growth in Energy Use and Its Causes

Since the Industrial Revolution, the emission rate for CO_2 has climbed hand-in-hand with the expansion of commercial energy use, since so much of the latter (currently 78%) is obtained from fossil-fuel sources. Barring an unforseen, massive switch to nuclear energy or renewable fuels,

CO_2 emission rates are expected to match commercial energy production rate increases as the developing world undergoes industrialization and as the economies of developed countries continue to expand. Indeed, a 2003 assessment by the European Union predicts that, over the first three decades of this century, carbon dioxide emissions will rise globally by an annual average of 2.1%, due to a 1.8% annual increase in energy usage. The fraction of fossil-fuel energy obtained from coal is expected to *increase* over this period—due to higher and higher prices for oil and natural gas as they become more scarce—thereby increasing the carbon intensity. The report predicts a cumulative increase in energy use for the United States of 50% and for the European Union of 18%. According to the report, energy use by developing countries will triple (corresponding to 4% annual, compounded growth), with the consequence that they will be responsible for 58% of CO_2 emissions by 2030, though they will still trail most industrialized countries in emissions per capita.

Currently the emissions of carbon dioxide amount to about 4 tonnes per person per year when averaged over the global population; this per capita annual emission is usually expressed as 1 tonne of *carbon*, and it is this reference to carbon that will be used henceforth. People in developed counties have much higher annual average emissions than do those in developing countries: 3 versus 0.5 tonnes of carbon per person. The United States leads in both total and per capita CO_2 emissions according to the bar graphs in Figure 5-6, where we have listed in order the top 20 emitter countries of carbon dioxide. The left set of bars indicates each country's percentage of total global emissions and the right-hand set shows the per capita emissions. Notice that the United States, Canada, and Australia have the highest per capita CO_2 emission rates, in part due to the high transportation requirements of these vast lands, compared to compact European countries. It is also true that in these three countries fossil-fuel energy is much cheaper than in European countries. Developed countries other than these three have remarkably similar per capita annual carbon emissions—about 2 tonnes—which perhaps is the value we might expect currently developing countries to achieve once they are fully developed. Alternatively, if developing countries can implement renewable energy technologies in constructing their economies, they would avoid the heavy fossil-fuel dependence and intense carbon dioxide emissions characteristic of all currently developed countries.

Because populations of different countries vary so much, their greenhouse gas emissions per capita or per dollar of GNP are no guide to their *total* emissions. Thus in Figure 5-6 we see that China and India make substantial contributions to the total global emissions since their populations are so large, even though their per capita emission rates are still quite modest. Both countries generate most of their electricity by burning coal. China's total CO_2 emissions actually fell by 7% between 1996 (its peak year) and 2000 because of radical reform in its energy industry and the Asian economic crisis.

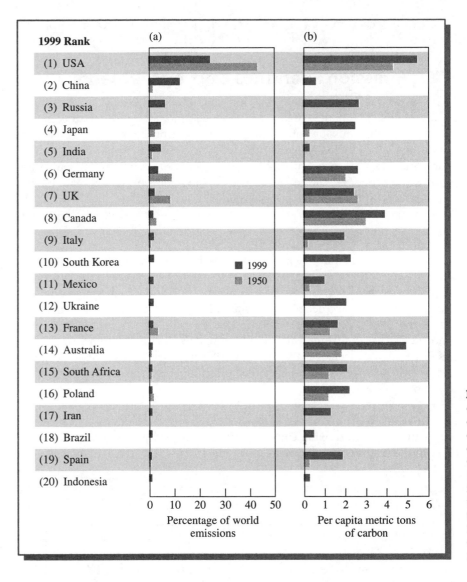

FIGURE 5-6 (a) Total annual emissions of CO_2, expressed as a percentage of the world total, and (b) per capita annual CO_2 emissions for the top 20 total emitter countries in 1999, with comparative data for 1950. [Source: Carbon Dioxide International Analysis Center (part of Oak Ridge National Laboratory).]

PROBLEM 5-2

Canada has massive supplies of heavy oil in tar sands, which are being used to make gasoline by combining them with natural gas. Assume that the empirical formulas of these three fuels are CH, CH_2, and CH_4, respectively, and that gasoline is made by hydrogenating the tar with hydrogen produced by extracting H_2 from natural gas. Combine the hydrogenation and hydrogen production equations to use all the H_2 and thereby deduce the overall reaction of CH and CH_4 with steam to produce gasoline and carbon dioxide.

CO_2 Emission Scenarios and Agreements

The Intergovernmental Panel on Climate Control, as part of its continuing study of global climate change, has produced a number of different scenarios concerning the emissions of greenhouse gases during the twenty-first century and has also attempted to predict the likely response of the climate system to such increase. In each emission scenario, it used estimates of population growth, of economic and technological development, of the energy mix, and of the likely emission controls that could be put in place.

Patterns of Growth in CO_2 Concentrations

Because carbon dioxide has such a long lifetime in the atmosphere, a century or more on average, the gas *accumulates* in air. Thus almost all the CO_2 emissions from the 1990s, for example, that did not find a temporary sink will remain in the air for decades to come, adding to the bulk of the emissions from the 1980s, the 1970s, and previous years. The actual carbon dioxide molecules that constitute this additional mass will change from year to year, as some CO_2 molecules leave the temporary sinks and enter the atmosphere while an equal number of such molecules from the air enter one or another temporary sink.

The growth pattern of the CO_2 concentration in air is determined mainly by the pattern of CO_2 emissions. Suppose, for example, that the same amount of carbon dioxide emissions was added to the air each year and did not find a temporary sink (Figure 5-7a, black line). The total amount of CO_2

FIGURE 5-7 CO_2 concentration related to its emission level. (a) Constant emissions of CO_2 produce a linearly increasing concentration of the gas. (b) Linearly increasing emissions produce a quadratically increasing concentration of the gas.

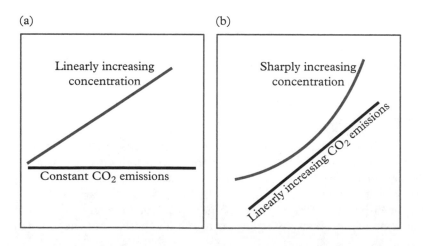

(a)

Linearly increasing concentration

Constant CO_2 emissions

(b)

Sharply increasing concentration

Linearly increasing CO_2 emissions

in air—and hence its concentration—would then increase annually by a constant amount. The carbon dioxide concentration increases linearly with time in this case (Figure 5-7a, green line). If the world could hold its carbon dioxide emissions constant at the year 2000 value, then the CO$_2$ concentration would increase linearly, by the current 1.6 ppm value, and would consequently become slightly greater than 500 ppm in 2100. Currently, global CO$_2$ emissions are growing slowly, increasing by only about 1% annually (see Figure 4-8), so the atmospheric carbon dioxide concentration is growing almost linearly with time (see insert in Figure 4-7).

Another scenario, which at some times has been more realistic than the situation just described, is that the CO$_2$ *emissions* were not the same each year, but *increased linearly*, that is, by a constant amount k each year. Thus, if the emissions one year amounted to A, the next year they were $A + k$, and the following year $A + 2k$, etc. (Figure 5-7b, black curve). In this case, if the fraction of the gas that enters the oceanic sink is constant, the growth in CO$_2$ *concentration* will be quadratic, much sharper than linear: the resulting plot of CO$_2$ concentration curves upward, as illustrated in Figure 5-7b (green curve). Indeed, in the decades preceding the mid-1970s, the CO$_2$ concentration did increase quadratically (see insert in Figure 4-7) since carbon dioxide emissions were increasing rapidly (see Figure 4-8). However, since then, the CO$_2$ concentration has increased in an approximately linear manner with time, reflecting the slower increase in emissions in this period, among other factors—including the fact that the fraction of emissions that find a temporary sink varies with time (as was discussed in Chapter 4).

IPCC Scenarios for CO$_2$ Emissions and Concentrations

In its 2001 report, the IPCC described a number of very different scenarios for greenhouse gas emissions for the rest of the century. The magnitudes of the emissions predicted for century's end vary dramatically: 5, 13.5, 20, and 29 Gt of C annually, ranging from 0.6 to 3.5 times the current value of about 8 Gt C/year. The carbon dioxide concentrations projected for 2100 for the IPCC scenarios range from 500 to more than 900 ppm, compared to today's 373 and the preindustrial 280-ppm levels. The most likely global temperature increases, relative to 1990, range from 2.0 to 4.5°C for the scenarios, although due to uncertainty in the sensitivity of the climate to CO$_2$, the range is best stated as 1.4 to 5.8°C.

Even with constant carbon dioxide emissions, at current levels or a few percent lower, the carbon dioxide concentration in the atmosphere will continue to grow. Some policymakers have promoted the idea that, through international agreements or allocation schemes, the world should control future CO$_2$ emissions so that the atmospheric level of the gas never exceeds some specific *concentration*. Although there is no consensus about what the most appropriate target is, for our discussion we shall use 550 ppm. This is twice the preindustrial value—in other words, a situation in which human actions have doubled the natural atmospheric carbon dioxide concentration.

One way in which global CO_2 *emissions* could rise and fall with time in order to eventually achieve the 550-ppm concentration target is shown by the curve in Figure 5-8a. In Figure 5-8b we show how the corresponding atmospheric CO_2 *concentration* would change with time for this emission scenario. The scenario was developed assuming that international agreement on CO_2 emissions can be achieved in the relatively near future. Consequently, it assumes modest growth in CO_2 releases until about 2070, at which point a slow decline would set in. The temperature increases—which track the CO_2 concentration curve

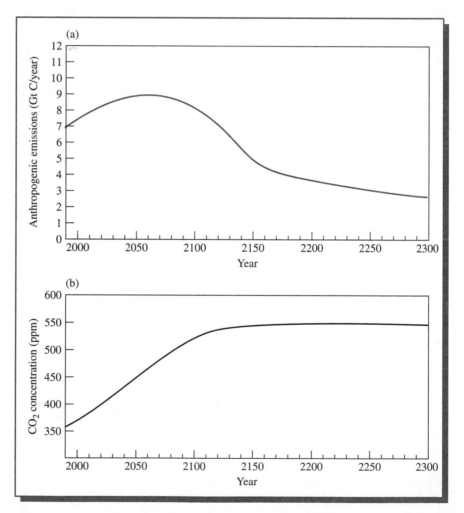

FIGURE 5-8 Approximate (a) annual CO_2 emission rates, (b) resultant atmospheric CO_2 concentrations to meet a 550-ppm stabilization target.

closely—by 2100 would be just under 2° (relative to that for the year 2000). The rise in sea levels would be reduced by about one-third if we embark soon on the scenario to never exceed the 550-ppm concentration of carbon dioxide.

Alternative scenarios to the one shown by the curves in Figures 5-8a and 5-8b, in which effective CO_2 controls are not implemented until several additional decades later, would eventually require sharper decline in emissions and would reach the 550-ppm limit sooner. Such alternative proposals allow more time to further develop replacement technologies, such as the solar energy techniques discussed later in this chapter, before we begin to end our reliance on fossil fuels. Such scenarios require the world to generate more emissions-free power than today's total power consumption by about mid-century, a major challenge to achieve. By the end of the century almost all power would have to be emissions-free. It is *not* possible to defer emission reductions indefinitely if the 550-ppm concentration target is to be achieved.

International Agreements on Greenhouse Gas Emissions

Faced with the prospect that increased CO_2 emissions over the next century could result in a significant increase in global air temperature with its resultant modifications of climate, some national governments and organizations have been debating how future emissions can be minimized while still allowing economies to grow.

The first agreement on greenhouse gas emissions was reached at the Rio Environmental Summit meeting in 1992; each developed country was to ensure that its CO_2 emission rate in 2000 would be no greater than it was in 1990. This target was met, in fact, by very few countries; most are currently emitting at levels well above their targets.

The second agreement was reached in negotiations held at Kyoto, Japan, in 1997. Thirty-nine industrialized nations agreed by 2008–2012 to decrease their collective CO_2-equivalent emissions by 5.2% compared to 1990 levels. The greenhouse gases affected by the Kyoto Accord are carbon dioxide, methane, nitrous oxide, hydrofluorocarbons, perfluorocarbons, and sulfur hexafluoride.

Under the agreement, the United States was due to cut its emissions to 7% less than its 1990 level, Canada and Japan by 6%, and the European Union collectively by 8% (with wide variations for individual countries within this unit). Some countries, such as Australia, were permitted to increase their emissions beyond 1990 levels. Emissions by developing countries were not controlled by the Kyoto Accord, since they were not significant players in emitting greenhouse gases in the past and therefore did not contribute much to current global warming.

As a result of the Kyoto agreement, the annual per capita CO_2 emissions in 2010 in developed countries would have decreased from 3.1 tonnes of carbon in 1997 to 2.8 tonnes, whereas, because of economic development, emissions in developing countries would probably have risen from 0.5 to 0.7 tonnes. The CO_2 concentration in air would have been a little over 1 ppm less than otherwise. However, the United States and Australia subsequently withdrew from the agreement and, indeed, the implementation status of the entire accord is now in doubt. Nevertheless, some U.S. states—mainly in New England—have decided on their own to limit greenhouse gas emissions.

The existing increase, by one-third, of the atmospheric CO_2 level and the temperature increase and climate modification that this probably caused resulted in large part from the industrialization and the increase in the standard of living that developed countries have achieved. Without a significant change in the methods by which energy is produced and stored, the same nations will continue to require about the same rate of CO_2 emissions in the future to maintain their economic growth.

As discussed previously, energy use in developing countries is growing much faster on average than in developed countries; if, as seems likely, these countries, too, rely mainly on fossil fuels, their CO_2 emissions will increase greatly. A growth rate of 4% in any quantity, if compounded, would lead to a doubling of usage in 17 years (see Problem 5-1). Consequently, if developing countries contribute half the new emissions by 2008 (see Figure 5-3) and keep increasing their emissions by 4% compounded per year, they will be collectively responsible for two-thirds of annual emissions by about 2030.

Rather than a procedure in which countries have CO_2 emission targets negotiated at international meetings, schemes have also been discussed that are based on allocations that could be traded between countries on the open market. In a manner similar to the way in which SO_2 emission rights currently are traded in the United States, countries that need to emit more than their collective CO_2 allocations could purchase unused allocations from countries with an excess. A bonus of this scheme is that it provides an incentive to develop and invest in cleaner technologies, since avoiding CO_2 emissions could be cheaper than purchasing additional rights—especially in the future when few nations will have excess emission capacity and the price of emission rights will rise.

The question of how CO_2 allocations can be made fairly to initiate the free-market CO_2 emission trading scheme is a perplexing one. In the simplest scheme, each country would be assigned an allocation based strictly on its (current) population. For example, if it was concluded that the current average annual emission of 1 tonne of carbon as CO_2 per capita could be sustained indefinitely, then this quantity would be allocated to a country for each of its residents. If it was decided to cut back current global emission levels, e.g., by one-quarter, then only 0.75 tonne per capita per year would be allocated, etc.

An immediate consequence of the per capita allocation method would be the annual transfer of substantial funds from all developed countries to developing and undeveloped countries, since, according to the data in Figure 5-6, the former all exceed the 1-tonne average, by factors ranging from two to five. Although this method would provide external funding so that developing countries could establish efficient energy infrastructures, it would likely not prove popular in developed countries since their energy costs would rise.

One alternative allocation scheme is based upon how much energy is required for industrial production by a country and how efficiently it uses energy. Thus a country's carbon dioxide allocation would be directly proportional to its gross national product (GNP). This allocation method rewards compact, energy-efficient developed countries at the expense of those—both developed and developing—that emit more CO_2 per unit of GNP. However, such a scheme would permit continued economic growth by developing countries, since their CO_2 allocations would track their economic growth. The global ratio of allowed carbon dioxide to dollar of GNP would have to decline with time if global emissions are to be controlled, since global GNP rises by several percent per year. Interestingly, the $CO_2/\$$ GNP ratio is much more independent of the level of economic development than is the ratio based on population; e.g., China emits about 1.0 kg of carbon dioxide for each dollar of production, compared to 0.9 kg for the United States, 0.5 kg for Japan, 1.0 kg for Germany, and 0.7 kg for India.

We conclude by commenting on the paradox that faces humanity today concerning the enhancement of the greenhouse effect. On the one hand, there exists the possibility that doubling or quadrupling the CO_2 concentration will have no measurable effect on climate, and that efforts taken to prevent such an increase not only would represent an economic burden for both the developed and the developing worlds, but would perhaps be wasted in the outcome. On the other hand, if the predictions of scientists who model the Earth's climate turn out to be realistic, but we do nothing to prevent further build-up of the gases, both present and future generations will collectively suffer from rapid and perhaps cataclysmic changes to the Earth's climate.

Minimizing Future Emissions of Greenhouse Gases

Carbon Intensity and Carbon Taxes

In devising strategies to minimize the amount of carbon dioxide emitted to the atmosphere in the future, scientists and policymakers take into account the fact that fossil fuels differ in the amount of the gas that they emit per unit amount of energy produced.

To a first approximation, the amount of heat released when a carbon-containing substance burns is directly proportional to the amount of oxygen it consumes. From this principle we can compare the amount of CO_2 released when different carbon-based fuels are combusted, and by this comparison we can decide which fuel is preferable from a global warming perspective. Consider the reactions of coal (mainly carbon), oil (essentially polymers of CH_2), and natural gas (essentially CH_4) with atmospheric oxygen, written so that the amount of carbon dioxide is identical in each case:

$$C + O_2 \longrightarrow CO_2$$

$$CH_2 + 1.5\,O_2 \longrightarrow CO_2 + H_2O$$

$$CH_4 + 2\,O_2 \longrightarrow CO_2 + 2\,H_2O$$

It follows from the stoichiometry of these reactions that, per mole of O_2 consumed and thus approximately per joule of energy produced, natural gas generates less carbon dioxide than does oil, which in turn is superior to coal, in a ratio of $1:1.33:2$. (The actual ratio is computed in Problem 5-3.) Unfortunately, some methane is lost to the atmosphere when natural gas pipelines leak; the greenhouse-enhancing effect of this methane may well override some of the advantage methane has in producing less CO_2 per joule upon combustion compared to oil and especially compared to coal (see Additional Problem 5).

PROBLEM 5-3

Given the thermochemical data below, determine the actual ratio of CO_2 per joule of heat released on the combustion of methane, CH_2, and elemental carbon (graphite) (ΔH_f values in kJ/mol): CH_4, -74.9; $CO_2(g)$, -393.5; $H_2O(l)$, -285.8; C(graphite), 0.0; CH_2, -20.6.

PROBLEM 5-4

The relative amounts of oxygen required to oxidize organic compounds to carbon dioxide and water can be deduced from calculating the *change* in the oxidation number (state) of the carbon atom in going from the fuel molecule to the product. Show that the ratio of oxygen required to combust C, CH_2, and CH_4 stands in the ratio of $2:3:4$ according to such a calculation.

Ton for ton, the potential of a fossil fuel to generate CO_2, thereby causing atmospheric warming, depends on its carbon content, so some policymakers believe that **carbon taxes,** i.e., taxes based on the amount of carbon contained in a fuel rather than on its total mass, should be instituted as a disincentive to use the less desirable fuels and to minimize overall fossil-fuel usage as well. The imposition of a tax on carbon also allows the market price for fuels to reflect their social as well as their economic costs. The tax rate could be set to rise with time in order to encourage the amount of abatement to increase. The possibility of switching to fuels that emit no carbon dioxide in their production or combustion, such as hydrogen gas produced by solar energy, is explored later in this chapter. In fact, the hydrogen-to-carbon ratio of the average global fuel mix has been continuously increasing over the last century and a half, as we moved from economies whose energy sources were dominated by wood (H/C ratio of about 0.1), to coal (1.0 ratio), to oil (about 2.0), and now to natural gas (4.0); this is the same direction as moving to lower CO_2/energy ratios, as implied above.

Sequestration of CO_2

In the future, CO_2 might be removed chemically from the exhaust gas of power plants that burn fossil fuels instead of being released to the atmosphere. The carbon dioxide gas so recovered would then be **sequestered;** i.e., it would be deposited in a location that would prevent its release into the air, although all such schemes are energy-intensive. For example, the CO_2 could be sequestered by burial in the deep oceans, where it would dissolve, or in very deep aquifers under land or under water, or in empty oil and natural gas wells or coal seams.

Dissolving carbon dioxide gas directly in seawater would produce *carbonic acid*, H_2CO_3, a weak acid that would increase the acidity of the ocean in the near vicinity:

$$CO_2(g) + H_2O(aq) \rightleftharpoons H_2CO_3(aq)$$

$$H_2CO_3(aq) \rightleftharpoons H^+ + HCO_3^-$$

The pH of ocean water would be lowered by tenths of a unit by the addition of large amounts of carbon dioxide, although much larger decreases of several pH units would occur near the points of injection.

This scheme of delivering carbon dioxide in massive amounts to the seas, labeled "ocean acidic" in Figure 5-9, has the potential to store many hundreds of gigatonnes of carbon dioxide for many hundreds of years. Even relatively shallow injection of the gas in the ocean, at 200–400 m depth, would produce a satisfactory result, provided that the seafloor there is slanted sufficiently to allow the dense, CO_2-rich water to be transported by gravity to greater depths. The carbon dioxide could be transported by pipeline from the shore to these depths. Simulations show that most of the gas would return to the surface and enter the atmosphere within a few decades if the CO_2-rich water is simply diluted by mixing with surrounding

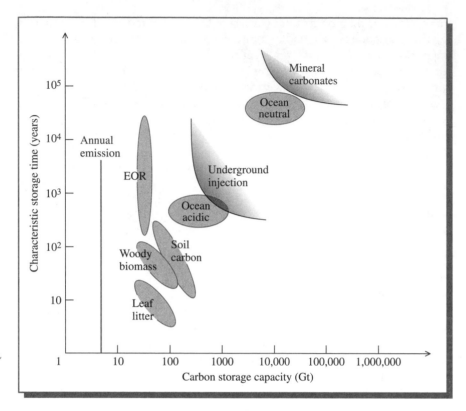

FIGURE 5-9 *Capacities and storage times for various CO_2 sequestration technologies. [Source: Adapted from K. S. Lackner, "A Guide to CO_2 Seques- tration," Science 300 (2003): 1677.]*

water, rather than sinking. Over a period of centuries, excess carbon dioxide would eventually return to the atmosphere, but presumably by that time alternative energy sources would have replaced fossil fuels and the atmos- pheric CO_2 problem would then be less serious.

Burial of carbon dioxide at intermediate depths of about 1000 m initially produces a clathrate, analogous to that involving methane and water. Since the CO_2–water clathrate is denser than pure water, it should sink to the deep ocean. However, in practice, droplets of liquid carbon dioxide are found to simply dissolve in the surrounding water.

Direct disposal to the deep ocean of CO_2 would require a pipeline to penetrate to a depth of 3000–5000 m, producing a pool of liquified carbon dioxide, denser than seawater at this depth. Some, perhaps just the surface, or all of the liquid carbon dioxide would be converted to the solid clathrate. The pool of liquid carbon dioxide would, probably over centuries, dissolve into the surrounding water. Unfortunately, sea life under this pool would be exterminated. There is also some fear that earthquakes or asteroid impact could destabilize the pool, resulting in the release of massive amounts of carbon dioxide gas into the air above.

Near the seafloor, dissolved carbon dioxide could eventually react with the solid **calcium carbonate,** $CaCO_3$, in sediments formed from seashells, etc. to produce soluble **calcium bicarbonate,** $Ca(HCO_3)_2$:

$$CO_2(g) + H_2O(aq) + CaCO_3(s) \longrightarrow Ca(HCO_3)_2(aq)$$

(This reaction is discussed in detail in Chapter 9.) For practical purposes, the CO_2, now chemically trapped in the bicarbonate form, would remain indefinitely in the dissolved state.

In an alternative scheme, labeled "ocean neutral" in Figure 5-9, calcium carbonate or some other suitable substance such as *calcium silicate* (a cheap, abundant mineral) would be reacted with carbon dioxide to transform it to aqueous calcium bicarbonate, which could be drained into ocean depths:

$$2\ CO_2 + H_2O + CaSiO_3 \longrightarrow SiO_2 + Ca(HCO_3)_2(aq)$$

Acidity problems associated with direct carbon dioxide dissolution in seawater are avoided in this way. Huge amounts of limestone or calcium silicate would be required for this form of sequestration, but the potential for CO_2 storage by this method is very great and the storage time of the gas is many thousands of years (see Figure 5-9).

PROBLEM 5-5

Calculate the mass, in tonnes, of calcium carbonate that is required to react with each tonne of carbon dioxide.

There have been suggestions that the CO_2 output from power plants could be pumped deep underground into regions where cracks and pores in common alkaline rocks such as calcium aluminosilicates could react with the gas in microorganism-catalyzed processes to produce calcium carbonate and thereby store the CO_2. Such carbonate minerals are known to be present in deep caves in Hawaii and elsewhere, so the process could well be feasible if the reactions occur quickly enough. Recently Norway has started to pump concentrated CO_2 gas into sandstone rocks located a kilometer below the North Sea; the pores in the rock were left empty by extraction of natural gas from them in the past. The gas could react with the rock and thus be immobilized.

Alternatively, surface rocks containing alkaline silicates could be crushed and then reacted with carbon dioxide to produce insoluble solid carbonates that could simply be buried in the ground. Unfortunately, direct carbonation reactions are slow unless the mineral is heated, a step that is costly in money and energy. In one indirect scheme, magnesium silicate rock is reacted with *hydrochloric acid*, HCl, to produce silicon dioxide and *magnesium chloride*, $MgCl_2$. Reaction of this salt with carbonic acid produces insoluble *magnesium carbonate*, $MgCO_3$, and re-forms the hydrogen chloride which, in principle, can be recycled, e.g.,

$$Mg_2SiO_4 + 4\,HCl \longrightarrow 2\,MgCl_2 + 2\,H_2O + SiO_2$$
$$MgCl_2 + H_2CO_3 \longrightarrow MgCO_3 + 2\,HCl$$

There are still energy costs and CO_2 production associated with such procedures, however.

In the short run, the easiest route to begin sequestration of carbon dioxide is probably to inject it into reservoirs containing crude oil or natural gas. The total capacity for carbon dioxide storage by such *enhanced oil recovery*, labeled EOR in Figure 5-9, is less than 100 Gt, however. This technology is already used to enhance the recovery of oil in some fields, although currently most of the CO_2 is again recovered and reused. In an interesting international project under construction, a 300-km-long pipeline will allow 5000 t/day of CO_2 to flow from a coal gasification plant in North Dakota to an oilfield in Saskatchewan, allowing more oil to be extracted from the latter and sequestering most of the carbon dioxide.

Depleted oil and gas reservoirs could be used to store carbon dioxide. These underground caverns are known to be stable, since they have held their original materials for millions of years. Carbon dioxide storage in coal seams that lie too far underground to be mined may also be feasible. Pumping CO_2 into the coal helps it release adsorbed methane, which then can be pumped to the surface and used. Several hundred gigatonnes of carbon dioxide could be sequestered in coal.

Much larger in volume and capacity than oil and gas reservoirs are *saline aquifers*, porous rocks containing salty water in large formations that lie far underground, well below fresh-water supplies. Carbon dioxide injected into such an aquifer initially remains a compressed gas or a liquid but slowly dissolves in the sometimes very alkaline brine. The brine is contained mainly in small pore spaces in the soil, so the dissolved carbon dioxide would be unlikely to leak from the aquifer. The stability of each aquifer to potential seismic instability and gas leakage would have to be individually assessed before it was used. The Norwegian energy company Statoil has already demonstrated the feasibility of this approach by storing annually about a million tonnes of carbon dioxide (a 9% impurity that must be removed from its crude natural gas) in a saline aquifer that lies 1000 m under the floor of the North Sea. Interestingly, Statoil found it cheaper to sequester CO_2 this way than to pay the $50 per tonne carbon tax the Norwegian government has instituted. The North Sea aquifers are sufficiently large to absorb all European emissions of carbon dioxide for many hundreds of years. Large saline aquifers are found in the

United States just below the lower Great Lakes, in southern Florida, in northeastern Texas, and in northern midwest states.

Another interesting suggestion for sequestering carbon dioxide involves the creation of giant balls of solid carbon dioxide (or dry ice) with insulated storage below $-79°C$ (the sublimation point of the solid CO_2) at the Earth's surface. If a very thick blanket of glass wool insulation is used, the release of gaseous carbon dioxide from the 400-m-diameter balls would not occur for many centuries.

The energy input required for the CO_2-concentrating phase of most of these so-called **carbon sequestration,** schemes would represent a substantial fraction, from one-third to one-half, of the output of the power plant to which it is connected. To be economically and technologically feasible, the dilute carbon dioxide in the emission gases from a combustion process must first be captured and concentrated. This is traditionally done by passing the air containing the gas through a solvent consisting of an amine, R_2NH, which converts it into an anion:

$$R_2NH + CO_2 \longrightarrow R_2NCO_2^- + H^+$$

Once the amine is saturated with the gas, heat is used to reverse the reaction and produce a concentrated stream of carbon dioxide. Scientists and engineers have tried to replace this energy-intensive process of stripping and concentrating the carbon dioxide by using membranes or other solvents, but no cheap method has yet been discovered.

Of course, only a fraction of the CO_2 produced by combustion comes from large, centralized sources such as power plants. It is not clear how carbon dioxide could be extracted from automobile exhaust and heating systems, which currently account for more than two-thirds of CO_2 emissions. A possible solution to this dilemma is to first transfer the fuel value of the coal or oil or natural gas to hydrogen, which would then be used to power the vehicles or furnaces, retaining the carbon dioxide at the central facility.

Removing CO_2 from the Atmosphere

A potential technique for extracting some of the carbon dioxide that is already dispersed in the atmosphere and depositing it in ocean depths is the *iron fertilization* proposal. Experiments indicate that large portions of the seas, especially the tropical Pacific and the Southern Oceans, lack plankton because they are very iron deficient. Artificially adding iron to these areas would result in massive blooms of plankton, some of which, in the Southern Ocean at least, would quickly descend into the deep oceans, thereby locking away the carbon dioxide that had been used in photosynthetic activity. Experiments are under way to test the feasibility of this approach. Results obtained so far indicate that, while fertilization by iron does occur, rather little of the phytoplankton sinks to the deep ocean, so sequestration by this approach would be limited. In addition, some scientists have pointed out that the decomposition of the phytoplankton consumes oxygen and encourages bacteria that produce methane and nitrous oxide, thereby increasing the concentration of these greenhouse gases in air. Other side effects of fertilization could produce additional negative environmental effects.

Carbon dioxide can also be removed from the atmosphere by growing plants specifically for this purpose. Some utility companies and some countries have proposed a scheme by which they are given credit to offset some of their CO_2 emissions by planting forests that would absorb and temporarily sequester carbon dioxide as they grew. In another proposal, carbon dioxide from a power plant would be used to grow vast amounts of algae, which then could be used as fuel for combustion. However, the crediting of carbon dioxide sequestering by growth of biomass is contro-versial. For example, the release of CO_2 from soil into the air that occurs when ground is cleared for growing trees can exceed the total carbon diox-ide absorbed by the new trees for a decade or more. Also, the carbon stored in trees would be released back into the atmosphere if the wood burned or rotted.

Reducing CO_2 Emissions by Improving Energy Efficiency

Some writers have noted that much of the current energy expended in both the industrial and domestic spheres could be saved by adopting the most *effi-cient* technologies for every purpose. For example, the use of low-wattage compact fluorescent light bulbs instead of incandescent bulbs would signifi-cantly reduce the amount of electrical energy used for lighting in homes; the *payback period*—until the much higher capital cost of these bulbs is more than met by savings in electrical costs—is a few years. Similarly, automobiles could be made much more energy efficient and thus use less gasoline to travel a given distance.

However, improving energy efficiency would *not* necessarily lead to a reduction in the demand for energy and a reduction in carbon dioxide emissions. The reason is that if energy-consuming equipment is made more efficient, the monetary cost for performing a given task drops, and there follows a natural tendency to use the equipment more, since it is so cheap to operate. For example, if you buy a very energy-efficient car, you will be able to afford to take more trips in it, since each one would be cheaper than with a "gas guzzler." Thus some policymakers believe that because of this rebound effect, energy savings and CO_2 reductions from efficiency would not be achieved in the long run by making energy-consuming devices more efficient. Increased efficiency would have to be accompanied by price increases on the fuel, presumably in the form of taxes, for it to reduce overall consumption.

Reducing Methane Emissions

Although our focus so far has been on reducing emissions of carbon dioxide to the atmosphere, global warming can also be combatted by decreasing the amount of methane that is released. The major opportunities for methane reduction are

• better maintenance of natural gas pipelines to reduce their leakage, a practice already under way in Russia;

• capture and combustion of the methane released by landfills, underground coal mines, and oil production; and

• changes in the techniques of rice production, by draining the field a few days before the plants flower, the point at which maximum emissions begin.

Review Questions

1. List some of the consequences, including those affecting human health, that may occur as a result of global warming in the future. Why might soil in some areas be too dry for agriculture, even though more rain falls on it?

2. Explain why sea levels are expected to rise as global air temperatures increase.

3. Define the term *commercial energy*. Upon what factors does the magnitude of its use in a country depend?

4. What is the equation relating exponential growth to the annual increase in a quantity?

5. What are the ultimate origins of coal, oil, and natural gas? Which fuel is in greatest reserve abundance?

6. What is the *Kyoto Accord*? What gas emissions are limited under it? Would the Kyoto agreement have halted global warming?

7. Describe the scheme whereby a nation's allocation of carbon dioxide emissions could be traded on a market. Describe two schemes by which initial allocations could be made.

8. What is a *carbon tax*, and what are the arguments in favor of it? Why do you think many people oppose it?

9. Describe several methods by which carbon dioxide emissions from power plants could be sequestered.

Green Chemistry Questions

1. The formation of polylactic acid (PLA) from biomass developed by Cargill Dow won a Presidential Green Chemistry Challenge Award.

(a) Which of the three focus areas (see the Introduction to Green Chemistry) for these awards does this award best fit into?

(b) List three of the twelve principles of green chemistry (see the Introduction to Green Chemistry) that are addressed by the green chemistry developed by Cargill Dow.

2. What are the environmental advantages of using PLA in place of petroleum-based polymers?

3. Why does the use of biodegradable polymers (to replace petroleum-based polymers) not offer complete solutions to energy and environmental problems?

Additional Problems

1. Write down a list of reasons that you think proponents of carbon dioxide allocations based strictly on a country's population would advance in support of their position. What objections can be raised to their position? Repeat this exercise for an allocation scheme based on GNP.

2. A sign was spotted outside a farmers' market with the slogan: "Help Stop Climate Change: Buy Local Produce." Explain the rationale behind this bit of advertising, and discuss whether taking the advice of this sign could indeed "help stop climate change."

3. The United States has pushed for the counting of the CO_2 that is naturally taken up by sinks in a country as a "credit" against the reductions in CO_2 emissions currently dictated by the Kyoto Accord. What would be the major sink that the United States would like to use to as CO_2 emission reduction credit? What are the arguments against the use of these sinks as credit toward CO_2 emission reductions?

4. Given that the density of dry ice (solid CO_2) is 1.56 g/cm^3, what diameter of dry-ice ball (in meters) would be produced from the 5 metric tonnes of CO_2 produced on average by each person in industrialized countries each year?

5. The replacement by natural gas of oil or coal used in power plants has been proposed as a mechanism by which CO_2 emissions can be reduced. However, much of the advantage of switching to gas would be canceled by methane escaping into the atmosphere from gas pipelines since it is 23 times as effective, on a molecule-per-molecule basis, in causing global warming as carbon dioxide is. Calculate the maximum percentage of CH_4 that can escape if replacement of oil by natural gas is to reduce the rate of global warming. [Hint: Recall that the heat energy outputs of the fuels are proportional to the amount of O_2 they consume.]

Further Reading

1. J. T. Houghton et al., *Climate Change 2001* (Cambridge: Cambridge University Press, 2001).

2. E. Claussen, ed., *Climate Change: Science, Strategies, & Solutions* (Arlington, VA: Pew Center on Global Climate Change, 2001).

3. H. Inhaber and H. Saunders, "Road to Nowhere," *The Sciences* (November/December 1994): 20–25. Argues that energy conservation leads to increased consumption.

4. T. R. Karl et al., "The Coming Climate," *Scientific American* (May 1997): 78–83.

5. F. Muller, "Mitigating Climate Change: The Case for Energy Taxes," *Environment* (March 1996): 13–20, 36–43.

6. P. Reiter, "Climate Change and Mosquito-Borne Disease," *Environmental Health Perspectives* 109 (supplement 1) (2001): 141–161.

7. M. I. Hoffert et al., "Advanced Technology Paths to Global Climate Stability: Energy for a Greenhouse Planet," *Science* 298 (2003): 981.

8. J. L. Sarmiento and N. Gruber, "Sinks for Anthropogenic Carbon," *Physics Today* 55 (2002): 30

9. K. Caldeira et al., "Climate Sensitivity Uncertainty and the Need for Energy Without CO_2 Emission," *Science* 299 (2003): 2052.

10. J. Haley, ed., *Global Warming: Opposing Viewpoints* (San Diego, CA: Greenhaven Press, 2002).

Websites of Interest

Log on to www.whfreeman.com/envchem3e/ and click on Chapter 5.

CHAPTER **6**

Renewable Energy, Alternative Fuels, and the Hydrogen Economy

In Chapters 2–5 we have seen how the atmosphere has been affected by emissions into it of pollutant gases such as sulfur and nitrogen oxides and greenhouse gases such as carbon dioxide and methane. The emphasis in this chapter is on alternative technologies under development that could reduce the anthropogenic production of such gases in the future while still allowing economic growth to occur. We begin by considering some possible solutions to the further buildup of atmospheric CO_2 by a partial switchover from fossil fuels to renewable energy, especially solar power. We then make an extensive survey of the various alternative fuels, including hydrogen, that may be more "greenhouse friendly" than those used at present and that also would be effective in reducing air pollution. The possibility of generating energy by nuclear power is discussed in Chapter 13.

A wind farm in Scotland. (Image State)

Renewable Energy

The Sun sends enough energy to the Earth to supply all of our conceivable energy requirements, about 10,000 times more than we use now and will in the future, if only we could trap it efficiently. In addition to being plentiful and reliable, it is **renewable energy**—energy that will not run out *and* whose capture and use do not result in the direct emission of greenhouse gases or other pollutants.

The world currently uses about 12 **terawatts** (TW, 10^{12} watts) of power, about 85% of which is generated by the burning of fossil fuels. Since 1 watt is 1 joule per second, and since there are 3.2×10^7 seconds in a year, our annual power consumption is about 3.8×10^{20} J, 380 EJ. Given that an average light bulb is rated at 60 W, we are using the equivalent of 200 billion light bulbs at a time, an average of about 35 per person, nonstop. Of course, this figure is an average for people in developed and developing countries; if we redo the calculation for North Americans, we are using about 200 60-W light bulbs for every man, woman, and child!

Hydroelectric Power

In fact, humans already harness considerable amounts of indirect solar energy in the form of **hydroelectric power.** In the hydrological cycle, the Sun's energy evaporates water from oceans, lakes, rivers, and the soil and transports the H_2O molecules upward in the atmosphere via winds. After the water molecules have condensed to raindrops, they still possess considerable potential energy owing to their elevation, only part of which is dissipated if they fall onto land or a water body that lies above sea level. We can harness some of this remaining potential energy by forcing the downward-flowing water to turn turbines and thereby generate electricity. Although there are small hydroelectric installations that use the flow of a river current, most large-scale facilities use dams and waterfalls, where the water pressure—and hence the power yield—is much greater. If all sites were exploited, the total amount of energy that could be obtained from hydroelectric sources is about 100 EJ/year; the current usage is about 20% of this total. Recall from Chapter 4, however, that the flooding of vegetation on land that is used to provide water storage for new dams can produce enough greenhouse gas emissions—mainly methane—to cancel the CO_2 emission savings from the use of hydroelectric power rather than fossil-fuel combustion. In addition, mercury from the flooded vegetation and soil leads to the poisoning of fish in the waterway (see Chapter 11). Other environmental problems associated with hydroelectric dams include silt build-up behind the turbines, eutrophication of the flooded waterways, and devastation to fish populations, such as salmon, that find their migratory routes blocked.

Wind Energy

An even larger amount of indirect solar energy, about 300 EJ annually, is potentially available as **wind power,** although only 0.05% of it currently is being tapped. Winds are air flows that result from the tendency of air masses that have undergone different amounts of heating by sunlight, and therefore have unequal pressures, to equalize those pressures. As a consequence of the local terrain, some geographical regions experience almost constant windy conditions.

The force of the wind can be exploited to do useful work and to generate electrical energy in the same way that the force of flowing water is used. For example, the strong, sustained winds in central North America were exploited by windmills to pump water and later to generate small amounts of electricity on individual farms until the middle of the twentieth century. Of course, windmills have been in use in Europe, especially in Holland, for centuries.

In recent decades, the large-scale generation of electricity by arrays of huge, high-tech windmills gathered in "wind farms" has become feasible. Wind power is currently the world's fastest-growing source of energy and could provide 10% of its electricity by 2020—the fraction and date of the European target. A 2003 EU report on energy predicts that 4% of the world's electricity will be produced by wind power by 2030. In 2001 alone, global wind-power capacity grew by one-third, and in the 1995–2001 period rose fivefold to 25,000 megawatts (MW) (see green curve, Figure 6-1).

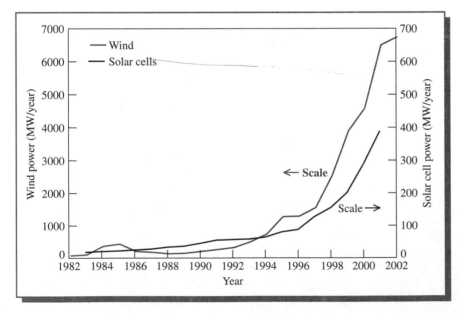

FIGURE 6-1 Growth in production of wind energy and solar-cell energy. [Source: L. R. Brown et al., *Vital Signs 2002* and *Vital Signs 2003* (New York: W.W. Norton, 2002, 2003).]

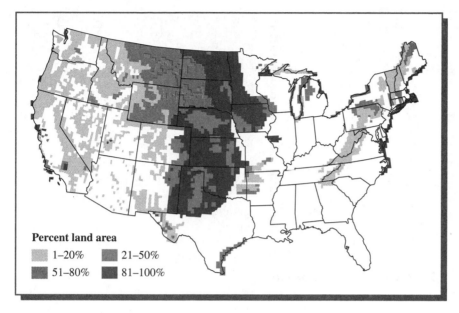

FIGURE 6-2 Percentage of land area estimated to have Class 3 or higher wind power in the contiguous United States. [Source: "Wind Energy Resource Atlas of the United States," Chapter 2: http://rredc.nrel.gov/wind/pubs/atlas/maps/chap2/2-10m.html.]

Percent land area
- 1–20%
- 21–50%
- 51–80%
- 81–100%

Wind power could be expanded to provide eventually up to perhaps one-fifth of the world's electricity. Currently, however, it accounts for less than 1%, a figure that also applies to the United States and Canada. The most elaborate wind farm installations are in Denmark and California; they will supply 20% of Denmark's electrical energy by 2005. The world's largest wind farm covers 130 km^2 in Oregon and Washington and will eventually involve 460 turbines. However, 90% of the U.S. potential for wind power lies in 12 states in the Midwest, ranging from North Dakota to Texas (see Figure 6-2), and the demand for electricity is centered far from most of these areas. Indeed, the United States has enough potential wind power to supply all its electricity now and in the foreseeable future.

Large growth in wind-power installations has occurred in Germany (currently the world leader in wind power), Spain, Denmark, the United States, and India, although there is the potential for this technology to be useful in many parts of the world. Indeed, the cost of generating electricity using modern windmill technology and feeding it into existing power grids is already competitive with conventional energy sources. As of the mid-1990s, most of the interest in constructing new facilities was in Europe and Asia, with relatively little activity planned for the Western Hemisphere. Europe has almost 90% of the world's current wind-energy capacity. Germany derives more than 3% of its electricity in this way and plans to increase that to 25% by 2025 as it phases out its nuclear power industry.

There is often public resistance to siting wind turbines in populated areas due to their visual unsightliness. For this reason, placing them on agricultural

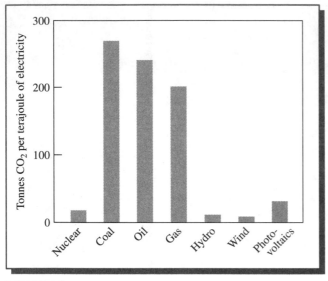

FIGURE 6-3 CO_2 emissions associated with different energy sources. [Source: "The Power to Choose," *New Scientist,* (6 September 1997): 18.]

land or even offshore is becoming popular. A study indicated that U.S. locations in New England, on Lake Erie, and off the coast of mid-Atlantic states could alone generate up to 20% of the U.S. electricity supply. In Europe, where offshore locations are popular, most are anchored in water 8–10 m deep. West-coast waters off North America are too deep for such placements, and those in the southeastern United States are too prone to hurricanes.

In terms of **energy payback**—the amount of time required to generate the energy used in constructing the machinery—that for wind is only 3–4 months. Carbon dioxide emissions from wind power are currently the smallest for any power source (see Figure 6-3). The most efficient and largest commercial wind turbines currently are about 2-MW units, three times the size of the models of the mid-1990s. Behemoths with huge 120-m blades are under development; they will deliver 5 MW.

Of all the forms of renewable energy, wind power is the most economical. In 2000 in the United States its cost per kilowatt hour (kWh) was about 5 cents (4 cents with federal subsidy), about the same as that from new coal plants, almost as low as that generated by natural gas, and less than a tenth of its cost 20 years earlier. As mentioned before, however, there would also be significant transmission costs if wind energy were to be expanded in the midwestern states. If the world switches eventually to a hydrogen economy, hydrogen generation powered by wind in this area could generate much of the U.S. supply. Currently the electrolyzers required for the process are expensive.

Biomass

The **biomass** produced by the worldwide process of photosynthesis constitutes a form of solar energy. The annual amount of energy currently produced from this source is about 55 EJ; a much larger amount is potentially available. The use of wood, crop residues, and dung (dried excrement from plant-consuming animals) has been a traditional source of energy in undeveloped countries, but its domestic and small-scale use is very polluting to the air and inefficient. Small-scale biomass burning is generally phased out in favor of commercial energy such as fossil fuels and electricity as a country's economy develops. Nevertheless, biomass was second only to hydroelectric power in the production of renewable energy in the United States in the late twentieth century.

Recently, technology has been developed to use biomass in large-scale installations that do not pollute the air. For example, wood-chip waste can be burned to produce steam. Alternatively, wood can be gasified, or digested by bacteria, and converted into alcohol fuels (see the Alternative Fuels section later in this chapter). Fast-growing trees on plantations could be used for this purpose, using land not suitable or needed for agriculture. Currently, crops such as corn and sugarcane are grown to produce ethanol for fuel, but often these facilities consume so much fossil fuel in their operation that little is saved overall in CO_2 emissions (also discussed later in the chapter).

Overall, the power density of photosynthesis (about 0.6 W/m^2) is too low for it to supply most of the world's energy needs. The density is low because the efficiency of conversion of sunlight to chemical energy in photosynthesis is very low, no more than 1–2%, even in the most productive areas. At today's consumption levels, the amount of land required to supply our energy needs by biomass equals that of all agricultural land currently developed, more than 10% of Earth's land surface.

Geothermal Energy

Geothermal power, though not solar-based, is another useful form of renewable energy. This heat energy emanates from beneath the Earth's surface, resulting from radioactive decay of elements there and from conduction from the molten core of the Earth. Geothermal energy occurs in the form of steam and hot water. About 8 EJ is available annually from this source. Currently, the United States (in California and Hawaii especially) and a number of developing countries such as the Philippines, Mexico, and Indonesia are the leading users of geothermal energy; in most of these cases, the heat is used to generate electricity. However, the use of geothermal energy for heating is widespread—58 countries worldwide employ it to some extent.

In the past, geothermal energy has been exploited only where unusual geological features result in the existence of steam or hot water close to the Earth's surface. A project under way in Achen, Germany, is attempting to overcome this limitation by sinking a borehole 2.5 km (1.1 miles) into the Earth's crust, where rock temperatures are about 80°C. Cold water will flow down the outer component of the pipe, being warmed as it descends, and then pumped back through an inner pipe to the surface where it will be used.

Per capita, Iceland leads the world in geothermal energy. As a result of the mid-Atlantic meeting of two tectonic plates deep under the island, geothermal energy is plentiful. For some years, hot water and heat in the major city, Reykjavik, has been supplied by hot water tapped from sources a few hundred meters below the ground and distributed by pipelines. Deeper geothermal reservoirs, from which steam can also be obtained, are tapped to

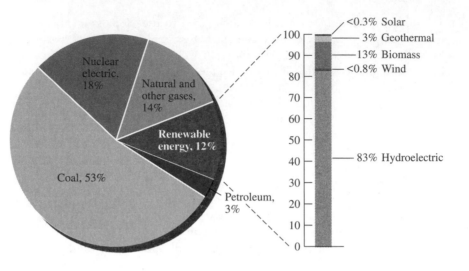

FIGURE 6-4 Sources of U.S. electric power (1997). [Source: "Bright Future—or Brief Flame—for Renewable Energy," *Science* 285 (1999): 678.]

generate some of the electricity. Eventually the plan is to tap more reservoirs and use the excess electricity to produce hydrogen. Iceland hopes to become the world's first economy based on hydrogen and to have its economy totally free of fossil fuels by about 2030.

As indicated in Figure 6-4, geothermal energy supplied more power than solar cells and wind power together in the United States in the late 1990s. Geothermal energy could potentially contribute up to 5% of North America's energy needs. One drawback to geothermal energy is the large quantity of hydrogen sulfide gas that it often releases when the hot fluid is tapped from below the Earth's surface, especially from deep sites.

Wave and Tidal Power

Wave power and **tidal power** can be obtained in many coastal regions of the world and is competitive economically in niche markets. It is estimated that about 20 EJ of power is potentially recoverable from waves and tides.

The source of the energy of tides is the gravitational influence of the Sun and Moon on the water's mass. In some locations, coastal currents generated by tides can be exploited to turn submerged turbines mounted on pipes that fit into holes drilled in the seafloor. Because water is so much denser than air, slow currents—about 10 km/hour are best—driving such "submerged windmills" efficiently generate electricity.

Tides cause large masses of water to be lifted and then lowered twice a day. If the tides in a coastal basin are generally high, a gate that can be opened or closed can be built across the basin. When the tide is coming in, the gate is left open so the water behind it rises. At high tide, the gate is closed. The dammed water leaving the basin turns a turbine, generating electricity.

Three tidal power plants are currently in operation, located in France, Nova Scotia, and Russia. These installations had high capital costs and can operate only twice daily. Although the energy produced is renewable and pollution-free, sedimentation occurs behind the dam gates, and tidal mudflats are often destroyed as a result of the operation.

Wave power at the sea's surface can be exploited. The machines based on an *oscillating water column* consist of a chamber that contains trapped air located just above the water surface. Wave power is generated by using the up-and-down motion of water that results from waves, which are caused by winds and thus are an indirect form of solar energy. The rising wave compresses air trapped in the chamber. The high-pressure air is then released through a valve, turning a turbine to produce electricity. As the wave recedes, air rushes back in through another valve, also spinning the turbine. Currently there are thousands of oceanic navigation buoys whose 60-W light bulbs are powered by this mechanism. Large-scale wave-power facilities are still in the future.

Types of Direct Solar Energy

The direct absorption of energy from sunlight and its subsequent conversion to more useful forms of energy such as electricity can occur by two mechanisms:

Thermal conversion Sunlight (especially its infrared component, which accounts for half its energy content) is captured as heat energy by some absorbing material. (An everday example of such a material is a shiny metal surface, which we know from experience becomes very hot when left in sunlight.) Solar energy is an excellent source of heat at temperatures near or below the boiling point of water, a category that accounts for up to half of total energy usage.

Photoconversion The absorption of photons associated with the ultraviolet, visible, and near-infrared components of sunlight brings about the excitation to higher energy levels of electrons in the absorbing material. The excitation subsequently causes a physical or chemical change (rather than a simple degradation to heat).

An example of *passive* solar technology—systems that use no continuous active intervention or additional energy source to operate them—is the use of solar box cookers in developing countries. In temperate climates, the design of buildings to absorb and retain (by insulation) a maximum fraction of the solar energy that falls on them in winter is another example.

Solar water heaters are used extensively in Australia, Israel, the southern United States, and other hot areas that receive lots of sunshine. They represent the biggest use of *active* solar technologies, which are defined as those that employ an additional energy source to operate them. Solar collectors located on the rooftops of private homes and apartment buildings, as well as some commercial establishments such as car washes, contain water that is circulated around a closed system by an electrically driven pump. Sunlight is absorbed by a black flat-plate collector, which transfers the heat to the water that flows over it and which is bounded on the outside by glass or a plastic window. The hot water is pumped to an insulated storage tank until it is required for bathing or laundry purposes or needed to supplement swimming pool water to heat it up.

In more elaborate installations, the hot water is passed through a **heat exchanger,** which is a system of pipes over which air is passed and thereby heated by thermal transfer. The hot air can be used immediately in winter to heat the rooms of the building. If not needed immediately, the heat can be stored in other materials such as rocks. Usually a backup system, in which water can be heated electrically or by burning fossil fuel, is incorporated to provide heat on cloudy days or in high-demand situations.

Using Thermal Conversion to Produce Electricity

By focusing the sunlight reflected by mirrors onto a receiver that contains a solid or a fluid, very high temperatures can be achieved. The hot fluid can be used to generate electricity by turning turbines.

As discussed in the next section, the fraction of thermal energy that can be extracted and converted to electricity from a mass of hot fluid at an absolute temperature T_h is limited by the *second law of thermodynamics* to be no greater than $(T_h - T_c)/T_h$, where T_c is the final absolute temperature of the cooling water. Consequently, it is advantageous to use a gas that has been heated to the highest possible temperature to maximize the amount of energy that is transformed to electricity rather than just degraded to waste heat. Indeed, temperatures of 1500°C have been achieved in steam heated by focusing sunlight. Generally, power plants need gas heated to 1200–1350°C at a pressure of 10–30 atm to operate. Simply focusing sunlight on tubes of air cannot achieve more than 700°C and 1 atm pressure. In one promising design, sunlight is focused by mirrors onto ceramic pins, which absorb the solar energy and heat up to 1800°C. Because they have a large surface area, the pins efficiently transfer the heat to air that flows around them. The hot, pressurized air (rather than steam) is then used to turn turbines and produce electricity.

The **solar thermal electricity** that results from power plants of this type may become competitive in price with conventional sources. This is particularly true if the waste heat, e.g., steam near the boiling point of water, can also be used for some purpose. This technique of using the waste heat from a heat-to-electricity conversion for a constructive purpose

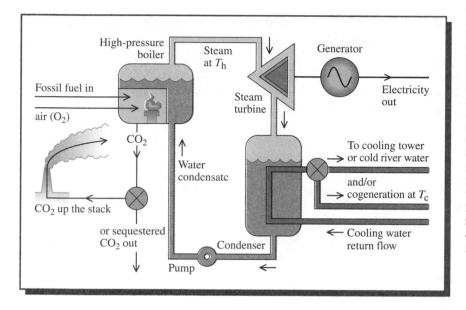

FIGURE 6-5 The generation of electricity from a steam turbine cycle. In solar thermal electricity generation, the water is heated by the Sun's rays rather than by burning a fossil fuel. T_h and T_c refer to steam and water temperatures, as discussed in the text. [Source: Modified from M. I. Hoffert et al., "Advanced Technology Paths to Global Climate Stability: Energy for a Greenhouse Planet," *Science* 298 (2002): 981.]

is called the **cogeneration** of energy. (It is a common feature of new power plants fueled by natural gas.) Unfortunately, power plants based on steam require large amounts of cooling water to condense the steam back into the liquid state as part of the system's cycle (see Figure 6-5), and in many areas that have abundant land and sunlight there is little water available for this purpose. Scientists have also pointed out that if the absorption of sunlight by such systems occurred on a massive scale, the Earth's albedo would be altered, with consequent effects on the climate that are difficult to predict.

Another way to use the very high temperature heat energy is to drive a thermochemical process such as the reduction of a metal oxide to the free metal (and oxygen gas). The metal could then be used to generate electricity in batteries or to react with water to form hydrogen fuel. In either case, the product is the metal oxide, which can be recycled for reuse. An example under development is the dissociation of zinc oxide, ZnO, into metallic zinc and oxygen at temperatures above 1700°C. Alternatively, the heat could be used to produce a combination of carbon monoxide and hydrogen from carbon dioxide and methane.

$$CO_2 + CH_4 \longrightarrow 2\,H_2 + 2\,CO \qquad \Delta H = +248 \text{ kJ/mol}$$

Reversal of this highly endothermic reaction produces heat that can be used to generate electricity, etc. without the net emission of greenhouse gases, since the methane and carbon dioxide products are collected and reused.

Limitations on the Conversion of Energy: The Second Law of Thermodynamics

In all processes that convert high-temperature heat into electricity, a portion of the original heat energy is inevitably lost as waste heat at a lower temperature. This loss is partially unavoidable as a consequence of the **second law of thermodynamics,** and applies to the production of solar thermal electricity as well as to other energy conversion processes.

According to the second law, **entropy** (or disorder) *must increase*—or at the least be unchanged—*when one type of energy is converted into another.* The law tells us that for any body at absolute temperature T that possesses an amount q of heat, the entropy S is a positive quantity given by the formula

$$S = q/T$$

Since high-quality (low-disorder) energy such as electricity has essentially zero entropy, clearly one cannot convert 100% of the heat into electricity, since the change ΔS in entropy associated with the conversion would be negative. However, if *some* of the initial heat energy q_h at the initial high temperature T_h is degraded to a smaller quantity q_c at a lower temperature T_c, then the entropy *change* ΔS for the process could be positive or zero:

$$\Delta S = \text{entropy of energy after conversion} - \text{entropy of energy before conversion}$$
$$= q_c/T_c + 0 - q_h/T_h$$

For the most complete conversion to electricity possible, $\Delta S = 0$; for this situation the equation can be rearranged to give the new relationship

$$q_c/T_c = q_h/T_h \qquad \text{or} \qquad q_c = q_h \, T_c/T_h$$

The amount of heat converted to electricity is $q_h - q_c$, which upon substitution for q_c is equal to

$$\text{heat converted} = q_h - q_h \, T_c/T_h$$
$$= q_h \, (1 - T_c/T_h)$$
$$= q_h \, (T_h - T_c)/T_h$$

Thus the maximum fraction of the original heat that can be converted to electricity is

$$\text{heat converted/initial heat} = (T_h - T_c)/T_h$$

In other words, the *maximum* yield of electricity increases as the *difference* in temperatures between the original heat source and the waste heat increases. Thus if the sunlight can be converted to heat at 1500°C (1783 K), and if the temperature of the waste heat could be held to 27°C (300 K), the fraction of energy that could be converted to electricity would be

$$(1783 - 300)/1783 = 0.83$$

In fact, the efficiencies calculated by the formula are somewhat overestimated when more than one physical phase is involved. For example, associated with the conversion cycle in traditional power plants is a step that condenses steam back into liquid water (Figure 6-5), a process in which entropy is decreased. Consequently, additional energy beyond that calculated must be degraded to waste heat to compensate.

PROBLEM 6-1

What is the maximum percentage of heat at 900°C that could be converted into electricity if the waste heat was produced as steam at 100°C?

PROBLEM 6-2

Electricity could be obtained by exploiting the thermal gradient between the surface and the deep waters of the ocean. The maximum gradient, about 20°, is achieved in tropical waters. What is the maximum percentage of the energy associated with this gradient that could be converted to electricity if the surface (cooling) water temperature is 25°C?

PROBLEM 6-3

In order to reach a conversion efficiency of 50%, to what minimum Celsius temperature must a heat source be raised if the waste heat has a temperature of 57°C?

Solar Cells

Electricity can be produced directly from solar energy by the photoconversion mechanism. This application exploits the **photovoltaic effect,** which is the creation of separated positive and negative charges in a material as a result of the excitation by light of an electron within the solid from its normal energy level to a higher, excited state. Both the excited electron and the location of the site of positive charge (the "hole") are mobile within the solid, so an electrical current could be made to flow in the material. The hole "moves" by means of the transfer of a *bonding* electron from an atom adjacent to the initial hole to the atom on which the hole is now located, thereby moving the position of the positive charge. Successive bonding-electron transfers of this type allow the hole to continue moving.

The material used for photovoltaic or solar cells is a semiconductor, which is a solid that has a conducting behavior intermediate between that of a metal (freely conducting) and an insulator (nonconducting). In semiconductors the bonds linking the atoms are relatively weak, so the separation in energy between the bonding and antibonding levels is relatively small (compared to that of an insulator). Consequently, the energy required to excite an electron from the least stable of the filled, bonding levels to the most stable of the empty, antibonding levels is small but finite. The most common semiconductor used in solar cells is elemental **silicon,** for which this *band gap* separating the energy levels is 124 kJ/mol, which corresponds to infrared light.

Silicon's light absorption ability extends from the band-gap energy of 124 kJ/mol through to energies associated with the visible region, so most of the photons of sunlight are absorbed. However, all the photon energy in excess of the 124 kJ/mol band gap is wasted by being converted into heat rather than promoting current flow. When this loss of energy is combined with that wasted by recombination of electrons and holes even in the purest single crystal silicon, only 28% of sunlight's energy is converted to electricity. Amorphous silicon has an efficiency of only slightly more than half this value—mainly because electron–hole combination occurs more readily—but it is now used extensively because it is so much less expensive to manufacture and can be produced in thin films.

Each solar cell provides only a tiny electrical current, so to generate electricity in useful quantities, many are joined together in a *solar array.* One problem with the electricity generated using solar cells is that it is *direct current* (dc) rather than the *alternating current* (ac) that is used in power grids and by most equipment and appliances. The dc electricity can be converted to ac, although with the loss of some power (as waste heat).

The cost of producing the solar cells and the problem of storing the electricity for use at night and on cloudy days is the greatest barrier to their increased use. As with other applications of solar energy, the capital cost in creating the infrastructure required to capture and use the "free" energy of the Sun is substantial. The cost of the encapsulation of cells, the wiring, and the construction of supporting structures is relatively high since each cell is inefficient; collectively this adds about as much to the total cost of a power system as the cost of the cells themselves. Currently, crystalline solar cells cost about $4/W to manufacture, so the system itself costs about $8/W; since the average home can be supplied by a 4-kW system, the cost of using cells to completely power a house is about $32,000. If solar cells can be made 20% efficient, this power demand could be met by about 20–25 m^2 of solar panels, i.e., an area about 5 m × 5 m.

Although the solar cells do not generate any carbon dioxide during their operation, their manufacture does consume significant amounts of energy and therefore causes substantial CO_2 emissions. Indeed, according to Figure 6-3, photovoltaics are the most CO_2-intensive of the various renewable energy forms. The energy and carbon dioxide payback periods for solar cells and their infrastructure currently are about three years but are expected to fall to one to two years when manufacturing techniques improve further. Once the cells are manufactured, however, their use to generate power in a home saves about as much carbon dioxide per year as is emitted from the family car. The Japanese government subsidized solar cells for tens of thousands of rooftops and, indeed, Japan now is the world's largest producer of such cells.

The cost of manufacturing solar cells has continued to fall with time, but electricity generated in this way is still not close to being competitive with conventional power generation methods. Currently, 90% of cells are made from crystalline silicon and the remaining 10% from thin-film amorphous silicon. To date, the silicon used for solar cells has been castoff or excess silicon from the semiconductor electronics industry. However, this supply will rapidly become inadequate if and when the computer industry revives from its recession.

Photovoltaic power may become attractive in hot, sunny locations such as the southwestern United States, where the peak power demand, driven by the need for air conditioning, coincides in time (summer afternoons) with the peak solar energy availability. Already, solar-cell power (plus storage) is cheaper than extending power grid lines a kilometer or more away from an existing network into a remote region, and it is competitive in cost with the use of diesel generators for this purpose. Portions of this textbook were written at a seaside location that is within sight of an offshore lighthouse powered by solar cells. The technology is also commonly used in water pumps, remote roadside telephones and signs, and satellites.

The use of solar cells in developing countries, most of which have sunshine in abundance, could obviate the need for the creation of power grids to carry electricity over long distances from source to user and represents the greatest potential market for expansion of photovoltaic power. Indeed, the majority of solar cells made in the United States are exported. Solar-cell electricity is already used to power water pumps, lights, refrigerators, and TVs in some developing countries. A mid-1990s survey found that 45% of new solar cells were used to electrify homes, villages, and water pumps; 36% were used for communications and other industrial applications, and 14% were used for the generation of grid-bound electricity. However, by 2000, more than 50% of new solar cells were connected to grid systems.

Like wind power, world production of solar cells for power production has increased sharply since the mid-1990s (see Figure 6-1; note the difference in scales for the two types of power). Although almost 400 mW of solar-cell capacity was shipped in 2001, it represents less than 10% of the wind-turbine power shipped in that year, a situation that has been true each year since the mid-1990s.

Could solar cells be used to supply *all* of the electricity needs of developed countries? Using today's efficiencies and materials, the energy needs of the United States could be met by a continuous array of solar cells covering a square of land about 160 km × 160 km (100 miles × 100 miles).

Although this area of 26,000 km^2 does not seem extraordinarily large, it should be realized that it is about a thousand times larger than the *total* area covered by all the solar cells manufactured through 1998 and would cost trillions of dollars to construct! An area about ten times larger—about 0.1% of the Earth's surface—could supply all the world's energy needs.

Conclusions About Renewable Fuels and Solar Energy

The 2003 assessment by the European Union for energy use in the year 2030 predicts that renewable energy, including wind, geothermal, and the direct forms of solar energy, will not keep pace with rising energy demand. Because people from rural regions of Asia and Africa will burn less firewood, due both to their migration to cities and to disappearing forests, renewable energy collectively will drop from the current 13% (only 2% of which is not from biomass burning) to only 8% of the global supply.

In these discussions some general features concerning the use of solar energy, as opposed to fossil-fuel and nuclear energy, have emerged, and others have also been reached by energy analysts. Many of these conclusions apply to all forms of renewable energy.

The **advantages** of solar energy appear to be that it

- is **free** and fantastically **abundant,**

- has **low environmental impact,**

- has **low operating costs,**

- does not require large, centralized suppliers and expensive distribution networks, and

- has **high public acceptance** as a "natural" form of energy.

The **disadvantages** of solar energy appear to be that it

- is **intermittent** in its availability, and thus requires that efficient storage or backup systems be constructed so that power can be supplied continuously;

- is **diffuse**—it provides a low density of energy per unit of surface collection area, so large areas of solar collectors are required to harvest the energy (one kilowatt requires about one square meter, on average);

- requires **high capital costs** to construct the energy collection and storage systems, offseting the free nature of the energy itself for many years until the investment is paid off; and

- receives little or **no economic (tax) or regulatory credit** from governments in recognition of the low amount of air pollution and greenhouse gas emissions it causes relative to fossil-fuel usage.

Alternative Fuels

We are all familiar with the widespread use of gasoline (called petrol in many English-speaking countries outside North America) to power motor vehicles. We are also aware that most air pollution problems in cities stem from emissions from gasoline engines (Chapter 2) and that fossil fuels produce large amounts of greenhouse gases when they are burned. Currently, about 18% of the total U.S. emissions of carbon dioxide into the atmosphere result from the operation of light-duty vehicles. About 60% of the U.S. use of petroleum is devoted to powering vehicles.

For these environmental reasons, attention is turning to the development of alternative fuel sources that are cleaner burning. Some of these alternatives are also, at least in principle, renewable in the sense that their production could be sustained indefinitely into the future without resulting in the accumulation of additional carbon dioxide. This is important, since a significant fraction of the world's energy is used for transportation. It is predicted that a billion cars will be on the roads by the middle of this century! In the following material we discuss the nature and properties of gasoline and of the major contenders—chiefly alcohols, natural gas, and hydrogen—for *fuels of the future*.

Gasoline

Petroleum, or *crude oil*, is a complex mixture of thousands of compounds, most of which are hydrocarbons; the proportions of the compounds vary from one oilfield to another. The most abundant type of hydrocarbon usually is the **alkane** series, which can be generically designated by the formula C_nH_{2n+2}. In petroleum, the alkane molecules vary from the simple methane, CH_4 ($n = 1$), to molecules having almost 100 carbons. Most of the alkane molecules in crude oil are of two structural types: a long, continuous chain of carbons or one main chain with short branches, e.g., 3-methylhexane. (See the Appendix if you are unfamiliar with the terminology and numbering systems of organic molecules.)

Petroleum also contains substantial amounts of **cycloalkanes,** mainly those with five or six carbons in a ring, such as the C_6H_{12} systems methylcyclopentane and cyclohexane:

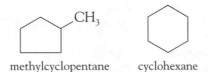

methylcyclopentane   cyclohexane

Petroleum contains some aromatic hydrocarbons, principally **benzene** and its simple derivatives in which one or two hydrogen atoms have been replaced by methyl or ethyl groups. Recall from Chapter 2 that **toluene** is benzene with one hydrogen replaced by one methyl group, and the *xylenes* are the three isomers with two methyl groups. Collectively, the benzene + toluene + xylene component in gasoline is called **BTX:**

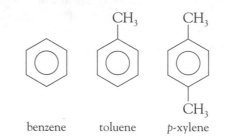

benzene   toluene   *p*-xylene

Gasoline often also contains some trimethylated benzenes and *ethylbenzene;* the mixture is then called BTEX.

PROBLEM 6-4

Deduce the structures of all the trimethylated benzenes. [Hint: For each of the three dimethylated benzenes, draw all the structures corresponding to placement of a third methyl group. Inspect each pair of structures you draw to eliminate duplicates.]

It is the BTX component of petroleum that is the most toxic to shellfish and other fish when an oil spill occurs in the ocean, whether from an oil tanker or an offshore oil well. Higher-molecular-weight hydrocarbons form sticky, tar-like blobs that adhere to birds and sea mammals as well as to rocks and other objects that the oil encounters.

Regular gasoline contains predominantly C_7 and C_8 hydrocarbons; diesel fuel contains mainly hydrocarbons with 9 to 11 carbon atoms. Generally speaking, the more carbon atoms in the alkane, the higher its boiling point and the lower its vapor pressure—and the lower its tendency to vaporize—at a given temperature. For this reason, gasoline destined for warm summer conditions is formulated with less of the smaller, more easily vaporized alkanes such as butanes and pentanes than that prepared for winters in cold climates. The presence of volatile hydrocarbons in gasoline is vital in cold climates so that automobile engines can be started.

In addition to hydrocarbons, petroleum also contains some sulfur compounds: **hydrogen sulfide** gas, H_2S, and organic sulfur compounds that are alcohol and ether analogs in which an S atom has replaced the oxygen. These substances are for the most part removed before the oil is sold for use. Small amounts of organic compounds containing oxygen or nitrogen also are present in crude oil. In general, the sulfur content (other than H_2S) of the fractions increases with boiling point, so the diesel-fuel fraction contains a higher percentage of sulfur than does gasoline, and the residue contains the highest concentration of sulfur of all, as well as most of the metals vanadium and nickel from the original crude oil, usually at levels of several parts per million. Sulfur that is present in fuels generally is converted during the fuel's combustion into *sulfur dioxide*, which is a serious pollutant if released into the air (Chapter 2).

Octane Enhancers

Gasoline that consists primarily of straight-chain alkanes and cycloalkanes has poor combustion characteristics when burned in internal combustion engines. A mixture of air and vaporized gasoline of this type tends to ignite spontaneously in the engine's cylinder before it is completely compressed and sparked, so the engine "knocks," with a resulting loss of power. Consequently, all gasoline is formulated to contain substances that will prevent knocking.

In contrast to unbranched alkanes, highly branched ones such as the octane isomer *2,2,4-trimethylpentane*, isooctane (illustrated below), have excellent burning characteristics. Unfortunately, they do not occur naturally in significant amounts in crude oil. The ability of a gasoline to generate power without engine knocking is measured by its **octane number.** To define the scale, isooctane is given the octane number of 100, and *n*-heptane is arbitrarily assigned a value of zero.

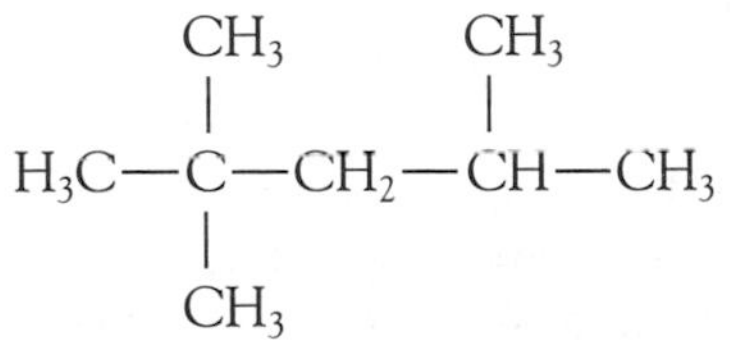

$$
\begin{array}{ccc}
 & CH_3 & CH_3 \\
 & | & | \\
H_3C-C-CH_2- & CH-CH_3 \\
 & | & \\
 & CH_3 &
\end{array}
$$

2,2,4-trimethylpentane ("isooctane")

Gasoline that has been produced simply by distillation of crude oil has an octane number of about 50, much too low for use in modern vehicles. However, when added to gasoline in small amounts, the compounds **tetramethyl lead,** $Pb(CH_3)_4$, and its ethyl equivalent prevent engine knocking and hence greatly boost the octane number of gasoline. For decades they were added worldwide to gasoline consisting predominantly of unbranched alkanes and cycloalkanes. However, these additives have now been largely phased out in most developed countries because of environmental concerns about lead, a topic discussed in Chapter 11.

In some European countries and in Canada, lead compounds were replaced by small quantities of an organic manganese compound called **MMT,** m*ethylcyclopentadienyl manganese tricarbonyl.* The use of MMT has

TABLE 6-1	Octane Numbers of Common Gasoline Additives
Compound	**Octane number**
Benzene	106
Toluene	118
p-Xylene (1,4-dimethylbenzene)	116
Methanol	116
Ethanol	112
MTBE	116

been controversial for health reasons (since manganese concentrations rise in air and soil as a result) and for technological reasons (since some car manufacturers claim it degrades emissions components on vehicles). MMT was banned in the United States until 1995 and is still little used there.

The alternative to using lead or manganese additives to boost octane ratings is to blend into gasoline significant quantities of highly branched alkanes or BTX or other organic substances such as MTBE (discussed later) that themselves have high octane numbers. A list of the common additives is shown in Table 6-1. Currently, most unleaded gasoline sold in the United States contains significant quantities of BTX (as high as 40% in the past) or of ethanol (especially in the Midwest) or of MTBE to boost the octane number. (For example, regular unleaded gasoline usually contains 0.2–2.5% MTBE. Because it must have a high octane rating, premium gasoline contains 2–9%.) Unfortunately, the BTX hydrocarbons are more reactive than the alkanes that they replace in causing photochemical air pollution, so in a sense, lead pollution has been reduced at the price of producing more smog. In addition, the use of BTX in unleaded gasoline in countries such as Great Britain in which few cars were equipped with catalytic converters resulted in the past in an increase in BTX concentrations in outdoor air. Benzene in particular is a worrisome air pollutant since at higher levels it has been linked to increases in the incidence of leukemia (see Chapter 2).

The **reformulated gasoline** used in 1995–2000 in North America contained a maximum of 1% benzene and 25% (by volume) of aromatics in total, with a minimum of 2% oxygen (by mass). The second phase of reformulated gasoline, which entered the U.S. market in 2000, reduces the benzene and BTX components even further and lowers the sulfur content to 30 ppm. The European Union plans to cut its maximum level of benzene in gasoline by 75%, effective in 2005.

The use of alcohols and compounds derived from them as additives or as oxygenated motor fuels in their own right is discussed in a later section.

Natural Gas and Propane (LPG)

In the developed world, the fuel called **natural gas** is used extensively. It is mainly methane but contains small amounts of ethane and propane. Normally the gas is transported by pipelines from its source to domestic consumers, who use it for cooking and heating, and to some utilities that burn it instead of coal or oil in power plants to produce electricity:

$$CH_4(g) + 2\ O_2(g) \longrightarrow CO_2(g) + 2\ H_2O(l) \quad \Delta H = -890\ kJ/mol$$

Unfortunately, where pipelines do not exist, the natural gas that is produced as a by-product of petroleum production at oil wells is simply wasted by venting or flaring it off, thereby adding to the atmospheric burden of greenhouse gases.

Highly **compressed natural gas** (CNG) is used to power some vehicles, especially in Canada, Italy, Argentina, the United States, New Zealand, and Russia. Due to the cost of converting a gasoline engine to accept natural gas as the fuel, the current use of CNG is mainly restricted to vehicles such as taxis and commercial trucks that are in almost constant service. For such vehicles, the additional capital cost of converting the fuel system is much less in the long run than the savings from the lower cost of the fuel. Because the compressed gas must be maintained at very high pressure to keep its storage volume reasonable, heavy fuel tanks with thick walls are required. In order to keep the weight and size of the tank reasonable, the driving range (before refilling) of CNG vehicles is usually considerably shorter than that of gasoline-powered vehicles.

Similar but somewhat less serious considerations apply to **propane,** C_3H_8, also a main component of **liquified petroleum gas** (LPG), in its use as a gasoline replacement in vehicles. The heat energy produced per gram of propane combusted, 50.3 kJ, is not quite as high as that of methane, 55.6 kJ. The heat released per gram by burning gasoline depends on the composition of the particular blend under consideration, but it is generally slightly less than that for propane. Both LPG and propane are readily liquefied under pressure, so they can be stored much more efficiently than can natural gas.

Compressed natural gas has environmental advantages and disadvantages as a vehicular fuel when compared to gasoline. Since methane molecules contain no carbon chains, neither organic particulates nor reactive hydrocarbons are formed and emitted into the air as a result of its combustion; however, a small amount of each pollutant type is formed from the ethane and propane components of commercial natural gas. Overall, regional air quality is improved by the use of natural gas rather than gasoline or diesel oil. However, the release of methane gas from pipelines during transmission or from tailpipes of vehicles due to incomplete combustion could lead to increased global warming, since methane is a potent greenhouse gas (Chapter 4). A massive conversion in North America to CNG as a vehicular fuel would be limited by supply problems for the gas, which is now used extensively for domestic heating and cooking and increasingly as the fuel in new electric power plants.

Some interesting proposals have been made recently to improve the performance of natural gas as a vehicular fuel. More efficient burning of the methane results if a small amount—about 15% by volume—of hydrogen gas is added to it. Alternatively, a smaller volume for storage of methane results if it is liquefied rather than simply compressed; however, more energy is expended in the process.

Although methane is useful as a fuel in its own right, attention has also been paid to its conversion to more *energy-dense* fuels such as methanol that are liquids at atmospheric temperatures and pressures.

Oxygenated Fuels: Methanol

Methanol, CH_3OH, and **ethanol,** C_2H_5OH, are both colorless liquids that burn easily in air to produce heat. Ethanol was used as a fuel for combustion engines as far back as the late 1800s. Many energy experts believe that alcohols will increase in importance as vehicular fuels in the near future. Indeed, these two alcohols have found niche domestic uses as fuels for many decades and have recently found increasing acceptance as gasoline additives or gasoline replacements for vehicles. In the late 1990s methanol fueling stations were under construction in parts of the United States such as California, where the majority of alcohol-fueled vehicles have been sold. Alcohols have the advantage over hydrogen and natural gas that they are liquids at ordinary pressures and temperatures and consequently are energy-dense fuels.

PROBLEM 6-5

Given that the enthalpies of combustion, per mole, of methanol and ethanol are -726 and -1367 kJ, and that the density of each is 0.79 g/mL, calculate the heat released by methanol and by ethanol (a) per gram and (b) per milliliter. From your results, comment on the superiority of one alcohol or the other with respect to energy intensity on a weight and on a volume basis. Are these alcohols superior or inferior to methane as fuels in terms of energy intensity per gram? How do they compare to gasoline, for which about 43 kJ are released per gram?

Although small amounts of both alcohols can be produced by a variety of different chemical or biochemical reactions, the large-scale production of methanol for fuel purposes would likely start with a fossil fuel, either coal or natural gas. In contrast, ethanol for fuel is produced from plant material—in Brazil from sugarcane and in the United States and Canada from the starch of grains and of corn—as discussed in detail later.

The conventional conversion of either coal or natural gas into methanol begins with the reaction of the fossil fuel with steam to produce a mixture of CO and H_2, often called **synthesis gas:**

$$C(s) + H_2O(g) \longrightarrow CO(g) + H_2(g)$$

$$CH_4(g) + H_2O(g) \longrightarrow CO(g) + 3\,H_2(g)$$

In the first process, steam is blown over white-hot coal, whereas in the second, methane gas is combined with steam that has been heated to about 1000°C. These methods produce synthesis gas from nonrenewable raw materials. An analogous mixture of hydrogen and carbon monoxide can also be obtained by heating renewable sources of biomass such as wood or the cellulosic component of garbage. The wood is first chipped and then gasified. The gaseous product is a mixture of CO, CO_2, and H_2. The processes for synthesis gas production are summarized in schematic form at the left side of Figure 6-6.

Methanol is synthesized from a 2:1 molar ratio of H_2 to CO in the presence of a catalyst:

$$2\,H_2 + CO \xrightarrow{\text{catalyst (Cu/ZnO)}} CH_3OH$$

Unfortunately, existing catalysts allow only a partial conversion (about one-fifth) of the gases into methanol for each pass of the gas mixture over the catalyst, and the processes are energy-intensive and require relatively high

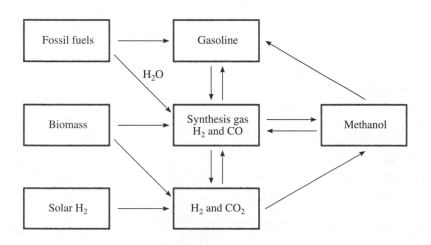

FIGURE 6-6 Scheme for the production of fuels in a hydrogen economy.

temperatures. Research is under way to develop catalysts that will operate at lower temperatures and thereby allow higher yields.

PROBLEM 6-6

The enthalpies of formation of $CO(g)$ and $CH_3OH(l)$, respectively, are -110.5 and -239.1 kJ/mol. Calculate the enthalpy of the reaction that forms methanol from synthesis gas. From your answer, predict whether the equilibrium amount of methanol obtained will increase or decrease as the temperature is lowered. Given your result, comment on the interest in developing low-temperature catalysts.

The correct 2:1 molar ratio of H_2 to CO required for the methanol synthesis reaction above is rarely obtained initially from the raw materials. For example, the reaction of steam with coal gives instead synthesis gas with a 1:1 ratio, and with natural gas a 3:1 ratio is obtained. The ratio can be adjusted to the required 2:1 by subjecting the mixture to the **water-gas shift reaction,** which is an equilibrium that can be written as

$$CO_2 + H_2 \underset{\xleftarrow{\hspace{1cm}}}{\overset{\text{catalyst}}{\xrightarrow{\hspace{1cm}}}} CO + H_2O$$

or as its reverse. Since running the reaction in the direction shown consumes H_2 and produces CO, and the opposite result is obtained by running the reaction in reverse, the initial 3:1 or 1:1 ratio of H_2 to CO can be altered to 2:1 by the partial conversion of the excess material, whether it is H_2 or CO, into the other, deficient material.

For example, consider the adjustment of the 3:1 ratio produced by the reaction of methane with steam to the required 2:1 ratio. Call the initial molar amount of CO produced a; then the initial amount of H_2 is $3a$. Since the hydrogen is initially in excess, some of it must be converted to CO; thus the appropriate direction for the shift reaction is indeed the forward direction written above. When this reaction achieves equilibrium, a molar amount x of hydrogen will have been consumed and an additional molar amount, x, of carbon monoxide will have been produced:

$$CO_2 + H_2 \longrightarrow CO + H_2O$$

from initial reaction: $3a$ a

at new equilibrium: $3a - x$ $a + x$

The value of x is obtained by requiring that the new, equilibrium ratio of H_2 to CO be 2:1:

$$\frac{3a - x}{a + x} = \frac{2}{1}$$

By algebraic manipulation of this equation, the ratio of x to a can be obtained:

$$x/a = 1/3$$

Thus the fraction $x/3a$ of the initial amount of H_2 from the natural gas that must be converted to CO is 1/9.

The two chemical reactions that when combined correspond to the conversion of methane in the correct 2:1 ratio are shown and added together below; it has been assumed for simplicity that $a = 1$:

$$CH_4 + H_2O \longrightarrow CO + 3\,H_2$$

$$1/3\,H_2 + 1/3\,CO_2 \longrightarrow 1/3\,CO + 1/3\,H_2O$$

sum $\qquad CH_4 + 2/3\,H_2O + 1/3\,CO_2 \longrightarrow 4/3\,CO + 8/3\,H_2$

When combined with a catalyst, the CO and H_2 from the sum of these reactions will yield 4/3 mole of CH_3OH, giving the net overall reaction

$$CH_4 + 2/3\,H_2O + 1/3\,CO_2 \longrightarrow 4/3\,CH_3OH$$

PROBLEM 6-7

In order to synthesize a compound with the empirical formula CH_3O (and no other products) starting from methane and steam, what ratio of H_2 to CO would be required? What fraction of the hydrogen gas produced from the reaction of methane and steam would have to be converted to carbon monoxide using the water-gas shift reaction to accomplish the transformation?

Research is currently under way to find a way to directly convert methane into methanol in a much more efficient manner than described above. Most of the difficulty stems from the fact that methane is a very unreactive substance: its C—H bond dissociation energy is highest of all the alkanes. Once one C—H bond is broken, however, the molecule becomes highly reactive because the other C—H bonds are weakened, and in the presence of oxygen it tends to oxidize completely to carbon dioxide rather than partially to a useful intermediate stage such as methanol.

Methanol can also be produced by combining carbon dioxide and hydrogen gases (see Figure 6-6) in the presence of a suitable catalyst:

$$CO_2(g) + 3\,H_2(g) \xrightarrow{\text{catalyst}} CH_3OH(l) + H_2O(l)$$

Since this reaction is only slightly exothermic, most of the fuel energy of the hydrogen is present in the methanol product. Based on equilibrium considerations, methanol formation would be favored by low temperatures and high pressures, and research has centered on finding a catalyst that will operate efficiently under such conditions without being deactivated. Low temperatures also prevent the formation of carbon monoxide rather than methanol.

Some of the massive quantities of CO_2 that are released annually into the atmosphere could be used as reactants in this process. Indeed, methanol produced in this manner could be considered a renewable fuel *provided* that the hydrogen is produced without the consumption of fossil fuel, e.g., by solar energy (see below).

Although methanol can be used as a vehicular fuel on its own, there are chemical reactions by which it can be converted to gasoline. Similarly, synthesis gas itself can be converted to gasoline, thereby allowing the production of this fuel from either natural gas or coal (Figure 6-6). Currently, neither of these processes, nor the production of fuel methanol itself, is sufficiently efficient that it can compete economically with gasoline produced from crude oil.

Oxygenated Fuels: Ethanol

Ethanol, also called *ethyl alcohol* or *grain alcohol*, can be obtained on a massive scale by the fermentation of plant crops that consist predominantly of carbohydrates, whether starch or sugars or cellulose. As in the fermentation processes that produce beer and wine, the ethanol that results from fermentation here is in a dilute solution (~10%), the major component of which is water. To convert this aqueous solution of ethanol into the almost-pure liquid alcohol that is useful as a fuel, the ethanol must be distilled from the water. The heat energy required to vaporize the ethanol from the solution is very substantial.

If the heat for distilling the ethanol is supplied from fossil-fuel combustion, then when it is added to the natural gas required to produce nitrogen fertilizer, etc., the total fossil-fuel energy used to produce the alcohol is comparable to the energy realized when the alcohol is subsequently used as a fuel. Indeed, the fossil-fuel energy input *exceeded* the alcohol energy output for North American corn-based operations in the past. Some energy analysts believe that this is still the case, notwithstanding increased efficiencies on farms and the use of agricultural by-products from the operation. Other analysts state that the process now produces somewhat more energy than the fossil fuel it consumes. (In some operations, the heat needed to distill the ethanol is provided by waste biomass, and fossil-fuel consumption is less of an issue. However, the particulate air pollution that accompanies biomass combustion prohibits its use in most areas.)

The use of gasoline containing 10% ethanol (using alcohol produced in modern facilities from corn to replace some of the crude oil used in gasoline) does result in some savings, about 2%, of total greenhouse gas emissions to the atmosphere when the entire cycle of corn growth (which extracts CO_2 from the air but emits some nitrous oxide from nitrogen fertilizer use), natural gas combustion to distill the dilute alcohol, and combustion of the alcohol is considered. Indeed, Canada plans a substantial increase in the production of fuel ethanol as one component in meeting its commitment to the Kyoto treaty. The dependence of both the United States and Canada on imported oil is also reduced somewhat by use of domestic ethanol. However, the production of ethanol from corn in North America is essentially the conversion of natural gas or coal into a convenient vehicular fuel, ethanol. There is probably a small saving of greenhouse gas emissions relative to producing the equivalent quantity of gasoline from conventional crude oil, and a more substantial one compared to producing the gasoline from sources such as tar sands and coal by energy-intensive operations.

Massive amounts of ethanol—over 10 billion liters annually—for use as vehicular fuel are currently produced from sugarcane in Brazil; unfortunately, the water pollution resulting from disposal of the by-products of this operation is massive. Smaller quantities of ethanol are obtained from cane in Zimbabwe and from corn and grain in some midwestern American states and recently also in Canada. As of the mid-1990s, more than 3.5 billion liters of ethanol for use as fuel were being produced and used annually from the starch of surplus corn and grain in the midwestern United States, and that amount is expected to increase substantially in the near future. Many farmers in the United States and Canada have provided major political support for the production and use of ethanol in gasoline, and in particular for the government subsidies required to make it economically competitive with petroleum.

Can ethanol produced from corn ever replace petroleum completely and worldwide? The growing area required to do this would be about twice the arable land used for all food crops today, so clearly the answer is no. (Indeed, in developed countries, current energy consumption rates exceed the energy generated in food crops.) Thus alcohol from fermentation of corn is unlikely to become a major fuel replacement for gasoline. Brazil produces enough ethanol from sugarcane to fulfil about one-fifth of its liquid fuel needs. The second-largest producer, the United States, generates less than 1% of its fuel needs from ethanol. Many of Brazil's vehicles operate on hydrous ethanol (95% C_2H_5OH) and the rest on a blend of 78% gasoline and 22% ethanol.

The reason why only the corn kernels, not the cob, leaves, and stalk, are used to produce ethanol currently is that only the kernels contain starch, whereas the rest of the plant is cellulose. It is necessary to break down the cellulose polymer into glucose before fermentation is possible. At present, enzymes capable of this degradation are difficult and expensive to produce, accounting for up to half the cost of producing the ethanol.

There exists even more potential for ethanol production from other sources of cellulose, particularly hardwood. The latter could be grown on land that is currently out of production or on marginal cropland; and waste from forestry operations could also be used. Short-rotation hardwood crops require substantially less fertilizers and pesticides than does corn, and such crops can produce much more biomass per unit area. *Lignin*, an unfermentable component of the wood, could be mechanically dewatered and then burned to provide the energy for the distillation step. Proponents of the use of cellulose to produce ethanol believe that the net output of carbon dioxide can be kept to one-fifth of that of the gasoline it replaces, mainly because the fossil-fuel input required to produce the wood is low compared to that for corn. The overall monetary cost of producing ethanol from cellulose would slightly exceed that for corn. The greatest potential for cost reduction lies in the development of enzymes used to degrade the cellulose to sugars and to ferment the sugars. Waste paper could also be used as a feedstock; research has produced a genetically engineered bacteria with yeast-like enzymes that efficiently convert paper mixed with dilute acids into ethanol.

Alcohols and Ethers as Vehicular Fuels

As fuels for vehicles, alcohols can be used *neat*, i.e., in pure form, or as a component in a solution that includes gasoline. Often these fuels are referred to by either M (for methanol) or E (for ethanol) combined with a subscript that indicates the percentage of alcohol in the liquid; thus the pure alcohols are M_{100} and E_{100}. Currently E_{100} is used mainly in Brazil, where about one-eighth of cars have been designed with engines that permit pure ethanol to be used as fuel. In North America, the "gasohol" currently sold is about 10% ethanol and 90% gasoline, i.e., E_{10}.

One disadvantage of methanol blends is that the pure alcohol is only soluble to the extent of about 15% in gasoline, corresponding to M_{15}; larger amounts of methanol form a second layer rather than dissolving. The inadvertent presence of water causes this unacceptable phase separation to occur at an even lower percentage of methanol. Additives such as *tertiary-butyl alcohol* (2-methyl-2-propanol) that are soluble in both methanol and gasoline prevent such separations from occurring. Looking at things from the other direction, gasoline is moderately soluble in methanol, so fuel blends such as M_{85} have been tested and are on sale in limited quantities. There are now vehicles on the road that use E_{85} fuel. Blends such as M_{85} and E_{85} do not have the problems with cold starts in vehicles that plague the pure alcohols, which, as discussed later, have much lower vapor pressures than gasoline and therefore generate little of the gaseous fuel required to start the engine (see Figure 6-7).

Some concern has been expressed about the safety of methanol as a vehicular fuel, given the toxicity of the compound. Methanol–water solutions have been widely used as windshield washer liquids in northern climates for many years without much environmental impact. However, the use of methanol as a fuel may be more dangerous as it would involve a very high concentration of the alcohol. Ethanol is much less toxic than either methanol or gasoline.

Another disadvantage of alcohols, especially methanol, as fuels is that the energy they produce per liter combusted is somewhat less than is generated by an equal quantity of gasoline; to travel the same distance, fuel tanks for alcohols would have to be larger. In principle, about 1.25 gal of ethanol are needed to

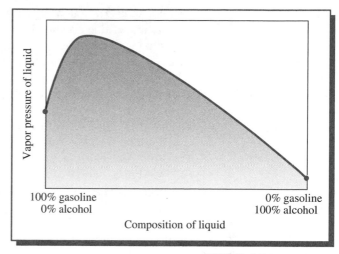

FIGURE 6-7 Variation in vapor pressure with composition for typical alcohol–gasoline mixtures.

generate the same amount of energy as is obtained with 1 gal of gasoline. In practice, however, the efficiency of combustion is greater with the alcohols, so the volume penalty would not be this large, and may be negligible. Another difficulty is that methanol cannot be used in conventional automobile engines because it reacts with and corrodes some engine and fuel tank components. However, alcohols possess some advantages: they are inherently high-octane fuels and, indeed, methanol is used to power all the cars at the Indy 500 races. Methanol has the added advantage that it does not produce a fireball when a tank-rupturing crash of racing cars occurs: it vaporizes less quickly than gasoline and, once formed, the vapor disperses more quickly.

One of the attractive features of oxygenated transportation fuels such as alcohols is that they result in lower emissions of many pollutants—specifically carbon monoxide, alkenes, aromatics, and particulates—compared to emissions from combustion of pure gasoline or diesel fuel, particularly from older vehicles that do not have catalytic converters. In North America, however, the turnover of vehicle fleets means that very few cars still on the road emit much CO. The reduction in urban ozone that would result from the lowered emissions of carbon monoxide and reactive hydrocarbons would be countered by increases due to the higher amounts of aldehydes (formaldehyde from methanol and acetaldehyde from ethanol) and vaporized alcohols that would be emitted into the air. This is particularly true for urban areas in which ozone formation is NO_X-limited rather than hydrocarbon-limited. However, NO_X emissions from burning methanol are lower than from ethanol, which in turn are lower than from gasoline, presumably because the flame temperature—which governs the amount of nitric oxide initially formed—increases in that order. Studies in Rio de Janeiro indicate that the concentration in air of the important pollutant *peroxyacetylnitrate* (see Chapter 3), which is formed readily from the acetaldehyde emissions, has increased due to the use of ethanol fuel; since Brazilian cars are not equipped with catalytic converters, this finding is not directly relevant to the North American situation. However, measurements in Albuquerque, New Mexico, have established that the use of ethanol as a gasoline additive increased the concentrations of pollutants such as PAN in the air of that city. It is curious that some proponents of ethanol as a fuel point to its ability to reduce CO emissions, which in fact is important only for cars without catalytic converters, but downplay the effects of aldehyde emissions by stating that they can always be minimized by use of catalytic converters!

Methanol can be used to produce **dimethyl ether,** CH_3OCH_3, which has been tested as a replacement for diesel fuel in trucks and buses:

$$2\,CH_3OH \longrightarrow CH_3OCH_3 + H_2O$$

This ether is nontoxic and degrades easily in the atmosphere, and in fact is used as a propellant in spray cans. Since its molecules contain no C—C bonds, soot-based particulate matter is produced in only very small quantities in its combustion (see Chapter 2) compared to amounts produced from diesel fuel. The NO_X emissions from dimethyl ether combustion are also lower than usually found for diesel engines.

MTBE as a Fuel Additive

Methanol is also used to produce the oxygenated gasoline additive MTBE, **methyl tertiary-butyl ether**

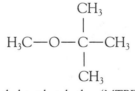

methyl *tert*-butyl ether (MTBE)

Ethanol can be used to produce the corresponding *ethyl tert-butyl ether* (ETBE), which also can be blended with gasoline, but it is more expensive than MTBE to produce. Their octane numbers are 116 (MTBE) and 118 (ETBE). MTBE is used in North American and European unleaded gasoline blends to increase their overall octane number, and also to reduce carbon monoxide (and unburned hydrocarbon) air pollution since, like the alcohols, it is an oxygenated fuel and generates less CO during its combustion than the hydrocarbons it replaces. Some gasoline for sale in American cities in the winter contains up to 15% MTBE; the U.S. Clean Air Act of 1992 requires that gasoline sold in winter months in 39 designated areas that have problems controlling CO pollution contain at least 2.7% oxygen by weight. Starting in January 1995, nine urban areas and certain other regions of the United States that had high ground-level ozone pollution problems were required to sell gasoline that contains at least 2.0% oxygen. Furthermore, in a very controversial measure, 15% of gasoline must contain oxygenates from renewable sources such as ethanol.

The advantages of using MTBE rather than ethanol to achieve the required oxygen level in gasoline are that it has a higher octane number and it does not evaporate as readily. However, as is the case with alcohols, its combustion can also produce more aldehydes and other oxygen-containing air pollutants than hydrocarbons.

The use of MTBE itself is not without controversy. It has an objectionable odor, and residents of some cities like Fairbanks, Alaska, where it has been used in oxygenated fuels, complain that it actually makes them sick. (The crude oil used in Alaska has characteristics that accentuate the MTBE odor problem.) MTBE has an odor something like turpentine that can be detected by humans at levels below those that cause health problems. The state of California has banned the use of MTBE in gasoline as of the end of 2003.

PROBLEM 6-8

Calculate the mass percentage of methanol, ethanol, or MTBE that must be present in gasoline to yield a mixture that is 2.0% oxygen by weight.

PROBLEM 6-9

Assuming that MTBE reacts in the atmosphere at the methyl group attached to the oxygen, use the principles of atmospheric reactivity developed in Chapter 3 to show that the first stable product in its decomposition sequence in air is an ester.

Another problem associated with MTBE is its contamination of well water, which has now occurred at many sites in the United States. Several U.S. states and the EPA have set action levels for MTBE in drinking water at a few tens of parts per billion, values that are exceeded in some supplies and at which the odor of the additive is apparent.

The main sources of MTBE in well water are leaking underground fuel tanks, leaking pipelines, spillage of gasoline at gas stations, and vehicle accidents. In contrast to the hydrocarbon components of gasoline, MTBE is rather soluble in water and therefore is quite mobile in soil and groundwater. It is also quite resistant to biological degradation because its carbon chains are very short and branched; its half-life is of the order of years.

In order to reduce the evaporation of volatile organic compounds (VOCs) from gasoline—since they contribute to the ozone problem—the maximum vapor pressure of gasoline sold in the United States in the summer months is also being regulated. To achieve the lower overall volatility, the amount of (highly volatile) butane in gasoline is reduced and replaced by MTBE or ETBE, both of which have low volatility. In contrast to these higher-molecular-weight ethers, ethanol and methanol as minor additives are quite volatile and actually increase the vapor pressure of gasoline. This behavior can be understood by reference to Figure 6-7, in which the vapor pressure of gasoline–ethanol mixtures is plotted against composition, with pure gasoline at the left side of the horizontal axis and pure ethanol at the right. As ethanol is added to gasoline, the vapor pressure of the mixture rises sharply, since the C_2 compound behaves in this hydrocarbon environment much like a low-molecular-weight—and therefore volatile—hydrocarbon. In contrast, as a pure liquid ethanol has a lower vapor pressure than gasoline since the hydrogen bonding between ethanol molecules is important in this situation and provides a "glue" that makes it difficult to break the molecules apart and vaporize them.

Biofuels

The *International Energy Agency* (IEA) has summarized studies on the greenhouse gas reduction characteristics of various **biofuels,** i.e., fuels such as ethanol that are derived from contemporary plant matter. It notes that significant nitrous oxide emissions result from the use of nitrogen-based fertilizers to grow the energy crops, such as corn, and has added its effects in determining equivalent carbon dioxide emissions. The IEA concludes that, even though its carbon dioxide output on combustion is reabsorbed in the growing of the subsequent corn crop, ethanol production from corn releases about 110% of the CO_2 released from the production and combustion of the energy-equivalent amount of gasoline. This percentage assumes that coal and electricity are used to supply distillation and other energy needs. This figure drops to less than 80% if natural gas is used in a cogeneration facility and to less than 60% if biomass is used to fuel the distillation step, etc. The recent advances in lowering fossil-fuel consumption during ethanol production mentioned previously have presumably lowered these percentages somewhat.

The biofuel technology that has the most promise for both producing useful energy and substantially lowering greenhouse gas emissions is the use of wood gasification to generate methanol or electricity. Gasification produces synthesis gas, from which methanol can be made (Figure 6-6), or the mixture can be burned to produce heat to generate electricity.

Another biofuel that has found some application, especially in Europe, is **biodiesel.** This material corresponds to a vegetable oil such as that from rapeseed that has been esterified and can be used in diesel engines without much modification. It produces less soot and sulfur dioxide emissions when it is combusted than the diesel fuel that it replaces. Because of the energy used in its production, and the nitrous oxide emissions associated with fertilizers, the CO_2 equivalent reductions that arise from its use are not very great and are comparable with that for corn-produced ethanol.

Hydrogen—Fuel of the Future?

Hydrogen gas can be used as a fuel in the same way as carbon-containing compounds; some futurists believe that the world will eventually have a hydrogen-based economy. Hydrogen gas combines with oxygen gas to produce water, and in the process it releases a substantial quantity of energy:

$$H_2(g) + 1/2\ O_2(g) \xrightarrow{\text{spark}} H_2O(g) \qquad \Delta H = -242\ \text{kJ/mol}$$

The idea that hydrogen would be the ultimate fuel of the future goes back at least as far as 1874, when it was mentioned by a character in the novel *Mysterious Island* by Jules Verne. Indeed, hydrogen has already found use as fuel in applications for which lightness is an important factor, namely, in powering the Saturn rockets to the moon and the U.S. space shuttles.

Hydrogen is superior even to electricity in some ways, since its transmission by pipelines over long distances consumes less energy than the transmission through wires of the same amount of energy as electricity, and since batteries are not required for local storage of energy.

Combusting Hydrogen

Hydrogen gas can be combined with oxygen to produce heat by conventional flame combustion or by low-temperature combustion in catalytic heaters. The combustion efficiency, i.e., the fraction of energy converted to useful energy rather than to waste heat, is approximately 25%, about the same as for gasoline. The main advantages to using hydrogen as a combustion fuel are its low mass per unit of energy produced and the smaller (but not zero) quantity of polluting gases its combustion produces, when compared to other fuels. BMW may begin marketing hydrogen-fueled combustion-engine cars by 2010.

Although it is sometimes stated that hydrogen combustion produces only water vapor and no pollutants, this in fact is not true. Of course, no carbon-containing pollutants, including carbon dioxide, are emitted. Since combustion involves a flame, however, some of the nitrogen from the air that is used as the source of oxygen reacts to form nitrogen oxides, NO_X. Some *hydrogen peroxide*, H_2O_2, is released as well. Thus hydrogen-burning vehicles are not really zero-emission systems. It is true that the lower flame temperature for the $H_2 + O_2$ combustion, compared to that for fossil fuels with oxygen, inherently produces less NO_X, perhaps two-thirds less. The nitrogen oxide release can be eliminated by using pure oxygen rather than air to burn the hydrogen; alternatively, it can be reduced even further by passing the emission gases over a catalytic converter or by lowering the flame temperature as much as possible.

Generating Electricity by Powering Fuel Cells with Hydrogen

Hydrogen and oxygen can be combined in **fuel cells** in order to produce electricity (a hydrogen technology also used in space vehicles). Fuel cells are similar in operation to batteries except that the reactants are supplied *continuously*. In the hydrogen–oxygen fuel cell, the two gases are passed over separate electrodes that are connected by an external electrical connection through which electrons travel and also by an electrolyte through which ions travel.

The components of a fuel cell, then, are the same as those of an electrolysis operation in which water would be split into hydrogen and oxygen, but the chemical reaction that occurs is exactly the opposite. Instead of electricity being used to drive the electrolysis reaction, it is produced.

Fuel cells have the advantage over combustion since a more useful form of energy is produced (electricity rather than heat) and the process produces no polluting gases as by-products. In principle, the only product of the reaction is water. At the catalytic surface of the first electrode, the H_2 gas produces H^+ ions and electrons, which travel around the external circuit to the second electrode, across which O_2 gas is bubbled (see Figure 6-8). Meanwhile, the H^+ ions travel through the electrolyte and recombine with the electrons and O_2 to produce water at the electrode. Although some of the reaction energy is necessarily released as heat— about 20%, due to requirements of the second law of thermodynamics—most of it is converted to electrical energy associated with the current that flows between the electrodes. Electric motors, whether in a fuel-cell or battery vehicle, are 80–85% efficient in converting electrical to mechanical energy. Real fuel cells overall are now about 50–55% efficient; 70% efficiency may be attained eventually. By contrast, internal combustion engines using gasoline are 15–25% efficient, diesels 30–35%.

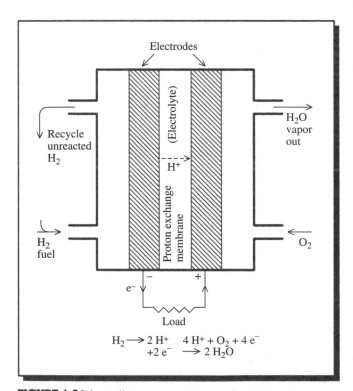

FIGURE 6-8 Schematic diagram of a hydrogen–oxygen fuel cell (PEMFC version).

Although commercial electric cars are currently powered by batteries, future improvements in technology could lead to their replacement by fuel cells. Mercedes-Benz has announced that commercial production of fuel-cell cars will begin about 2010; it unveiled a prototype vehicle of this type in 1996.

Many other auto manufacturers are currently pursuing the development of electric cars that use fuel cells. Prototype buses running in Vancouver and Chicago use innovative fuel cells for their power source. The electrolyte used in the fuel cells of these vehicles is a hair-thin (about 100 μm) synthetic polymer that acts as a proton exchange membrane. The membrane, when moist, conducts protons well since it incorporates sulfonate groups. It also keeps the hydrogen and oxygen gases from mixing. The electrodes of such **polymeric-electrolyte-membrane fuel cells,** labeled PEMFC in Figure 6-9, are graphite with a small amount of platinum dispersed as nanoparticles in thin layers (about 50 μm thick) on its surface. Each cell, which operates at about 80°C, generates about 0.8 V of electricity, so many must be stacked together in order to provide sufficient power for the vehicle. In the current versions of the buses, compressed hydrogen is stored in tanks under the roof of the bus.

Some incentives to develop vehicles that use fuel cells powered by hydrogen are

• to reduce urban smog, which is partially driven by emissions from gasoline and diesel engines;

• to reduce energy consumption, since fuel cells are much more efficient in producing motive power than are combustion engines; and

• to reduce carbon dioxide emissions, since fuel cells powered by hydrogen are carbon-free.

Some analysts point out that the cost of improving air quality and CO_2 emissions by switching the transportation system to such vehicles over the next few decades is much higher than alternative strategies, such as scrapping old cars, improving vehicle fuel efficiency, reducing NO_X emissions from power plants, and capturing and sequestering CO_2 emissions from power plants.

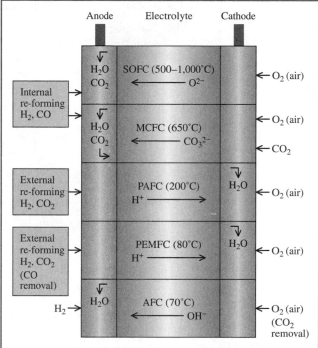

FIGURE 6-9 Operating characteristics of various fuel cells. [Source: B. C. H. Steele and A. Heinzel, "Materials for Fuel Cell Technologies," *Nature* 414 (2001): 345.]

Obtaining Fuel-Cell Hydrogen from Liquid Fuels

Because of the limited practicality of transporting hydrogen in individual cars and trucks, there is active research in designing systems that allow it to be extracted as needed from liquid fuels, which are much more convenient to transport. For example, in the near future, the hydrogen may instead be obtained as needed from liquid methanol by on-board decomposition of the latter to hydrogen gas using the reverse of the methanol-formation reaction discussed earlier:

$$CH_3OH \longrightarrow 2\,H_2 + CO$$

In the General Motors version of this process, the *re-former* unit operates at 275°C and uses a copper oxide/zinc oxide catalyst to promote the reaction. The water-gas shift reaction is subsequently used to react the CO in the synthesis gas with steam and provide additional H_2 gas, giving the overall reaction

$$CH_3OH + H_2O \longrightarrow 3\,H_2 + CO_2$$

A similar on-board process has been developed recently that converts gasoline or diesel fuel into carbon dioxide and hydrogen. Hydrogen can be obtained from gasoline or methanol with about 75% efficiency, and this may be the source of hydrogen in fuel-cell vehicles. Unfortunately, the current PEMFC and alkaline fuel cells (see Problem 6-10), as well as one based on phosphoric acid, all require relatively pure hydrogen, free especially of carbon monoxide—a gas that is formed in the re-former process and is difficult to eliminate completely. Carbon monoxide bonds to the sites of the catalyst (e.g., platinum) intended to promote the fuel-cell reaction and blocks the catalytic activity there. Concentrations of CO greater than 20 ppm in the hydrogen gas slow down most fuel cells appreciably. Perhaps a CO-tolerant electrode catalyst, possibly one incorporating a second metal or a metal oxide as well as platinum, will be developed in the future to overcome this problem by oxidizing the adsorbed CO to carbon dioxide. Hydrogen that is virtually free of carbon monoxide could be produced from methanol by an oxidative steam re-forming process at 230°C:

$$4\,CH_3OH + O_2 + 2\,H_2O \longrightarrow 10\,H_2 + 4\,CO_2$$

Since oxygen is involved, however, not all the fuel value of methanol is captured in the hydrogen product.

PROBLEM 6-10

An alkaline electrolyte can be used in the H_2–O_2 fuel cell to replace the acidic environment mentioned previously (see Figure 6-9). Assuming that the reaction of O_2 with water and electrons produces hydroxide ions, OH^-, and that these ions travel to the other electrode, where they react with hydrogen to give up electrons and produce more water, deduce the two balanced half-reactions and the balanced overall reaction for such a fuel cell. (The *alkaline fuel cell*, labelled AFC in Figure 6-9, is used in space shuttles and Apollo spacecraft to provide electricity.)

Fuel cells may also be used in the near future in small electric power plants, partly because pollutant emissions from them are so small compared to those from fossil-fuel combustion (e.g., only about 1% of the NO_X). Indeed, **phosphoric acid**–based fuel cells (PAFC in Figure 6-9) have been operating since the early 1990s in some hospitals and hotels to generate power. The most promising fuel cells for power plants involve a **molten carbonate** salt—e.g., potassium and lithium carbonates, plus additives, at 650°C—as the electrolyte. (The *molten-carbonate fuel cell* is labeled MCFC in Figure 6-9.) The hydrogen gas reacts with **carbonate ions,** CO_3^{2-}, to produce carbon dioxide, water, and electrons at the anode, while the carbon dioxide reacts with electrons and oxygen from air to re-form the carbonate ions at the cathode. The carbon dioxide must be recycled from the anode back to the cathode during the cell's operation. The hydrogen is produced on-site by reaction of methane with steam, so CH_4 is the actual source of the fuel energy here. By-product waste heat can be recovered and used in a cogeneration mode, and the dc electricity produced by the fuel cell is converted to ac for distribution. MCFC efficiency approaches 50%.

Like the molten-carbonate fuel cell, a cell based on **solid oxides** (see SOFC, Figure 6-9) is much more tolerant of carbon monoxide impurities in the fuel hydrogen gas than are the other fuel cells. The electrolyte in the solid-oxide fuel cell is a ceramic mixture of oxides of zirconium and yttrium. The **oxide anion,** O^{2-}, produced from oxygen gas carries the charge between the electrodes, travels through the solid from cathode to anode as the fuel is consumed, and forms water at the anode. The high operating temperature (up to 1000°C) of the solid-oxide cell allows fuel to be reformed internally and to form hydrogen ions without the use of expensive catalysts, so methane or other hydrocarbons can be used as the fuel instead of hydrogen. However, carbon deposits tend to form at the anode and stick to it at the high operating temperatures that are involved with these units.

The solid-oxide fuel cell, like the molten-carbonate one, is practical for central power plants and its fuel efficiency is also high. Both types suffer from practical problems with electrodes, such as carbon deposition. The latter problem for solid-oxide fuel cells can be overcome by operating at a lower temperature.

PROBLEM 6-11

For the molten-carbonate fuel cell, obtain balanced half-reactions for the processes at the two electrodes and add them together to determine the overall reaction.

Electric Cars Powered by Batteries

An alternative to vehicles that use fuel cells are those powered by batteries. Some *electric cars* have already been produced and most use a number of the same sort of lead-acid batteries that gasoline-powered vehicles have traditionally employed singly to operate the starter motor. In the future, electric cars will probably use nickel-cadmium, nickel-metal hydride, and lithium-based batteries. The practical difficulties that discourage widespread adoption of such vehicles are their high cost, low driving range between battery charges, the length of the battery recharge period, and the weight of the batteries. Like fuel-cell systems, they have the attraction of zero emissions during operation, little operating noise, and low maintenance costs. Of course, pollution is emitted into the environment when the electricity required for these cars is generated in the first place. Some researchers have predicted that lead pollution stemming from the manufacture, handling, disposal, and recycling of lead-acid batteries would raise lead emissions into the environment by an amount that would exceed the levels that were associated with leaded gasoline. Critics of this analysis have pointed out that the data used are flawed and that not all the lead lost in the processing steps will be emitted into the environment rather than disposed of properly.

Even electric vehicles are not really pollution-free if a fossil fuel is burned to generate the electricity to charge the battery, since the fossil fuel's combustion in any power plant yields NO_X that is released into the atmosphere.

The turn of the century saw the introduction of **hybrid** combustion/electric powered vehicles, such as the Toyota Prius, into the market. Such vehicles overcome the requirement of long and frequent recharging of all-battery systems, since the battery is recharged continuously when the (small) gasoline engine is in operation and is producing excess power. During braking, recovered kinetic energy is channeled back to the battery. Motive power for the vehicle is supplied by both the gasoline and electric motors, the proportion depending on the driving situation. Hybrid vehicles are highly fuel efficient, so they emit much less carbon dioxide, and also much less nitrogen oxides, carbon monoxide, and VOCs than conventional vehicles. The batteries involved are nickel-metal hydride, which are lighter and more compact than lead batteries of equivalent potential.

Other Uses for Fuel Cells

The first wave of consumer products powered by fuel cells will likely be on the market in the next year or two. Laptop computers will be powered by fuel cells rather than rechargeable batteries. These fuel cells will use methanol rather than hydrogen directly as fuel and will contain removable cartridges containing the alcohol. The advantage for the user of fuel cells over battery power is a much longer working time before the power runs out. Using methanol or natural gas rather than H_2 directly as the fuel in fuel cells does avoid the problem of the generation and storage of hydrogen. One problem with methanol is the generation of by-product carbon monoxide, which poisons the catalyst, as discussed previously. Diluting the methanol with water lessens this problem but cuts the power output of the cell. In addition, neither methanol nor other liquid fuel that could in principle be used directly instead of H_2 in a fuel cell reacts fast enough to produce the required electrical current in a vehicle, although recent research on the use of dilute solutions of methanol indicates that these problems may be overcome eventually.

Storing Hydrogen

In rocketry applications, hydrogen is stored as a liquid, as is the oxygen. Since hydrogen's boiling point of only 20 K ($-253°C$) at 1 atm pressure is so low, a large amount of energy must be expended in keeping it very cold, in addition to the energy used to liquefy it. This drawback effectively limits the applications of liquid hydrogen to a few specialized situations in which its lightness (low density) is the most important factor.

Hydrogen could be stored as a compressed gas, in much the same way as is done for methane in the form of natural gas. However, compared to CH_4, hydrogen has a drawback: a much greater amount of H_2 gas needs to be stored in order to release the same amount of energy. Compared with methane, the combustion of one mole of hydrogen consumes only one-quarter of the oxygen and consequently generates about one-quarter of the energy, even though both occupy equal volumes under the same pressure (ideal gas law). Thus the "bulky" nature of hydrogen gas limits its applications (see Problem 6-14).

It is instructive to compare the volumes of hydrogen under different conditions required to fuel a hydrogen-fuel-cell car (assuming 50% efficiency) to travel 400 km (240 miles), approximately the distance one can obtain in an efficient gasoline-powered car with a tank capacity of 40–50 L. The amount of hydrogen required is 4 kg, which occupies

- 45,000 L, or 45 m^3, e.g., a balloon having a 5-m diameter or a cube 3.6 m on each side, if it exists as a gas at normal atmospheric pressure, or

- 225 L (about 60 gal, equal to about five normal-sized gasoline tanks) as a gas compressed to about 200 atm (routinely achievable), or

- 56 L as a liquid (or solid) maintained at $-252°C$ (at 1 atm pressure), or

- 35–75 L if stored as a metal hydride, if effective systems can be developed, as discussed below.

A practical and safe way to store hydrogen for use in small vehicles may be in the form of a metal hydride. Many metals, including alloys, absorb large amounts of hydrogen gas reversibly—as a sponge absorbs water. The molecular form of hydrogen becomes dissociated into atoms at the surface of the metal as it is absorbed and forms metal hydrides by incorporating the small atoms of hydrogen in "holes" in the crystalline structure of the metal. Thus the hydrogen exists as atoms, not molecules, within the lattice, which expands slightly to incorporate them. For example, titanium metal absorbs hydrogen to form the hydride of formula TiH_2, a compound in which the density of hydrogen is twice that of liquid H_2! Heating the solid gradually releases the hydrogen as a molecular gas, which then can be burned in air or oxygen to power the vehicle.

Research continues to find a light metal alloy that can efficiently store hydrogen without making the vehicle excessively heavy. Even existing metal hydride systems are lighter than the pressurized tanks needed to store liquid hydrogen. Most industrial research now centers on metal systems. Practical considerations require that an alloy to store hydrogen

- be capable of quickly and reversibly absorbing hydrogen,

- not become brittle after many repeated cycles of absorption and desorption,

- operate in the pressure and temperature ranges of $1-10$ atm and $0-100°C$,

- not be so dense that it weighs down the vehicle excessively (a concentration of hydrogen of at least 6.5 mass % is the U.S. Department of Energy target), and

- not require a huge volume (at least 62 kg H/m^3, equivalent to 4 kg in 65 L, is the target).

Lanthanum–nickel alloys derived from $LaNi_5$ have all these characteristics except one: they are too heavy (mass % < 2), a deficiency shared by all known metal hydrides that operate near ambient temperature. Many lighter hydrides and alloys, such as MgH_2 and Mg_2NiH_4, are known, but they do not operate reversibly under moderate conditions. Research on systems formed by the lighter metals continues but has not yet been successful in producing alloys that fulfil all five conditions listed.

One of the practical difficulties in using hydrogen as a fuel is its tendency to react over time with the metal in pipelines or storage containers in which it is used. This reaction embrittles the metal and it deteriorates, eventually forming a powder. Recent progress has been made in overcoming this difficulty by using composite materials rather than simple metals as the structural materials for storage and transport facilities.

Some research in the past reported that tiny fibers made of graphite, a light material, can store up to three times their weight in hydrogen between the graphite layers and would be a safe, lightweight storage mechanism for hydrogen. However, research in the late 1990s involving extensive hydrogen storage in carbon nanotubes has not proven to be reproducible. One of the difficulties is the very tiny (milligram) samples of carbon nanotubes that are available for experimentation. Some researchers believe that, if the nanotubes are broken so that they have an open end, hydrogen can enter the tube. Other experiments indicate that only a single layer of H_2 gas conventionally adsorbed to the outside of the tubes is actually stored, a concentration too small (<2% by mass) to be useful.

Overall, the problem of devising a practical, economical, and safe way of storing hydrogen has not yet been achieved, and in the eyes of some analysts, "no breakthrough is yet in sight," despite much interest and research activity. It may be that the weight requirements associated with all practical methods of storing hydrogen will limit its use to large vehicles such as buses and airplanes.

As mentioned in the discussion of fuel cells, in some applications it may be more feasible to transport and store hydrogen in the form of an energy-dense liquid such as methanol and to use it as required for power generation. Toluene, C_7H_8, has also been proposed as a long-distance hydrogen carrier; it could then be dehydrogenated when the hydrogen is required.

Another way to temporarily store hydrogen is by means of the alkali salt (lithium or sodium) of the **borohydride ion,** BH_4^-. For their prototype minivan that runs on fuel cells, Chrysler uses a 20% solution of sodium borohydride in water to store hydrogen, which is released when the solution is pumped over a ruthenium catalyst, prompting the redox reaction of the H^- in borohydride with the H^+ in water to produce H_2:

$$BH_4^- + 3\,H_2O \longrightarrow 4\,H_2 + H_2BO_3^-$$

The density of hydrogen in the borohydride solution is comparable to that in liquid hydrogen.

The possibility of storing hydrogen in clathrates in water—much like methane in water clathrates—may be feasible. Although hydrogen molecules are too small to be efficiently trapped at low pressures, it has been discovered recently that two or four H_2 molecules can be stored in each ice clathrate at room temperature at enormous pressures, about 2000 atm. Once formed, though, the clathrate can be stored at liquid nitrogen temperatures at reduced pressure.

PROBLEM 6-12

Calculate the mass of titanium metal required to absorb each kilogram of hydrogen and form TiH_2 in a "tankful" of hydrogen. Repeat the calculation for magnesium if the hydride has the formula MgH_2. Which metal is superior for storage of hydrogen from a weight standpoint?

PROBLEM 6-13

Assuming that the energy released by combustion of H_2 is proportional to the amount of oxygen it consumes, estimate the ratio of heat released by one mole of methane compared to one mole of hydrogen gas.

PROBLEM 6-14

Using the thermochemical information in the H_2 combustion equation, calculate the enthalpy (heat) of combustion of hydrogen per gram, and by comparing it to that of methane (see page 271), decide which fuel is superior on a weight basis. By comparing the actual energy released by combustion per mole of gas—and hence per molar volume—decide which fuel is superior on a volume basis.

Producing Hydrogen

It is important to realize that hydrogen is not an energy *source*, since it does not occur as the free element in the Earth's crust. Hydrogen gas is an **energy vector** or carrier only; it must be produced, usually from water and/or methane, with the consumption of large amounts of energy and/or of other fuels. The industrial infrastructure that would be required to produce enough hydrogen to fuel all the vehicles in the United States is enormous, since it would require about as much energy as the current electric power capacity.

The most expensive commercial way to produce hydrogen is by electrolysis of water, using electricity generated by some energy source:

$$2\,H_2O(l) \xrightarrow{\text{electricity}} 2\,H_2(g) + O_2(g)$$

Unfortunately, about half the electrical energy is inadvertently converted to heat and therefore wasted in this process.

A hope for the future is that solar energy collected by photovoltaic collectors or wind power will be economically efficient in providing electricity to generate hydrogen. Currently there are prototype plants in Saudi Arabia and in Germany that use electricity from solar energy to produce hydrogen, a process about 7% efficient in converting the hydrogen. The stored energy is later recovered by reacting the hydrogen with oxygen. Excess electricity from hydroelectric or nuclear power or wind-power installations—i.e., power generated but not required immediately for use—could be used to produce hydrogen by electrolysis of water.

Even better than the use of solar electricity to electrolyze water would be the direct decomposition of water into hydrogen and oxygen by absorbed sunlight, but no practical, efficient method has yet been devised to effect this transformation. One of the difficulties in using sunlight to decompose water is that H_2O does not absorb light in the visible or UV-A regions; thus some substance must be found that can absorb sunlight, transfer the energy to the decomposition process, and finally be regenerated. The substances proposed to date for this purpose are very inefficient in converting sunlight. In addition, since the light-absorbing substances and others required are not 100% recoverable at the end of the cycle, they must be continuously resupplied—thus the hydrogen that is produced is not really a renewable fuel.

One catalyst that has been found to convert sunlight into hydrogen by electrolyzing water is **titanium dioxide,** TiO_2. A small potential is applied to the electrode in the cell's operation. Titanium dioxide is stable to sunlight (unlike many other potential light-absorbing materials) and cheap, but pure, TiO_2 absorbs only ultraviolet light. By blending carbon into TiO_2 so that C replaces some of the oxide ions, the efficiency in producing hydrogen gas is increased eightfold, to more than 8% of the Sun's energy, because the addition of carbon extends absorption into the visible region (to 535 nm).

PROBLEM 6-15

Determine the longest wavelength of light that has photons capable of decomposing liquid water into H_2 and O_2 gases, given that for this process $\Delta H = +285.8$ kJ/mol of water. In which region of the spectrum does this wavelength lie? [Hint: Recall from Chapter 1 the relationship between reaction heat and light wavelength.]

In principle, the thermal conversion of sunlight into heat can produce temperatures hot enough to decompose water into hydrogen and oxygen. Research in Israel, using a *solar tower* of mirrors to concentrate sunlight by a factor of 10,000 and thereby produce temperatures of about 2200°C in a reactor, has succeeded in splitting about one-quarter of water vapor at low pressures into H_2 and O_2.

Hydrogen gas can be produced by reacting a fossil fuel such as coal or petroleum or natural gas with steam to form hydrogen and carbon dioxide. The energy value of the fuel is transferred from carbon to the hydrogen atoms of water; chemically speaking, the reduced status of the carbon is transferred to the hydrogen. The net reactions, assuming carbon to be mainly graphite, are:

$$C + 2\,H_2O \longrightarrow 2\,H_2 + CO_2$$

$$CH_4 + 2\,H_2O \longrightarrow 4\,H_2 + CO_2$$

Notice that as much carbon dioxide is produced in this way as would be obtained by combustion of the fossil fuels in oxygen. As discussed previously, the actual conversions occur in two steps: first the fossil fuel reacts with steam to yield carbon monoxide and some hydrogen (Figure 6-6). Then the CO/H_2 synthesis gas mixture and additional steam are passed over a suitable catalyst to obtain additional hydrogen and complete the oxidation of the carbon by the water-gas shift reaction driven in the direction shown here:

$$CO + H_2O \xrightarrow{\text{catalyst}} CO_2 + H_2$$

It is interesting to note that at the turn of the twentieth century and for several decades thereafter, the synthesis gas produced when coal reacts with steam was itself used as the fuel in many municipal street-lighting systems around the world.

Hydrogen gas could be produced in a renewable way from biomass grown for this purpose. Some research indicates that aqueous solutions of both glucose and glycerol can be decomposed at moderate temperatures (225–265°C) and pressures (27–54 atm) with a platinum-based catalyst to produce hydrogen and carbon dioxide (Figure 6-6).

Associated with every conversion of one fuel to another are energy losses, mainly to waste heat, some of which are dictated by the second law of thermodynamics and therefore cannot be avoided. The energy of natural gas can be transferred to hydrogen with an efficiency of about 72%; transfer from coal is 55–60% efficient. Thus, if the resulting fuel is used only to generate heat, significantly less CO_2 is emitted if the original fossil fuel is burned rather than being first converted to hydrogen.

Finally, it should be mentioned that hydrogen is considered to be a dangerous fuel due to its high flammability and explosiveness; it ignites more easily than do most conventional fuels. On the positive side, however, spills of liquid hydrogen rapidly evaporate and rise high into the air. (Some of the fears surrounding hydrogen stem from the 1930s incident in which the airship *Hindenburg* was destroyed in a catastrophic fire. However, it was the thin layer of aluminum encasing the hydrogen gas that initially was ignited, not the H_2 itself.)

Conclusions Concerning Hydrogen and Other Alternative Fuels

The alternative fuel currently available with the greatest potential for greenhouse gas reduction, compared to the use of reformulated gasoline, is propane (LPG). If the effect of methane leakage from compressed natural gas usage over a long period of time is prorated, its use also reduces the greenhouse gas emissions. Unfortunately, both LPG and CNG are associated with somewhat higher NO_X emission rates. Nevertheless, their ozone-forming potential—especially that of CNG—is significantly less than that of reformulated gasoline, as indeed is that of methanol and methanol blends.

The relative advantages and disadvantages of a number of alternative fuels and technologies, according to an academic analysis published in 2000, are compared to those of nonreformulated gasoline in Table 6-2. Use of vehicles burning compressed natural gas or any of the electric vehicles (battery, hybrid, or fuel cell) lower air pollution significantly, whereas the use of diesel fuel makes it considerably worse. The CNG and battery vehicles analyzed were those with driving ranges of only 160 km (100 miles), however. All the alternatives listed offer some relief from the global warming potential of gasoline; however, the highest-rated alternative, ethanol, was assumed to be produced without combustion of any fossil fuels, which we know from our discussions is unrealistic.

In summary, technology is currently being developed to produce and use the new vehicle fuels hydrogen, natural gas, propane, and alcohols. Each possesses some advantages and some disadvantages compared to gasoline. It is clear that the main current disadvantage of alcohols is the monetary cost of production; this factor applies even more strongly to hydrogen. A large part of the expense of producing these fuels lies in the inefficiency of the chemical and physical transformation steps that are required to form them. In particular, the steps of the chemical transformations usually require heat that, in practice, is supplied by burning a fossil fuel, and much of the heat is wasted rather than being incorporated into the products. Furthermore, the reactions themselves rarely go to 100% conversion to the desired product. Even when only physical transformations are required—such as the compression or liquefaction of hydrogen—the energy costs are significant.

TABLE 6-2	Advantages and Disadvantages of Alternative Fuels and Vehicles[a]					
Item	Reformulated diesel	CNG	Ethanol (from biomass)	Hybrid vehicle	Electric vehicle Battery	Fuel cell
O_3, NO_X, VOCs	−3	+1	0	+1	+2	+1
PM	−1	0	0	+1	+2	+1
Air toxics	−4	+1	0	+2	+2	+2
Fuel cycle emissions	0	+2	+1	+1	0	+1
Overall air pollution	−8	+4	+1	+5	+6	+5
Global warming	+3	+2	+6[b]	+3	+1	+3

Source: Based on Table 1 of L. Lave et al., Environmental Science and Technology, 34 (2000): 3598–3605.
[a] Relative magnitudes (arbitrary scale) of advantages (+ scores) and disadvantages (− scores) of alternative fuels and vehicles in reducing air pollution and global warming, compared to reformulated gasoline in conventional engine vehicles.
[b] Does not include CO_2 emissions from fossil fuels used in producing ethanol.

Review Questions

1. Define the term *renewable energy* and list several forms of it. Which form is growing the fastest?

2. Describe the difference between the two methods of absorbing energy from sunlight. What is the difference between active and passive systems?

3. What is meant by *solar thermal electricity*, and how is it generated? What is meant by the term *cogeneration*?

4. State the *second law of thermodynamics*. According to this law, what formula gives the maximum fraction of heat that can be transformed into electricity?

5. Define the *photovoltaic effect*.

6. List four advantages and four disadvantages of solar energy.

7. What is the most important class of hydrocarbons present in crude oil?

8. What is meant by the *BTX fraction* of gasoline? Is it toxic?

9. What is meant by *engine knocking*?

10. How is the *octane number* rating scale for fuels defined? Name a few fuels that have octane numbers greater than 100.

11. List several ways in which the octane number of fuels can be increased by the addition of other compounds to straight-chain alkane mixtures.

12. What is the main component of *natural gas*? Write out the balanced chemical equation illustrating its combustion.

13. Why is natural gas considered to be an environmentally superior fuel to oil or coal? What phenomenon involved in its transmission by pipeline might offset this advantage?

14. What is meant by the term *CNG*? What are the advantages and disadvantages of fueling vehicles with CNG?

15. What is the *water-gas shift reaction*? Describe the methods by which methanol can be produced in volume for use as a fuel. What does M_{85} mean?

16. What do *MTBE* and *ETBE* stand for? What is the structure of MTBE? What is meant by E_{10} fuel?

17. What are the advantages and disadvantages in regard to air pollution of using alcohol fuels?

18. Describe the method used in producing ethanol in volume for use as a fuel. What are the potential feedstocks for this process?

19. What are the very energy-intensive steps involved in production of fuel ethanol? Why isn't ethanol a fully renewable fuel?

20. Describe the three ways in which hydrogen can be stored in vehicles for use as a fuel and discuss briefly the disadvantages of each method.

21. Does the burning of hydrogen really produce no pollutants? Under what conditions do no pollutants form?

22. What is the difference between an *energy source* and an *energy vector* (carrier)? Into which category does H_2 fall?

23. Describe how a hydrogen fuel cell works, and write balanced half-reactions for its operation in acidic media. What other types of fuel cells exist?

24. Describe the production of hydrogen gas by electrolysis. Can solar energy be used for this purpose? Why isn't water decomposed directly by absorption of sunlight?

Additional Problems

1. Given that an average of 342 W of sunlight energy falls on each square meter of Earth, that the surface area of a sphere is $4\pi r^2$, and that the Earth's radius r is about 5000 km, calculate in joules the total amount of sunlight received annually by the Earth. What percentage of this quantity needs to be captured in order to provide our current commercial energy needs?

2. Consider the use of methanol, CH_3OH, as an oxygenated liquid fuel for suitably modified cars. **(a)** By writing the balanced chemical equation for its combustion in air, determine whether it is more similar to coal, oil, or natural gas in terms of the joules of energy released per mole of CO_2 produced. **(b)** Determine the balanced equation by which methanol can be produced by reacting elemental carbon (coal) with water vapor, given that CO_2 is the only other product in the reaction. **(c)** Does the combined scheme of parts **(a)** and **(b)** represent a way of using coal but producing less carbon dioxide per joule than by its direct combustion? Explain your answer.

3. Deduce the fraction of the CO or H_2 produced by the reaction of coal with steam that must be converted to H_2 or CO, respectively, by the water-gas shift reaction in order to obtain the 2:1 ratio of hydrogen to carbon monoxide that is required to synthesize methanol. Deduce also the net reactions of conversion of coal to methanol.

4. Deduce the balanced reaction in which synthesis gas is formed by combining methane and carbon dioxide. From enthalpy of formation data given in Problems 5-3 and 6-6, deduce the enthalpy change for this reaction. By applying Le Châtelier's principle, deduce whether the conversion of the gases to carbon monoxide and hydrogen will be favored by low or by high pressures, and by low or by high temperatures. Combining these results with those obtained in Problem 3, determine the fraction of the total carbon dioxide resulting from the production and combustion of methanol synthesized from this synthesis gas that would be

renewable, i.e., recycled from the consumption of carbon dioxide in the process.

5. Contact several of the new car dealerships in your area to discover which vehicles currently on sale can use one or more of the alternative fuels: CNG, LPG (propane), M_{85}, E_{100} or E_{85}, hydrogen, or electricity. For each fuel and vehicle, inquire about the average kilometers or miles per liter or gallon of fuel; from this information and using fuel prices obtained from local service stations, estimate the driving cost per kilometer or mile for each vehicle and fuel combination. Are any of the combinations competitive in cost with gasoline?

6. Contact a garage in your area that converts existing gasoline-fueled vehicles into those that accept CNG or LPG (propane). Determine the common conversion price, and what the likely kilometers-per-liter or miles-per-gallon performance for the new fuel in a common vehicle will be. From the cost of the fuels in your area, estimate the distance that the vehicle must be driven before the cost of the conversion has been met by savings on the fuel.

7. The Bay of Fundy, located between the Canadian provinces of Nova Scotia and New Brunswick, has the highest tides in the world. The difference between high and low tides can be as much as 16 m. A total of 14 billion tonnes of seawater flow into and out of Minas Basin, a part of the Bay of Fundy, during each tide. The energy tapped as tidal power comes from the change in potential energy of this water as it falls in the Earth's gravitation field. Given that the potential energy of a mass m at height h in a gravitational field is given by $m \times g \times h$, where g is the gravitational constant, calculate the amount of energy that corresponds to a tidal drop of 16 m in the Minas Basin.

8. One way of considering the direct environmental impact of the combustion of various fuels is to look at the amount of heat produced per unit of CO_2 generated. Determine and compare the values of kilojoules of heat per mole of CO_2 produced in the case of the combustion of methanol, ethanol, and n-octane. Use the following molar enthalpies of combustion (ΔH_c, in kJ/mol): methanol, -726; ethanol, -1367; and n-octane, -5450. Comment on the values obtained.

9. The use of biomass fuels, including scavenged wood and wood-chip waste, has been proposed as a way of reducing CO_2 emissions, even though such fuels would produce a large amount of CO_2 per unit of heat produced. Explain the rationale behind this idea.

Further Reading

1. J. A. Turner, "A Realizable Renewable Energy Future," Science 285 (1999): 687–689.

2. R. Edwards, "The Big Break" [wave power], New Scientist (3 October 1998): 30–33.

3. K. S. Betts, "The Wind at the End of the Tunnel," Environmental Science and Technology (1 July 2000): 306A–312A.

4. N. Fell, "Deep Heat" [geothermal energy], New Scientist (22 February 2003): 40–42.

5. D. Pimentel, "Ethanol Fuels: Energy Balance, Economics, and Environmental Impacts Are Negative," Natural Resources Research 12 (2003): 127–134.

6. W. Hoagland, "Solar Energy," Scientific American (September 1995): 170–173.

7. P. Hoffmann, Tomorrow's Energy: Hydrogen, Fuel Cells, and the Prospects for a Cleaner Planet (Cambridge, MA: MIT Press, 2001).

8. G. Hoogers, "Fuel Cells: Power for the Future," Physics World (August 1998): 31–36.

9. C. Hanish, "Powering Tomorrow's Cars," Environmental Science and Technology (1 November 1999): 458A–462A.

10. N. L. C. Steele and D. T. Allen, "An Abridged Life-Cycle Assessment of Electric Vehicle Batteries," Environmental Science and Technology (1 January 1998): 40A–46A.

Websites of Interest

Log on to www.whfreeman.com/envchem3e/ and click on Chapter 6.

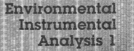

Environmental Instrumental Analysis 1

Instrumental Determination of NO$_X$ by Chemiluminescence

In the preceding chapters, we have seen that nitrogen oxides play a leading role in atmospheric chemistry, both in the stratosphere and at ground level. In this box, we see how the concentration of NO and NO$_2$ gases in environmental air samples can be determined using a sophisticated, modern method of analysis.

The process by which chemicals react to produce light is termed **chemiluminescence.** If conditions for a particular reaction are well known and can be controlled in an analytical instrument, chemiluminescence can be used as a sensitive and selective means of determining the concentration of components in the reaction. A few well-known chemical reactions that involve chemiluminescence are the basis for methods in tropospheric and stratospheric environmental analysis. One of the most common is the detection of nitric oxide (NO) and nitrogen dioxide (NO$_2$).

The chemiluminescence reaction that produces light in this method is the gas-phase reaction of NO with ozone (O$_3$):

$$NO + O_3 \longrightarrow NO_2{}^* + O_2$$

In the NO detector this reaction takes place in a small steel reaction vessel under controlled conditions. The excited-state nitrogen dioxide created in this reaction, designated NO$_2{}^*$, very quickly returns to the ground state by giving off light in the red and infrared range of the spectrum (λ = 600 to 2800 nm)

$$NO_2{}^* \longrightarrow NO_2 + light$$

The reaction is dependent on pressure and temperature. A constant low pressure is maintained in the reaction vessel by use of a vacuum pump that constantly evacuates the chamber. Typical cell pressures are approximately 1 to 100 torr. The amount of light produced in the reaction is proportional to the amount of whichever reactant is *not* in excess in the reaction chamber. *If* ozone is provided in excess (from an ozone generator), then the light output reflects changes in NO concentration. The light created by the reaction is detected by a photomultiplier tube (PMT) whose output is fed to a computer system. The computer software correlates the amount of light produced with the quantity of reactant by referring to the relation between light intensity and NO concentration, which it has obtained from previous calibration experiments. In instruments of this kind the PMT signal is often integrated over short time periods (e.g., 10 sec) using a "photon counting" system that improves sensitivity.

The schematic diagram of this instrument, shown in the figure at the top of page 300, includes the reaction chamber, ozone generator, PMT, computer, and vacuum pump. Gas from the atmosphere is drawn directly into the reaction chamber and immediately mixed with an excess of O$_3$ (i.e., more O$_3$ than NO). Typical sampling volumes are 1,000 standard cubic centimeters per minute (sccm). Some instruments have gold-plated surfaces inside to prevent surface reactions and increase the light-collection abilities of the chamber. The light produced is detected by the PMT, which is mounted immediately adjacent to the

(continued on p. 300)

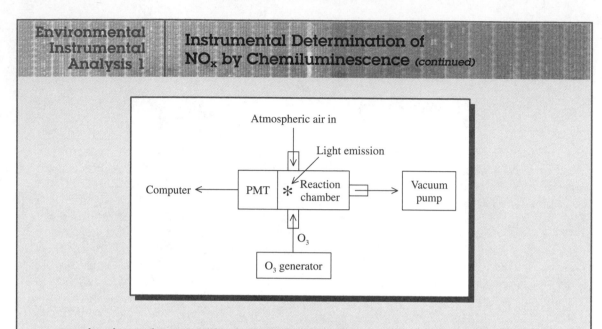

reaction chamber and separated by a transparent window or a filter that can block out light from interfering reactions.

The same instrument can be used to determine NO$_2$ by incorporating a chemical reduction step for the incoming air in which NO$_2$ is reduced to NO by a hot metal catalyst before entering the reaction chamber. NO is then determined as before, but now the signal includes input from the presence of the air's

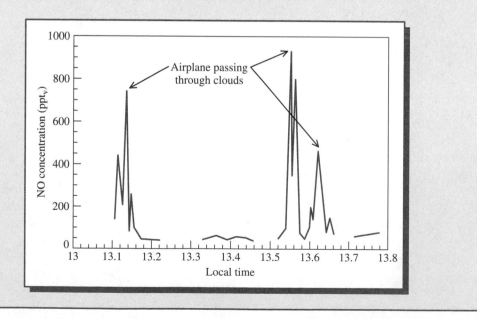

of the air's NO *and* NO$_2$. If alternating signals are generated in a short time period—one with the atmospheric air stream that has been reduced and the other without reduction—by means of a switching valve, then the concentration of both nitrogen oxides can be determined:

[NO] = signal from unreduced air flow

[NO$_2$] = signal from reduced air flow
          − signal from unreduced air flow

Instruments of this kind have been used in airplane-based stratospheric and tropospheric sampling projects by the National Atmospheric and Space Administration and the National Center for Atmospheric Research. Similar instrumentation has been used for tropospheric studies of urban pollution using laboratory-based smog chambers and for indoor air-pollution studies.

The figure at the bottom of page 300 shows the variation in concentration of NO recorded as an airborne scientific expedition equipped with this kind of detector flew through tall, vertically developed, and actively raining cumulonimbus clouds.

The researchers were flying level at 9.3 km altitude while sampling the tropospheric air over the Pacific Ocean west of Hawaii. The regions of the flight path involving clouds are noted. [NO] in parts per trillion by volume is plotted on the y-axis and local time is plotted on the x-axis. These data demonstrate the production of NO in electrically active clouds.

Reference: B. A. Ridley, M. A. Carroll, and G. L. Gregory, "Measurements of Nitric Oxide in the Boundary Layer and Free Troposphere over the Pacific Ocean," *Journal of Geophysical Research* 92(D2) (1987): 2025–2047.

<table>
<tr><td></td><td></td></tr>
</table>

Environmental Instrumental Analysis 2	Instrumental Determination of Atmospheric Methane

In the preceding chapters, we have seen that methane is an important gas not only in chemical reactions in the stratosphere and at ground level, but also in producing global warming. In this box, the determination of the concentration of methane in environmental air samples is described.

The determination of the concentration of methane in the atmosphere can be accomplished either by analyzing atmospheric gas samples in the laboratory or by using more rugged field instrumentation for direct analysis where the sample is taken, e.g., in a rain forest, on a research vessel at sea, or from an airplane. Both methods usually rely on gas chromatography (GC) as the means of separating the complex mixture of atmospheric components and a very sensitive instrument called the flame ionization detector (FID) to actually measure each of the separated analytes (components).

The power of gas chromatography stems from its ability to separate individual components of a mixture injected into the chromatographic column and to identify the analytes as they exit the column one by one. The GC separation process occurs as the gaseous compounds, under the influence of the column's temperature and chromatographic surface, and a flowing carrier gas or mobile phase undergo a series of nondestructive interactions with the chromatographic surface as they pass through the column. Since each compound to be separated—methane and other atmospheric components in this case—interacts in a slightly different manner with the column's

chromatographic surface, the result is a different transit time for each compound and, therefore, an individual *exit time* or *retention time*.

High-resolution capillary GC columns, designed to separate literally hundreds of compounds from a single mixture, have a very small inside diameter, less than 0.53 mm. Many different chromatographic surfaces are available; each is designed to separate different families of analytes. For the paraffin family, of which methane is the first member, nonpolar chromatographic surfaces or phases are used. In a sample of air collected near a rain forest slash-and-burn site, many other hydrocarbons in addition to methane are routinely detected, e.g., isoprene (C_5H_8), ethane, propane, and β-pinene. The table details the concentration of these compounds 30 m above the ground in the biosphere near Manaus, Brazil, in July and Au-

	Concentration (ppb_v)
Compound	30 m above ground
Methane	1657
Ethane	1.17
Isoprene	2.56
β-pinene	0.03

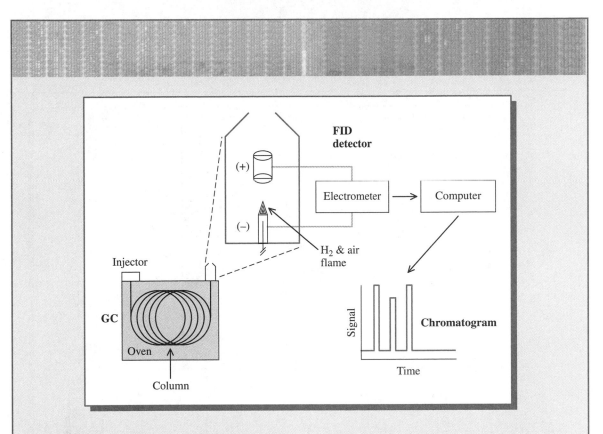

gust of 1985. Tethered balloons were raised and lowered to predetermined heights and a radio-controlled valve system was used to collect samples of biospheric gases in Teflon bags. These gas samples were transferred to stainless-steel canisters and then returned for GC/FID analysis to the National Center for Atmospheric Research.

The GC/FID system, shown above, contains three basic parts: injector, chromatographic column, and detector. The chromatographic column is plumbed into the GC's injector, where the mixture is initially introduced into the column. The column—whose temperature is controlled by a small oven—terminates in the base of the FID, where each exiting analyte is detected.

The FID response to hydrocarbons is based on the detection of carbon-containing ions created in the combustion processes of the FID's flame, a flame created by burning a hydrogen/air mixture. This extremely sensitive system can detect CH_4 in the picogram (10^{-12} g) range. This translates into parts per billion by volume in atmospheric gas samples.

The result of all chromatographic methods is, in one form or another, a graphical representation of the data called a **chromatogram.** This is most often a plot of retention time versus detector signal. Careful

(continued on p. 304)

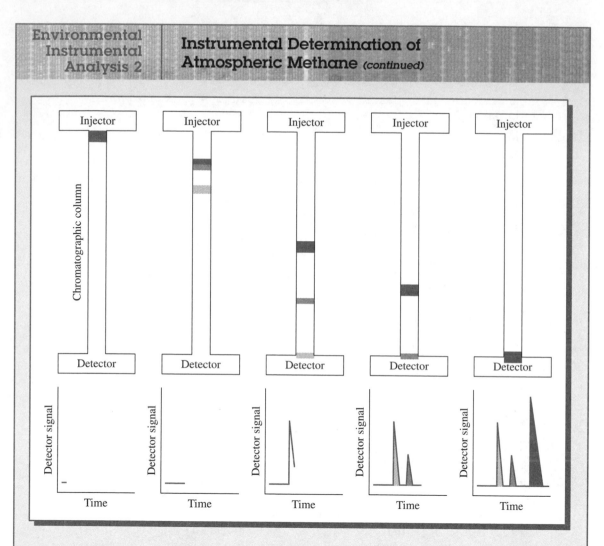

Instrumental Determination of
Atmospheric Methane (continued)

analysis of the chromatogram yields the identity, based on known standards, and quantity of each of the components in the original atmospheric mixture.

The figure above shows a schematic of the chromatographic process with the injector—where the sample is introduced—at the top of the column and the detector at the bottom. A chromatograph "in progress" is presented at five different steps in the chromatographic process in the lower panels.

Reference: P. R. Zimmerman, J. P. Greenberg, and C. E. Westberg, "Measurements of Atmospheric Hydrocarbons and Biogenic Emission Fluxes in the Amazon Boundary Layer," *Journal of Geophysical Research* 93(D2) (1988): 1407–1416.

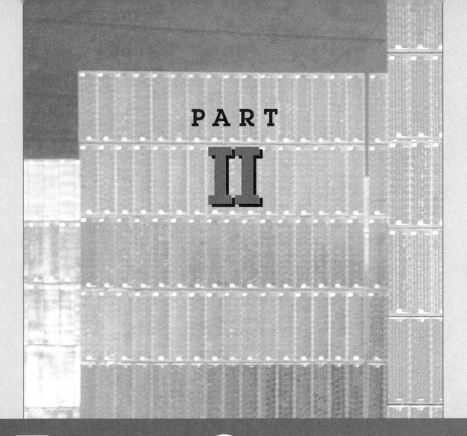

PART
II

TOXIC ORGANIC COMPOUNDS

Pesticides

Background

The term **synthetic chemical** is used to describe substances that generally do not occur in nature but have been synthesized by chemists from simpler substances. The great majority of commercial synthetic chemicals are organic compounds, and most use petroleum or natural gas as the original source of their carbon.

In Chapters 7 and 8 the environmental consequences of widespread use of synthetic organic chemicals are discussed. Our emphasis will be on those substances whose toxicity has led to concerns regarding human health, especially with respect to cancer and to birth defects, as well as to the well-being of lower organisms.

In this chapter we discuss pesticides, considering the environmental problems associated with their use and some general principles of toxicology that apply to their effects on human health. In Chapter 8, nonpesticidal organic compounds of environmental concern are described.

Some cormorants in the Great Lakes region have severe birth defects—such as the crossed bill that this cormorant has—owing to exposure before birth to organochlorine chemicals in the food chain of the mother birds. (Courtesy of International Joint Commission, Detroit/Ottawa)

Types of Pesticides

Pesticides are substances that kill or otherwise control an unwanted organism. The various categories of pesticides are listed in Table 7-1. All chemical pesticides share the common property of blocking a vital metabolic process in the organisms to which they are toxic. In this chapter we first discuss **insecticides**—substances that kill insects—and then **herbicides**—compounds that kill plants. Also mentioned in

TABLE 7-1	Pesticides and Their Targets
Pesticide type	**Target organism**
Acaricide	Mites
Algicide	Algae
Avicide	Birds
Bactericide	Bacteria
Disinfectant	Microorganisms
Fungicide	Fungi
Herbicide	Plants
Insecticide	Insects
Larvicide	Insect larvae
Molluscicide	Snails, slugs
Nematicide	Nematodes
Piscicide	Fish
Rodenticide	Rodents
Termiticides	Termites

passing are some **fungicides,** substances used to control the growth of various types of fungus, especially to protect stored seeds before planting. Collectively, these three categories represent the great bulk of the *1 billion kilograms* of pesticides used annually in North America.

About half of the use of pesticides in North America involves agriculture, although worldwide the figure rises to 85%. Almost all commercial food crops are now produced with the use of insecticides, herbicides, and fungicides, except, of course, in organic farming. Indeed, the current ability in developed countries to produce and harvest large amounts of food on relatively small amounts of land with a relatively small input of human labor has been made possible by the use of pesticides. Currently, the greatest U.S. use of insecticides occurs in the growing of cotton, whereas the majority of herbicides are used in the growing of corn and soybeans. Depending upon the main crops they produce, countries vary in what types of pesticides they consume in large amounts. For example, herbicides account for three-quarters of pesticide use in Malaysia, whereas insecticides are the most widely used pesticides in India and the Philippines, as are fungicides in Colombia.

Some 80–90% of American domestic households contain at least one synthetic pesticide; typical examples are weed killers for the lawn and garden, algae controls for the swimming pool, flea powders for use on pets, and sprays to kill insects such as cockroaches.

Almost since their introduction, synthetic pesticides have been a concern because of the potential impact on human health of eating food contaminated with these chemicals. Half the foods eaten in the United States contain measurable levels of at least one pesticide. For that reason, many have been banned or restricted in their use. Nevertheless, a 1993 report by the U.S. National Academy of Science pointed out that pesticide regulation to date has not paid enough attention to the protection of human health, especially that of infants and children, whose growth and development are at stake. Children, kilogram for kilogram, eat more food than adults and tend to eat more food (such as apples, grapes, and carrots) with higher pesticide levels than do adults, and their internal organs—including the brain—are still developing and maturing, making them more vulnerable to any negative effects these chemicals may have. In addition, small children play on floors and lawns and put many objects in their mouths, increasing their exposure to pesticides used in homes and yards.

Some scientists dismiss these concerns by emphasizing that living plants themselves manufacture insecticides in order to discourage insects and fungi from consuming them, and consequently we are exposed in our food supply to much higher concentrations of these "natural" pesticides than to synthetic ones. Natural pesticides are not necessarily less toxic than synthetic ones, as we shall see.

Traditional Insecticides

Insecticides of one type or another have been used by societies for thousands of years. A principal motivation for using insecticides is to control disease: human deaths due to insect-borne diseases through the ages have greatly exceeded those attributable to the effects of warfare. The use of insecticides has greatly reduced the incidence of diseases transmitted by insects and the rodents that harbor them: malaria, yellow fever, bubonic plague, sleeping sickness, and, recently, illness from the *West Nile virus* scarcely exhaust the list of these scourges. The other principal motivation for insecticide use is to prevent insects from attacking food crops: even with extensive use of pesticides, about one-third of the world's total crop yield is destroyed by pests or weeds during growth, harvesting, and storage. People also try to control insects such as the mosquito and the common fly simply because their presence is annoying.

The earliest recorded usage of pesticides was the burning of sulfur to fumigate Greek homes around 1000 B.C. **Fumigants** are pesticides that enter the insect as an inhaled gas. The use of **sulfur dioxide, SO_2,** from the burning of solid *elemental sulfur*, sometimes by incorporating the element in candles, continued at least into the nineteenth century. Sulfur itself, in the form of dusts and sprays, was also used as an insecticide and as a fungicide; it is still employed as a fumigant against powdery mildew on plants. Inorganic fluorides, such as *sodium fluoride*, NaF, were used domestically to control ant populations. Both sodium fluoride and *boric acid* were used to kill cockroaches in infested buildings. Various oils, whether derived from petroleum or from living sources such as fish and whales, have been utilized for hundreds of years as insecticides and as *dormant sprays* to kill insect eggs.

The use of arsenic and its compounds to control insects dates back to Roman times. It was employed by the Chinese in the sixteenth century and became quite widespread from the late nineteenth century until the Second World War. *Paris green*, a copper–arsenic salt, was a popular insecticide introduced in 1867 into the United States. Other arsenic-containing salts have also been employed. They all operate as stomach poisons and kill insects that ingest them. Arsenic compounds continued to be heavily used as insecticides into the early 1950s and are discussed in more detail in Chapter 11.

Unfortunately, inorganic and metal-containing organic pesticides are usually quite toxic to humans and other mammals, especially at the dosage levels that are required to make them effective pesticides. Mass poisonings have occurred as a result of the use of some mercury-based fungicides, as will be discussed in Chapter 11. In addition, toxic metals and semimetals, such as the arsenic commonly used in such pesticides, are not biodegradable: once released into the environment, they will remain indefinitely in the water, wildlife, soil, or sediments and may enter the food supply if liberated from these sites.

Organochlorine Insecticides

Many organic insecticides developed during and after World War II have largely displaced these inorganic and metal-organic substances, as will be discussed. Usually only small amounts of the organic compounds are required to be effective against the target pests, so smaller amounts of chemicals enter the environment. Given a dose of each large enough to act as a pesticide, the organic substances are generally much less toxic to humans than are the inorganic and metal-organic pesticides. Finally, organic pesticides were initially thought to be biodegradable, although, as we shall see, this has certainly not been found to be true in many cases.

In the 1940s and the 1950s the chemical industries in North America and western Europe produced large quantities of many new pesticides, especially insecticides. The active ingredients in most of these pesticides were **organochlorines,** which are organic compounds that contain chlorine. Many organochlorines share several notable properties:

- stability against decomposition or degradation in the environment;

- very low solubility in water, unless oxygen or nitrogen is also present in the molecules;

- high solubility in hydrocarbon-like environments, such as the fatty material in living matter; and

- relatively high toxicity to insects but low toxicity to humans.

As an example of an organochlorine pesticide, consider the compound **hexachlorobenzene** (HCB), C_6Cl_6. It is stable, easy to prepare from chlorine and benzene, and for several decades after World War II it found use as an agricultural fungicide for cereal crops. Since it is extremely persistent and is still emitted as a by-product in the chemical industry and in combustion processes, it remains a widespread environmental contaminant. It is of concern because it causes liver cancer in laboratory rodents and perhaps also in humans. Our current daily exposure to HCB is not sufficient to pose a significant health hazard, even though it is estimated that 99% of Americans have detectable levels of the chemical in their body fat.

HCB

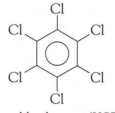

hexachlorobenzene (HCB)

Hexachlorobenzene is one of the 12 chemicals, the "dirty dozen," listed by the United Nations Environmental Program as **Persistent Organic Pollutants,** or POPs, which are being banned or phased out by international agreement (see Table 7-2). Each country signing the Stockholm Treaty

TABLE 7-2	U.N.'s Persistent Organic Pollutants (POP) "Dirty Dozen" and Their Status in Several Countries						
POP	U.S.	Canada	U.K.	Mexico	China	India	Use
DDT	X	X	X	R	R	R	Mosquitoes
Aldrin	X	X	X	X	OK	OK	Termites
Dieldrin	X	X	X	X	OK	R	Crops
Endrin	X	X	X	X	OK	X	Rodents
Chlordane	R	X	X	OK	R	OK	Termites
Heptachlor	R	X	X	X	OK	OK	Soil insects
Hexachloro-benzene	X	X	X				Fungicide
Mirex	X	X			R	R		Ants, termites
Toxaphene	X	X	X	X	OK	X	Ticks, mites
PCBs*	X	R	R	OK			Many uses
Dioxins*	BP	BP	BP	BP	BP	BP	
Furans*	BP	BP	BP	BP	BP	BP	

Codes: *not pesticides; X = banned or no registered uses; R = severely restricted uses only; OK = not restricted; BP = by-product, not produced deliberately.

that bans these POPs must develop its own implementation plan. Other compounds with the same negative characteristics as the dirty dozen may be added later to the list. HCB is already banned, or has no registered uses, in the United States, Canada, and the United Kingdom.

Moth Balls

Another chlorinated benzene, the 1,4 isomer of **dichlorobenzene,** is used as an insecticidal fumigant, i.e., an airborne insecticide. It is one type of domestic *moth ball* and is also used in solid and liquid room deodorizers. Although it is a crystalline solid, it has an appreciable vapor pressure. Consequently, enough of it will vaporize to act as an effective insecticide in the immediate area around the solid. The same compound has also been used as a soil fumigant and a pesticide. However, it is an animal carcinogen and accumulates to some extent in the environment. It may well be the chemical responsible for the greatest carcinogenic risk of all indoor VOCs (volatile organic compounds).

Pesticides in Water

The solubilities of trace substances such as HCB in liquids and solids are often expressed on a "parts per" scale, rather than on a mass or moles per unit volume basis. However, the "parts per" scales for condensed (nongaseous) media express the ratio of the *mass* of the solute to the *mass* of the solution, *not* the ratio of moles or molecules that is used for gases. Since the mass of 1 liter of a natural water sample is very close to 1 kilogram, the HCB solubility of 0.0062 mg of solute per liter corresponds to 0.0062 mg of solute per 1000 g of solution. Multiplying both numerator and denominator by 1000, we conclude that 0.0062 mg/L is equivalent to 0.0062 g of solute per 1 million grams of solution, i.e., to 0.0062 ppm. In general, the value for the ppm solubility of any trace substance in water is the same as its value in units of milligrams per liter or micrograms per gram.

PROBLEM 7-1

For aqueous solutions, (a) convert 0.04 μg/L to the ppm and ppb scales and (b) convert 3 ppb to the μg/L scale.

The pollution of aquatic environments is not merely a question of the concentration of pollutant actually in *solution*, and the small values for the solubilities of organochlorines in water may be deceptive on this score. Most organochlorine compounds are much more soluble in organic media than in water. In bodies of water such as rivers and lakes, organochlorines are much more likely to be bound to the surfaces of organic particulate matter suspended in the water and on the muddy sediments at the bottom than to be dissolved in the water itself. From these sources they enter living organisms such as fish. For reasons discussed in detail later, their concentration in fish is often thousands or millions of times greater than that dissolved in polluted drinking water. It is due to this phenomenon that concentrations of organochlorines have often reached dangerous levels in many species. Many organochlorine insecticides have been removed from use as a consequence. For humans, the amount of organochlorines ingested by eating a single lake fish is generally greater than the total organochlorines acquired in a lifetime of drinking water from the same lake!

DDT

DDT, or *para-dichlorodiphenyltrichloroethane,* has had a tumultuous history. It was hailed as miraculous in 1945 by Sir Winston Churchill because of its use in the war effort. It was very effective against the mosquitoes that carry malaria and yellow fever, against body lice that can transmit typhus, and against plague-carrying fleas. The World Health Organization has estimated that malaria-reduction programs, one component of which was the use of DDT, have saved the lives of more than 5 million people.

Unfortunately, DDT was widely overused, particularly in agriculture, which consumed 70–80% of its production, and forestry. As a result, its environmental concentration rose rapidly and it began to affect the reproductive abilities of birds that indirectly incorporated it into their bodies. By 1962 DDT was being called an "elixir of death" by Rachel Carson in her influential book *Silent Spring* because of its role in decreasing the populations of birds such as the bald eagle, whose dietary intake of the chemical was very high. Further details concerning the history of DDT are given in Box 7-1.

DDT's Structure

Structurally, a DDT molecule is a substituted ethane. At one carbon, all three hydrogens are replaced by chlorine atoms, while at the other, two of the three hydrogens are replaced by a benzene-like ring. Each ring contains a chlorine atom at the para position, i.e., directly opposite the ring carbon that is joined to the (shaded) ethane unit:

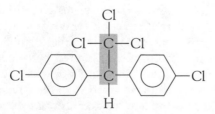

(DDT): *para*-dichlorodiphenyltrichloroethane

BOX 7-1 | The History of DDT

Before DDT, the only insecticides available were those such as arsenic compounds, which were very toxic and persistent, and those extracted from plants, which quickly lost their effectiveness once exposed to the elements. Thus DDT seemed at first to be the ideal insecticide: it was not very toxic to humans but highly toxic to insects; the fact that it was persistent represented a further advantage. DDT was discovered to be an insecticide in 1939 by Paul Müller, a chemist working on the development of various chemicals to fight agricultural insects for the Swiss firm Geigy. Müller was awarded the Nobel Prize in Medicine and Physiology in 1948 in recognition of the many civilian lives DDT saved after the war.

Products containing DDT were marketed within Switzerland beginning in 1941. Since Switzerland was a neutral country in World War II, its government informed both the Allies and the Axis countries about the discovery and uses of DDT; however, it was only the Allies who realized its potential for wartime use to combat infestations of disease-carrying insects in hot climates.

By the end of World War I, more than five million deaths had been caused by typhus. To avoid a repetition of such disasters, an incipient epidemic of typhus in Naples, Italy, was thwarted by spraying all the civilians and the occupying Allied troops with DDT. Outbreaks of typhus in other parts of Europe, including the concentration camps at Dachau and Bergen-Belsen, were dealt with in the same way by the Allied troops as they advanced. Aerial spraying with DDT to combat biting insects was carried out by the Allies at Guadalcanal and elsewhere in the Pacific before their troops invaded the islands. DDT was also used to combat mosquitoes that car-

ried malaria in various parts of Europe, both during and after the war.

Once World War II ended, DDT began to be used not only for public health purposes in hot climates but also extensively in developed countries to control insect pests attacking agricultural crops. Initially it was used on fruit trees and on vegetable crops and subsequently in the growing of cotton. Eventually some insect populations became resistant to DDT and its effectiveness decreased. This phenomenon led farmers to apply greater and greater amounts of the insecticide, particularly on cotton fields.

Within the scientific community, reservations about DDT as the "perfect insecticide" began to be heard almost as soon as it first went into use. In particular, it was known that DDT in soil persisted for several years and could become concentrated in a food chain. The general public became aware of environmental problems associated with DDT upon the publication in 1962 of Rachel Carson's book *Silent Spring*. In it, she discussed the decline of the American robin in certain regions of the United States, due to its consumption of earthworms laden with the DDT used in massive amounts to combat Dutch elm disease. Carson's book stimulated widespread public concern about DDT and other pesticides. Through a series of legal hearings in the United States instigated by lawyers and scientists working with the Environmental Defense Fund, DDT was eventually banned or severely restricted in its use in most states. In 1973 the Environmental Protection Agency banned all DDT uses except those essential to public health. Similar bans were instituted by Sweden in 1969 and later in most other developed countries. DDT is still being used in developing countries to control disease or combat agricultural insects.

Many animal species metabolize (convert by reactions into other substances) DDT by the elimination of HCl; a hydrogen atom is removed from one ethane carbon and a chlorine atom from the other, thereby creating a derivative of ethene called **DDE** (**dichlorodiphenyldichloroethene**):

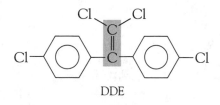

DDE

Substances that are produced by the metabolism of a chemical are called **metabolites;** thus DDE is a metabolite of DDT. The chemical DDE is also produced slowly in the environment by the degradation of DDT under alkaline conditions and by DDT-resistant insects that detoxify DDT by this transformation. Unfortunately, in some birds DDE interferes with the enzyme that regulates the distribution of calcium, so contaminated birds produce eggs that have shells (calcium carbonate) too thin to withstand the weight of the parents sitting on them to hatch them.

PROBLEM 7-2

The structures shown for DDT and DDE have both the ring chlorines in the para position and are sometimes labeled p,p'-DDT and p,p'-DDE, where the p and p' prefixes refer to the chlorine positions in the first and second rings. Deduce the structures and the appropriate labels for all the other unique isomers of both DDT and DDE. Note that the two rings are equivalent, so that, e.g., o,m'-DDT is the same compound as m,o'-DDT.

DDE in Body Fat

DDT's persistence made it an ideal insecticide: one spraying gave protection from insects for weeks to years, depending on the method of application. Its persistence is due to its low vapor pressure and consequent slow rate of evaporation, to its low reactivity with respect to light and to chemicals and microorganisms in the environment, and to its very low solubility in water. Like other organochlorine insecticides, DDT is soluble in organic solvents and therefore in the fat of animal tissue. DDT and/or its degradation products have been found in all birds and fish that have been analyzed, even those living in deserts or ocean depths. We all have some DDT (to the extent of about 3 ppm for North American adults) stored in our body fat.

In humans, most ingested DDT is slowly but eventually eliminated. Most of the "DDT" stored in human fat is actually the DDE that was present in the food that was eaten, having previously been converted from DDT originally in the environment. Unfortunately, DDE is almost non-degradable biologically and is very fat-soluble, so it remains in our bodies for a long time.

Bans on DDT and Other Persistent Pollutants

For environmental reasons, DDT is now banned from use in most Western industrialized countries. Its use had been declining anyway as resistant insect populations evolved that could metabolize DDT to the noninsecticidal DDE and thus render it inactive. The United Nations includes DDT on its list of 12 Persistent Organic Pollutants (Table 7-2). Of the dozen, only DDT will not be totally banned. Its use will be restricted to the control of disease and as an intermediate in the production of *dicofol*, a miticide whose structure is identical except that DDT's remaining hydrogen atom on the ethane unit is replaced by a hydroxy group, —OH. Although many groups pressured the United Nations to include a total ban on DDT, others were strongly opposed to a complete ban, given that it so effective in small amounts against malaria, a disease that kills 1 million children annually in Africa. Groups opposing a ban argued that the health hazards of DDT to humans are minuscule compared to the great benefits it can provide. Switching to an alternative would probably be beyond the financial means of many poor countries and would require more frequent applications, since alternatives are not as persistent and are usually more acutely toxic. Groups supporting a total, immediate ban countered that some countries such as Mexico had already eliminated malaria without the use of DDT and that most mosquitoes responsible for carrying the disease were already resistant to DDT in parts of the world such as India. According to the final treaty, countries can ask to continue using DDT against malaria until effective *and* affordable alternatives are available.

DDT and DDE still enter the environment everywhere as a result of long-range air transport—a topic discussed at length near the end of this chapter—from developing countries where DDT is still in use to control malaria and typhus and for some agricultural purposes. Currently 11 countries in Africa, 7 in Asia, and 5 in Latin America use DDT for disease control (see also Table 7-2).

Thanks to restrictions and bans, the environmental concentrations of DDT and DDE in developed countries dropped substantially in the early and middle years of the 1970s and have now become stabilized at small but nonzero levels. As a result of the decline in DDE levels, bald eagles have made a comeback around Lake Erie and elsewhere. Similarly, the population of Arctic peregrine falcons, a bird that was driven to near-extinction due to the effects of DDE, has now recovered to such an extent that it has been removed from the list of endangered species in the United States.

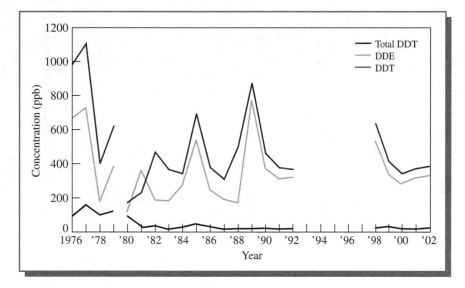

FIGURE 7-1 Concentrations of DDT, DDE, and their sum (total DDT) in 65-cm coho salmon from Lake Ontario. [Source: Alan Hayton, Ontario Ministry of the Environment.]

The concentration of DDT has been found to decline much more rapidly than that of the metabolite DDE, as illustrated in Figure 7-1 for levels in coho salmon from a large polluted water body, Lake Ontario. Indeed, the metabolite levels have shown no consistent sign of a decline in recent decades.

The concentration of DDT in humans has also declined drastically, as illustrated in Figure 7-2a, where its level in human breast milk of Swedish women from 1967 to 1997 is illustrated. The 1997 level is only 1% of that of 1967. The curve drawn through these points is exponential, so the decline is kinetically first-order, with a half-life of four years. The DDE levels in the same women (Figure 7-2b) also began to decline exponentially beginning in the early 1970s, although, as expected, the half-life is longer, six years. Data for women in North America show similar declines over this period.

According to recent research, one probable effect of DDT when levels were much higher in North America than they are today—the early 1960s, for example—was that the risk of mothers delivering preterm babies was greater, by a factor of up to 4 or 5, the higher their DDT exposure, as measured by their blood DDE level. Given that these effects begin to appear for blood DDE levels greater than 10 ppb, and that the average DDE concentration in North American women is now much lower, this risk is probably not one that women in developed countries face today. However, DDE levels in women in countries still using DDT for malaria control often fall in the affected range.

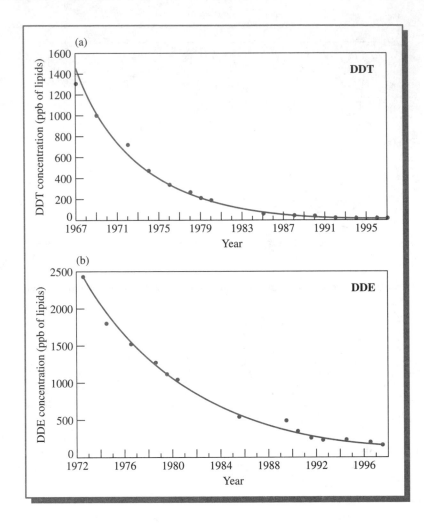

FIGURE 7-2 Trends in (a) DDT and (b) DDE concentrations in breast milk of Swedish women. [Source: K. Noren and D. Meironyte, "Certain Organochlorine and Organobromine Contaminants in Swedish Human Milk in Perspective of Past 20–30 Years," *Chemosphere* 40 (2000): 1111.]

The Accumulation of Organochlorines in Biological Systems

Many organochlorine compounds are found in the tissues of fish in concentrations that are orders of magnitude higher than those in the waters in which they swim. Hydrophobic (water-hating) substances like DDT are particularly liable to exhibit this phenomenon. There are several reasons for this **bioaccumulation** of chemicals in biological systems.

Bioconcentration

In the first place, many organochlorines are inherently much more soluble in hydrocarbon-like media, such as the fatty tissue in fish, than they are in water. Thus, when water passes through a fish's gills, the compounds selectively diffuse from the water into the fish's fatty flesh and become more concentrated there: this process (which also affects organisms besides fish) is called **bioconcentration**. The **bioconcentration factor,** BCF, represents the equilibrium ratio of the concentration of a specific chemical in a fish relative to that dissolved in the surrounding water, provided that the diffusion mechanism represents the only source of the substance for the fish. BCF values occur over a very wide range and vary not only from chemical to chemical but also, to a certain extent, from one type of fish to another, particularly because of variations in the abilities of different fish to metabolize a given substance.

The BCF of a chemical can be predicted, to within a factor of about 10, from a simple laboratory experiment: the chemical is allowed to equilibrate between the liquid layers in a two-phase system made up of water and **1-octanol,** $CH_3(CH_2)_6CH_2OH$, an alcohol that has been found experimentally to be a suitable surrogate for the fatty portions of fish. The **partition coefficient,** K_{ow}, for a substance S is defined as

$$K_{ow} = [S]_{octanol}/[S]_{water}$$

where the square brackets denote concentrations in molarity or ppm units. Largely as a matter of convenience, the value of K_{ow} is often reported as its base-10 logarithm, since its magnitude sometimes is quite large. For example, for DDT (see Table 7-3), K_{ow} is about 1,000,000, i.e., 10^6, so log $K_{ow} = 6$. Experimentally, the bioconcentration factor for DDT lies in the range of 20,000 to 400,000, depending upon the type of fish. The K_{ow} value of a compound is a fairly reliable approximation to the BCF values found for fish. The approximation typically breaks down when the molecules are too large to diffuse into the fish.

TABLE 7-3	Selected Data for Some Pesticides	
Pesticide	Solubility in H_2O (ppm)	log K_{ow}
HCB	0.0062	5.5–6.2
DDT	0.0034	6.2
Toxaphene	3	5.3
Dieldrin	0.1	6.2
Mirex	0.20	6.9–7.5
Malathion	145	2.9
Parathion	24	3.8
Atrazine	35–70	2.2–2.7

Data from K. Verschueren, *Handbook of Environmental Data on Organic Chemicals* (New York: Van Nostrand Reinhold, 1996).

In general, the higher its octanol–water partition coeffficient K_{ow}, the more likely a chemical is to be bound to organic matter in soils and sediment and ultimately to migrate to fat tissues of living organisms. However, log K_{ow} values of 7 or 8 or higher are indicative of chemicals with such strong adsorption to sediments that they are actually unlikely to be mobile enough to enter living tissue. Thus it is chemicals with log K_{ow} values in the 4–7 range that bioconcentrate to the greatest degree.

PROBLEM 7-3

For HCB, log K_{ow} = 5.7. What would be the predicted concentration of HCB due to bioconcentration in the fat of fish that swim in waters containing 0.000010 ppm of the chemical?

Biomagnification

Fish also accumulate organic chemicals from the food they eat and from their intake of particulates in water and sediments onto which the chemicals have

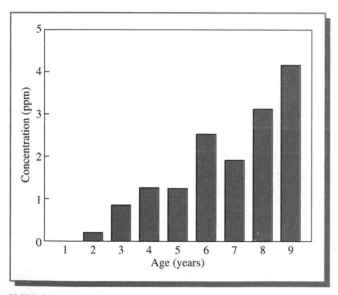

been adsorbed. In many such cases, the chemicals are not metabolized by the fish: the substance simply accumulates in the fatty tissue of the fish, where its concentration increases with time. For example, the concentration of DDT in trout from Lake Ontario increases almost linearly with the age of the fish, as illustrated in Figure 7-3. The average concentration of many chemicals also increases dramatically as one proceeds up a **food chain,** which is a sequence of species, each one of which feeds on the one preceding it in the chain. The **food web,** incorporating interlocking food chains, for the Great Lakes is illustrated in Figure 7-4. Over a lifetime, a fish eats many times its weight in food from the lower levels of the food chain, but it retains rather than eliminates most organochlorine chemicals from these meals.

FIGURE 7-3 Variation with age of DDT concentration in Lake Ontario trout caught in the same year. [Source: "Toxic Chemicals in the Great Lakes and Associated Effects" (Ottawa: Government of Canada, 1991).]

A chemical whose concentration increases along a food chain is said to be **biomagnified.** In essence, the biomagnification results from a sequence of bioaccumulation steps that occur along the chain. The difference between bioconcentration from water and biomagnification along a food chain is illustrated symbolically in Figure 7-5. Notice in Figure 7-4 the herring gull's higher level of DDT compared with the fish below it in the

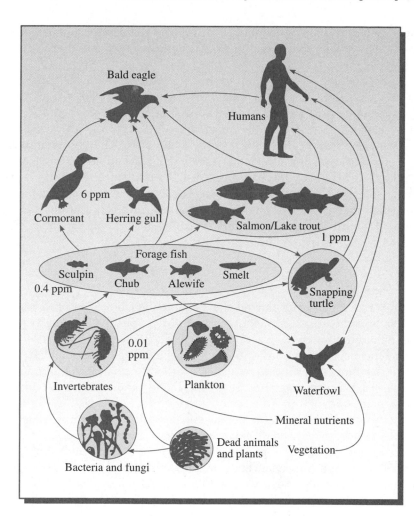

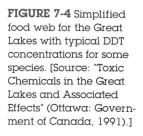

FIGURE 7-4 Simplified food web for the Great Lakes with typical DDT concentrations for some species. [Source: "Toxic Chemicals in the Great Lakes and Associated Effects" (Ottawa: Government of Canada, 1991).]

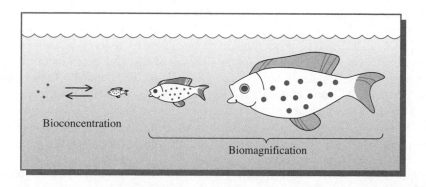

FIGURE 7-5 Schematic representation of the two modes of bioaccumulation that operate in biological matter present in a body of water.

chains. Fish at the top of the aquatic part of the chain bioaccumulate DDT rather effectively, so that even higher concentrations are found in the birds of prey that feed on them.

As an example of biomagnification, consider that the DDT concentration in seawater in Long Island Sound and the protected waters of its southern shore at one time was as high as 0.000003 ppm, but it reached 0.04 ppm in the plankton, 0.5 ppm in the fat of minnows, 2 ppm in the needlefish that swim in these waters, and 25 ppm in the fat of the cormorants and osprey that feed on the fish, for a total biomagnification factor of about 10 million. It is by such mechanisms that DDE levels in some birds of prey became so great that their ability to reproduce successfully was impaired. The bioaccumulation of organochlorines in fish and other animals is the reason why most of the human daily intake of such chemicals comes from our food supply rather than from the water we drink.

Analogs of DDT

A number of compounds having the same general molecular structure as DDT are found to display similar insecticidal properties. This similarity arises from the mechanism of DDT action, which is due more to its molecular *shape* than to chemical interactions with specific species. The shape of a DDT molecule is determined by the two tetrahedral carbons in the ethane unit and by the two flat benzene rings. In insects, DDT and other molecules with the same general size and three-dimensional shape become wedged in the nerve channel that leads out from the nerve cell. Normally, this channel transmits impulses only as needed via sodium ions. But a continuous series of Na^+-initiated nerve impulses is produced when the DDT molecule holds the channel open. As a consequence, the muscles of the insect twitch constantly, eventually exhausting it with convulsions that lead to death. The same process does not occur in humans and other warm-blooded animals since DDT molecules do not exhibit such binding action in nerve channels.

Examples of other molecules with DDT-like action include *DDD* (sometimes called TDE), *para-dichlorodiphenyldichloroethane*, which is an environmental degradation product of DDT: it differs only in that one chlorine from the $—CCl_3$ group is replaced by a hydrogen. Since the overall shapes and sizes of DDT and DDD are similar, their toxicity to insects is as well. In the past, DDD was itself sold as an insecticide, but it has also been discontinued because it bioaccumulates. Notice that DDE, unlike DDT and DDD, is based on a *planar* $C＝C$ unit rather than a $C—C$ linkage with tetrahedral groups at each end. Thus, whereas DDD is a DDT-like insecticide, DDE is not, since its three-dimensional shape is very different: DDE is flat rather than propeller-shaped, so it does not become wedged in the insect's nerve channels.

Scientists have devised analogs to DDT that have the same general size and shape, and consequently possess the same insecticidal properties, but are more biodegradeable and thus do not present the bioaccumulation problem associated with DDT. The most important of these analogs is **methoxychlor:**

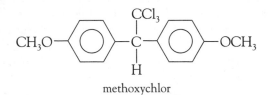

methoxychlor

The *para*-chlorines of DDT are replaced in methoxychlor by methoxy groups, —OCH_3, which are approximately the same size as chlorine but react much more readily. These reactions produce water-soluble products that not only degrade in the environment but are excreted rather than accumulated by organisms. Methoxychlor is still used both domestically and agriculturally to control flies and mosquitoes.

PROBLEM 7-4

Draw the molecular structure of DDD.

PROBLEM 7-5

Methyl groups are approximately the same size as chlorine atoms, but hydrogen atoms are significantly smaller. Would you expect insecticidal properties for DDT molecules in which (a) the —CCl_3 group is replaced by —$C(CH_3)_3$ and (b) the para chlorines are replaced by hydrogens?

Other Organochlorine Insecticides

Toxaphene

During the 1970s, after DDT had been banned, the insecticide that replaced it in many agricultural applications, such as the growing of cotton and soybeans, was **toxaphene.** It is a mixture of hundreds of similar substances, all of which result when the hydrocarbon *camphene*, produced from chemicals extracted from pine trees, is partially chlorinated:

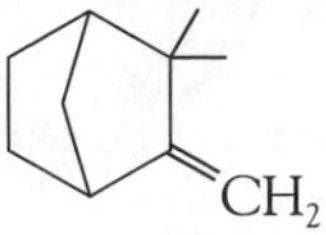

camphene

Toxaphene became the most heavily used insecticide (1966–1976) in the United States before restrictions were placed on its use in 1982 and a total ban was imposed in 1990. Over 100 million kilograms of the insecticide were used in the United States. It was banned in Canada in 1983 and is now outlawed in most developed and developing countries. More than 85% of toxaphene use in the United States occurred in the southeastern cotton-growing states, though some was used as a herbicide in the growing of peanuts and soybeans. Figure 7-6 shows how toxaphene has since spread over North America by air transport after evaporation.

Toxaphene is extremely toxic to fish. Indeed, it was first used in the 1950s in North America to rid lakes of undesirable fish. However, it was found to be so persistent that the lakes could not be successfully restocked for years thereafter!

Toxaphene bioaccumulates in fatty tissues and causes cancer in test rodents; consequently it is listed as a Persistent Organic Pollutant (see Table 7-2). Although it is now banned in developed countries as well as some developing ones, toxaphene is still being deposited in bodies of water remote from its point of use because of long-range transport by air from the developing countries that still make restricted use of it.

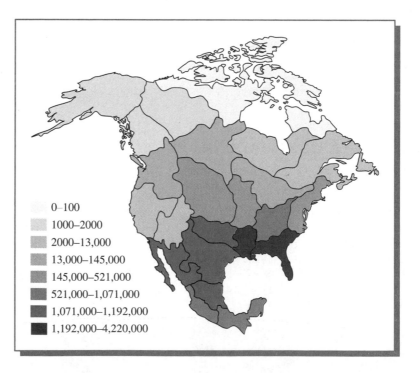

FIGURE 7-6 The calculated spread of toxaphene from the southeastern United States across North America. [Source: K. E. Betts, "Modeling Continent-Wide Contamination," *Environmental Science and Technology* 35 (2001): 481A.]

0–100
1000–2000
2000–13,000
13,000–145,000
145,000–521,000
521,000–1,071,000
1,071,000–1,192,000
1,192,000–4,220,000

Although only a small percentage of toxaphene production occurred in the Great Lakes basin, its concentration in fish in relatively clean Lake Superior exceeds that of any other organochlorine. For a while it was thought that toxaphene might be produced inadvertently as a by-product in pulp and paper mills in the area that used chlorine to bleach paper, since the wood itself might have provided the necessary hydrocarbons. However, this is now known by experiment not to be the case. The levels of toxaphene in the fish of all the other Great Lakes decreased by large factors in the decade following its ban. The near-constant toxaphene levels in the fish of Lake Superior mimic the slowly changing concentration of the chemical dissolved in its waters. Lake's Superior's colder water—and hence slower volatilization rate—and slower sedimentation rate compared to the other Great Lakes is thought to be the main reason why it is so slow to rid itself of its toxaphene loading.

Hexachlorinated Cyclohexane

During World War II the derivative of cyclohexane having one of the two hydrogens on each carbon replaced by chlorine, namely, *1,2,3,4,5,6-* **hexachlorocyclohexane,** was discovered to be an effective insecticide against a wide variety of insects. In fact, there are eight isomers with this formula; they differ only in the relative orientations of the chlorine atoms bonded to different carbons. (The structure below is not intended to illustrate the chlorine orientations—only their points of attachment).

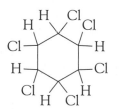

1, 2, 3, 4, 5, 6-hexachlorocyclohexane

(The compound is also known as *benzene hexachloride* or BHC, not to be confused with hexachlorobenzene.)

A commercial mixture of most of the hexachlorocyclohexane isomers was used to control mosquitoes and in agricultural applications after World War II. Its use has been restricted since the 1970s because of its toxicity and tendency to bioaccumulate. Only one of the eight isomers actually kills insects, the gamma isomer, which is sold separately under the name *Lindane*. It was the active ingredient in several commercial medical preparations used to rid children of lice and scabies and to treat seeds and seedlings.

Chlorinated Cyclopentadiene Insecticides

Cyclopentadiene, shown at left below, is an abundant by-product of petroleum refining. As its name implies, there are two double bonds in each molecule. When fully chlorinated (diagram at right), it can be combined with one of several other organic molecules to produce a whole series of insecticidal compounds with properties, including environmental persistence, that at first made them superficially attractive.

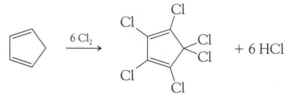

cyclopentadiene   perchlorocyclopentadiene

Most cyclodiene insecticides that were commercially important have now been branded as Persistent Organic Pollutants by the United Nations Environmental Program and are listed in Table 7-2. They were used to control soil insects, fire ants, cockroaches, termites, grasshoppers, locusts, and other insect pests.

The cyclodiene pesticides, starting with **aldrin** and **dieldrin,** arrived on the market in about 1950. Given their persistence, their potential toxicity, their tendency to accumulate in fatty tissues, and the suspicion that dieldrin was causing excess mortality of adult bald eagles, the use of almost all of these compounds has now either been banned or severely restricted in North America and most western European countries. Nonetheless, dieldrin and DDT were the most common POPs still detectable in food in 2002. Some of the compounds are still in use elsewhere (see Table 7-2).

Agricultural uses of dieldrin, mainly to combat soil insects, and use in buildings to control termites were largely prohibited in North America by the mid-1980s. It was used extensively in tropical countries to control the tsetse fly and is still used in some countries to kill termites. It continues to enter water systems in many areas by percolating from waste disposal sites. A recent Danish study has connected dieldrin levels in women with their risk of breast cancer.

Endosulfan (not on the U.N. list) is still used extensively throughout the world as an insecticide for both domestic and agricultural applications. Its bioconcentration and environmental persistence are much lower than those of other cyclodienes, since it is more reactive:

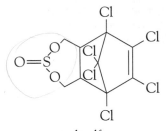

endosulfan

If two perchlorocyclopentadiene molecules are chemically combined, the resultant molecule, known as **mirex,** also acts as an insecticide; it is particularly effective against the fire ant found in the southeastern United States:

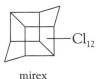

mirex

(All 10 carbon atoms are bonded to chlorine, but for clarity the individual chlorine atoms are not shown; only the total is displayed in the formula.) Mirex was also sold under the name *dechlorane* as a flame retardant additive for synthetic and natural materials. Most mirex/dechlorane use occurred in the 1960s. Although banned since the mid-1970s, mirex is presently classified as a Persistent Organic Pollutant by the United Nations.

For the most part, the chlorinated cyclodiene pesticides are chemical products of the past. Their use has been phased out or at least severely restricted because of environmental and human health considerations.

Principles of Toxicology

Toxicology is the study of the harmful effects to living organisms of substances that are foreign to them. The substances of interest include both synthetic chemicals and those that exist naturally in the environment. In toxicology, the effects are normally determined by injecting animals with the substance of interest and observing how the health of the animal is affected. By contrast, in **epidemiology** scientists do not run experiments in a lab but instead determine the health history of a selected group of human beings and attempt to relate differences in disease rates, etc. to differences in substances to which they have been accidently exposed.

Toxicological data concerning the harmfulness of a substance to an organism, such as an organochlorine pesticide or a heavy metal, are gathered most easily by determining its **acute toxicity,** which is the rapid onset of symptoms—including death at the extreme limit—following the intake of a dose of the substance. For example, experiments show that it takes only a few tenths of 1 microgram of the most acutely toxic synthetic compound—the dioxin to be discussed in Chapter 8—to kill most rodents within a few hours after it is administered orally to them.

Although the acute toxicity of a substance is of interest when we are exposed accidently to pure chemicals, in **environmental toxicology** we are usually more concerned about **chronic** (continuous, long-term) **exposures** at relatively low individual doses of a toxic chemical that is present in the air we breathe, the water we drink, or the food we eat. Generally speaking, any effects—whether cancer, birth defects, etc.—of such continuing exposures are also long-lasting and therefore also classified as chronic.

The same chemical may give rise to both acute and chronic effects in the same organism, although usually by different physiological mechanisms. For example, a symptom of acute toxicity in humans of exposure to many organochlorines is a skin irritation that leads to **chloracne,** a persistent, disfiguring, and painful analog to common acne, and there is the fear that persistent exposure to much lower individual doses than those that produce skin disease could eventually lead to cancer.

Dose–Response Relationships

Most of the quantitative information concerning the toxicity of substances is obtained from experiments performed by administering doses of the substances to animals, although recently tests on certain bacteria have been used to determine whether a substance is likely to be a carcinogen or not. Owing to practical considerations including cost and time, most experiments involve acute rather than chronic toxicity, even though it is the latter that usually is of primary interest in environmental science. To determine directly the effects of continuous low-level exposures over long periods would require a very large number of test animals and a long project time. The practical alternative is to evaluate the effects using high doses—when the effects are substantial and clear-cut—and then extrapolate the results down to environmental exposures. Unfortunately, there is no assurance that such extrapolations are always reliable, since the cellular mechanisms that produce the effects at high and low doses could differ.

The **dose** of the substance administered in toxicity tests is usually expressed as the mass of the chemical, usually in milligrams, per unit of the test animal's body weight, usually expressed in kilograms, thus giving units of milligrams per kilogram. The division by body weight is necessary because the toxicity of a given amount of a substance usually decreases as the size of the individual increases. (Recall that the maximum recommended doses for medicines such as headache remedies are smaller for children than for adults, primarily because of the difference in body masses.) It is also assumed that toxicity values obtained from experiments on small test animals are approximately transferable to humans provided the differences in body weight are taken into account. Normally the toxicity of a substance increases with increasing dose, although exceptions are known.

PROBLEM 7-6

If a dose of a few tenths of a microgram of a certain substance is sufficient to kill a mouse, approximately what mass of the substance would be fatal to you? What average level of the substance would have to be present in the water you drink for you to receive a fatal dose from this source in a week? Note that your weight in kilograms is that in pounds divided by 2.2.

Individuals differ significantly in their susceptibility to a given chemical: some respond to it even at very low doses whereas others require a much higher dose before they respond. For this reason scientists have created **dose–response relationships** for toxic substances, including environmental agents. A typical dose–response curve for acute toxicity is illustrated in Figure 7-7a. The dose is plotted on the (horizontal) x axis, and the cumulative percentage of test animals that display the measured effect (e.g., death) when given a particular dose is shown on the (vertical) y axis. For example, in Figure 7-7a, about 60% of the test animals were affected by a dose of about 4 mg/kg.

Because the range of doses on such graphs often exceeds an order of magnitude, and because the effects at the low end of the concentration scale are often important in environmental decision making and cannot be seen clearly using linear scales, the dose–response plot is often recast using a logarithmic scale for doses. Usually an S-shaped or sigmoidal-shaped curve results from this transformation—see Figure 7-7b.

Most often, the response effect on test animals that is used to construct dose–response curves is death. The dose that proves to be lethal to 50% of the population of test animals is called the LD_{50} value of the substance; its determination from a dose–response curve is illustrated by the dashed lines in Figure 7-7a. The *smaller* the value of LD_{50}, the *more potent* (i.e., more toxic) the chemical, since less of it is required to affect the animal. A chemical much less toxic than that illustrated in Figure 7-7b would have a sigmoidal curve shifted to the right of the one shown.

Many sources quote values for the LOD_{50}, the **lethal oral dose**, when the chemical has been administered orally to the test animals, as opposed to dermal or some other means of administration. For example, the LOD_{50} value for DDT for rats is about 110 mg/kg. As mentioned previously, the

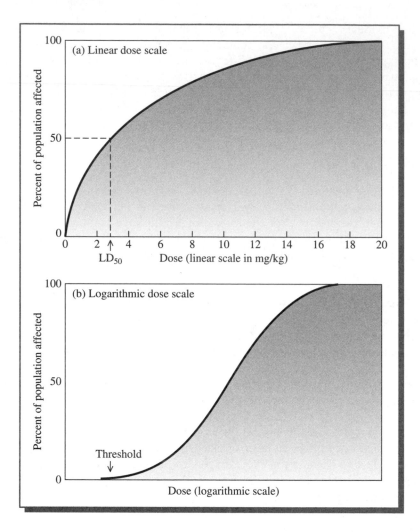

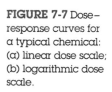

FIGURE 7-7 Dose–response curves for a typical chemical: (a) linear dose scale; (b) logarithmic dose scale.

presumption is usually made that LD_{50} and LOD_{50} values are approximately transferable between species. In the case of DDT, for example, humans are known to have survived doses of about 10 mg/kg, so presumably the LOD_{50} value for humans is greater than 10 mg/kg. However, we have no *direct* evidence that the 110 mg/kg value for rats is also valid for humans.

Of much more concern than the acute toxicity of DDT is its ability to cause chronic effects in humans, such as cancer. Although DDT is not traditionally considered to be a human carcinogen, some small-scale epidemiological studies found that the higher the concentration of DDE in a woman's blood, the more likely she was to have contracted breast cancer. However, later full-scale studies in the United States have failed to confirm this association between breast cancer and DDE. Thus it does not seem that lifetime exposure to DDT is an important cause of breast cancer. However, these studies do not address the issue raised recently of whether exposure during the teenage years, when breasts are developing rapidly, might not be a factor in developing breast cancer decades later.

The range of LD_{50} and LOD_{50} values for acute toxicity of various chemical and biological substances is enormous, spanning about ten powers of 10. Approximate LD_{50} values for some interesting common substances and for the pesticides commonly encountered in domestic products are listed in Table 7-4. Substances that have LD_{50} values of many *grams* per kilogram body weight, such as sugar, are classified as *hardly toxic* or practically nontoxic. All substances are toxic at sufficiently high doses; as the Renaissance-era German philosopher Paracelsus observed, all things are poison, and it is the dose that differentiates a poison from a remedy. Terms such as *moderately toxic, slightly toxic,* and *highly toxic* are sometimes used as descriptors, but there are no agreed-upon definitions for these terms.

TABLE 7-4	LD$_{50}$ Ranges for Some Common Natural and Synthetic Substances, Including Pesticides		
	LD$_{50}$ (approximate) (mg/kg)	LD$_{50}$ (approximate) (g/kg)	Natural substances	Synthetic substances
	>10,000	>10	Sugar	
	1,000	1	Salt; ethanol; pyrethrins	Malathion; atrazine; HCB; mirex; glyphosate; aspirin
	100	10^{-1}	Caffeine; rotenone	DDT; 2,4-D; toxaphene; dimethoate; carbaryl; 2,4,5-T; paraquat; cyanazine; codeine; Tylenol
	10	10^{-2}		Dichlorvos; denitrothion; carbofuran; diazinon; NaCN; As$_2$O$_3$
	1	10^{-3}	Nicotine	Parathion; aldicarb; strychnine
	10^{-1}	10^{-4}	Rattlesnake toxin	
	10^{-2}	10^{-5}	Aflatoxin-B	
	10^{-3}	10^{-6}		2,3,7,8-TCDD
	10^{-4}	10^{-7}		
	10^{-5}	10^{-8}	Tetanus and botulism toxins	

(Left margin, vertical: Direction of increasing toxicity)

In the dose–response curves for some substances there exists a dose below which none of the animals are affected; this is called the **threshold** and is illustrated in Figure 7-7b. The highest dose at which no effects are seen lies slightly below it and is called the **no observable effects level** (NOEL), although sometimes the two terms are used interchangeably. It is difficult to determine the threshold or NOEL level: it may be that if more animals were involved in a particular study, effects at low doses might be uncovered that are not apparent with only a small number of test animals. Most toxicologists believe that for toxic effects other than carcinogenesis there is probably a nonzero threshold for each chemical. A few scientists hold the controversial view that for some substances the curve in Figure 7-7b actually falls below the zero or NOEL level for very low concentrations before returning to zero at zero dose, indicating that tiny amounts of these substances could have a positive rather than a negative effect on health.

Experiments involving test animals are also used to determine how carcinogenic a compound is. However, the simple **Ames test** that uses bacteria can be used fairly reliably to distinguish compounds likely to be human carcinogens from those that are not.

Risk Assessment

Once toxicological and/or epidemiological information concerning a chemical is available, a **risk assessment** analysis can be performed. This analysis tries to answer quantitatively the questions "What are the likely types of toxicity expected for the human population exposed to a chemical?" and "What is the probability of each effect occurring to the population?" Where necessary, risk assessment also tries to determine permissible exposures to the substance in question.

In order to perform a risk assessment on a chemical, it is necessary to know

- **hazard evaluation** information; i.e., the types of toxicity (acute? cancer? birth defects?) that are are expected from it;

- quantitative **dose–response** information concerning the various possible modes of exposure (oral, dermal, inhalation) for it; and

- an estimate of the potential **human exposure** to the chemical.

For chronic exposures, threshold or NOEL dose information is normally expressed as milligrams of the chemical per kilogram of body weight per day. In determining the threshold level for the most sensitive members of the human population, it is common to divide the NOEL from animal studies by a safety factor, typically 100. The resulting value is called the maximum **acceptable daily intake,** ADI, or maximum daily dose; the U.S. EPA instead refers to it as the **toxicity reference dose,** RfD. Note that the ADI or RfD value does *not* represent a sharp dividing line separating absolutely safe from absolutely unsafe exposures, since the transferability of toxicity information from animals to humans is not exact, and in many cases the safety factor is quite generous. Some scientists have suggested dividing the NOEL by a further factor of 10 in order to protect very susceptible groups such as children. Indeed, the 1996 Food Quality Protection Act in the United States requires that the EPA set limits for residues of pesticides on foods 10 times lower than what is considered to be safe for an adult.

PROBLEM 7-7

The NOEL for a chemical is found to be 0.010 mg/kg/day. What would the ADI or RfD value for adults be set at? What mass of the compound is the maximum that a 55-kg woman should ingest daily?

As mentioned, in performing a risk assessment an attempt is made to estimate the exposure of the affected population. For example, for chemicals whose mode of exposure is primarily through drinking water, regulatory agencies such as the U.S. EPA consider a hypothetical average person who drinks about 2 L of water daily and whose body weight averages 70 kg (154 lb) through life. If the ADI (or RfD) is 0.0020 mg/kg/day, then for the 70-kg person, the mass of substance that can be consumed per day is $0.0020 \times 70 = 0.14$ mg/day. Thus the maximum allowable concentration of the chemical in water would be 0.14 mg/2 L = 0.07 mg/L = 0.07 ppm. Of course, if there are other significant sources for the substance, they must be taken into account in determining the drinking water standard. Also, exposure to several chemicals of the same type (e.g., several organochlorines) might lead to additive effects, so the standard for each one should presumably be lowered to take this into consideration.

The EPA regulates exposure to carcinogens by assuming that the dose–response relationship has no threshold and can be linearly extrapolated from zero dose to the area for which experimental results are available. The maximum daily doses are then determined by assuming that each person receives the dose every day over his or her lifetime and that this exposure should not increase the likelihood of cancer to more than one person in every million.

In determining regulations to control risk, usually no consideration is given to the economic costs involved. Many economists believe that because regulations cost money to implement, and because society may decide that it has only limited resources that it is willing to spend on regulation, a cost-benefit analysis should be used to help decide which substances to regulate. Associated with this line of thinking is the idea that regulations should show a positive payback: the benefits from regulation should exceed the costs. Of course it is difficult to place a specific monetary value on environmental benefits in many cases. For example, is the $200,000 average cost for saving a life by regulating the chloroform content in water (see Chapter 10) a reasonable amount to pay? If so, would it still be reasonable if the cost instead were 10 or 100 or 1000 times this amount?

The Distribution of Environmental Pollutants

When a persistent chemical, such as DDT, is released into the environment, we find that later some of it has dissolved in natural bodies of water, some is in the air, some is present in soil and sediments, and some is located in living matter. A constant interchange of the chemical occurs among these various phases. It is possible to estimate the amount and concentration of the chemical in each phase once the release of the chemical has ended and sufficient time has passed for equilibrium among the phases to be achieved. Even when equilibrium conditions are not yet in place, it is of value to determine the phases where the chemical will ultimately be concentrated.

Recall from your previous background in chemistry that in calculations involving substances participating in *chemical* reactions we algebraically combine experimental values of equilibrium constants with information concerning initial concentrations in order to determine equilibrium concentrations. A somewhat analogous procedure can be applied to determine the distribution of a substance when by *physical* processes it has achieved equilibrium between several phases. The condition that equilibrium has been achieved in its distribution is that the **fugacity,** f, of the substance, which is defined as *its tendency to escape from a given physical phase*, is equal for all phases. Fugacity has units of pressure, e.g., atmospheres or kilopascals. Thus when all the DDT in the environment has been distributed among air, water, sediment, biota, etc., the concentration in each phase is such that its tendency to escape from any phase (and enter any other) has the same value for all phases.

As you might expect, the fugacity of a substance in a given phase is proportional to its concentration, C, in that phase:

$$f = C/Z$$

where Z is the *fugacity capacity constant* for the substance and the phase. Generally, the higher the value of Z, the greater the tendency of a chemical to concentrate in that phase. (These capacity constants are analogous to the equilibrium constants used in chemical reaction calculations.) If we use x to denote the phase of interest, then

$$f_x = C_x/Z_x$$

We can determine the concentration in each phase by rearranging the equation, to give

$$C_x = f_x Z_x$$

At equilibrium, the f_x values for all phases are identical, equal to f, say. Thus, if we know f, we can determine the concentration in each phase from the simplified equation

$$C_x = f Z_x$$

As in chemical equilibrium problems, we usually know the total number of moles, n_{total}, of the material. As in chemical problems, it is useful to state the mass conservation condition: the sum of the equilibrium number of moles, n_x, present in each phase x must add up to n_{total}. By definition, each n_x is equal to the concentration C_x times the volume V_x for the phase:

$$n_x = C_x V_x$$

Substitution of the next-to-last equation into the last one gives

$$n_x = f Z_x V_x$$

When we sum the n_x values over all phases X of interest, we must obtain the total number of moles. Thus

$$n_{total} = f \Sigma Z_x V_x$$

Rearrangement of this equation allows us to calculate the value of the system fugacity:

$$f = n_{total} / \Sigma Z_x V_x$$

An Example of a Fugacity Calculation

As an example of how fugacity calculations are carried out in practice, consider the distribution of 1 mol of DDT among three phases: air, water, and sediment in a model compartment of Earth (Figure 7-8). As discussed later, we take the volume of air to be 10^{10} m^3, the water volume to be 7×10^6 m^3, and the volume of accessible sediment to be 2×10^4 m^3. The values of the Z_x constants for DDT, in units of mol/atm m^3, are determined from experimental data to be

for the air phase, 40.3
for the water phase, 3.92×10^4
for the sediment phase, 2.25×10^9

In the evaluations of Z_x values from experimental data, a temperature of 25°C is usually assumed for simplicity. The Z_x values for sediments (and biota) are assumed to be proportional to the octanol–water partition coefficients K_{ow} discussed earlier in the chapter.

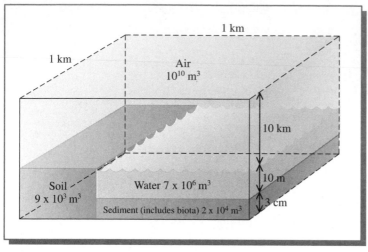

1 km

1 km

Air
10^{10} m^3

10 km

10 m

3 cm

Soil
9×10^3 m^3

Water 7×10^6 m^3

Sediment (includes biota) 2×10^4 m^3

FIGURE 7-8 Model world parameters used in fugacity calculations.

After substitution of the values for Z_x and V_x, the value of the fugacity in this case is

$$f = 1.0/(40.3 \times 10^{10} + 3.92 \times 10^4 \times 7 \times 10^6 + 2.25 \times 10^9 \times 2 \times 10^4)$$

$$= 1.0/(4.03 \times 10^{11} + 2.74 \times 10^{11} + 4.5 \times 10^{13})$$

$$= 2.19 \times 10^{-14} \text{ atm}$$

The concentration of the chemical and the amount of it can now be computed for each phase:

$$C_x = f Z_x \quad \text{so}$$

$$[\text{DDT in air}] = 2.19 \times 10^{-14} \times 40.3 = 8.8 \times 10^{-13} \text{ mol/m}^3$$

$$[\text{DDT in water}] = 2.19 \times 10^{-14} \times 3.92 \times 10^4 = 8.6 \times 10^{-10} \text{ mol/m}^3$$

$$[\text{DDT in sediment}] = 2.19 \times 10^{-14} \times 2.25 \times 10^9 = 4.9 \times 10^{-5} \text{ mol/m}^3$$

Notice the preferential concentration of DDT in sediment, which is hydophobic due to its carbon content.

The *amounts* in each phase are given by the fZV values, i.e., the concentrations multiplied by the volumes. Thus the number of moles of DDT

$$\text{in air} = 8.8 \times 10^{-13} \times 1 \times 10^{10} = 0.0088 \text{ mol}$$

$$\text{in water} = 8.6 \times 10^{-10} \times 7 \times 10^6 = 0.0060 \text{ mol}$$

$$\text{in sediments} = 4.9 \times 10^{-5} \times 2 \times 10^4 = 0.98 \text{ mol}$$

Thus we see that, with air, water, and sediment available, 98% of the DDT will be found in sediments and about 1% in both air and water. Notice that the concentration of DDT in water is greater than in air, but the total amount of it in air exceeds that in water because the air volume is so much larger. This sort of switch in ordering between amount and concentration in different phases is common for pollutant chemicals.

Model World Parameters for Fugacity Calculations

The volumes for the various phases used in these calculations are based on a model "world" whose components are able to be in equilibrium with each other. Since only concentrations are obtained in the calculations, it is important that only the *relative* volumes, not their absolute values, be appropriate. The model world is a square 1 km by 1 km whose parameters are assumed to be the average values for the real Earth. The atmosphere is taken to be 10 km high, which is a reasonable approximation to the troposphere. The air volume then is $(1000 \text{ m} \times 1000 \text{ m}) \times (10,000 \text{ m}) = 10^{10} \text{ m}^3$. The 1-km square is assumed to be 70% covered by water and 30% by soil. The average water depth is taken to be 10 m, which is relatively shallow since we are interested only in the part that achieves equilibrium with the air. Thus the water volume is $0.7 \times (1000 \text{ m} \times 1000 \text{ m}) \times 10 \text{ m} = 7 \times 10^6 \text{ m}^3$. The sediment in equilibrium with this water is assumed to be only 3 cm deep, giving it a volume of $0.7 \times 1000 \text{ m} \times 1000 \text{ m} \times 0.03 \text{ m} = 2.1 \times 10^4 \text{ m}^3$. In addition to air, water, and sediment, the model usually also includes soil, whose effective volume is $9 \times 10^3 \text{ m}^3$, plus 35 m^3 of solids suspended in the water, and about 3.5 m^3 of biota such as fish. The Z values for biota are usually of the same order of magnitude as those for sediment, so the concentration of a given chemical in biota is close to that in sediment.

PROBLEM 7-8

The Z values for hexachlorobenzene are 4×10^{-4} in air, 9.5×10^{-5} in water, and 2.3 in sediment (and biota). Using the model world volumes given, calculate the equilibrium concentrations when 1 mol of hexachlorobenzene is distributed among air, water, and sediment.

Organophosphate and Carbamate Insecticides

Organophosphate Insecticides

The largest use of **organophosphate** (OP) insecticides is in agriculture (see Figure 7-9), but they also find many domestic uses. Organophosphates generally are nonpersistent; in this respect, they represent an environmental advantage over organochlorines since they do not bioaccumulate in the food chain, presenting a chronic exposure and health problem to us later. However, they are generally much more acutely toxic to humans and other mammals than are organochlorines. Many organophosphates represent an acute danger to the health of those who apply them and to others who may come into contact with them. Exposure to these chemicals by inhalation, swallowing, or absorption through the skin can lead to immediate health problems. However, organophosphates metabolize relatively quickly and are excreted in the urine.

After application, organophosphates decompose within days or weeks and thus are seldom found to bioconcentrate in food chains. However, because they have a wide range of uses in homes, on lawns, in commercial buildings, and in agriculture, almost everyone is regularly exposed to them. There is some evidence that OPs cause chronic as well as acute

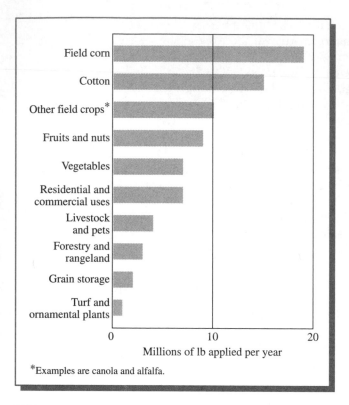

Field corn

Cotton

Other field crops*

Fruits and nuts

Vegetables

Residential and
commercial uses

Livestock
and pets

Forestry and
rangeland

Grain storage

Turf and
ornamental plants

0   10   20

Millions of lb applied per year

*Examples are canola and alfalfa.

FIGURE 7-9 Consumption of organophosphate pesticides by various crops in the United States. [Source: B. Hileman "Reexamining Pesticide Risk," *Chemical and Engineering News* (17 July 2000), 34.]

health problems. Children in the United States generally have their greatest exposure to organophosphates through the food they eat, with exposure from drinking water at 1–10% of that from food. A 2003 report found that preschool children (in Seattle) who consumed mainly organic fruits, vegetables, and juices had much lower exposure to organophosphates, as measured by metabolites in their urine, than children with conventional diets. A number of recent studies, none of them large enough to be statistically definitive, have found links between indoor use of insecticides — organophosphates especially — and childhood leukemia and brain cancer. The U.S. EPA has placed OPs in its highest-priority group in its current re-examination of pesticides.

Structurally, all organophosphate pesticides contain a central pentavalent phosphorus atom to which are connected

- an oxygen or sulfur atom doubly bonded to the P atom,

- two methoxy ($—OCH_3$) or ethoxy ($—OCH_2CH_3$) groups singly bonded to the P atom, and

- a longer, more complex, characteristic R group singly bonded, usually through an oxygen or sulfur atom to the phosphorus unit, that differentiates one organophosphate from another.

The three main classes of the organophosphates are illustrated in Figure 7-10. Those containing a P=S unit are converted inside the insect into the corresponding molecules with a P=O unit, producing a more toxic substance. The P=S form is used initially because it penetrates the insect more readily and is more stable than the corresponding P=O compound. These organophosphates decompose fairly rapidly in the environment because oxygen in the air alters P=S bonds to P=O. Water molecules add to and thereby split P—O bonds in all the organophosphates, ultimately yielding nontoxic substances such as *phosphoric acid*, $O=P(OH)_3$ and its ions, and alcohols.

Dichlorvos is an example of a organophosphate molecule containing no sulfur. It is a relatively volatile insecticide and is used as a domestic fumigant released from impregnated fly strips hung from ceilings and light fixtures. The

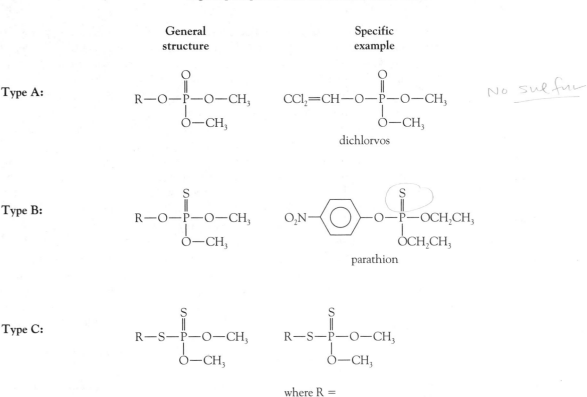

FIGURE 7-10 General structures and examples of organophosphate insecticides. In some molecules (e.g., parathion) the methoxy groups are replaced by ethoxy groups.

chemical slowly evaporates, and its vapor kills flies in the room. Plastic is impregnated with dichlorvos for use in flea collars. It is relatively toxic to mammals; its LOD_{50} is 25 mg/kg in rats. (Note the -os ending of the name of this insecticide. The commercial names of organophosphate insecticides often have this or a -fos ending, signaling their nature.)

Parathion (Figure 7-10) is an example of an organophosphate in which only the doubly bonded oxygen has been replaced by sulfur. It is very toxic (LOD_{50} = 3 mg/kg) and is probably responsible for more deaths of agricultural field workers than any other pesticide. Since it is nonspecific to insects, its use can inadvertently kill birds and other nontarget organisms. Bees, too, which are often economically valuable, are indiscriminately destroyed by parathion. It is now banned in some Western industrialized countries but is still widely used in developing countries. The background levels of parathion in homes in the agricultural community in the United States have dropped by an order of magnitude since it was banned in 1991. The structure of *fenitrothion* is very similar to that of parathion but it is much less toxic to mammals ($LOD_{50} =$ 250 mg/kg). It has been used extensively as a spray against spruce budworm in evergreen forests in eastern Canada, though not without some controversy.

Diazinon, commonly used for insect control in homes (against ants and roaches), gardens and lawns (including grub control), shrubs, and on pets, and once thought to be relatively safe ($LOD_{50} = 300$ mg/kg), also has a P=S structure. Because diazinon is toxic to birds, its use has been partially restricted for some time. There is now some evidence that children treated for diazinon poisoning suffer persistent neurobehavioral problems. Consequently, the residential uses of diazinon in North America are being phased out.

Chlorpyrifos ($LOD_{50} = 135$ mg/kg), another member of this P=S organophosphate group, was commonly used in households to control cockroaches, ants, termites, and other insects. Indeed, it was the insecticide most commonly sprayed by exterminators to kill cockroaches. However, it has been withdrawn in the United States for domestic use by its manufacturer due to health concerns, especially involving fetal and childhood exposure, following experiments on rats, and due to unintentional poisonings. Its uses had been restricted as of the late 1990s by the EPA, as had that of **methyl parathion,** which is parathion with methyl rather than ethyl groups. Indeed, most indoor uses of organophosphates have been eliminated by regulatory action in recent years. A recent survey linked the use of indoor insecticides, but not herbicides or outdoor insecticides, especially during pregnancy, to an increase in leukemia among the children born following fetal exposures.

Malathion is the most important example of the organophosphate type in which two oxygens have been replaced by sulfur, giving a P=S and one P—S unit. Upon exposure to oxygen, malathion is converted to *malaoxon*, which has one sulfur replaced by oxygen. Introduced in 1950, malathion is not particularly toxic to mammals ($LOD_{50} = 885$ mg/kg) but it is nevertheless fatal to many insects since they metabolize it in a different way. However, if improperly stored, it can be converted to an isomer that is almost 100 times as toxic; it was responsible for the death of five spray workers and the sickness of thousands more during a malaria-eradication program in Pakistan in 1976.

Malathion is still used in domestic fly sprays and to protect agricultural crops. In combination with a protein bait, low concentrations of malathion have been sprayed from helicopters over several areas in the United States (California, Florida, Texas) to combat infestations of the Mediterranean fruit fly, a dangerously destructive pest. As in California, aerial spraying of malathion in Chile to combat the fruit fly has also proven to be controversial. It has also been sprayed in parts of New York City (1999) and Florida (1990) to protect against St. Louis encephalitis, carried by mosquitoes. For decades, the entire city of Winnipeg, Manitoba, has been sprayed with malathion several times each summer to keep down the mosquito population and recently to protect against the West Nile virus.

Dimethoate ($LOD_{50} = 250$ mg/kg) belongs to the same structural group as malathion; it is often used for insect control on food crops, including those grown in backyards. Yet another group member, *azinphos-methyl* ($LOD_{50} = 5$ mg/kg), is an insecticide widely used by professionals on fruits and vegetables. However, its use in the United States has been restricted somewhat, since it can pose acute risks to young children exposed to it through their diets and to agricultural workers because its LOD is quite low. According to a survey in the mid-1990s, azinphos-methyl and another organophosphate of this same group, *phosalone*, together with the fungicide *diphenylamine*, were the most common pesticides present in residues in and on apples grown commercially in Canada.

The organophosphates are toxic to insects because they inhibit enzymes in the nervous system; thus they function as nerve poisons. In particular, organophosphates disrupt the communication carried between cells by the **acetylcholine** molecule. This cell-to-cell transmission cannot operate properly unless the acetylcholine molecule is destroyed after it has executed its function. Organophosphates block the action of the enzymes whose job it is to destroy the acetylcholine, by selectively bonding to them. The presence of the insecticide molecule has the effect of suppressing the continued transmission of impulses between nerve cells that is essential to the coordination of the organism's vital processes, and death ensues. (At the atomic level, it is the phosphorus atom of the organophosphate molecule that attaches to the enzyme and stays bound to it for many hours.)

Carbamate Insecticides

The mode of action of **carbamate** insecticides is similar to that of the organophosphates; they differ in that it is a carbon atom rather than a phosphorus that attacks the acetylcholine-destroying enzyme. The carbamates, introduced as insecticides in 1951, are derivatives of **carbamic acid,** H_2NCOOH. One of the hydrogens attached to the nitrogen is replaced by an alkyl group, usually methyl, and the hydrogen attached to the oxygen is replaced by a longer, more complex organic group, symbolized below simply as R:

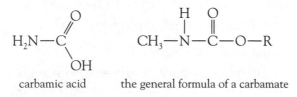

carbamic acid   the general formula of a carbamate

Like the organophosphates, the carbamates are short-lived in the environment since they react with water and decompose to simple, nontoxic products. (The reaction with water involves the splitting of one of the single bonds to the central carbon.) They are more attractive for some applications since their dermal toxicity is rather low.

Important examples of the carbamate pesticides are **carbofuran** (LOD_{50} = 8 mg/kg), **carbaryl** (LOD_{50} = 307), and **aldicarb** (LOD_{50} = 0.9); the latter is very toxic indeed to humans. Carbaryl is a widely used lawn and garden insecticide and has a low toxicity to mammals, but it is particularly toxic to honeybees.

Health Problems of Organophosphates and Carbamates

In summary, the organophosphates and carbamates solve the problem of environmental persistence and accumulation associated with organochlorine insecticides, but sometimes at the expense of dramatically increased acute toxicity to the humans and animals who encounter them while the chemicals are still in the active form. These less persistent insecticides—together with the pyrethroids mentioned below—largely replaced organochlorines in residential uses. Organophosphates and carbamates are a particular problem in developing countries, where widespread ignorance about their hazards and failure to use protective clothing—due to ignorance or to the heat—have led to sickness and many deaths among agricultural workers. The types of pesticides used in developing countries are also more likely to be highly toxic, even banned elsewhere for that reason. Carbofuran, for example, is used to control the Andean weevil in potato crops grown in Ecuador, with the result that crop yields are much improved, but at the expense of pesticide poisoning of some of the potato farmers. Estimates by the United Nations and the World Health Organization put the number of persons who suffer acute illnesses from short-term exposure to pesticides in the millions annually; 10,000–40,000 die each year from the poisoning, about three-quarters of these in developing countries. About 99% of the deaths from pesticide poisonings occur in developing countries. Even so, about 20,000 people receive emergency medical care in the United States annually for actual or suspected poisoning from pesticides; about 30 people annually die from it.

Natural and Green Insecticides, and Integrated Pest Management

Pesticides from Natural Sources

As pointed out earlier, many plants can themselves manufacture certain molecules for their own self-protection that either kill or disable insects. Chemists have isolated some of these compounds so that they can be used to control insects in other contexts. Examples are nicotine, rotenone, the pheromones, and juvenile hormones.

One group of natural pesticides that has been used by humans for centuries is the **pyrethrins.** The original compounds (the general structure is illustrated below) were obtained from the flowers of a certain species of chrysanthemum.

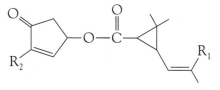

general pyrethrin structure

In the form of dried, ground-up flower heads, pyrethrins were used in Napoleonic times to control body lice; they are still used in flea sprays for animals. They are generally considered to be safe to use. Like organophosphates, they paralyze insects, though they usually do not kill them. Unfortunately, these compounds are unstable in sunlight. For that reason, several synthetic pyrethrin-like insecticides that are stable outdoors—and therefore can be used in agricultural applications—have been developed by chemists. Synthetic pyrethrins are called **pyrethroids,** though they usually are given names ending in *-thrin* to denote their origin (e.g., *permethrin*). Pyrethroids are now also a common ingredient in domestic insecticides, as any visit to a garden center will testify. They have been used in Mexico to spray houses in which malaria control is still necessary.

Rotenone, a complex natural product derived from the roots of certain bean plants, has been used as a crop insecticide for over 150 years and for centuries to paralyze and/or kill fish. The compound enters fish via their gills and disrupts the respiratory system. It is also highly efficient against insects and is decomposed by sunlight. Rotenone is widely used in hundreds of commercial products, including flea-and-tick powders and sprays for tomato plants. Notice in Table 7-4 that pyrethrins and rotenone, which are natural substances, have about the same acute toxicity as some synthetic ones, such as malathion, even though they are often marketed as safer, natural pesticides. Unfortunately, there is some evidence that chronic exposure to rotenone can contribute to the onset of Parkinson's disease.

Integrated Pest Management

In recent years, **integrated pest management** (IPM) strategies have been developed. They combine the best features of various feasible methods of pest control—not just the use of chemicals—into a long-range, ecologically sound plan to control pests so that they do not cause economic injury. Generally speaking, a unique plan is developed for each area and crop, with chemicals used only as a last resort and when the monetary cost of their use will be more than recovered by increased crop yield. The six pest control methods that can be combined are

- chemical control—the use of chemical pesticides, both synthetic and natural;

- biological control—reducing pest populations by the introduction of predators, parasites, or pathogens;

- cultural control—introducing farm practices that prevent pests from flourishing;

- host-plant resistance—using plants that are resistant to attack, including plants adapted by genetic engineering to have greater resistance;

- physical control—using nonchemical methods to reduce pest populations; and

- regulatory control—preventing the invasion of an area by new species.

Green Chemistry: Insecticides That Target Only Certain Insects

Insecticides such as the organophosphates and carbamates interrupt the function of specific enzymes that are common to most insects (and to humans). They are thus toxic to a wide array of insect species and are known as **broad-spectrum insecticides.** Although it may be an advantage to kill more than one species with a single pesticide, this often leads to the demise of beneficial insects such as pollinators (bees) and natural enemies (lady bugs and praying mantises) of insects that are pests.

An approach to limiting the environmental effect of an insecticide is to develop insecticides that are toxic to only certain species, i.e., the target organism. One way to accomplish this is to find a biological function that is unique to the target organism and develop an insecticide that interrupts only that function. The Rohm and Haas Company of Philadelphia, Pennsylvania, won a Presidential Green Chemistry Challenge Award in 1998 for the development of *Confirm*, *Mach 2*, and *Intrepid*. The active ingredients in these insecticides are members of the **diacylhydrazine** family of compounds (Figure 7-11a) and are effective in controlling caterpillars.

Caterpillars are the larval stage of insects such as moths and butterflies, and during the larval stage they must shed their cuticle to grow. The concentration of *20-hydroxyecdysone* (Figure 7-11b), which is produced by the caterpillar and is a member of the steroid family of compounds, increases during the molting process. As a result of its presence, the caterpillar ceases to feed and sheds its cuticle. The concentration of this natural compound then decreases and the caterpillar resumes feeding. The diacylhydrazines present in the commercial products Confirm, Mach 2, and Intrepid mimic 20-hydroxyecdysone; however, their concentrations do not decline, and consequently the insect never resumes feeding. The insect thus dies of starvation or dehydration. These insecticides target only insects that go through molting stages during their growth; thus most insects will be unaffected.

Confirm and Intrepid are classified as **reduced-risk pesticides** by the U.S. EPA. This classification program was started in 1993. To be placed in this category, a pesticide must meet one or more of the following requirements:

1. it reduces pesticide risks to human health;

2. it reduces pesticide risks to nontarget organisms;

3. it reduces the potential for contamination of valued environmental resources; or

4. it broadens the adoption of IPM (integrated pest management, discussed in the preceding section) or makes it more effective.

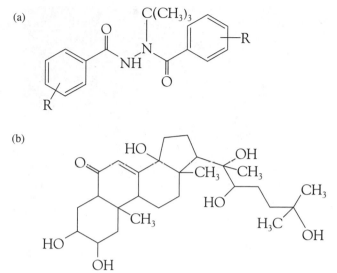

FIGURE 7-11 (a) Diacyl-hydrazine pesticides; (b) 20-hydroxyecdysone.

The diacylhydrazene insecticides certainly meet the first two of these requirements. In order to encourage the development of lower-risk pesticides, the EPA rewards the developers of pesticides that contain active ingredients that meet the EPA reduced-risk criteria with expedited review. (http://www.epa.gov/oppfead1/fqpa/rripmpp.htm).

Green Chemistry: A New Method for Controlling Termites

Termites invade over 1.5 million homes in the United States annually and cause about $1.5 billion in damage. Traditional treatments for termites involve treating the soil around the affected structure with 100–200 gal of pesticide solution to create an impenetrable barrier. This process may result in groundwater contamination, accidental worker exposure, and detrimental effects to beneficial insects.

Dow AgroSciences in Indianapolis won a Presidential Green Chemistry Challenge Award in 2000 for its development of *Sentricon*. In contrast to the traditional control of termites, Sentricon employs monitoring stations to first detect the presence of termites prior to the use of any insecticide. The

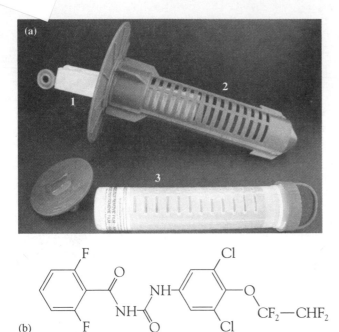

monitoring stations (Figure 7-12a) consist of pieces of wood (1) contained in perforated plastic tubes (2), which are placed in the ground around the structure. If termites are detected in any of the monitoring stations, the wood pieces are replaced by a perforated plastic tube (3) containing the bait. Bait stations may also be placed in the structure. The bait consists of a mixture of cellulose and the pesticide **hexaflumuron.** Hexaflumuron interrupts the molting process of termites and thus is not harmful to most beneficial insects. Termites that have ingested the bait return to their nests and share the bait by trophallaxis, thus spreading the insecticide throughout the colony. Once the colony has been decimated, the bait is replaced with wood and monitoring resumes.

Hexaflumuron was the first pesticide to be classified as a reduced-risk pesticide by the U.S. EPA. Hexaflumuron is significantly less toxic, and is used in quantities 100 to 1000 times less, than traditional pesticides (see Table 7-5) employed for termite control.

FIGURE 7-12 (a) Sentricon monitoring/baiting station; (b) Hexaflumuron structure. [Source: Photo by Michael Cann.]

Herbicides

Herbicides are chemicals that destroy plants. They are usually employed to kill weeds without causing injury to desirable vegetation; e.g., to eliminate broad-leaf weeds from lawns without killing the grass. The agricultural use of herbicides has replaced human and mechanical weeding in developed countries and has thereby sharply reduced the number of people employed

TABLE 7-5	Toxicities of Traditional Termiticides vs. Hexaflumuron[a]	
Pesticide (compound type)	Acute oral LD_{50} (mg/kg)	Acute dermal LD_{50} (mg/kg)
Chlorpyrifos (organophosphate)	135–163	2000
Permethrin (pyrethroid)	430–4000	>4000
Imidacloprid (chloronicotinyl)	424–475	>5000
Fipronil (pyrazole)	100	>2000
Hexaflumuron	>5000	>2000

[a] Typical quantity applied: traditional pesticides, 750–7000 g; hexaflumuron, 2–5 g.

in agriculture. Herbicides are also used to eliminate undesirable plants from roadsides, railway and powerline rights-of-way, etc., and sometimes to defoliate entire regions. Ever since the late 1960s, herbicides have been the most widely used type of pesticide in North America. As of the early 1990s, about half the herbicide used in the United States is applied to corn, soybean, and cotton crops. More recently, however, the application of insecticides to cotton has been significantly reduced by the introduction of cotton genetically modified to incorporate resistance to insects.

In ancient times, armies sometimes used salt or a mixture of brine and ashes to sterilize land that they had conquered, intending to make it uninhabitable by future generations of the enemy. In the first half of the twentieth century several inorganic compounds were used as weed killers—principally **sodium arsenite,** Na_3AsO_3; *sodium chlorate,* $NaClO_3$; and *copper sulfate,* $CuSO_4$. The latter two belong to a large group of salts formerly used as herbicidal sprays that kill plants by the rather primitive action of extracting the water from them, while at the same time leaving the treated land still capable of supporting agriculture.

Eventually, organic derivatives of arsenic replaced its inorganic compounds as herbicides since they are less toxic to mammals. Inorganic and metal-organic herbicides have been largely phased out because of their persistence in soil. Completely organic herbicides now dominate the market; their utility is based partially on the fact that they are much more toxic to certain types of plants than to others, so they can be used to eradicate the former while leaving the latter unharmed.

Atrazine and Other Triazines

One modern class of herbicides is the **triazines,** based on the symmetric aromatic structure shown below, which has alternating carbon and nitrogen atoms in a six-membered benzene-like ring:

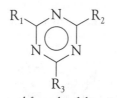

the general formula of the triazines

In herbicidal triazines, R_1 = Cl and R_2, R_3 = amino groups. In these triazines, one carbon atom in the ring is bonded to chlorine and the other two to amino groups, which are nitrogen atoms singly bonded to hydrogens and/or carbon chains.

The best-known member of this group is **atrazine,** which was introduced in 1958 and has been used since in huge quantities to destroy weeds in corn fields. Indeed, atrazine is the most heavily used herbicide in the United

States, accounting for 40% of all weed killers applied in the country—and probably the world—including use on 75% of corn crops. In atrazine, R_2 is —NH—CH$_2$CH$_3$ and R_3 is —NH—CH(CH$_3$)$_2$.

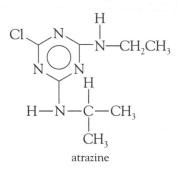

atrazine

It usually is applied to cultivated soils, at the rate of a few kilograms per hectare or one kilogram per acre, in order to kill grassy weeds, mainly in support of corn and soybean cultivation.

Biochemically, atrazine acts as a herbicide by blocking photosynthesis in the plant in the photochemical stage that initiates the reduction of atmospheric carbon dioxide to carbohydrate. Higher plants, including corn, tolerate triazines better than do weeds, since they rapidly degrade them to nontoxic metabolites. However, if the triazine concentration builds up in a soil—e.g., because of lack of moisture to degrade it—a stage can be reached where no plants will grow. In high concentrations, atrazine has been used to eliminate all plant life, e.g., to create parking lots. Some weeds are becoming atrazine-tolerant. The main ecological risk from its wide-spread use is the death of sensitive plants in water systems close to agricultural fields. Canada has set 2 ppb as the maximum concentration in water for protection of aquatic life. Some controversial recent research regarding the effects of low levels of atrazine on wildlife are discussed in Chapter 8.

While in soil, atrazine is degraded by microbes. One such biochemical reaction results in the replacement of chlorine by a hydroxyl group, —OH, yielding a metabolite that is not toxic to plants. The other microbial pathways involve the loss of either the ethyl group or the isopropyl group from an amino unit, with its replacement by hydrogen; these metabolites are toxic to plants.

Atrazine is moderately soluble (30 ppm) in water. During rainstorms it is readily desorbed from soil particles and dissolved in the water moving through the soil. In waterways that drain agricultural land on which atrazine is used its concentration typically is found to be a few parts per billion. Usually atrazine is detectable in well water in such regions. The possible risks of atrazine to amphibians are discussed in Chapter 8.

Although it persists in most soils only for a few months, once it or its metabolites enter waterways, atrazine's half-life is several years. For example, in the Great Lakes its half-life is about two to five years, whereas it is less than half a year in Chesapeake Bay, presumably because the water is warmer so metabolism is faster there.

Unfortunately, atrazine is not removed by typical treatments of drinking water unless carbon filtration is used. However, less than 0.25% of the population in the U.S. corn belt states consume atrazine at greater than the 3 ppb, its **maximum contaminant level,** MCL. The MCL values for substances are the maximum permissible concentrations of substances dissolved in the water of any public system in the United States and are based on average annual concentrations. Currently the U.S. EPA is re-evaluating the potential risk of atrazine to humans and the environment. In the European Union, the health advisory level for atrazine and its metabolites has been set at 0.5 ppb. Several European countries have banned atrazine as they would any pesticide that exceeds a level of 0.1 ppb in drinking water, regardless of whether or not it has been proven to be a human health risk.

Since atrazine's measured BCF is less than 10, bioaccumulation does not represent a significant problem. Atrazine is not a very acutely toxic compound (its LOD_{50} is about 2000 mg/kg). However, some preliminary surveys on the health of farmers and other individuals exposed to it in high concentrations show disturbing links to higher cancer rates and a higher incidence of birth defects. No definitive studies linking atrazine use to human health problems have as yet been reported. Nevertheless, the U.S. EPA has listed it as a "possible human carcinogen" and has directed states to devise plans to protect groundwater from herbicide contamination.

In certain American agricultural regions, atrazine use has been banned outright. For a few years the triazine called *cyanazine*—which has the same chemical formula as atrazine except that one hydrogen in the isopropyl group is replaced by a cyanide group—became quite popular as an agricultural herbicide, but its manufacturer now has volutarily phased it out of production because of questions about its effect on human health. Other triazines with similar uses are *simazine* and *metribuzin*.

Chloroacetamides

In some regions where soybeans and corn are grown intensively, atrazine has yielded its status as the herbicide of choice to one of the **chloroacetamides,** which are derivatives of *chloroacetic acid,* $ClCH_2COOH$, in which the —OH group is replaced by an amino group. The most prominent herbicides of this type are **alachlor, metolachlor,** and **acetochlor.** These three compounds differ only in minor variations in the complex organic groups R_1 and R_2 attached to the amino nitrogen. Alachlor is a carcinogen in animals, and metolachlor is suspected of being one as well. The EPA has proposed that the use of alachlor, metolachlor, atrazine, and simazine all be carefully managed in areas where they are used intensively, since they represent a significant risk to groundwater.

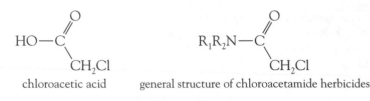

chloroacetic acid   general structure of chloroacetamide herbicides

Generally, the concentrations of these herbicides in waterways that drain agricultural land peak in May and are nondetectable by the end of summer; however, all are somewhat toxic to fish. Alachlor and its degradation products have been detected in groundwater that lies under corn fields. Metolachlor is known to degrade in the environment by the action of sunlight and of water.

Atrazine and its metabolite *DEA*, and metolachlor were the agricultural herbicides most often detected in streams and shallow groundwater in both urban and agricultural areas, according to an investigations by the U.S. Geological Survey in the early-mid 1990s. Domestic herbicides found most often were the triazines simazine and *prometon*. Insecticides found in highest concentrations—principally carbaryl and the organophosphates diazinon, malathion, and chlorpyrifos—were higher in urban than rural regions, presumably because of domestic usage. More than 95% of the streams and 50% of the groundwater samples were found to contain at least one pesticide at detectable levels. Research in Switzerland has found levels of atrazine, alachlor, and other agricultural pesticides that exceed drinking water standards in rainwater. Presumably the pesticides evaporated from farm fields.

Glyphosate

Glyphosate is an example of a *phosphonate*, a class of compounds that are structurally similar to organophosphates except that one oxygen of the four that surround phosphorus is missing and is replaced by an organic group—in this case a methylene group, —CH_2—, attached to the simple amino acid *glycine:*

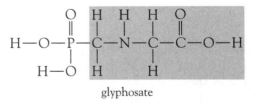

glyphosate

Glyphosate is widely used as a herbicide, e.g., as the commercial product *Roundup*. It is rather nontoxic: its LD_{50} values are high for both oral and dermal routes of exposure, although acute ingestion of or exposure to large quantities of it is fatal. Dermal and oral absorption is low, and it is eliminated essentially unmetabolized. Glyphosate is nonresidual and there is no evidence that it bioaccumulates in animal tissue or is carcinogenic or teratogenic. The same is true of its initial breakdown product, the substance corresponding to cleavage of the rightmost NH—CH$_2$ bond in the structure on the facing page.

Glyphosate operates by inhibiting the synthesis of amino acids containing the aromatic benzene ring, which in turn prevents protein synthesis from occurring. Although it kills almost all plants, some strains of soybeans have been genetically altered using biotechnology so that they are resistant to glyphosate and consequently it can be used as a weed killer in the growth of the crop. Its advantages in growing soybeans are that it replaces several different herbicides and that only one application is required, although the total volume of herbicide used is not reduced substantially. Its greater tendency to remain adsorbed on soil means that it has a lesser tendency to occur in runoff and subsequently in water supplies than do the herbicides atrazine and alachlor that it replaces. Thus the evidence gathered so far indicates that glyphosate is a relatively benign herbicide.

Phenoxy Herbicides

Phenoxy weed killers were introduced at the end of World War II. Environmentally, the by-products in commercial products containing such herbicides are often of greater concern than the herbicides themselves, as we shall see in Chapter 8. For that reason, we begin by discussing the chemistry of phenol, the fundamental component of these compounds.

Phenols are mildly acidic; in the presence of concentrated solutions of a strong base like NaOH, the hydrogen of the OH group is lost as H$^+$ (as occurs with any common acid) and the *phenoxide anion*, C$_6$H$_5$O$^-$, is produced in the form of its sodium salt:

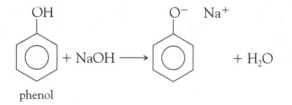

The O$^-$Na$^+$ group is a reactive one, and this property can be exploited to prepare molecules containing the C—O—C linkage. Thus if an R—Cl molecule is heated together with a salt containing the phenoxide ion, NaCl is eliminated and the phenoxy oxygen links the benzene ring to the R group:

$$C_6H_5O^-Na^+ + Cl—R \longrightarrow C_6H_5—O—R + NaCl$$

Such a reaction is the most direct commercial route to the large-scale preparation of the herbicide, introduced in 1944, whose well-known commercial name is *2,4,5-T*. Here (in the reaction immediately above) the R group is acetic acid, CH$_3$COOH, minus one of its methyl group hydrogens (so R = —CH$_2$COOH and the Cl—R reactant is Cl—CH$_2$COOH). Then, according to the reaction, we obtain C_6H_5—O—CH_2COOH, **phenoxy-acetic acid,** as an intermediate in the production of the actual herbicides.

In the commercial herbicides, some of the five remaining hydrogen atoms of the benzene ring in phenoxyacetic acid are replaced by chlorine atoms.

Note that the numbering scheme for the benzene ring begins at the carbon attached to the oxygen.

The **2,4-D** compounds (**2,4-dichlorophenoxyacetic acid**) is produced by a strategy that does not involve first forming the corresponding chlorinated phenol. It is used to kill broad-leaf weeds in lawns, golf course fairways and greens, and agricultural fields. In contrast, **2,4,5-T (2,4,5-trichlorophenoxyacetic acid)** is effective in clearing brush, e.g., on roadsides and powerline corridors.

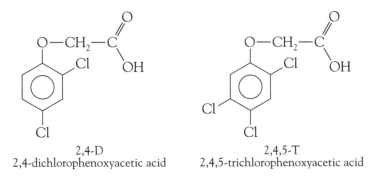

2,4-D
2,4-dichlorophenoxyacetic acid
    2,4,5-T
2,4,5-trichlorophenoxyacetic acid

The herbicide MCPA is just 2,4-D with the chlorine in the 2 position replaced by a methyl group, CH_3. The herbicides called *dichlorprop, silvex,* and *mecoprop* have structures identical to 2,4-D, to 2,4,5-T, and to MCPA, respectively, except that they have a methyl group replacing one hydrogen atom of the —CH_2– group in the acid chain; thus they are phenoxy herbicides that are based on *propionic acid,* CH_3—CH_2—COOH, rather than on acetic acid. The herbicide *dicamba* is the same as 2,4-D with a methoxy group at the 5 position of the benzene ring; it is often used as a weed killer in corn fields.

Huge quantities of 2,4-D and its closely related analogs just described are used in developed countries for the control of weeds in both agricultural and domestic settings. In some communities their continued use on lawns has become controversial because of their suspected effects on human health. In particular, farmers in the midwestern United States who mix and apply large quantities of 2,4-D to their crops are found to have an increased incidence of the cancer known as non-Hodgkin's lymphoma.

Summary

Some common organic pesticides, including those currently on sale for domestic use, are listed in Table 7-6, along with their maximum contaminant-level values in U.S. drinking water supplies. Also indicated in the table is their ranking for human health impact on the general public (according to a survey of reported illnesses in California in the late 1980s). Notice that both herbicides and insecticides, including those designed to be safer than those used in previous decades, are represented in the health impact ratings. In general, there is no pesticide that is completely safe. However, the elimination of all synthetic pesticides would lead to an increase in the transmission of disease by insects and an increase in the cost of food, both of which would affect human

TABLE 7-6	Common Pesticides and Their Properties			
Pesticide	Type[a]	Domestic use?	U.S. maximum water contaminant level (ppb)	California health impact rating
Alachlor	H; chloracetamide		2	
Atrazine	H; triazine		3	
Carbofuran	I (systemic); carbamate		40	
Carbaryl (Sevin)	I; carbamate	Yes		
Chlordane	I (especially soil); cyclodiene		2	
Chlorpyrifos	I (especially houshold); Type B organophosphate	Yes		3
Diazinon	I (soil, trees); Type B organophosphate	Yes		4
Dichlorvos	I (fly strip fumigant); Type A organophosphate	Yes		
Dimethoate	I (systemic); Type C organophosphate	Yes		
Dinoseb	I (fruit spray); dinitrophenol		7	
2,4-D, mecoprop, dicamba	H; phenoxy- acetic acids	Yes	70	
Endrin	I; cyclodiene		2	
Gluteraldehyde	H	Yes		10

(continued on p. 354)

TABLE 7-6	Common Pesticides and Their Properties *(continued)*			
Pesticide	Type[a]	Domestic use?	U.S. maximum water contaminant level (ppb)	California health impact rating
Glyphosate	H; organo-phosphate	Yes	700	8
Heptachlor	I (soil); cyclodiene		1	
Lindane	I	Yes	2	
Metaldehyde	I (mollusks); aldehyde polymer	Yes		
Methoxychlor	I; DDT-like		40	
Malathion	I; Type C organophosphate	Yes		6
Propetamphos	I (household pests); organophosphate	Yes		7
Propoxur	I (including roaches, wasps); carbamate	Yes		
Pyrethrins/ piperonyl butoxide (several-thrins)	I	Yes		9

[a] H = herbicide; I = insecticide.

health adversely. Any decision about discontinuing the production and use of a given pesticide must consider whether cheap, safer alternatives are available and, if not, what the consequences of both action and inaction are. The quandary about whether to ban the use of DDT in tropical developing countries is an excellent illustration.

When a new pesticide, or indeed any other synthetic chemical, is about to be introduced into the market, many environmental groups and some government agencies have proposed that we should err on the side of being too cautious and only allow its introduction if there are no signs that significant problems could arise. They propose that in such situations, to prevent possible harm to the health of humans and other organisms, we should employ what is now known as the **precautionary principle.** One definition of this principle was given at the 1992 Rio Conference on Environment and Development: "Where there are threats of serious or irreversible damage, lack of full scientific certainty shall not be used as a reason for postponing cost-effective measures to prevent environmental degradation." Opponents of the use of this principle point out that it is impossible to anticipate all possible consequences, positive or negative, of introducing a new substance and that consequently we could become frozen into inaction.

Review Questions

1. What are the three main categories of pesticides? What types of organisms are killed by each category?

2. What is meant by the term *fumigant*?

3. Name three important properties shared by organochlorine pesticides.

4. Draw the structure of DDT and state what the initials stand for.

5. What units are usually used to state the concentrations of trace contaminants in water?

6. What were the main uses of DDT? Explain why it is no longer used in many developed countries and why some developing countries wish to continue using it.

7. Explain how DDT functions as an insecticide.

8. Draw the structure of DDE. Is it a pesticide or not? Explain.

9. Explain what is meant by the terms *bioconcentration* and *bioconcentration factor* (BCF).

10. Explain what is meant by the term *biomagnification* and how it differs from *bioconcentration*.

11. Write the defining equation for the *partition coefficient* K_{ow}. How is it related to a compound's BCF? What is octanol supposed to be a surrogate for in this experiment?

12. Describe one analog of DDT that works in the same fashion but does not bioaccumulate.

13. In general terms, explain what toxaphene is and why it is no longer in use.

14. Draw the structure of cyclopentadiene. Name at least three insecticides produced from it.

15. Define the terms *acute toxicity* and *dose*.

16. Sketch a typical dose–response curve relationship for a toxic chemical using (**a**) a linear, and (**b**) a logarithmic scale for doses.

17. Define the terms LD_{50} and LOD_{50}.

18. Define *fugacity* and state how its value is related to the various phases in which a pollutant exists.

19. What are the general structures of the three main subclasses of organophosphate insecticides? Give the name of one insecticide in each subclass. Explain how organophosphates function as insecticides.

20. In what way are organophosphate insecticides considered superior to organochlorines as pesticides? In what way are they more dangerous?

21. What is the general structure of carbamate insecticides? Name one example.

22. What are five of the pest control methods used in pest control management?

23. What is the function of a herbicide? Name a few "old-fashioned" insecticides that contained metals.

24. What is the general structure of a triazine herbicide? Name two examples.

25. What is the general structure of chloroacetamide herbicides? Name one example.

26. What is meant by the *maximum contaminant level* (MCL) of a water pollutant?

27. What is the formula of glyphosate? What are its advantages over other herbicides?

28. What is phenol? Draw its structure and that of 2,4-dichlorophenol.

29. Draw the structures and write out the names of the two most important phenoxy herbicides.

30. What is meant by the *precautionary principle*?

Green Chemistry Questions

1. The development of the insecticides Confirm, Mach2, and Intrepid won a Presidential Green Chemistry Challenge Award.

(a) Which of the three focus areas (see the Introduction to Green Chemistry) for these awards does this award best fit into?

(b) List one of the twelve principles of green chemistry (see the Introduction to Green Chemistry) that are addressed by these new insecticides.

2. What environmental advantage do Confirm, Mach2, and Intrepid offer compared to conventional pesticides?

3. **(a)** What is a U.S. EPA *reduced-risk pesticide*?

(b) Which categories under the reduced-risk criteria do Confirm, Mach2, and Intrepid meet?

4. How do Confirm, Mach2, and Intrepid act to target only specific insects?

5. The development of the Sentricon system won a Presidential Green Chemistry Challenge Award.

(a) Which of the three focus areas (see the Introduction) for these awards does this award best fit into?

(b) List two of the twelve principles of green chemistry (see the Introduction to Green Chemistry) that are addressed by the Sentricon system.

6. What environmental advantages does the hexaflumuron/Sentricon system offer compared to conventional termite control pesticides/methods?

7. Which categories under the reduced-risk criteria does the hexaflumuron/Sentricon system meet?

Additional Problems

1. The threshold/NOEL level found for a particular chemical from animal studies is 0.004 mg/kg body weight/day. The only source for the chemical is freshwater fish, where it occurs at an average level of 0.2 ppm. What is the maximum average daily consumption of such fish that would keep exposure level below the ADI or RfD for the compound?

2. An approximate mathematical fit to the form of the dose–response curve of Figure 7-7a is

$$R = 1 - e^{-d}$$

where R is the fractional response and d is the dose.

(a) Plot R versus d for values of d ranging from 0 to 5 on both linear and logarithmic scales for d. (Be sure to include some small values of d, from 0.01 to 0.10, in the logarithmic plot to ensure that the form of the curve near zero is displayed.) Do the forms of the curves resemble those in Figures 7-7a and 7-7b?

(b) Both from your graphs and by solving the equation above analytically, find the dose corresponding to LD_{50}.

(c) Does the function R have a nonzero threshold at low doses? Can you confidently predict the answer to this from inspecting your logarithmic dose–response curve?

3. In fugacity calculations, the Z values for dieldrin are 4×10^{-4} in air, 2.0 in water, and 2×10^{-5} in sediment (and biota). Using the model world volumes in the text, calculate the equilibrium concentrations and amounts when 1 mol of dieldrin is distributed between air, water, and sediment.

4. The fat content of breast milk, while varying somewhat from mother to mother, with the stage of development of the nursing baby, and with other factors, generally is on the order of 4.2 g/100 mL for mature "hindmilk." Based on Figure 7-2, calculate the mass of DDT that would have been ingested by a typical breast-fed Swedish infant in 1969 upon consuming 250 mL of breast milk.

5. The overall bioconcentration factor (BCF) for a particular substance in a particular aquatic species (not just in the fat tissues) can be estimated as the value of K_{ow} for that substance times the fraction of body fat in the species of interest. Rainbow trout (5.0% body fat) taken from a particular lake were tested and found to contain 22 ppb parathion in their tissues. Use the information in Table 7-3 to determine the concentration of parathion in this lake.

6. In the case of effects on aquatic species, the toxicity of a compound is most conveniently expressed as LC_{50}, the concentration that is lethal to half of the test population in a stated period of time. The organophosphate pesticide Azinphosmethyl has a 96-h LC_{50} of 3 ppb for rainbow trout. In an unfortunate incident, a total of 200 g of this pesticide was sprayed on a field, and a subsequent heavy rain washed 35% of the applied pesticide into a nearby small lake with a surface area of 30,000 m^2 and an average depth of 0.5 m. Assuming there were rainbow trout in this lake, was it likely that a significant number of fish were killed?

Further Reading

1. V. Turusov et al., "DDT: Ubiquity, Persistence, and Risks," *Environmental Health Perspectives* 110 (2002): 125.

2. B. Hileman, "Reexamining Pesticide Risk," *Chemical and Engineering News* (17 July 2000): 34.

3. G. Santoro, "Silent Summer," *Discover* (July 2000): 76.

4. K. Noren and D. Meironyte, "Certain Organochlorine and Organobromine Contaminants in Swedish Human Milk in Perspective of Past 20–30 Years," *Chemosphere* 40 (2000): 1111.

5. D. P. Tierney et al., "Predicted Atrazine Concentrations in the Great Lakes: Implications for Biological Effects," *Journal of Great Lakes Research* 25 (1999): 455.

6. G. M. Williams et al., "Safety Evaluation and Risk Assessment of the Herbicide Roundup and Its Active Ingredient, Glyphosate, for Humans," *Regulatory Toxicology and Pharamacology* 31 (2000): 117.

Websites of Interest

Log on to www.whfreeman.com/envchem3e/ and click on Chapter 7.

CHAPTER 8

Nonpesticide Toxic Organic Compounds of Environmental Concern

As we have seen in Chapter 7, compounds used as pesticides are also somewhat toxic to humans and can bioaccumulate and cause environmental problems. But sometimes it is the highly toxic trace impurities in commercial lots of such substances that are the principal concern regarding human health. In this chapter we shall analyze how such hazardous by-products, especially dioxins, arise in the environment, not only from pesticide manufacture but from other anthropogenic processes as well. Also considered are PCBs, industrial chemicals of widespread environmental concern with respect to both their own properties and those of their contaminants. As we shall see, the toxicity mechanisms by which these contaminants, the PCBs themselves, and dioxins operate are all quite similar. We shall also look at a series of other toxic organic products that have become air and water pollutants and at how persistent substances become distributed around the world by air currents. Finally, we conclude by considering environmental chemicals that have disruptive effects on our reproductive systems.

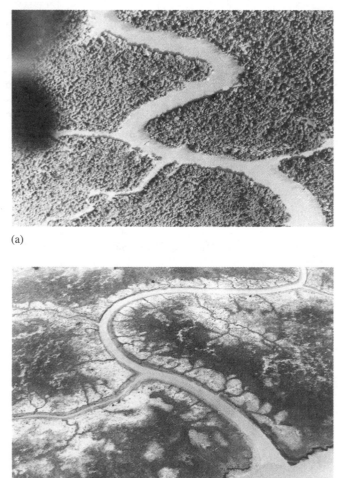

(a)

(b)
A forest in South Vietnam (a) before and (b) after spraying with Agent Orange. (AP)

Dioxins

Dioxin Production in the Preparation of 2,4,5-T

Traditionally, the industrial synthesis of the herbicide *2,4,5-T* (discussed in Chapter 7) starts with *2,4,5-trichlorophenol*, which itself is produced by reacting NaOH with the appropriate *tetrachlorobenzene*. The OH group replaces one chlorine atom in the process. Unfortunately, during this synthesis there occurs an additional reaction that converts a very small portion of the trichlorophenol product into "dioxin." In this side reaction, two *trichlorophenoxy anions* react with each other, resulting in the elimination of two chloride ions:

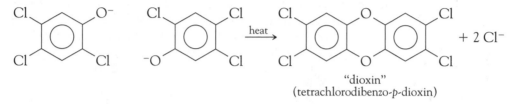

"dioxin"
(tetrachlorodibenzo-*p*-dioxin)

In this process a new six-membered ring is formed that links the two chlorinated benzene rings. This central ring has two oxygen atoms located para (i.e., opposite) to each other, as in the simple molecule *1,4-dioxin* or *para-dioxin* (*p*-dioxin):

1,4-dioxin

Although the molecule labeled "dioxin" is correctly known as a *tetrachlorodibenzo-p-dioxin*, it has become popularly known simply as dioxin, with the understanding that it is the most toxic of a class of related compounds.

The side reaction that produces dioxin as a by-product is kinetically second-order in chlorophenoxide. In other words, the rate of the reaction depends on the square (or second power) of the ion's concentration. Consequently, the rate of dioxin production increases dramatically as the initial chlorophenoxide ion concentration increases. In addition, the rate of this side reaction increases rapidly with increasing reaction temperature. Therefore, the extent to which the trichlorophenol, and consequently the commercial herbicide, become contaminated with the dioxin by-product can be minimized by controlling concentration and temperature in the preparation of the original trichlorophenol. Today, the contamination of commercial 2,4,5-T by this dioxin can be kept to about 0.1 ppm by keeping both the phenoxide concentration and the temperature low. Nevertheless, its manufacture and use in North America were phased out in the mid-1980s because of concerns about its dioxin content, however small.

A 1:1 mixture of 2,4-D and 2,4,5-T, called *Agent Orange,* was used extensively as a defoliant during the Vietnam War. Since the mixture contained dioxin levels of about 10 ppm, it is clear that the production of the trichlorophenol used for 2,4,5-T preparation was not carefully controlled to minimize contamination. As a result, the soil in southern Vietnam is contaminated by dioxins. The consequences of this contamination for the residents, and for the American troops that were exposed while spraying was under way, are still controversial. The potential human health consequences of exposure to dioxin are discussed in a later section.

Environmental contamination by dioxin also occurred as the result of an explosion in a chemical factory in Seveso, Italy, in 1976. The factory produced 2,4,5-trichlorophenol from tetrachlorobenzene, as described above. On one occasion the reaction was not brought to a complete halt before workers left for the weekend. The reaction continued unmonitored, and the heat released by the reaction eventually resulted in an explosion. Since the trichlorophenol had been heated to a high temperature, a large amount of dioxin—probably several kilograms—was produced. The explosion effectively distributed the toxin in the environment and many wildlife deaths resulted from the contamination. Although a large number of humans, both adults and children, were also exposed to the chemical as a result of the explosion, no serious health effects were found for many years. Recent studies, however, have established that the rates of several types of cancer have increased in people who lived in the zones most exposed to dioxin from the explosion. Specifically, the risk of contracting breast cancer increased in proportion to the dioxin exposure, as measured by the level of the substance in women's blood samples taken soon after the explosion.

Dioxin Numbering System

The nomenclature and numbering system used for ring compounds like the dioxins is a little unusual. Since the central dioxin ring is connected on either side to benzene rings, the three-ring unit is properly known as **dibenzo-p-dioxin.** The chlorine substitution on the outer rings also should be indicated, so the dioxin shown below is a **tetrachlorodibenzo-p-dioxin,** or **TCDD:**

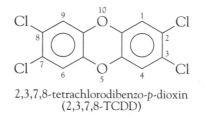

2,3,7,8-tetrachlorodibenzo-*p*-dioxin
(2,3,7,8-TCDD)

The numbering scheme for the ring carbons in dioxins takes into account the fact that carbons shared between two rings carry no hydrogen atoms and so need not be numbered. Thus C-1 is the carbon next to one shared by the rings, and the numbering follows a direct path from there. By convention, the oxygen atoms are also part of the numbered sequence in this scheme, although their locations are not used in naming any of this family of compounds. The C-1 is chosen to give the lowest possible number to the first substituent; if there is a choice after this criterion has been applied, then that which gives the lowest number to the second substituent is used, etc. Applying these rules, the dioxin shown above is named 2,3,7,8-TCDD, or to give it its full name, **2,3,7,8-tetrachlorodibenzo-p-dioxin.** No wonder it is simply called dioxin in the press!

There are actually 75 different chlorinated dibenzo-p-dioxin compounds when one includes all the possible structures with one to eight chlorines, given that a number of isomers exist for most of these eight types. Different members of a chemical family that differ only in the number and position of the same substituent are called **congeners.**

All dioxin congeners are planar: all carbon, oxygen, hydrogen, and chlorine atoms lie in the same plane. For convenience, we refer to the benzene carbon atoms closest to the central dioxin ring as alpha carbons and the outer ones as beta carbons:

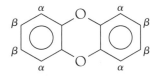

This unsubstituted dibenzo-p-dioxin molecule has two types of symmetry that are useful to consider when describing substitution patterns. First there is lateral or *left-right* symmetry: the carbon atom labeled β at the top of the left-side ring is equivalent to the β carbon at the top of the right-side ring, and similarly for the two β carbons at the bottom. The dioxin ring also has *up-down* symmetry: the carbon atom labeled β at the top of the left-side ring is equivalent to the β carbon at the bottom of that ring, and similarly for the two β carbons of the right-side ring. Thus all four β carbons are, in fact, equivalent in the unsubstituted dioxin. Similarly, the four α carbons all denote equivalent positions. Consequently, there are only two unique *monochlorodibenzo-p-dioxins:* due to the equivalence of the four α positions, those that would be numbered 4-, 6-, and 9-chlorodibenzo-p-dioxin are all equivalent to the 1- molecule. Similarly, 3-, 7-, and 8-chlorodibenzo-p-dioxin are all equivalent to the 2- molecule because of the equivalence of the β positions. Some or all of the equivalences can be lost when multiple substitution occurs.

PROBLEM 8-1

By drawing the structures and comparing them, decide whether 1,3-, 2,4-, 6,8-, and 7,9-dichlorodibenzo-p-dioxins are all unique compounds or whether they are all the same compound. Are 1,2- and 1,8-dichlorodibenzo-*p*-dioxins unique compounds? Using a systematic procedure, deduce the structures of all unique dichlorodibenzo-*p*-dioxins, keeping in mind that before substitution the two rings are equivalent and that the molecule has up-down symmetry.

Chlorophenols as Pesticides

In addition to their use as starting materials in the production of herbicides, chlorophenols find use as wood preservatives (fungicides) and as slimicides. The most common wood preservative, in use since 1936, is **pentachlorophenol** (PCP, but not the angel dust compound known by the same initials); all the benzene hydrogens have been substituted in this compound:

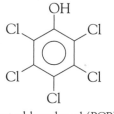

pentachlorophenol (PCP)

Commercial PCP is not pure pentachlorophenol but is significantly contaminated with *2,3,4,6-tetrachlorophenol*. This mixture has many pesticidal uses: it is used as a herbicide (e.g., as a preharvest defoliant), an insecticide (termite control), a fungicide (wood preservation and seed treatment), and a molluscicide (snail control). Some trichlorophenol isomers and some tetrachlorophenol isomers are also sold as wood preservatives.

Unfortunately, if wood treated with such preservatives is eventually burned, a fraction of the chlorophenols can react to eliminate HCl, thereby producing members of the chlorinated dioxin family. Thus **octachlorodibenzo-*p*-dioxin,** OCDD, is produced as an unwanted by-product in the incomplete combustion of pentachlorophenol products:

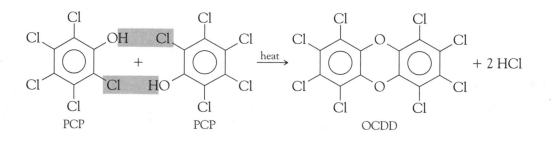

OCDD is the most prevalent dioxin congener found in human fat and in many environmental samples.

Indeed, pentachlorophenols are one of the largest chemical sources of dioxins in the environment; however, the main dioxin they contain, OCDD, is not particularly toxic, as discussed in a later section. Commercial supplies of chlorinated phenols themselves are contaminated with various dioxins.

PROBLEM 8-2

In naming OCDD and pentachlorophenol, no numbers are used to specify the positions of the chlorine substituents. Why is that not necessary here, whereas it is required in, e.g., 2,3,7,8-TCDD?

In general, any two phenol molecules that each have a chlorine on a carbon atom next to a carbon with an OH group can combine to produce a dibenzo-*p*-dioxin molecule. The two phenols that combine need not be identical but simply need to make contact when they have been heated sufficiently to facilitate HCl elimination and dioxin formation. Similarly, coupling of phenoxide anions can occur with Cl⁻ elimination, as discussed previously in the case of 2,4,5-T synthesis. The methodology of problem solving that can be used to deduce the chlorophenolic origin of environmental dioxins is discussed in Box 8-1.

PROBLEM 8-3

(a) Deduce the structures and the correct numbering for the two tetra-chlorophenol isomers that exist in addition to the 2,3,4,6 isomer mentioned in the text. (b) For each of these two isomers, deduce the structure and names of the dioxin(s) that would result if two molecules of that isomer were to react together.

PROBLEM 8-4

Deduce which dioxin(s) would be produced in side reactions if 2,4-D were to be synthesized from 2,4-dichlorophenol.

Detecting Dioxins in Food and Water

As a consequence of their widespread occurrence in the environment and their tendency to dissolve in fatty matter, dioxins bioaccumulate in the food chain. More than 90% of human exposure to dioxins is attributable to the food we eat, particularly meat, fish, and dairy products. Typically, dioxins and furans (a group of compounds resembling the dioxins in structure, which

BOX 8-1 | Deducing the Probable Chlorophenolic Origins of a Dioxin

The chlorophenolic source of dioxins found in environmental samples can be deduced by reversing the logic used in the text to deduce which dioxin would be produced by the coupling of two specific chlorophenols.

Consider the congener 1,2,7,8-tetrachlorodibenzo-*p*-dioxin; it could have been formed by elimination of two HCl molecules from two chlorophenol molecules in the following two ways (here T stands for trichlorophenol).

dioxin structure came from the chlorophenol on the right side of the molecule and the bottom oxygen from the chlorophenol on the left, leads to the possibility that the trichlorophenol molecules that combined were the 2,4,5 and the 2,3,6 congeners. Thus a 1,2,7,8-tetrachlorodibenzo-*p*-dioxin molecule in the environment could have arisen by combination of a 2,4,5-trichlorophenol molecule with either a 2,3,4- or a 2,3,6-substituted congener.

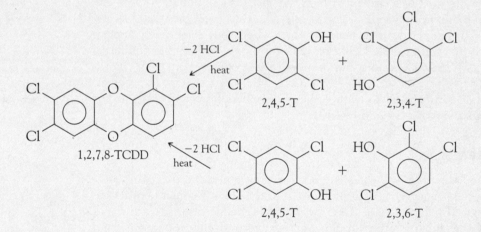

If it is assumed that the oxygen atom at the top of the dioxin congener originates with the chlorophenol congener on the left side of the dioxin molecule, then the bottom oxygen must come from the chlorophenol on the right side of the dioxin molecule; with this set of assumptions, the original reactants must have been 2,4,5- and 2,3,4-trichlorophenol. (Notice in the diagram above that the chlorine atoms eliminated must have arisen from positions adjacent to the oxygen atoms.) The alternative possibility, that the oxygen atom at the top of the

Unfortunately, some dioxins undergo rearrangement of substituents during their formation, so such a "retrosynthesis" approach is not an infallible guide to the origin of dioxins discovered in the environment.

PROBLEM 1

Deduce the two possible combinations of polychlorophenol molecules that, when coupled together through loss of two HCl molecules, would produce a molecule of 1,2,9-trichlorodibenzo-*p*-dioxin.

we'll discuss later) are present in fish and meat at levels of tens or hundreds of picograms (pg, or 10^{-12} g) per gram of food; in other words, they occur at levels of tens or hundreds of parts per trillion. However, the bulk of dioxins and furans in nature are not present in biological systems: attachment to soil and to sediments of rivers, lakes, and oceans provides their most common sinks.

The ability of chemists to detect TCDD and other organochlorines in environmental samples has improved by orders of magnitude over the past few decades. In the early 1960s, when Carson's *Silent Spring* was published, the lower limit for analysis of DDT and such compounds was the parts-per-million level. Ten years later, detecting such substances at the parts-per-billion level was possible but not common or easy. By 1990, parts-per-trillion detection was possible in soil or biota samples, and parts-per-quadrillion was possible for water samples. Today, a few labs detect some substances at limits up to 1000 times lower than these! At these levels, many organochlorines are found in *every* environmental sample, no matter how "clean" the environment it came from. By the late 1990s, researchers at the Centers for Disease Control in Atlanta were able to detect as little as 10^{-16} g of TCDD in human serum samples.

The potential impact on human health of exposure to dioxins is documented later, following a discussion of the properties of PCBs and furans, two types of chemicals with which dioxins share many properties.

PROBLEM 8-5

Given its formula and Avogadro's constant (6.02×10^{23} molecules/mol), deduce how many molecules are present in 10^{-16} g of TCDD.

PCBs

The well-known acronym **PCB** stands for **polychlorinated biphenyl,** a group of industrial organochlorine chemicals that became a major environmental concern in the 1980s and 1990s. Although not pesticides, they found a wide variety of applications in modern society because of certain other properties they possess. Since the late 1950s, over 1 million metric tons of PCBs have been produced, about half in the United States and the rest mainly in France, Japan, and the former Soviet bloc. Like many other organochlorines, they are very persistent in the environment and they bioaccumulate in living systems. As a result of careless disposal practices, they have become a major environmental pollutant in many areas of the world. More than 95% of the entire U.S. population have detectable concentrations of PCBs in their bodies. Due both to their own toxicity and to that of their furan contaminants, PCBs in the environment have become a cause for concern because of their potential impact on human health, particularly with regard to growth and development.

In the following material we consider what PCBs are, how they are made, what they are used for, and how they become contaminated and released into the environment.

The Structure of PCB Molecules

Biphenyl molecules consist of two benzene rings linked by a single bond formed between two carbons that have each lost their hydrogen atom:

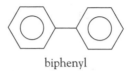

biphenyl

Like benzene, if biphenyl is reacted with Cl_2 in the presence of a *ferric chloride* catalyst, some of its hydrogen atoms are replaced by chlorine atoms. The more chlorine initially present and the longer the reaction is allowed to proceed, the greater the extent (on average) of chlorination of the biphenyl molecule. The products are polychlorinated biphenyls, PCBs. The reaction of biphenyl with chlorine produces a mixture of many of the 209 congeners of the PCB family; the exact proportions depend on the ratio of chlorine to biphenyl, the reaction time, and the reaction temperature. An example of a PCB molecule is

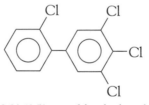

2,3′,4′,5′-tetrachlorobiphenyl

Although many individual PCB compounds are solids, the mixtures are liquids or solids with low melting points. Commercially, individual PCB compounds were not isolated; rather they were sold as partially separated mixtures, with the average chlorine content in different products ranging from 21 to 68%.

PROBLEM 8-6

The general formula for any PCB congener is $C_{12}H_{10-n}Cl_n$, where n ranges from 1 to 10. Calculate the average number of chlorine atoms per PCB molecule in a mixture of congeners that is 60% chlorine by mass, a common value for commercial samples.

The Numbering Systems for PCBs

The numbering scheme used for individual PCB congeners begins with the carbon joined to the other ring; it is given the number 1, and the other carbons around the ring are numbered sequentially. As illustrated, the positions in the second ring are also numbered 1 through 6, starting with the ring-joining carbon, but are distinguished by primes. By convention, the 2' position in the second ring lies on the same side of the C—C bond joining the rings as the 2 position in the first ring, etc.

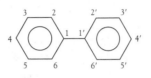

In most instances, the two rings in a chlorinated biphenyl molecule are not equivalent, since the patterns of substitution differ. The unprimed ring is chosen to be the one that will give a substituent with the lowest-numbered carbon. Using all these rules, we can deduce that the name of the PCB molecule shown on page 366 is *2,3',4',5'-tetrachlorobiphenyl*.

Very rapid rotation occurs around carbon–carbon single bonds in most organic molecules, including the C—C link joining the two rings in biphenyl and in most PCBs. Thus it is not normally possible to isolate compounds corresponding to different relative orientations of the two rings in a PCB molecule. For example, 3,3'- and 3,5'-*dichlorobiphenyl* are not individually isolatable compounds, since one is constantly being converted into the other and back again by rapid rotation about the C—C bond linking the rings:

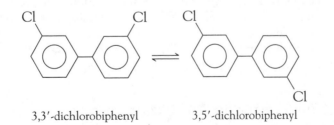

3,3'-dichlorobiphenyl   3,5'-dichlorobiphenyl

The label used for the compound is the one that has the lowest number for the second chlorine, so the molecule shown is called the 3,3' isomer. Although the rings rotate rapidly with respect to each other, the energetically optimum orientation is the one with the rings coplanar or close to it, except, as we shall see later, when large atoms or groups occupy the 2 and 6 positions.

Using a systematic procedure, draw the structures of all unique dichloro-biphenyls, assuming first that free rotation about the bond joining the rings does *not* occur. Then deduce which pairs of structures become identical due to free rotation.

Commercial Uses of PCBs

All PCBs are practically insoluble in water but are soluble in hydrophobic media, such as fatty or oily substances. Commercially, they were attractive because they are chemically inert liquids and are difficult to burn, have low vapor pressures, are inexpensive to produce, and are excellent electrical insulators. As a result of these properties, they were used extensively as the coolant fluids in power transformers and capacitors. Later they were also employed as plasticizers, i.e., agents used to keep plastic materials such as PVC products more flexible; in carbonless copy paper; as deinking solvents for recycling newsprint; as heat transfer fluids in machinery; as waterproofing agents; and even further uses.

Because of their stability and extensive use, together with careless disposal practices, PCBs became widespread and persistent environmental contaminants. When their accumulation and harmful effects were recognized, **open uses,** i.e., those for which their disposal could not be controlled, were terminated. Although North American production of PCBs was halted in 1977, the substances remain in use in some electrical transformers currently in service. As these are gradually decommissioned, their PCB content is usually stored in order to prevent further contamination of the environment. In the United States, the EPA expects a 90% reduction of PCB use in electrical equipment by 2006. Canada has proposed a phase-out of all PCB uses by 2008. In some locales, stored PCBs are being destroyed by incineration, using techniques discussed in Chapter 12. Previously, PCB-containing transformers and capacitors were often just dumped into landfills and their PCB content was allowed to leak into the ground. Indeed, PCBs were inadvertently released into the environment during their production, use, storage, and disposal.

PCBs Cycling Among Air, Water, and Sediments

If released into the environment, PCBs persist for many years because they are so resistant to breakdown by chemical or biological agents. Although their solubility in water is very slight—indeed, they are more likely to be adsorbed onto suspended particles in the water than dissolved in it—the tiny amounts of PCBs in surface waters are constantly being volatilized and subsequently redeposited on land or in water after traveling in air for a few days. By such mechanisms, PCBs have been transported worldwide. There are measurable background levels of PCBs even in polar regions and at the bottom of oceans. Indeed, the ultimate sinks for PCBs that are mobile are in the deep sediments of oceans and large lakes. This environmental load of PCBs will continue to be recycled among air, land, and water, including the biosphere, for decades to come, as analyzed in greater detail near the end of this chapter. Unfortunately, only about 30% of manufactured PCBs are currently found in the environment, so much of the production lingers in storage or old electrical equipment and may ultimately be released.

A quantitative measure of the recycling of PCBs within a water body is provided by the **mass balance** of current annual inputs and outputs of these compounds. The balance for a very large, relatively clean water body—Lake Superior—is illustrated in Figure 8-1. Although it is the least polluted of the Great Lakes, Superior's burden of PCBs in its water and sediments is substantial. Currently, almost all the input of PCBs occurs from the air, with relatively little added from industries or via tributary rivers (Figure 8-1a). Overall, Lake Superior is now gradually exhaling its historical load of PCBs into the air, the output to air being much greater than the annual input from the atmosphere. The variation in PCB concentrations with time in Lake Superior is illustrated in Figure 8-1b. Little of Superior's PCBs are now being lost to sediments; about as much is redissolved from them as is deposited onto them each year.

By contrast, the mass balance of PCBs in Lake Ontario, another of the Great Lakes, is quite different from that of Lake Superior. The PCB concentration in Lake Ontario water substantially exceeds that of Lake Superior, since it is located in a much more industrialized area. In Lake Ontario the greatest current input comes from land-based sources such as waste dumps that still leach PCBs into the lake and its tributaries. About the same quantity of PCBs is present in the water flowing out of the lake. Approximately equal amounts are lost annually to sediments and to the atmosphere; about one-third of such losses are canceled by new inputs from the sediments and the air.

PROBLEM 8-8

The PCB concentration in Lake Michigan is declining according to a first-order rate law with a rate constant of 0.078/year. If the PCB concentration in Lake Michigan averaged 0.047 ppt in 1994, what will it be in 2010? In what year will the concentration fall to 0.010 ppt? What is the half-life period of PCBs in this lake? [Hint: Recall that for first-order processes, the fraction of any sample that still remains after time t has passed is $f = e^{-kt}$ where k is the rate constant.]

Because of their persistence and their solubility in fatty tissue, PCBs in food chains undergo biomagnification; an example is shown in Figure 8-2.

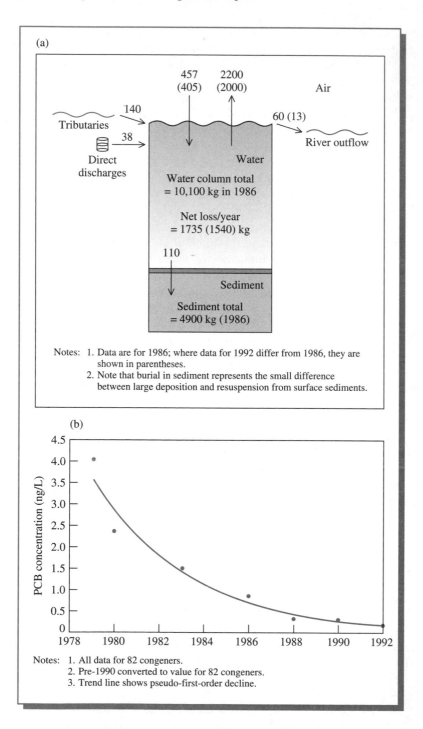

(a)

457   2200
(405)   (2000)   Air

140
Tributaries →

38
Direct discharges →

Water

Water column total
= 10,100 kg in 1986

Net loss/year
= 1735 (1540) kg

110

Sediment

Sediment total
= 4900 kg (1986)

60 (13) →
River outflow

Notes: 1. Data are for 1986; where data for 1992 differ from 1986, they are
           shown in parentheses.
        2. Note that burial in sediment represents the small difference
           between large deposition and resuspension from surface sediments.

(b)

PCB concentration (ng/L)

1978   1980   1982   1984   1986   1988   1990   1992

Notes: 1. All data for 82 congeners.
        2. Pre-1990 converted to value for 82 congeners.
        3. Trend line shows pseudo-first-order decline.

FIGURE 8-1 (a) Mass balance of PCBs in Lake Superior, in kilograms per year. (b) The trend for 1979–1992 of PCB concentration in the water column of Lake Superior. [Source for (a) and (b): *The State of Canada's Environment 1996* (Ottawa: Government of Canada, 1996).]

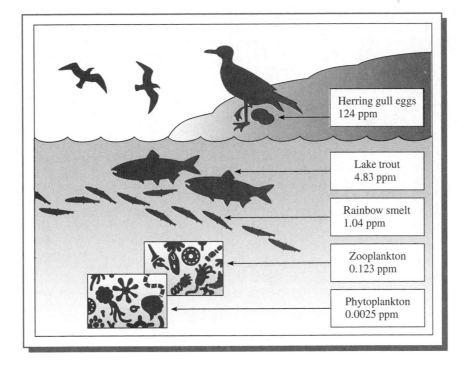

FIGURE 8-2 The bioconcentration of PCBs in the Great Lakes aquatic food chain. [Source: *The State of Canada's Environment* (Ottawa: Government of Canada, 1991).]

Herring gull eggs
124 ppm

Lake trout
4.83 ppm

Rainbow smelt
1.04 ppm

Zooplankton
0.123 ppm

Phytoplankton
0.0025 ppm

Note that the ratio of PCBs in the eggs of herring gulls in the Great Lakes was 50,000 times that in the phytoplankton in the water at the time of these measurements. The good news is that the average levels of PCBs in such eggs have fallen with time in many locations, as the data in Figure 8-3a illustrate for gull colonies in Lake Ontario around Toronto. Since the concentrations are plotted in the figure on a logarithmic scale, the data are seen to fit two intersecting straight-line curves. The PCB concentrations in eggs were decreasing exponentially with time with a half-life of about five years but are now declining with a half-life of seven years. This complicated behavior arises because of the continuing source of PCBs to the system. PCB levels in fish at the top of the Lake Ontario food chain have also declined since the 1970s, but the current rate of decrease is slow and erratic (see Figure 8-3b).

The relative concentrations of the congeners of a PCB mixture begin to change once they enter the environment. Microorganisms in soils and sediments and large organisms such as fish both preferentially metabolize congeners having relatively few chlorine atoms. Thus the relative concentrations of the more heavily chlorinated congeners increase with time, since they are degraded much more slowly. Thus, e.g., between 1977 and 1993, the proportion of PCB molecules with 4 or 5 chlorine atoms

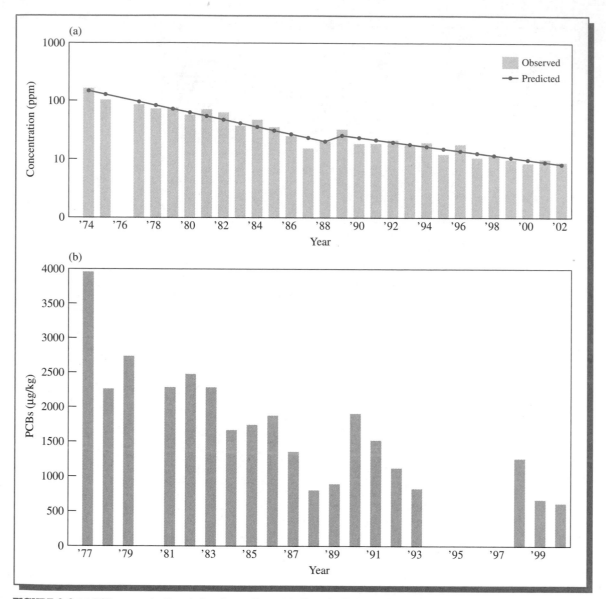

FIGURE 8-3 (a) PCB concentrations in herring gull eggs in Toronto Harbor, 1974–2002. The predicted curves correspond to exponential decay in those time periods. [Source: Dr. Chip Weseloh, Environment Canada.] (b) PCB concentrations in 65-cm coho salmon from Lake Ontario. [Source: Ontario Ministry of the Environment.]

decreased by 6% each in trout in Lake Ontario, whereas those with 7 or 8 chlorines increased by 7% and 4%, respectively.

PCB Contamination by Furans

Strong heating of PCBs in the presence of a source of oxygen can result in the production of small amounts of **dibenzofurans.** These compounds are structurally similar to dioxins; they differ only in that they are missing one oxygen atom in the central ring. The furan ring contains five atoms, one oxygen and four carbon atoms that participate in double bonds:

furan

The *dibenzofurans* (DFs) have a benzene ring fused to opposite sides of the furan ring:

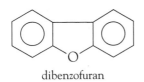

dibenzofuran

As with dioxins, all chlorinated dibenzofuran congeners are planar; i.e., all C, O, H, and Cl atoms lie in the same plane. They are formed from PCBs by the elimination of the atoms X and Y bonded to two carbons that are ortho to those that link the rings and that lie on the same side of the C—C link between the rings:

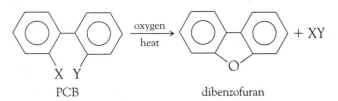

The atoms X and Y can both be chlorine, or one can be hydrogen and the other chlorine, so the molecule eliminated can be Cl_2 or ClH (i.e., HCl). A more detailed analysis of the nature of the specific furans that result from particular PCB congeners is given in Box 8-2.

Most of the chlorine in the original PCB molecule is still present in the dibenzofuran; *polychlorinated dibenzofurans* are known commonly as **PCDFs.** The numbering scheme for substituents is the same as that for

BOX 8-2 | Predicting the Furans That Will Form from a Given PCB

In deducing the nature of the polychlorinated dibenzofuran (PCDF) that would be formed from a particular PCB, it should be remembered that free rotation occurs about the single bond joining the two rings in the original biphenyl in all PCBs at the elevated temperatures of the reaction. Thus HCl elimination in 2,3′-dichlorobiphenyl gives both 4- and 2-chlorodibenzofuran.

At the high temperatures of this reaction, some interchange of the adjacent substituents in the 2 and 3 positions (ortho and meta) of any given ring can occur as a prelude to HCl elimination; in particular *chlorine can move from an ortho to a meta position, and hydrogen from meta to ortho, preceding HCl elimination.* For example, when 2,6,2′,6′-tetrachlorobiphenyl (see next page) is heated in air, some of its molecules lose a pair of ortho chlorines to give a dichlorodibenzofuran, and some first interchange Cl and H in one ring to eliminate HCl and produce a trichlorodibenzofuran. Free rotation about the C—C bond does *not* occur *after* the interchange, as presumably the elimination occurs immediately.

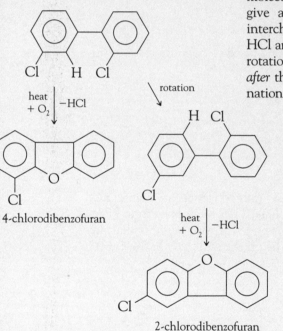

4-chlorodibenzofuran

2-chlorodibenzofuran

dioxins (PCDDs); note, however, that by convention the numbering starts next to the carbon that forms the single C—C bond *opposite* the oxygen:

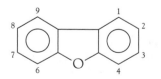

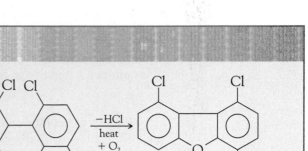

2,6,2′,6′-tetrachlorobiphenyl

1,4,9-trichlorodibenzofuran

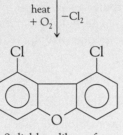

1,9-dichlorodibenzofuran

PROBLEM 1

For each PCB shown below, deduce which furans would be expected to be produced by Cl_2 or HCl elimination when the PCB is heated in air. Write the correct name for each PCDF.

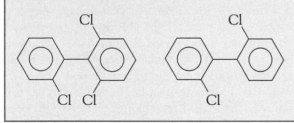

PROBLEM 2

Recently it has been discovered that upon strong heating in air PCBs can also react by elimination of two ortho hydrogen atoms (one on each ring) as H_2. Decide which, if any, additional PCDFs will be produced if the PCBs in Problem 1 can eliminate H_2.

While there exist 75 different chlorine-substituted dibenzo-p-dioxins, there are 135 dibenzofuran congeners, since the symmetry of the ring system is lower for furans. In particular, although the furans have the same left-right symmetry as dioxins, they do not have their up-down symmetry.

PROBLEM 8-9

Draw the structures of all the 16 unique dichlorodibenzofurans and deduce the numbering required in their names.

Almost all commercial PCB samples are contaminated with some PCDFs, but it usually amounts to only a few ppm in the originally manufactured liquids. However, if the PCBs are heated to high temperatures and if some oxygen is present, conversion of PCBs to PCDFs increases the level of contamination by orders of magnitude. The furan concentration in used PCB cooling fluids is found to be greater than in the virgin materials, presumably due to the moderate heating that the fluid undergoes during its normal use. Furan production also occurs if one attempts to burn PCBs with anything but an unusually hot flame.

Other Sources of Dioxins and Furans

In addition to the sources discussed, polychlorinated dibenzofurans and dibenzo-*p*-dioxins are also produced as the by-products of a myriad of processes, including the bleaching of pulp, the incineration of garbage and hospital waste, the recycling of metals, and the production of common solvents such as *tri-* and *per-chloroethene*.

Pulp and Paper Mills

Pulp and paper mills that use chlorine to bleach pulp are dioxin and furan sources. These contaminants, among many other chlorinated compounds, result from the reaction of the chlorine with some of the organic molecules released by the pulp. The tan color of the pulp that has undergone the initial stages of processing is due to the light-absorbing properties of the *lignin* component of the original wood fibers. A generalized structure for lignin is shown in Figure 8-4. In order to make white paper, the residual (~10%) lignin present after initial processing must be removed, usually by bleaching the pulp with oxidizing agents. If you examine the generalized structure of lignin in Figure 8-4, you can observe several sites of monosubstituted *phenols* and *phenolic ethers*, and ortho-substituted *phenyl diethers*. From these structural components it is not difficult to imagine how lignin can serve as a precursor to furans and dioxins when it reacts with chlorinating agents such as elemental chlorine.

More furans than dioxins are formed in the bleaching of pulp by elemental chlorine. The furan congeners of highest concentrations in the pulp are 1,2,7,8-TCDF and the more toxic 2,3,7,8-TCDF. Unfortunately, the most abundant dioxin produced by the pulp-and-paper bleaching

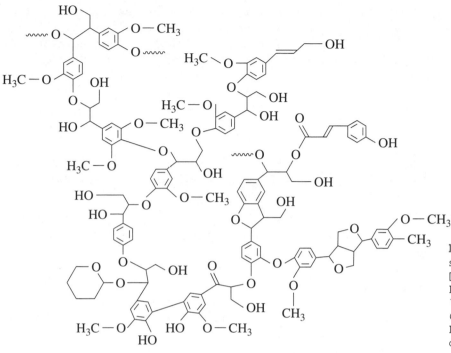

FIGURE 8-4 Generalized structure of lignin. [Source: M. C. Cann and M. E. Connelly, *Real-World Cases in Green Chemistry* (Washington, D. C.: American Chemical Society, 2000).]

process is the highly toxic 2,3,7,8-TCDD congener. The paper and the effluent contain dioxins at parts-per-trillion levels, which resulted in total releases in the past, in North America, of several hundred grams of 2,3,7,8-TCDD annually.

Because of the problems of producing dioxins and furans, the use of **elemental chlorine,** Cl_2, as a bleaching agent for paper was banned in the United States as of April 2001. Most pulp and paper mills there and in other developed countries changed their bleaching agent from elemental chlorine to **chlorine dioxide,** ClO_2, from which the furan and dioxin output is much smaller, even undetectable in many cases. The difference is due to the mechanism by which the compounds attack the pulp's residual lignin. Elemental chlorine reacts to insert chlorine as a substituent on the aromatic rings in lignin, yielding products that are soluble in alkali and can then be washed away. Experiments suggest that during the oxidation of the lignin, two of the component benzene rings can couple together to form a dibenzofuran or dibenzo-*p*-dioxin system that subsequently is chlorinated and in the process becomes detached from the lignin system. In contrast, chlorine dioxide destroys the aromaticity of the benzene rings by free-radical processes and therefore produces fewer chlorinated products that contain aromatic rings. Some mills now produce paper pulp without any use of chlorine compounds. Ozone, hydrogen peroxide, and even high-pressure oxygen are the alternative bleaching agents used in these **totally chlorine free** (TCF) pulp mills. Mills that still use chlorine to bleach now remove contaminants from wastewater by treatments such as reverse osmosis (see Chapter 10).

Green Chemistry: H_2O_2, an Environmentally Benign Bleaching Agent for the Production of Paper

TCF bleaching agents for paper such as *hydrogen peroxide*, *ozone*, and *diatomic oxygen* have been developed. While TCF agents eliminate the formation of dioxins and furans, these methods in general are problematic because the oxidizing agents are not as strong as elemental chlorine or chlorine dioxide. Thus they generally require longer reaction times and higher temperatures (more energy input), and they lead to significant breakdown of the cellulose fibers, which weakens the paper, requiring more wood to produce the same amount of paper.

Terry Collins of Carnegie Mellon University earned a Presidential Green Chemistry Challenge Award in 1999 for his development of compounds known as **tetraamido-macrocyclic ligands** (TAML, Figure 8-5), which enhance the strength of hydrogen peroxide as an oxidizing agent. From an environmental viewpoint, hydrogen peroxide is a particularly enticing reagent since its by-products are water and oxygen, which are about as environmentally benign as one could hope for. The use of TAML in conjunction with hydrogen peroxide reduces the temperature (i.e., the energy requirements) and reaction times normally required for bleaching with hydrogen peroxide, thus making hydrogen peroxide a viable alternative for this process.

The TAML ligands can be modified by varying the alkyl groups (R) on the right side of the structure in Figure 8-5. Changing these groups influences the lifetime of the catalysts. In uses such as the bleaching of paper it is important for the TAML catalysts to decompose in a relatively short time so that they do not become a burden to the environment. However, they must last long enough to fulfill their role as catalysts for hydrogen peroxide. TAML catalysts not only offer significant promise for the bleaching of paper, they are also being considered for use in laundry applications, the disinfection of water, and the decontamination of biological warfare agents such as anthrax.

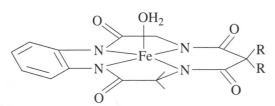

FIGURE 8-5 Tetraamido-macrocyclic ligands (TAML): activators for hydrogen peroxide. [Source: M. C. Cann and M. E. Connelly, *Real-World Cases in Green Chemistry* (Washington, D.C.: American Chemical Society, 2000).]

Fires and Incineration as Sources of Dioxins and Furans

Fires of many kinds, including forest fires and those in incinerators, release various congeners of the dioxin family into the environment; these chemicals are produced as minor by-products from the chlorine and organic matter in the fuel. Dioxin production seems unavoidable whenever combustion of organic matter occurs in the presence of chlorine, unless steps are taken to ensure complete combustion by using very high flame temperatures. Some environmentalists worry particularly about the dioxin emissions when the chlorine-containing plastic PVC is incinerated or involved in other fires. Indeed, recent research on the combustion of newspapers indicates that chlorinated dioxin and furan production rises as the amount of salt or PVC present also increases. In many environmental samples of combustion products, several dozen different dioxin congeners are found, all in comparable amounts. Congeners with relatively high numbers of chlorine substituents usually are the most common.

Incinerators now are the largest anthropogenic source of dioxins in the environment. Dioxins and furans are formed in the postcombustion zone of incinerators, where the temperature is much lower (250–500°C) than in the flame itself (see Chapter 12). They are formed during the oxidative degradation of the graphite-like structures in the soot particles that were produced by the incomplete combustion of the waste. Trace metal ions in the original waste probably catalyze the process. The small amounts of chlorine in the waste provide more than enough to partially chlorinate the furans and dioxins. Recent research indicates that the dibenzofuran and dibenzo-p-dioxin ring systems are formed at high temperatures (>650°C) and that chlorination progressively occurs when the temperature cools below 650°C and gradually is reduced to 200°C.

Characteristically, incineration produces a greater mass of furans than of dioxins. The yields of specific dioxin congeners increase with the degree of chlorination through to OCDD, whereas the peak production of furans occurs with four to six chlorines. In contrast to waste incineration, industrial coal combustion generates little dioxin because the coal burns much more completely, generating little soot to decompose later into dioxins and furans.

Chlorine Content of Dioxin and Furan Emissions

The profile of estimated annual global PCDF and PCDD emissions for the congeners with 4 to 8 chlorines, i.e., those believed to be toxic, is shown in Figure 8-6a. As discussed previously, furans outnumber dioxins, and furans peak at molecules with four chlorines, whereas the dioxin peak is less pronounced and occurs at about six chlorines. Furans with these intermediate amounts of chlorine have toxicities similar to that of 2,3,7,8-TCDD,

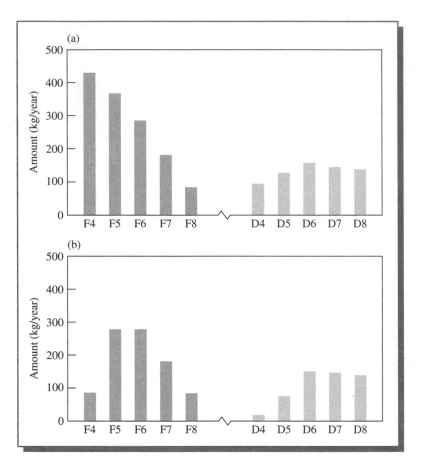

FIGURE 8-6 Annual PCDD and PCDF (a) emissions and (b) deposition rates after reactions with the hydroxyl radical, OH. The letters F and D represent furans and dioxins; the numbers indicate the number of chlorine atoms per molecule. [Source: J. I. Baker and R. A. Hites, "Is Combustion a Major Source of Polychlorinated Dibenzo-p-Dioxins and Dibenzofurans to the Environment?" *Environmental Science and Technology* 34 (2000): 2879.]

whereas fully chlorinated dioxin molecules have low toxicities. Consequently, the threat to human health from furans in the environment may even exceed that from dioxins.

The masses of the dioxin and furan compounds that are eventually deposited from air onto soil and sediments, the primary mechanism by which dioxins eventually enter the food chain, is shown in Figure 8-6b. The loss in mass between emission and deposition is greater the *fewer* the number of chlorines present. This occurs because the principal loss mechanism is attack by addition at an unsubstituted carbon by the *hydroxyl free radical*, OH (followed by atmospheric oxidation of the resulting radical, as expected from the principles discussed in Chapter 3), and the rate of this initial reaction is greater the fewer the chlorines present. The actual amount of OCDD, the octachloro congener of dioxin, that is deposited greatly exceeds the estimate from this graph, which is based mainly on combustion sources. Some scientists believe that the discrepancy arises because much additional

OCDD is created in water droplets in air by the sunlight-initiated photochemical decomposition of PCP, pentachlorophenol, which eventually results in coupling two PCPs to produce OCDD.

Very small concentrations of dioxins—particularly highly chlorinated ones—were present in the environment in the preindustrial era, presumably as a result of forest fires, volcanoes, etc. Indeed, forest fires are still probably Canada's largest source of dioxins. According to analyses of soils and lake sediments, the greatest anthropogenic input of dioxins and furans to the environment in developed countries began in the 1930s and 1940s and peaked in the 1960s and 1970s. The principal sources were combustion and incineration, the smelting and processing of metals, the chemical industry, and existing environmental reservoirs. Inadvertent production of dioxins continues today, but at a slower rate—about half of the maximum, according to some sediment samples. The decrease in emissions resulted from deliberate steps taken by industrialized nations to reduce the production and dispersion of these toxic compounds. In particular, dioxin emissions from large sources in the United States declined by 75% from 1987 to 1995 alone, primarily due to reductions in air emissions from municipal and medical waste incinerators. New regulations should increase the reduction to 95%. However, uncontrolled combustion nonpoint sources such as rural backyard trash burning in barrels—especially when plastics such as PVC are included in the mix—have not been brought under control.

Once created, dioxins and furans are transported from place to place mainly in the atmosphere. Eventually they are deposited and can enter the food chain, becoming bioaccumulated in plants and animals. As previously mentioned, our exposure to them arises almost entirely through the foods we eat. In the next section we try to answer the question of what effects, if any, this exposure has on our health.

The Health Effects of Dioxins, Furans, and PCBs

Over a billion dollars has been spent on research to determine the extent to which dioxins, furans, and PCBs cause toxic reactions in humans. Nevertheless, conclusions about this issue are still tentative and controversial. Evidence about toxicity is derived from two sources: toxicological experiments on animals that have been deliberately exposed to the chemicals and epidemiological studies of humans who have been accidentally exposed.

It is generally agreed that most PCBs are *not* acutely toxic to humans: the LD_{50} values of most congeners are large. In high doses, PCBs cause cancer in test animals, and consequently they are listed as a "probable human carcinogen" by the U.S. EPA. However, studies of humans exposed to them have produced inconsistent results. Most groups of people who have been exposed to relatively high concentrations of PCBs—for example, as a result of their employment in electrical capacitor plants—have not experienced a higher overall death rate. (See, however, the comments about human cancer later in this chapter.) The most common human reaction is **chloracne,** a chemically induced dermataological response on the face to exposure to many types of organochlorine compounds.

Inadvertent PCB Poisonings

The most dramatic effects yet observed on human health from exposure to PCB mixtures occurred when two groups of people, one in Japan in 1968 and the other in Taiwan in 1979, unintentionally consumed PCBs that had accidentally been mixed with cooking oil. In the Japanese incident, and probably in the Taiwanese case as well, the PCBs had been used as a heat exchanger fluid in the deodorization process for the oil. Since the PCBs had been heated, their level of PCDF contamination was much greater than occurs in freshly prepared commercial PCBs. The thousands of Japanese and Taiwanese people who consumed the contaminated oil suffered health effects far worse than has been found for workers at PCB manufacturing and handling plants, even though the resulting PCB levels in their bodies were about the same. From this difference, it has been concluded that the main toxic agents in the poisonings were the PCDFs, and that they and dioxins were collectively responsible for about two-thirds of the health effects, with the PCBs themselves responsible for the remainder. Indeed, studies on laboratory animals indicate that the furans involved in these incidents are more than 500 times as toxic on a gram-for-gram basis than pure PCBs. Cognitive development was measured by IQ scores of children born to the most highly exposed Taiwanese mothers—even if birth occurred long after the consumption of the contaminated oil. Scores were found to be significantly lower than for children of unexposed mothers or for children born before the accident occurred. Interestingly, children whose fathers but not mothers had consumed the oil showed no detrimental effects. Further effects of this incident are discussed later, in the Environmental Estrogens section.

An incident comparable to the Asian cooking oil poisonings occurred in early 1999 in Belgium. Several kilograms of a mixture of PCBs that had previously been heated to a high temperature—converting a tiny fraction of the PCBs to furans—was put into a 80,000-kg batch of animal fat, which was then mixed with animal feed and shipped to about 1000 farmers. Poultry producers subsequently noticed a sudden drop in egg production and in egg hatchability, and high concentrations of dioxins were found in chicken meat. The contaminated food was pulled from the market and destroyed.

Effects of in Utero Exposure to PCBs

Research has recently been initiated to discover whether humans who have been exposed to PCBs through their diets are subject to reproductive problems. Sandra and Joseph Jacobson and their co-workers at Wayne State University in Detroit have spent more than two decades studying the offspring of people from the area around Lake Michigan, including children whose mothers regularly eat fish from the lake and who, as a result, are expected to have elevated levels of PCBs in their bodies. They have discovered statistically significant differences in children born to women with high levels of PCBs in their bodies; these differences are present not only at birth but persist to the age of at least eleven years.

The *prenatal* (i.e., before birth) exposure of the infants to PCBs was determined by analyzing the blood of their umbilical cord for these chemicals after birth. Because analytical techniques in the early 1980s were not sufficiently sensitive to detect the amounts of PCBs in all the umbilical cord samples, exposure for many infants was estimated from the PCB levels in their mother's blood and breast milk. The *postnatal* exposure of the children was assessed by analyzing their mother's breast milk and also by analyzing blood samples from the children at the age of four years.

The Jacobsons discovered that, at birth, the children of mothers who had transmitted the highest amounts of PCBs to the children *before* birth had, on average, a slightly lower birth weight and a slightly smaller head circumference; they were also slightly more premature than those born to women who passed along lower amounts. The severity of these deficits was larger the greater their prenatal exposure to PCBs. When tested at the age of seven months, many of the affected children displayed small difficulties with visual recognition memory, again with the extent of the problem increasing with prenatal transmission of PCBs. At the age of four years, the lower body weight observed at birth for highly exposed infants still lingered. More serious was the observation that the four-year-olds displayed progressively lower scores on several tests of mental functioning (with respect to verbal and memory abilities) the greater their prenatal PCB exposure (see Figure 8-7a).

At eleven years of age the effects of the prenatal PCB exposure were still apparent: the IQ scores of the part of the group that had been the most highly exposed before birth averaged 6 points below the others; the most affected mental functionings were memory and attention span. However, prenatal PCB exposure at any but the highest levels did not appear to have affected the IQ of the eleven-year-olds (see Figure 8-7b).

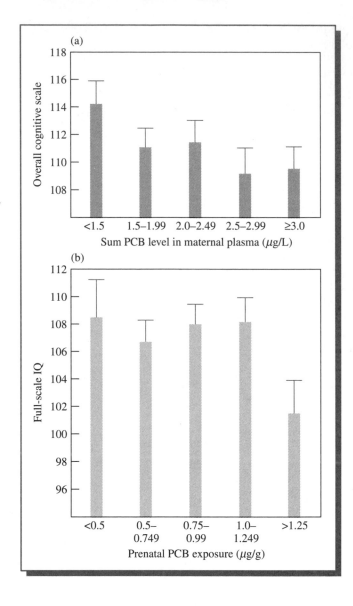

FIGURE 8-7 The effect on the intelligence of children of receiving PCBs prenatally: (a) overall cognitive abilities at $3\frac{1}{2}$ years of age, (b) full-scale IQ at 11 years. [Sources: (a) S. Patandin et al., "Effects of Environmental Exposure to Polychlorinated Biphenyls and Dioxins on Cognitive Abilities in Dutch Children at 42 Months of Age" *Journal of Pediatrics* 134 (1999): 33. (b) J. L. Jacobson et al., "A Benchmark Dose Analysis of Prenatal Exposure to Polychlorinated Biphenyls," *Environmental Health Perspectives* 110 (2002): 393.]

Interestingly, at both four and eleven years of age the children's total body content of PCBs, which is determined mainly from the breast milk they consumed as infants and to a lesser extent by the fish in their diet rather than from any prenatal transmission of the chemicals, was *not* the relevant factor in determining these physical and mental deficits. Rather, it was the smaller amounts of PCBs that had been transmitted from mother to fetus that were important. Thus PCBs appear to interfere with the proper prenatal development of the brain and with the mechanisms that ultimately determine physical size.

The study by the Jacobsons is one of the clearest examples available concerning the influence of toxic environmental chemicals on the health of human beings. It is important to realize that the highest levels of PCBs to which the children in this group were exposed before birth are not much greater than those to which the majority of unborn children in the general population were subjected a few decades ago. And while the chemicals did not produce gross birth defects in the children, they did result in small and consistent deficits of several kinds.

Studies of infants in North Carolina and in upper New York State have produced similar results to those found by the Jacobsons. Nevertheless, some doubt has recently been cast on the Jacobsons' results from an analysis that pointed out the difficulties in establishing the original in utero exposures. A more recent study in the Netherlands has overcome these analytical difficulties, since the ability to detect very low levels of chemicals such as PCBs in blood serum has improved dramatically. The ongoing Dutch study has supported the Jacobsons' findings. It found that prenatal exposure to PCBs is more important than postnatal exposure and produces lower birth and growth weight and lesser cognitive abilities in young children, although the latter deficit was found for all but the least exposed fraction of children, rather than only the most highly exposed fraction. Subtle cognitive deficits have also been found in children in northern Quebec, whose PCB concentrations are high due to the long-range air transport and subsequent deposition and entry into the food chain of the compounds (as discussed later in the chapter). Dutch researchers also found that PCBs and dioxins transmitted to babies during gestation and via breast milk weakened their immune systems, contributing to more infections in the first few years of life.

The Toxicity Patterns of Dioxins, Furans, and PCBs

Recent studies have shown that single doses of 2,3,7,8-TCDD administered to pregnant laboratory animals cause reproductive effects in their offspring. These results have raised some alarm about the potential effects of dioxins on human reproduction. In this connection, many scientists are worried about the dangers posed by environmental chemicals such as dioxins, furans, PCBs, and other organochlorines that can affect sex hormones, as discussed later in this chapter.

Test results from studies on animals indicate that the toxicity of dioxins, furans, and PCBs depends to an extraordinary degree on the extent and pattern of chlorine substitution. The following generalization can be made: the very toxic dioxins are those with four beta chlorine atoms and few if any alpha chlorines. Thus the most toxic is 2,3,7,8-TCDD, which has the maximum number (four) of beta chlorines and no alpha chlorines:

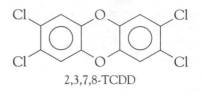

2,3,7,8-TCDD

Dioxin congeners that have three beta chlorines but no (or only one) alpha chlorine are appreciably toxic, but less so than the 2,3,7,8 compound. Fully chlorinated dioxin, that is, octachlorodibenzo-*p*-dioxin (OCDD), has a very low toxicity since all the alpha positions are occupied by chlorine:

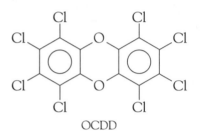

OCDD

Similarly, mono- and dichloro dioxins are usually not considered highly toxic, even if the chlorines are present in beta positions.

PROBLEM 8-10

Predict the order of relative toxicities of the following three dioxin congeners, given that, for systems not too dissimilar to TCDD, the presence of an alpha chlorine reduces the toxicity less than does the absence of a beta chlorine:

2,3,7-trichlorodibenzo-*p*-dioxin

1,2,3-trichlorodibenzo-*p*-dioxin

1,2,3,7,8-pentachlorodibenzo-*p*-dioxin

It is generally assumed that the toxicity pattern for furans is similar though not identical to that for dioxins in that the most toxic congeners have chlorines in all the beta positions. However, the most toxic furan, the 2,3,4,7,8 congener, does have one chlorine atom in an alpha position.

According to animal tests, the most toxic PCBs are those having *no* chlorine atoms (or at most one) in the positions that are ortho to the carbons that join the rings, that is, on the 2,2',6, and 6' carbons. Without ortho chlorines, the two benzene rings can easily adopt an almost coplanar configuration and rotation about the C—C bond joining the rings is rapid. However, because of the large size of chlorine atoms, they get in each other's way if they are present in both ortho positions on the same side of the two rings; this interaction forces the rings to twist away from each other and prevents the rings from adopting the coplanar geometry:

Consequently, PCB molecules with chlorines in three or four of the ortho positions cannot adopt a coplanar geometry.

If the rings are not kept from coplanarity by interference between chlorine atoms, and if certain meta and para carbons have chlorines attached, then the PCB molecule can readily attain the coplanar geometry that happens to be similar in size and shape to 2,3,7,8-TCDD. Such PCB molecules are found to be highly toxic. Apparently 2,3,7,8-TCDD and other molecules of its size and shape readily fit into the same cavity in a specific biological receptor; the complex of the molecule and the receptor can pass through cell membranes and thereby initiate toxic action. By comparing molecular models, it is not difficult to see, e.g., that the most toxic PCB, namely 3,3',4,4',5'-pentachlorobiphenyl is almost the same size and shape as 2,3,7,8-TCDD. It is believed that some of the toxic effects observed in the cooking-oil incidents were due to coplanar PCB congeners.

Only a very small fraction of commercial PCB mixtures correspond to coplanar PCBs having no ortho chlorines. Although individually less toxic, PCBs with one ortho chlorine and with chlorines in both para and at least one meta position contribute substantially to the overall toxicity of PCB mixtures since they are far more prevalent than those having no ortho chlorines.

In humans, the more highly chlorinated furans, dioxins, and PCBs are stored in fatty tissues and are neither readily metabolized nor excreted. This persistence is a consequence of their structure: few of them contain hydrogen atoms on adjacent pairs of carbons at which hydroxyl groups, OH, can readily be added in the biochemical reactions that are necessary for their elimination. In contrast, those compounds with few chlorines always contain one or more such adjacent pairs of hydrogens and tend to be excreted after hydroxylation rather than stored for a long time.

The TEQ Scale

Since most organisms, including humans, possess a mixture of many dioxins, furans, and PCBs stored in their body fat, and since all act qualitatively in the same way, it is useful to have a measure of the *net* toxicity of any mixture. To this end, scientists often report concentrations of these organochlorines in terms of the equivalent amount of 2,3,7,8-TCDD that, if present alone, would produce the same toxic effect. An international **toxicity equivalency factor,** or

TEQ, has been devised that rates each dioxin, furan, and PCB congener's toxicity relative to that of 2,3,7,8-TCDD, which is arbitrarily assigned a value of 1.0. Recently *polybrominated biphenyls* have been added to this scale.

A summary of the TEQ values for some of the more toxic dioxins, furans, and PCBs is given in Table 8-1. As an example, consider an individual who ingests 30 pg (picograms) of 2,3,7,8-TCDD, 60 pg of 1,2,3,7,8-PCDF, and 200 pg of OCDD. Since the TEQ factors for these three substances are, respectively, 1.0, 0.05, and 0.001, the intake is equivalent to

$$(30 \text{ pg} \times 1.0) + (60 \text{ pg} \times 0.05) + (200 \text{ pg} \times 0.001) = 33.2 \text{ pg}$$

Thus, even though a total of 290 pg of dioxins and furans were ingested by this person, it is equivalent in its toxicity to an intake of 33.2 pg of 2,3,7,8-TCDD.

TABLE 8-1	Toxicity Equivalence Factors (TEQ) for Some Important Dioxins, Furans, and PCBs
Dioxin or furan or PCB	**Toxicity equivalency factor**
2,3,7,8-Tetrachlorodibenzo-*p*-dioxin	1
1,2,3,7,8-Pentachlorodibenzo-*p*-dioxin	0.5
1,2,3,4,7,8-Hexachlorodibenzo-*p*-dioxin	
1,2,3,7,8,9-Hexachlorodibenzo-*p*-dioxin	0.1
1,2,3,6,7,8-Hexachlorodibenzo-*p*-dioxin	
1,2,3,4,6,7,8-Heptachlorodibenzo-*p*-dioxin	0.01
Octachlorodibenzo-*p*-dioxin	0.001
2,3,7,8-Tetrachlorodibenzofuran	0.1
2,3,4,7,8-Pentachlorodibenzofuran	0.5
1,2,3,7,8-Pentachlorodibenzofuran	0.05
1,2,3,4,7,8-Hexachlorodibenzofuran	
1,2,3,7,8,9-Hexachlorodibenzofuran	
1,2,3,6,7,8-Hexachlorodibenzofuran	0.1
2,3,4,6,7,8-Hexachlorodibenzofuran	
1,2,3,4,6,7,8-Heptachlorodibenzofuran	
1,2,3,4,7,8,9-Heptachlorodibenzofuran	0.01
Octachlorodibenzofuran	0.001
3,3′,4,4′,5-Pentachlorobiphenyl	0.1
3,3′,4,4′,5,5′-Hexachlorobiphenyl	0.01

Source: TEQ data are drawn mainly from the Canadian Environmental Protection Act Priority Substance List. Assessment Report No. 1 (1990).

PROBLEM 8-11

Using the TEQ values in Table 8-1, calculate the number of equivalent picograms of 2,3,7,8-TCDD that correspond to an intake of 24 pg of 1,2,3,7,8,9-hexachlorodibenzo-*p*-dioxin, 52 pg of 2,3,4,7,8-pentachlorodibenzofuran, and 200 pg of octachlorodibenzofuran.

The TEQ values for environmental samples are sometimes reported in the media as if they represent the concentration of 2,3,7,8-TCDD itself. However, this compound often is not even the dominant contributor to the TEQ toxicity. As discussed previously, combustion of organic matter produces relatively few toxic dioxins; the TEQ from such sources is often dominated by penta- and hexa-chlorinated furans. Similarly, the TEQ arising from use of chlorination in bleaching paper is dominated by toxicity from tetrachlorinated furans.

Dioxins, Furans, and PCBs in Food

About 95% of human exposure to dioxins and furans is from the presence of the compounds in food. A bar graph showing the TEQ values for contamination of various types of foods purchased in U.S. supermarkets in the 1990s is shown in Figure 8-8. Notice that fresh-water fish contained the highest

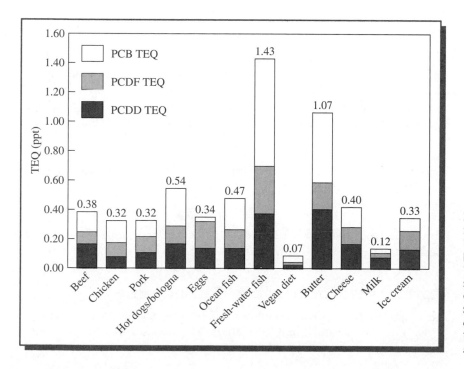

FIGURE 8-8 TEQ values for foods collected from U.S. supermarkets. [Source: A. Schecter et al., "Levels of Dioxins, Dibenzofurans, PCB and DDE Congeners in Pooled Food Samples Collected in 1995 at Supermarkets Across the United States," *Chemosphere* 34 (1997): 1437.]

levels of both PCB and furan toxicity. The composite vegan diet (i.e., all vegetable, fruit, and grain with no animal products at all) has a very low TEQ compared to that from animal-based components. A 2003 report by the U.S. Institute of Medicine recommended that girls should reduce their consumption of animal products in order to reduce the amount of dioxins that build up in their body fat and that could subsequently affect any children they might have.

PROBLEM 8-12

Given that the average TEQ of animal-based foods was about 0.4 pg of TCDD equivalent per gram when the data in Figure 8-8 were collected, and that the LD_{50} for 2,3,7,8-TCDD is about 0.001 mg/kg body weight, what mass of animal-based food would you have had to consume to ingest a fatal dose of it?

In the mid-1980s the average total concentration of all dioxins and furans in the fat tissue of adult North Americans was about 1000 ppt. However, because highly chlorinated and therefore less toxic congeners predominated, the TEQ value was much less: about 40 ppt of 2,3,7,8-TCDD equivalent. By the 1990s the TEQ in human fat had fallen to about 15 ppt. The maximum TEQ levels, about 75 ppt, were reached in the 1970s. The variation with time of stored dioxins and furans fits a model in which the daily TEQ dose amounted to about 0.5 pg/kg body weight in the early decades of the twentieth century, rose to over 6 pg/kg in the 1940s to 1970s, and has now declined again to 0.5 pg/kg.

The average North American adult's body contains about 15 kg of fat, so his/her total body burden of 2,3,7,8-TCDD equivalents now amounts to about 0.2 μg. Given that the average residence time t_{avg} of dioxins and furans in the human body is about 7 years, and using the relationship from Chapter 4 connecting t_{avg} to the total amount C and the input rate R, that is,

$$t_{avg} = C/R$$

then the average human rate of intake of 2,3,7,8-TCDD equivalents is calculated to be

$$R = C/t_{avg} = 0.2 \ \mu g/7 \ \text{years} = 0.03 \ \mu g/\text{year}$$

This value, which corresponds to about 1 pg/kg body weight/day, is close to the intake estimated over the last few years from the model discussed previously.

Dioxins as Probable Human Carcinogens

Although there is little argument about the *relative* acute toxicities of various dioxin and furan congeners, the *absolute* risk to humans is *very* controversial. As noted earlier in the chapter, the amount of

2,3,7,8-TCDD per kilogram of body weight required to kill a guinea pig is extraordinarily small—about 1 μg—making it the most toxic synthetic chemical known for that species. However, the LD_{50} required to kill many other types of animals is hundreds or thousands of times this amount: e.g., the LOD_{50} for hamsters is 1200 μg/kg; for frogs, 1000 μg/kg; for rabbits and mice, 115 μg/kg; for monkeys, 70 μg/kg; and for dogs it may be as low as 30 μg/kg. A 2000 EPA report on dioxin suggests that humans fall in the middle range for acute susceptibility to dioxins; humans given doses of 100 μg/kg suffered no apparent ill effects beyond chloroacne.

Scientists are more worried about the long-term effects of exposure to dioxins than about their acute toxicities. The Seveso study discussed previously was the first to show an increased rate of cancer among people exposed accidentally to TCDD. A recent study of American workers who were employed in industries that produced chemicals contaminated with 2,3,7,8-TCDD indicates that exposure at relatively high levels may cause cancer. The current theory concerning the action of dioxin predicts that there should be a threshold below which no toxic effects will occur, and recent studies of workers exposed to 2,3,7,8-TCDD support this hypothesis.

Another widespread exposure of humans to 2,3,7,8-TCDD occurred in the early 1970s in and around Times Beach, Missouri. Waste oil containing PCBs and 2,3,7,8-TCDD from the manufacture of 2,4,5-trichlorophenol was used for dust control on gravel roads. Some horses died due to exposure in an arena where the dioxin contamination was particularly high, and some children became ill. A decade later, widespread contamination of the soil in the town was discovered. In 1997 over 200,000 tonnes of soil from this town and 26 other affected sites in eastern Missouri that had 2,3,7,8-TCDD levels of 30–200 ppb were excavated and incinerated to remediate the problem. (The town of Times Beach was abandoned in 1982, partly due to a severe flash flood.) Although exposure to the chemical seems to have negatively affected their immune systems, a major study in 1986 did not find evidence of increased disease prevalence in the exposed group of Times Beach residents. Less formal studies and anecdotal evidence, however, indicate problems with seizures and congenital abnormalities, so the issue remains controversial.

In 2000 the U.S. EPA issued a draft report concerning the health risks of dioxins. The report concluded that 2,3,7,8-TCDD is a (known) human carcinogen—although this characterization was a point of controversy among members of the expert committee that reviewed the report—and that the mixtures of dioxins to which people are exposed is a "likely human carcinogen." The *International Agency for Research on Cancer* of the *World Health Organization* had previously classified TCDD as a known human carcinogen. Experiments had indicated that 2,3,7,8-TCDD was the most potent multisite carcinogen known in test animals. The EPA estimates that the most sensitive and most highly exposed Americans stand at least a 1-in-1000 chance of developing cancer from dioxin. This risk is 10 times higher than that stated in the 1994 assessment by the EPA. The report notes that noncancer effects of dioxin are at least as important as cancer.

However, there is currently no clear indication that the incidence of cancer has risen in the general American population due to dioxins, though this could well be due to the inability to relate effects to exposure at current levels. In addition to cancer, the report concludes that dioxins adversely affect the endocrine and immune systems and the development of fetuses, as we'll discuss later.

For furans, direct evidence of human susceptibility is available from the incidents of PCB-contaminated cooking-oil consumption mentioned previously. The most common symptoms observed in these groups were chloracne and other skin problems. Unusual pigmentation occurred in the skin of babies born to some of the mothers who had been exposed. The children also often had low birth weight and experienced a rather high infant-mortality rate. Children who directly consumed the oil showed retarded growth and abnormal tooth development. Many of the victims also reported aching or numbness in various parts of their bodies and frequent bronchial problems. Other than chloracne, such symptoms are not observed in workers who are occupationally exposed to PCBs and whose body burdens of PCBs are comparable to those who consumed the contaminated oil, but in which the PCDF concentration is orders of magnitude lower. However, some children of these workers had mild cases of the less serious problems seen in the poisoned group.

In animal studies, a threshold of about 1000 pg (i.e., 1 ng) of 2,3,7,8-TCDD equivalent per kilogram of body weight per day is observed with respect to the cancer-causing ability of dioxins and furans. In determining the maximum tolerable human exposure to such compounds, many governments apply a safety factor of 100, resulting in a guideline for maximum exposure of 10 pg/kg/day averaged over a lifetime. Currently, the average American ingests only about one-tenth of this amount from animal fats in his/her food supply. Exposure levels near the guideline limit are expected for persons consuming large amounts of fish that have elevated dioxin and furan levels.

Human Exposure to Dioxins, Furans, and PCBs

There continues to be vigorous debate in scientific, industrial, and medical communities regarding the environmental dangers of dioxins, furans, and PCBs. In one camp are those who feel that the dangers from these chemicals have been wildly overstated in the media and by some special-interest groups. They point to the very low concentrations of these substances in the environment, to the possibility that there is a threshold below which they have no effect on human health, to the lack of human fatalities resulting from them, and to the enormous economic costs associated with instituting further controls and cleanup measures. At the other extreme are persons who point to the substantial biomagnification and high toxicity per molecule of these substances and to their presence in almost all environments. They consider the detrimental effects such as cancer and birth deformities caused by these chemicals in wildlife to be "warning canaries" that signal potential ill effects in humans. Discovering where the truth lies between these opposing viewpoints presents a challenge even for environmental science students, to say nothing of the public at large!

Polynuclear Aromatic Hydrocarbons (PAHs)

The Molecular Structure of PAHs

There is a series of hydrocarbons whose molecules contain several six-membered benzene-like rings connected by the sharing of a pair of adjacent carbon atoms between adjoining **fused** rings. The simplest example is **naphthalene,** $C_{10}H_8$:

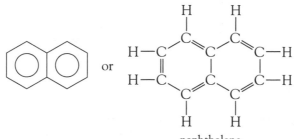

naphthalene

Notice that there are ten, not twelve, carbon atoms in total and that there are only eight hydrogen atoms, since the shared carbons have no attached hydrogens. As a compound, naphthalene is a volatile solid whose vapor is toxic to some insects. It has found use as one form of "moth balls," the other being *1,4-dichlorobenzene*.

Conceptually, there are two ways to fuse a third benzene ring to two carbons in naphthalene; one results in a linear arrangement for the centers (the "nuclei") of the rings while the other is a branched arrangement:

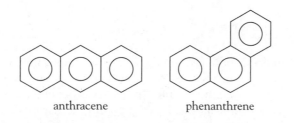

anthracene   phenanthrene

The resulting molecules, **anthracene** and **phenanthrene,** are PAH pollutants associated with incomplete combustion, especially of wood and coal. They are also released into the environment from the dumpsites of industrial plants that convert coal into gaseous fuel and from the refining of petroleum and shale. In rivers and lakes they are found mainly attached to sediments rather than dissolved in the water; both are subsequently partially incorporated by fresh-water mussels.

PROBLEM 8-13

Draw the full structural diagram for phenanthrene, showing all atoms and bonds explicitly.

PROBLEM 8-14

By determining their molecular formulas, show that these molecules are not additional isomers of $C_{14}H_{10}$:

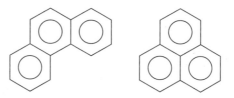

PROBLEM 8-15

Using a systematic procedure, deduce the structural formulas for the five unique isomers of $C_{18}H_{12}$, which contains four fused benzene rings.

In general, hydrocarbons that display benzene-like properties are called **aromatic;** those that contain fused benzene rings are called **polynuclear** (or polycyclic) **aromatic hydrocarbons,** or **PAHs** for short. Like benzene itself, many PAHs possess unusually high stability and a planar geometry. Other than naphthalene, they are not manufactured commercially. However, some PAHs are extracted from coal tar and used in commerce.

PAHs as Air Pollutants

PAHs are common air pollutants and are strongly implicated in the degradation of human health in some cities. Typically, the concentration of PAHs in urban outdoor air amounts to a few nanograms per cubic meter, although it reaches 10 times this amount in very polluted environments. PAHs are formed when carbon-containing materials are incompletely burned. Elevated PAH concentrations in indoor air are typically due to the smoking of tobacco and the burning of wood and coal.

PAHs containing four or fewer rings usually remain as gases if they are released into air. After spending less than a day in outside air, such PAHs are degraded by a sequence of free-radical reactions that begin, as expected from our previous analysis of air chemistry (Chapter 3), by the addition of the OH radical to a double bond.

In contrast to their smaller analogs, PAHs with more than four benzene rings do not exist for long in air as gaseous molecules. Owing to their low vapor pressure, they condense and become adsorbed onto the surfaces of suspended soot and ash particles. PAHs are found mainly on particles of submicron, i.e., respirable, size; consequently, PAHs can be transported into the lungs by breathing.

Figure 8-9 illustrates the concentrations of the most abundant PAHs in an air sample taken in Sweden. The four leftmost PAHs in the figure have three or four fused rings and, as expected, occur mainly as gases. The

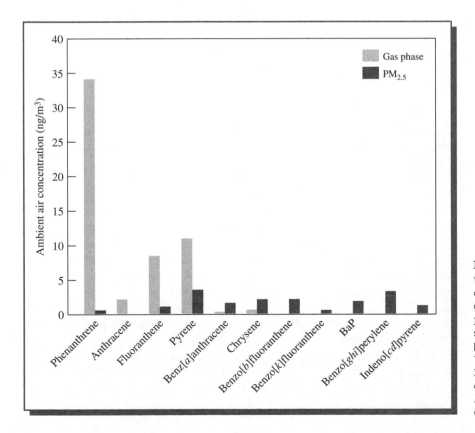

FIGURE 8-9 The distribution of individual PAHs attached to fine particles (PM$_{2.5}$) and in the gas phase in an ambient air sample from Sweden. [Source: C.-E. Bostrom et al., "Cancer Risk Assessment for Polycyclic Aromatic Hydrocarbons," *Environmental Health Perspectives* 110 (supplement 3) (2002): 451.]

remaining ones, which in contrast are found mainly associated with fine particulate matter, are larger. Even PAHs with two to four rings adsorb onto particles in the wintertime, since their vapor pressure decreases sharply at lower temperatures. The role of vapor pressure in determining the long-range air transport of pollutants such as PAHs is discussed later in this chapter.

Soot itself is mainly graphite-like carbon; it consists of a collection of tiny crystals (crystallites), each composed of stacks of planar layers of carbon atoms, all of which occur as fused benzene rings. **Graphite** is the ultimate PAH: its parallel planes of fused benzene rings each contain a vast number of carbon atoms.

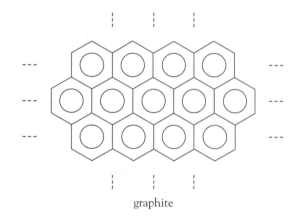

graphite

There are no hydrogen atoms in graphite except at the periphery of the layers. The surfaces of soot particles are excellent adsorbers of gaseous molecules.

PAHs are introduced into the environment from a number of sources: the exhaust of gasoline and especially diesel combustion engines, the "tar" of cigarette smoke, the surface of charred or burned food, the smoke from burning wood or coal, and other combustion processes in which the carbon of the fuel is not completely converted to CO or CO_2. Although PAHs constitute only about 0.1% of airborne particulate matter, their existence as air pollutants is of concern since many of them are carcinogenic, at least in test animals. Vehicle exhaust, especially from diesel engines, older gasoline-powered cars, and all vehicles in which the engine has not warmed up are the major contributors to PAH levels in cities. Aluminum smelters are a source of PAHs since their heated graphite anodes deteriorate over time, releasing the hydrocarbons. A recent survey in Taiwan of PAH emissions from exhaust stacks found the following trends for different types of restaurants, the differences presumably arising from the style of cooking employed:

Chinese   >>   Western   >   fast-food   >   Japanese

PAH levels were particularly high in the air near the World Trade Center buildings in New York for many days after they had been destroyed in September 2001. However, the heat generated by the flames in the fires carried most of the smoke aloft. In addition to PAHs, the high temperature and anaerobic conditions within the piles of debris also resulted in the oxidation of metals and organic substances by chlorine, producing a wide variety of contaminants.

PAHs as Water Pollutants

Polycyclic aromatic hydrocarbons are also serious water pollutants. PAHs enter the aquatic environment as a result of spills of oil from tankers, refineries, and offshore oil drilling sites. In drinking water the PAH level typically amounts to a few nanograms per liter and usually is an unimportant source of these compounds for humans. The larger PAHs bioaccumulate in the fatty tissues of some marine organisms and have been linked to the production of liver lesions and tumors in some fish. PAHs, PCBs, and the insecticide mirex are thought to play a role in the devastation of the populations of beluga whales in the St. Lawrence River; since discharges of these pollutants to the river have decreased substantially in recent years, the health of the belugas may eventually improve.

PAHs are generated in substantial quantity in the production of such coal tar derivatives as *creosote*, a wood preservative used especially on railway ties. Up to 85% of the 200 compounds in creosote are PAHs, including some carcinogenic ones. In addition to concerns about PAH exposure to workers installing the ties, gardeners who buy used railway ties for landscape design and people burning discarded ties are at risk. The leaching of PAHs from the creosote used to preserve the immersed lumber of fishing docks and the like represents a significant source of pollution to crustaceans such as lobsters.

Formation of PAHs During Incomplete Combustion

The mechanism of PAH formation during combustion of organic materials is complex but apparently is due primarily to the repolymerization of hydrocarbon fragments that are formed during the **cracking** (i.e., splitting into several parts) of larger fuel molecules in the flame. Fragments containing two carbon atoms are particularly prevalent after cracking and partial combustion have occurred. Two C_2 fragments can combine to form a C_4 free-radical chain, which could add another C_2 to form a six-membered ring. Such reactions occur quickly if one of the original C_2 fragments is itself a free radical.

The repolymerization reaction occurs particularly under oxygen-deficient conditions. Generally the PAH formation rate increases as the oxygen-to-fuel ratio decreases. The fragments often lose some hydrogen, which forms water after combining with oxygen during the reaction steps. The carbon-rich fragments combine to form the polynuclear aromatic hydrocarbons, which are the most stable molecules that have a high C-to-H ratio.

Since neither methane, the main component of natural gas, nor methanol molecules contain any C—C bonds with which to form C_2 units, their combustion as fuels produces very little PAH or other soot-based particulates.

Carcinogenic Properties of PAHs

The most notorious and common carcinogenic PAH is **benzo[a]pyrene,** BaP, which contains five fused benzene rings:

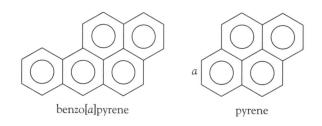

benzo[a]pyrene   pyrene

The molecule is named as a derivative of *pyrene*, which has the structure shown at the right. Conceptually, if an additional benzene ring is added at the pyrene bond labeled *a*, the benzo[a]pyrene molecule is obtained.

Benzo[a]pyrene is a common by-product of the incomplete combustion of fossil fuels, of organic matter (including garbage), and of wood. It is a carcinogen in test animals and a probable human carcinogen. BaP is worrisome since it bioaccumulates in the food chain: its log K_{ow} value is 6.3, comparable to that of many organochlorine insecticides (see Table 7-3). It is considered to be one of the top 15 organic carcinogens in drinking water, as listed in Table 8-2, along with the other 14 compounds in this category, most of which are pesticides.

A second example of a carcinogenic PAH is the four-ring hydrocarbon **benz[a]anthracene,** which is anthracene with another benzene ring fused to the *a* bond:

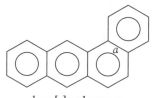

benz[a]anthracene

TABLE 8-2	Top Organic Carcinogens in U.S. Drinking Water[a]	
Chemical	MCL in ppb	Cancer risk per 100,000 people
Ethylene dibromide	0.05	12.5
Toxaphene	3	9.6
Vinyl chloride	2	8.4
Heptachlor	0.4	5.2
Heptachlor epoxide	0.2	5.2
Hexachlorobenzene	1	4.6
Benz[a]pyrene	0.2	4.2
Chlordane	2	2.0
Carbon tetrachloride	5	1.9
1,2-Dichloroethane	5	1.3
PCBs	0.5	0.5
Pentachlorophenol	1	0.3
Di(2-ethylhexyl) phthalate	6	0.2
Dichloromethane (methylene dichloride)	5	0.1

Source: A. H. Smith et al., "Arsenic Epidemiology and Drinking Water Standards," *Science* 296 (2002): 2146.
[a]Chemicals ordered by their cancer risk, were they to be consumed at their MCL (maximum contaminant level).

Some PAHs with certain of their hydrogen atoms replaced by methyl groups are even more potent carcinogens than are the parent hydrocarbons.

The relative positions in space of the fused rings in PAHs play a major role in determining their level of carcinogenic behavior in animals. The PAHs that are the most potent carcinogens each possess a **bay region** formed by the branching in the benzene ring sequence: the organization of carbon atoms as a bay region imparts a high degree of biochemical reactivity to the PAH, as explained in Box 8-3.

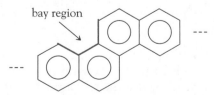

BOX 8-3 | More on the Mechanism of PAH Carcinogenesis

Research has established that the PAH molecules themselves are not carcinogenic agents; rather they must be transformed by several metabolic reactions in the body before the actual cancer-causing species is produced.

The first chemical transformation that occurs in the body is the formation of an epoxide ring across one C=C bond in the PAH. The specific epoxide of interest to the carcinogenic behavior of benzo[a]pyrene is

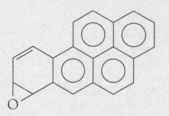

A fraction of these epoxide molecules subsequently add H_2O, to yield two —OH groups on adjacent carbons:

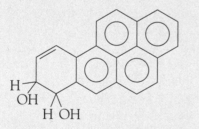

The double bond (shown in the structure) that remains in the same ring as the two —OH groups subsequently undergoes epoxidation, yielding the molecule that is the active carcinogen:

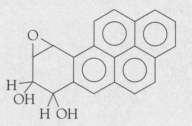

By adding H^+, this molecule can form a particularly stable cation that can bind to molecules such as DNA, thereby inducing mutations and cancer.

The metabolic reactions of epoxide formation and H_2O addition are part of the body's attempt to introduce —OH groups into hydrophobic molecules like PAHs to make them more easily dissolved in water and then eliminated. For BaP and other PAHs that possess a bay region, one of the intermediate products in this multistep process can be diverted instead into the formation of a very stable cation that induces cancer.

PROBLEM 8-16

Based on the bay region theory, would you expect naphthalene to be a carcinogen? How about anthracene or phenanthrene? What about the PAH called benzo[ghi]perylene (shown at the top of the facing page)?

benzo[*ghi*]perylene

Environmental Levels of PAHs and Human Cancer

Has exposure to PAHs been demonstrated to produce cancer in humans? The answer is both yes and no. For over 200 years it has been known that prolonged exposure in occupational settings to very high levels of coal tar, the principal toxic ingredient of which is benzo[*a*]pyrene, leads to cancer in humans. In 1775 the occurrence of scrotal cancer in chimney sweeps was associated with the soot lodged in the crevices of the skin of their genitalia. Modern workers in coke oven and gas production plants likewise experience increased levels of lung and kidney cancer due to this PAH.

The evidence for cancer induction in the general public, whose exposure is at levels that are orders of magnitude lower than in these occupational environments, is less clear-cut. The main cause of lung cancer is the inhalation of cigarette smoke, which contains many carcinogenic compounds in addition to PAHs; the deduction from health statistics of the much smaller influences of pollutants such as PAHs from other sources is difficult to accomplish. Some scientists speculate that the higher death rate from lung cancer in cities as compared to rural areas in many countries is due in part to breathing carcinogenic air pollutants like PAHs, although other factors such as a higher smoking rate also contribute.

Many cities in developing countries have chronic problems with carbon-based particulate air pollution. For example, serious indoor and outdoor air pollution, which arises primarily from the unvented burning of coal and biomass for cooking and heating and consists primarily of PAHs, sulfur dioxide, and particulate matter, is reputed to be responsible for over 1 million deaths annually in China. The rate of lung cancer in Chinese women is higher than for men, possibly due to higher exposures to PAHs from coal burning and from cooking-oil fumes.

Diesel engine exhaust has been labeled a "probable human carcinogen." It contains not only PAHs but also some of their derivatives containing the nitro group, —NO_2, as a substituent; these substances are even more active carcinogens than the corresponding PAHs. For example, *nitropyrene* and *dinitropyrene* are responsible for much of the mutagenic character of diesel exhaust, i.e., its ability to cause mutations that could ultimately produce cancer. These compounds are formed within the engine by the reaction of pyrene with NO_2 and

N_2O_4. There is also evidence that PAHs are nitrated by some of the constituents of photochemical smog.

As discussed in Chapter 2, the emission of particulates from heavy-duty diesel engines can be controlled by means of trap devices placed in the exhaust stream. These devices temporarily retain the solids, including soot, in the exhaust and eventually oxidize them further. Some scientists have worried that, during their residence in the traps, PAHs could undergo reaction to produce even greater quantities of nitrated PAHs. While this apparently does occur, tests indicate that the total mutagenic activity of a given amount of engine exhaust is actually decreased, since most of it is oxidized.

The particulate matter containing PAHs is traceable not only to smoke from the burning of coal, but also to the exhaust of diesel-fueled vehicles and of motor scooters with two-stroke engines. For example, the air in Bombay, India, is of such poor quality that breathing it for a day is said to be equivalent in toxicity to smoking 10 cigarettes!

For most nonsmokers in developed countries, the greatest exposure by far to carcinogenic PAHs arises from their diet, rather than directly from polluted air, water, or soil. As expected from their mode of preparation, charcoal-broiled and smoked meat and fish contain some of the highest levels of PAHs found in food. However, leafy vegetables such as lettuce and spinach can constitute an even greater source of carcinogenic PAHs due to the deposition of these substances from air onto the leaves of the vegetables while they are growing. Unrefined grains also contribute significantly to the total amount of PAHs ingested from food.

Other Toxic Organics of Environmental Concern

Insect Repellants

Throughout history, people have attemped to find substances that were effective in repelling insects, especially mosquitoes and other biting insects. The ideal repellant would not only be effective for long periods but would also be nontoxic and nonirritating to humans and would not be oily or have an offensive odor. The most effective natural repellants contain eucalyptus oil as their active ingredient. The most successful insect repellent is the synthetic compound known popularly as *DEET*, which is an acronym for *N,N-diethyl-3-methyltoluamide*:

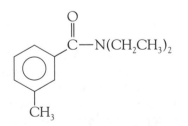

DEET

Some health agencies have warned against using DEET in high-concentration formulations, due to skin irritation problems resulting from application at these concentrations. There are conflicting reports concerning the safety of using DEET for children; some scientists believe that its use can produce seizures in some children.

PBDEs: A New Persistent Pollutant

Highly brominated organic compounds are common commercial fire retardants. These compounds function as fire retardants because, when heated, they release free bromine atoms that react with the free radicals of combustion to produce nonradicals and thereby quench the fire. Large amounts of these fire retardants are used worldwide, and because of their persistence, they now are accumulating in the environment and have even been detected in the Arctic.

Brominated diphenyl ethers, especially, at levels of 5–30% are incorporated into polyurethane foam, textiles, ready-made plastic products, and certain electronic equipment to prevent them from ever catching fire. From a conceptual viewpoint, the molecular structure of diphenyl ether is analogous to that of biphenyl, except that an ether oxygen atom joins the two benzene rings. Bromine atoms can occupy any of the 10 other positions on the rings, analogous to the chlorine atoms in PCBs, again giving 209 possible congeners. Molecules with fewer than four bromines are generally not found in commercial mixtures.

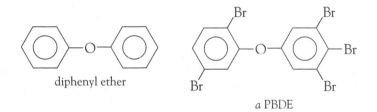

diphenyl ether

a PBDE

As a class, these compounds are known as *polybrominated diphenyl ethers,* **PBDEs.**

PBDE molecules are of particular environmental concern because some migration of them has occurred from commercial products into the environment, where they are now widely distributed. Like PCBs, they are persistent and lipophilic, they bioaccumulate, and some of them are toxic. PBDEs have recently been detected in U.S. sewage sludge (much of which is destined to be spread on agricultural land), in some fish caught in the wild, and even in sperm whales, which normally feed only in deep ocean waters. Also like PCBs, the commercial products are mixtures of congeners, not pure compounds, although the number of congeners in each product is relatively small. Unlike most PCBs, PBDEs are solids under ambient conditions rather than liquids.

The toxicity of PBDEs decreases as the number of bromines increases. Consequently, the least toxic PBDE is the fully brominated congener *decabromo diphenyl ether*. It is the almost exclusive ingredient (>97%) in *Deca*, the commercial mixture that is the predominant PBDE product on the market, which is used as the flame retardant in plastic components in computers and TV housings. Some scientists suspect that the PBDEs in Deca, while not themselves highly toxic, may degrade in the environment by loss of some bromine, thereby increasing their toxicity.

The commercial product called *Octa* consists mainly of congeners with six or seven bromine atoms and is used in thermoplastics.

A third product, *Penta*, is a mixture mainly of PBDEs having four or five bromine atoms. It is used as a fire retardant in polyurethane foams, such as those used in furniture upholstery and padding in vehicles. It is tetra- and pentabromodiphenyl ethers that are most widely distributed in the environment and that are the most bioaccumulative and the most toxic. The Penta mixture constitutes up to 30% by mass of some polyurethane foams, products that easily deteriorate by outdoor weathering and break into small, easily transportable fragments that can eventually find their way into natural waters. From this source PBDE molecules can enter the aquatic food chain.

PBDEs with a large amount of bromine (more than six Br atoms per molecule) probably do not bioaccumulate because they are not readily incorporated by organisms; instead they bind to particles and accumulate in sediments. However, congeners with four to six bromines are taken up by organisms and have the potential to bioaccumulate in the food chain. They may in the future represent a danger to human health due to exposure to them in food, especially fish.

Although few data are available, the concentration of PBDEs in human breast milk is known to have risen sharply in the 1990s in both U.S. and Canadian women—by more than a factor of 10 from 1982 to 2002 for the latter—and is approaching that of PCBs. The main human health concern is that PBDEs having relatively few bromine atoms may affect hormone and liver systems and interfere with neurodevelopment.

Because of environmental concerns, the European Union has banned the Penta and Octa products. Ironically, the human-body concentrations in Europeans appear to be much lower than in North Americans. Some scientists in North America have called for a ban on PBDEs such as that instituted in Europe, since they believe that these compounds will become a serious environmental problem if nothing is done. They believe that there is now enough evidence to invoke the Precautionary Principle (Chapter 7) and reduce the environmental risk from these substances. The counterargument is that these compounds are vital in reducing losses of human lives and property in fires. Late in 2003, U.S. manufacturers of Penta and Octa announced that they would phase out production of these mixtures by the end of 2004.

Two other brominated organic compounds are also used as fire retardants. Indeed, the largest-volume retardant of all is TBBPA, *tetrabromobisphenol*-A.

This substance is less likely than PBDEs to leach or volatilize into the environment since it is covalently bonded to the plastic, whereas PBDEs are only physically dissolved in it. The other fire-retardant product is the half-brominated cyclic hydrocarbon HBCD, h*exabromocyclododecane*.

Polybrominated biphenyls (PBBs) are also used as fire retardants but are now banned in some countries. A 1973 industrial accident in Michigan resulted in the widespread contamination of the food supply there with PBBs.

Perfluorinated Sulfonates

All the organic compounds discussed previously in this chapter act either as hydrophobic (water-repelling) or as *oleophobic* (oil-repelling) substances, but not as both. There exists a small class of organic compounds that will dissolve in neither of these classes. **Fluorinated surfactants** are compounds that consist of molecules and ions having a long perfluorinated carbon tail; that is, a hydrocarbon chain in which each hydrogen atom has been replaced by a fluorine atom. The best-known example of such a molecule is *perfluorooctane sulfonate* (PFOS):

$$CF_3(CF_2)_7 - \overset{\displaystyle O}{\underset{\displaystyle O}{\overset{\|}{\underset{\|}{S}}}} - OH$$

Notice the similarity in structure to sulfuric acid: one —OH group in the latter has been replaced by the unbranched 8-carbon octane chain, which is perfluorinated. This substance was used to make the 3M product *Scotchgard*, a fabric protector that, because of the characteristics of the perfluoro chain, repelled both water and oily spills and potential stains. Other compounds based upon PFOS were used in fire-fighting foams, pesticide formulations, cosmetics, lubricants, grease-resistant coatings for paper products, adhesives, and paints and polishes.

Recently the 3M company voluntarily decided to phase out the production of PFOS because it persists long enough in the environment to eventually be detected in human blood samples. Although it is not very toxic, its concentration in some wildlife had reached levels of concern to some scientists.

Environmental Estrogens

In the last two decades a new threat to the health of wildlife, and possibly of humans, exposed to synthetic organic chemicals in the environment has been identified. It has been established that certain compounds can affect the reproductive health of higher organisms by contributing to infertility in various ways, and they may also increase the rate of cancer in reproductive organs. Much of the public interest in this issue was stirred by the publication of the book *Our Stolen Future* by Theo Colburn and her associates in 1996. However, there remains great uncertainty and debate in the scientific community about whether or not there are significant risks to human health from environmental levels of these compounds.

Mechanism of Action of Environmental Estrogens

The chemicals in question interfere with the systems of the organism that work by transmitting extremely low concentrations, the parts-per-trillion level, of chemical messenger molecules called **hormones.** The hormones flow through the bloodstream from the point of production and storage to their target organs, including those involved in sexual reproduction in both females and males. The arrival of the hormone at a receptor is a signal for the cell to initiate action of some type. Most of the concern about humans centers on interference with **estrogens,** the female sex hormones (which also are present in males, but at a smaller concentrations). Sex hormones, including estrogens, contain the characteristic four-ring steroid structure (see Figure 8-10). They are produced from cholesterol in the ovaries of females and the testes of males in response to signals from the brain and other organs.

Hormone actions are initiated by the binding of the hormone to a specific receptor within a cell. The resulting hormone–receptor complex binds to specific regions of DNA in the cell nucleus, an action that in turn determines the action of the genes. Certain environmental chemicals can also bind to the hormone receptor and thereby either mimic or block the action of the hormone itself. In particular, the "promiscuous" estrogen receptor will bind to a number of compounds, even ones that bear little structural resemblance to estrogen itself. However, most such compounds bind to the receptor with only a small fraction of the strength of estrogen. Estrogen itself binds to its main receptor by hydrogen bonding with its two —OH groups and by attractive van der Waals forces from its ring system to amino acid side chains. A second estrogen receptor has recently been discovered. In addition to the interfering mechanisms discussed, some compounds can accelerate the breakdown of natural hormones, and it has recently been established that some promote the conversion of male hormones into female ones.

As a class, substances that interfere with the **endocrine system** of hormone production and transmission are often referred to as **environmental estrogens** (or *ecoestrogens*). More general names for this class of substances have also been suggested: synthetic *hormonally active agents* (HAAs), *endocrine disrupting chemicals* (EDCs), *environmental hormones*, and *xenoestrogens*.

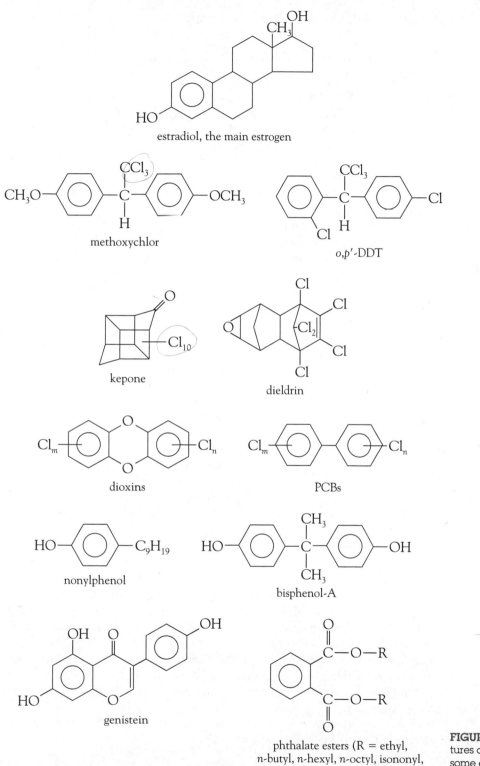

estradiol, the main estrogen

methoxychlor

o,p'-DDT

kepone

dieldrin

dioxins

PCBs

nonylphenol

bisphenol-A

genistein

phthalate esters (R = ethyl,
n-butyl, n-hexyl, n-octyl, isononyl,
isodecyl, benzyl butyl, 2-ethylhexyl)

FIGURE 8-10 The structures of estradiol and some environmental estrogens.

The Chemicals That Operate as Environmental Estrogens

Although an ecoestrogen must attach itself to, or block, the hormone receptor to be effective, there is not a strong resemblance in overall structure between synthetic substances that have been identified as hormonally active and the natural sex hormones, nor is there much structural similarity of the synthetic ones to each other. It is true that many molecules identified as environmental estrogens contain one or more hydroxyl groups, as does estrogen. Oxygen-containing organochlorine insecticides that are known to act hormonally include *methoxychlor* and *kepone* (Figure 8-10). However, other environmental estrogens are organochlorine molecules, including some PCBs, dioxins, and insecticides (Figure 8-10) that contain no oxygen atoms. In metabolizing them, however, the body itself attaches hydroxyl groups to some of the nonchlorinated carbon atoms in such molecules. Some of the resulting hydroxylated organochlorines are more hormonally active agents than the original compounds. The same is true of polycyclic aromatic hydrocarbons when they are hydroxylated.

Most nonchlorinated environmental estrogens are alcohols, often with the phenolic ring that is present in the steroidal hormones. **Nonylphenol** (Figure 8-10) is one important example. **Octylphenol** (which has a C_8 rather than a C_9 chain attached to the phenol ring) is even more hormonally active. Both these alkylphenols occur in the environment—including drinking water—as a result of the breakdown of larger *ethoxylate* molecules. This type of molecular breakdown occurs in sewage treatment plants, for example. The ethoxylates are used in detergents, spermicides, paints, and some plastics. They are commonly used as emulsifying agents in pesticides, and the nonylphenols produced by their decomposition enter the food chain via sprayed fruits and vegetables. The European Union is moving toward banning the use of ethoxylates, as Norway has done already for applications for which alternatives exist.

Another phenolic environmental estrogen of concern is **bisphenol-A** (Figure 8-10), a widely used substance that is polymerized industrially into polycarbonate plastics and some epoxy resins. Some of this compound is leached if the food containers made of the resin are heated for sterilization purposes. The plastics industry states that migration of bisphenol-A from plastic food containers does not occur under normal cleaning and use conditions, although some consumer groups claim evidence that some, albeit small, amounts of leaching from polycarbonate plastics such as baby bottles does occur. Although resins containing the compound are used to line aluminum beer and soda cans, bisphenol-A apparently does not leach from these containers. Bisphenol-A could also potentially leach from dental sealants that are made from resins prepared from it, although the initial evidence that this occurs is now in doubt.

Another hormonally active diol is **genistein,** a molecule whose structure (Figure 8-10) is similar to that of estrogen. Genistein is produced naturally by plants, rather than being a synthetic product. It is a member of a group of natural chemicals called *flavonoids*. Genistein is found in significant concentrations in wood products and in soybeans and soy-based food products. Researchers have discovered that genistein from pulp-mill effluent may cause feminizing and other reproductive effects in fish that swim in the effluent-fed waters. Both nonylphenol and genistein have been found to produce effects at very low doses in test animals. It is not clear yet whether genistein overall has a positive or a negative effect on human health. However, some scientists are worried about the levels of genistein ingested by babies fed with soy milk.

Phthalate esters (Figure 8-10) are widely used as plasticizers in common plastics, from which they can leach into the environment, since they are not chemically bonded to the polymer. They too have estrogenic action. Human exposure to **di-2-ethylhexyl phthalate** (DEHP) occurs via food packaging. The chemical is also present in many other plastics found around the home, including some used by children, where phthalate plasticizers can constitute up to 45% of the weight of the object. The presence of DEHP in plastic medical devices such as intravenous PVC bags is also of concern since leakage of small amounts the plasticizer into the patient occurs during medical procedures.

Concern about health effects of phthalates other than DEHP is small at present according to many scientists. The European Commission has banned the use of phthalate softeners in PVC toys meant for children under 3 years of age, since these children tend to suck and chew on their toys, particularly rattles and teethers, and would thereby extract and ingest some of the phthalate esters. However, a scientific panel convened by the U.S. Consumer Product Safety Commission concluded that the most common plasticizer used in PVC toys, *diisononyl phthalate*, does not pose a risk to humans.

The compounds discussed above are currently thought to be the most significant environmental estrogens uncovered to date. However, the estrogenic potential of many synthetic substances is not yet known. The U.S. EPA in 1999 began an extensive process of screening potential endocrine disruptors.

Effects of Environmental Estrogens on Wildlife

The most devastating consequences of environmental estrogens commonly are not observed in the mammals that originally ingest them. Rather, these effects result from the transfer of the estrogen from the mother to the fetus or egg. This disrupts the hormone balance in the recipient and causes reproductive abnormalities or produces changes that will result in cancer when the offspring grows to adulthood. During its development the fetus is particularly sensitive to fluctuations in hormone levels. For that reason, exposure to low levels of natural or environmental hormones can result in physiological changes that do not occur in adults exposed at the same levels.

The most famous example of the environmental effects of hormone-like chemicals on wildlife involves alligators in Lake Apopka, Florida. In 1980 massive amounts of DDT and its analogs were spilled into the lake. In the mid-1980s Professor Louis Guilette, Jr., of the University of Florida at Gainesville found that very few alligator eggs were hatching—and few hatchlings survived of those that were born—thereby threatening the future population of the colony. Furthermore, the eggs that did hatch produced alligators with abnormal reproductive systems, which therefore were unlikely to be able to reproduce. The ratio of natural estrogen to the male sex hormone testosterone was greatly elevated in the young alligators. Presumably as a consequence, the penises of male alligators were reduced in size compared to the norm. Apparently these effects were caused by **DDE** (Chapter 7), the metabolite of DDT, which has been found by research to inhibit binding of male hormones to their receptor.

Another Florida-based example involves the nearly extinct Florida panthers that live in the Everglades. Their reproductive difficulties may result from the consumption of raccoons whose levels of DDE and other endocrine disruptors are high as a consequence of their consumption of contaminated fish. Some researchers have linked reproductive problems, such as embryo mortality and deformities, of birds in the Great Lakes area to the hormonal activity of pollutants such as PCBs and dioxins.

Tests have revealed that the most estrogenic component of commercial DDT is not the main ingredient (~77%), the p,p'-DDT isomer (Chapter 7), but rather the minor o,p' isomer (~20%) component (Figure 8-10), which has one of the ring chlorines in the ortho position and one in the para. Some scientists have speculated that women who are directly exposed to o,p'-DDT by spraying may be at much higher risk of subsequently developing breast cancer than those in the developed world, whose main exposure is through DDE in their diets.

Research reported in 2002 and 2003 indicated that environmental concentrations, i.e., ppb levels, of the herbicide **atrazine** (Chapter 7) could modify the balance of hormones in just-hatched frogs and thereby affect their sexual development. The effect of atrazine was to increase the levels of an enzyme that converts the male hormone testosterone to the female one, estrogen. About 20% of the male tadpoles developed into hermaphrodites, having both testes and ovaries. Because of this, atrazine may be contributing to the worldwide decline in the population of the amphibian. This hormonal action indicates yet another way, in addition to attaching to or blocking a hormone receptor, that environmental chemicals could upset hormonal balances. However, research reported from some other groups has failed to duplicate the initial findings at very low atrazine concentrations, so the issue of whether this is a significant environmental problem is as yet unresolved.

Researchers have also found that both natural estrogen, secreted by women as part of their monthly cycles, and the synthetic derivative used in birth control pills are present in wastewater and sewage effluent and can cause feminization of males in some species of fish. Such feminization was encountered in the 1990s in some British waterways and was initially blamed on the presence of nonylphenol discharges in the water. A recent survey of estrogens in coastal waters found much higher levels in shallow bays that receive input from sewage than in the open oceans.

Effects of Environmental Estrogens on Humans

Much of the human evidence concerning the possible effects of estrogen mimics on developing fetuses was obtained from the experience of women who took the synthetic estrogen DES (diethylstilbestrol) in the 1948–1971 period to prevent miscarriage. Many of the daughters of these women are sterile and a small fraction of them have developed a rare vaginal cancer as a consequence of their prenatal exposure to DES. The male offspring have an increased incidence of abnormalities in their sexual organs, have decreased average sperm counts, and may have an increased risk of testicular cancer as a consequence of this exposure, but their fertility is not affected.

Although there is good evidence that high concentrations of environmental estrogens have caused reproductive problems in wildlife and laboratory animals, it is not certain that comparable effects occur in humans at levels to which we are exposed. The decline in male sperm counts and the increase in the rate of testicular cancer, which are often cited as examples of these effects, may not, in fact, be general phenomena. Both sperm counts and testicular cancer rates vary significantly between geographical regions, and the variations are not closely linked to differences in pollution levels. However, the current average dioxin level in humans is less than an order of magnitude greater than the lowest dose found by experiments to cause reproductive problems in the offspring of rats.

Recent research indicates that environmental estrogens in humans can affect sexual characteristics: girls who were exposed prenatally to high levels of DDE reach puberty on average almost a full year before those with the lowest exposures. Boys were not affected in this way. Perhaps the most dramatic effect of an environmental hormone on human reproduction is the recently discovered effect of dioxins and furans in influencing the male: female sex ratio, i.e., the ratio—normally about 0.51—of boys to girls at birth. Epidemiological evidence from the dioxin-exposed group in Seveso, Italy, and the Taiwanese PCB-poisoned-oil group indicates that males who were exposed to high concentrations of dioxins, or to furans, dioxins, and PCBs, during adolescence are, in adulthood, much less likely to father boys than girls. Most studies of men exposed *as adults* to high levels of dioxins did not show a similar result, and the same is largely true for men who were over age 20 and for all females at the time of the Taiwanese incident. Recent research from Austria indicates that workers occupationally exposed to TCDD when under 20 years of age later fathered many more girls than boys, whereas initial dioxin exposure at a later age had no effect on the sex ratio.

Thus it appears that exposure during puberty to environmental hormones from the dioxin-furan-PCB family of compounds can have measurable lifelong effects on subsequent reproduction characteristics of human males. Interestingly, higher-than-normal exposure of males to industrial PCB mixtures through the diet can result in a *higher* male:female sex ratio, perhaps because some PCB congeners exhibit estrogenic and some antiestrogenic or androgenic behavior, the net effect depending upon the ratio of one to the other.

Some scientists discount *any* adverse health effects of synthetic chemicals acting as environmental estrogens by pointing out that we all ingest much greater quantities of plant-based estrogen mimics called **phytoestrogens,** including genistein. Common sources of these natural chemicals include all soy products, broccoli, wheat, apples, and cherries. Indeed, there is some evidence that phytoestrogens have a *protective* effect against some types of cancers. It is true that phytoestrogens are quickly metabolized by the body and perhaps do not survive long enough to exert effects on a developing fetus, for example, whereas many synthetic environmental estrogens are stored in body fat rather than being metabolized.

Another puzzle in the environmental estrogen story relates to their dose–response behavior. In contrast to most toxic substances for which the response increases with the dose, eventually leveling off, the response curves for action by estrogen and estrogen mimics have an inverted-U shape (see Figure 8-11). The greatest effects are produced at low dosages; large dosages shut down the responding system to some extent. For example, the effects of atrazine on frogs are found by some researchers to operate only at very low concentrations, not at higher levels.

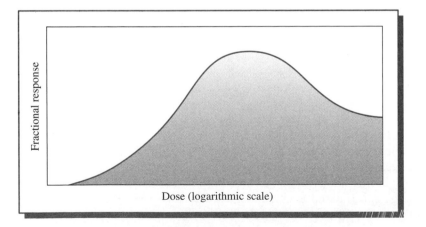

FIGURE 8-11
Dose–response curve (schematic) for estrogen and its mimics.

As an added complication, some estrogenic compounds are found to increase estrogenic activity in some tissues and block it in others! The relevance of these findings and many other unanswered questions in the environmental estrogen story will only be sorted out once much more research is done in this fascinating new chapter concerning the effects of low environmental levels of toxic organic chemicals on living organisms.

The Long-Range Transport of Atmospheric Pollutants

At first glance, it seems amazing to discover that relatively nonvolatile organochlorines and PAHs can eventually travel thousands of kilometers by air from their point of release and end up contaminating relatively pristine areas of the world such as the Arctic. Recently, some quantitative understanding of this **long-range transport of atmospheric pollutants** (LRTAP) has been reached using principles of physical chemistry.

By a global fractionation (or distillation) process, pollutants travel at different rates and are deposited in different geographical regions depending on their physical properties. Most persistent organic pollutants have sufficient volatility to evaporate—often rather slowly—at normal environmental temperatures from their temporary locations at the surface of soil or water bodies. However, because the vapor pressure of any chemical increases exponentially with temperature, evaporation is favored in tropical and semitropical areas, so these geographic regions are rarely the *final* resting places for pollutants. In contrast, cold air temperatures favor the condensation and adsorption of gaseous compounds onto suspended atmospheric particles, most of which are subsequently deposited onto the Earth's surface. Thus the Arctic and Antarctic regions are the final resting places for relatively mobile pollutants that are not deposited at lower latitudes because of their high volatility. Unfortunately, these compounds degrade even more slowly in these regions because temperatures are so cold.

Examples of pollutants that migrate to polar regions are the highly *chlorinated benzenes*; PAHs having three rings; and PCBs, dioxins, and furans that have only a few chlorines (see Table 8-3). Substances with even greater volatility, such as naphthalene and the less chlorinated benzenes, are not deposited even at the cold temperatures of polar regions; consequently they continue their worldwide travels more or less indefinitely until they are chemically destroyed, usually by reaction initiated by collision with the hydroxyl radical.

As implied in Table 8-3, the mobility of a chemical increases as the vapor pressure of its condensed form (as measured by that of the supercooled liquid at 25°C) increases. In addition, mobility increases as the condensation

TABLE 8-3	Predicted Mobilities of Persistent Airborne Pollutants			
Global transport behavior	Low mobility	Relatively low mobility	Relatively high mobility	High mobility
Property				
Vapor pressure of liquid at 25°C in pascals[a]	10^{-4}		10^{-2}	1
Condensation temperature	30°C		−10°C	−50°C
K_{oa}	10^{10}		10^8	10^6
Examples				
PAHs	>4 rings	4 rings	3 rings	1–2 rings
Chlorobenzenes	—	—	5–6 Cl	0–4 Cl
PCBs	9–8 Cl	4–8 Cl	1–4 Cl	0–1 Cl
PCDDs/PCDFs	4–8 Cl	2–4 Cl	0–1 Cl	—
Pesticide examples	Mirex	DDT Toxaphene Chlordane	HCB Dieldrin Hexachloro-cyclohexane	Mothballs

Source: Adapted mainly from F. Wania and D. Mackay, "Tracking the Distribution of Persistent Organic Pollutants," *Environmental Science and Technology* 30 (1996): 390A–396A.
[a]For the supercooled liquid.

temperature of the vapor form of the pollutant gas decreases. Thus substances that do not condense until the temperature drops to −30°C or lower eventually accumulate in polar regions, where such air temperatures are common. Substances with condensation temperatures below −50°C remain airborne indefinitely, since not even polar regions sustain such temperatures for long.

DDT is an intermediate case on these transport scales. It does evaporate sufficiently rapidly (supercooled liquid vapor pressure is 0.005 Pa), but its relatively high condensation temperature of 13°C (55°F) means that much of it becomes permanently deposited at mid-latitudes (especially in the winter) and only a small fraction of it migrates to the Arctic.

Although PCBs are predicted by the model to be deposited mainly in temperate areas rather than migrating en masse to the Arctic, the migration that does occur is great enough that animals there are quite contaminated by these chemicals. The world record for PCB contamination,

90 ppm, is held by polar bears in Spitsbergen, Norway. Even breast milk is higher in PCBs for women who live in far northern areas than in more temperate ones, a result partially of their high-fat diet, since organochlorines are known to accumulate in such a medium. A computer analysis of wind patterns shows that the largest source of dioxins for Inuit (Eskimo) people in the Canadian Arctic is incinerators in the U.S. Midwest.

PROBLEM 8-17

DDE has a 25°C vapor pressure (for its supercooled liquid) of 0.0032 Pa and a condensation temperature of −2°C. Is DDE more volatile than DDT or less? Predict whether a larger or smaller portion of the fraction that does vaporize will be deposited at polar latitudes compared to DDT itself.

Owing to the variations in air temperature during their transport, most molecules of mobile pollutants experience several successive cycles of evaporation and condensation as they migrate gradually toward colder climates. This "grasshopper effect" is illustrated in Figure 8-12 for a pulse of a relatively mobile pollutant that was emitted near the equator at time t_0. At a later time t_1, the majority of the pollutant mass is still present in tropical regions, but at a subsequent time t_2 it has moved mainly to the subtropics. Whether it eventually ever moves (hops) from temperate and subpolar regions to polar ones (at a later time t_6) depends upon whether its mobility is sufficiently high.

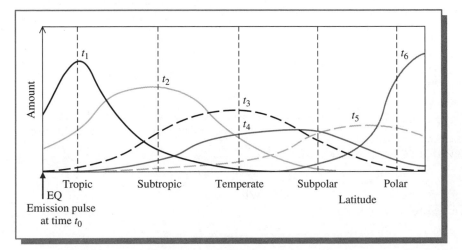

FIGURE 8-12 Calculated variation with time in the geographic distribution of an airborne pollutant released at the equator. [Source: F. Wania and D. Mackay, "Tracking the Distribution of an Airborne Pollutant," *Environmental Science and Technology* 30 (1996): 390A.]

Review Questions

1. Using structural diagrams, write the reaction by which 2,3,7,8-TCDD is produced from 2,4,5-trichlorophenol.

2. Draw the structure of 1,2,7,8-TCDD. What is the full name for this dioxin?

3. What, chemically speaking, was *Agent Orange*, and how was it used?

4. Draw the structure of *pentachlorophenol*. What is its main use as a compound? What is the main dioxin congener that it could produce?

5. What does *PCB* stand for? Draw the structural diagram of the 3,4′,5′-trichloro PCB molecule.

6. What were the main uses for PCBs? What is meant by an *open use*?

7. Draw the structure of a representative polychlorinated dibenzofuran congener.

8. Other than the chlorophenols and PCBs, what are some of the other sources of dioxins and furans in the environment? What is currently the largest anthropogenic source of dioxins?

9. From what medium—air, food, or water—does most human exposure to dioxin come about? Why is this so?

10. What molecules can be eliminated from a PCB molecule when it is heated to moderately high temperatures?

11. Are PCBs acutely toxic to humans? What is the basis for health concerns about them? Recount recent evidence that shows that PCBs can affect human development.

12. Are all dioxin congeners equally toxic? If not, what pattern of chlorine substitution leads to the greatest toxicity? What is the most toxic dioxin?

13. What is meant by a *coplanar PCB*? What structural features give rise to noncoplanarity?

14. What does *TEQ* stand for? Why is it used?

15. Is dioxin carcinogenic to humans or not? Discuss the evidence for and against.

16. What does *PAH* stand for? Draw the structures of two examples.

17. In what processes are PAHs commonly formed?

18. By means of a structural diagram, show what is meant by the *bay region* present in certain PAHs. How is the presence of this region related to the health effects of PAHs?

19. What does *PBDE* stand for? Draw the structure of any PBDE. What are some of the uses of this class of compound?

20. Draw the structure of a *perfluorinated sulfonate*. What are such substances used for?

21. Which three physical properties are used to predict the ultimate deposition zone of volatile chemicals?

22. Define the term *environmental estrogen*. Give two chloroorganic and two nonchloroorganic examples.

23. Recount some of the evidence that environmental estrogens affect the health of wildlife and of humans.

24. What is a *phytoestrogen*?

25. What does *LRTAP* stand for?

Green Chemistry Questions

1. The development of TAML catalyst for hydrogen peroxide by Terry Collins won a Presidential Green Chemistry Challenge Award.

(a) Which of the three focus areas (see the Introduction to Green Chemistry) for these awards does this award best fit into?

(b) List three of the twelve principles of green chemistry (see the Introduction to Green Chemistry) that are addressed by the chemistry developed by Collins.

2. What environmental advantages does the TAML–hydrogen peroxide method of bleaching pulp have over the use of elemental chlorine?

Additional Problems

1. The partitioning of PCBs among air, water, and sediments can be estimated by the fugacity model discussed in Chapter 7. The Z values for a typical environmental PCB are 4×10^{-4} in air, 0.03 in water, and about 10,000 in sediment (and biota). Using the model world volumes in Chapter 7, calculate the equilibrium concentrations when 1 mol of PCBs is distributed among air, water, and sediment.

2. Deduce which combination(s) of two different tetrachlorophenol isomers would produce the following hexachlorodibenzo-*p*-dioxins: **(a)** the 1,2,3,7,8,9 isomer, **(b)** the 1,2,4,6,8,9 isomer, and **(c)** the 1,2,3,6,7,9 isomer. [Hint: See Box 8-1]

3. Deduce which dioxins would probably result from the low-temperature combustion of a commercial sample of PCP.

4. In the purification of wastewater contaminated by pentachlorophenol and 2,3,5,6-tetrachlorophenol using ultraviolet light, it was noticed that OCDD and 1,2,3,4,6,7,8-heptachlorodibenzo-*p*-dioxin were formed. Deduce whether the latter was formed by the coupling of a molecule of each of the phenols or by photochemical dechlorination of OCDD. What potential flaw exists in treating water by UV light, given the nature of the products that are formed?

5. *Paraquat* is a well-known herbicide, which has acquired a sort of fame owing to its use to destroy crops of marijuana. By using information gathered using a search engine on the World Wide Web, develop a short essay on the uses and dangers of Paraquat, including the issue of poisoning by it in developing countries.

6. Using mechanical ball-and-stick or computer-generated molecular models, construct structures for **(a)** 2,3,7,8-TCDD, **(b)** dibenzofuran, and **(c)** biphenyl. Place chlorines on the dibenzofuran and biphenyl models so that the space filled by the carbon, oxygen, and chlorine atoms overlaps that of the dioxin as much as possible without occupying much of the space associated with the alpha positions. Do the resulting congeners represent the most toxic furans and biphenys according to TEQ values?

7. Professor Craig Stow of the University of Wisconsin has fit mathematical functions to the experimental variations with time (of the type illustrated by either of the straight-line curves in Figure 8-3) in PCB concentrations C in herring gull eggs found in the various Great Lakes. He established the following best fits to the data:

$$\text{Lake Erie:} \quad C = 56.04e^{-0.065t}$$
$$\text{Lake Michigan:} \quad C = 71.85e^{-0.33t} + 22.51$$
$$\text{Lake Huron:} \quad C = 35.59e^{-0.13t} + 0.01e^{+0.49t}$$

Here t is the year after 1978 and C is the PCB concentration in ppm for the eggs.

(a) What is the half-life of PCBs in Lake Erie?

(b) What is the predicted steady-state concentration of PCBs in gull eggs from Lake Michigan? In what year will a concentration only 10% above this value be attained? Does inspection of the data in Figure 8-3 lead you to believe that eggs from Lake Ontario are also tending toward a nonzero steady state concentration?

(c) What is the minimum PCB concentration predicted for eggs from Lake Huron? In what year would that minimum level have been achieved? If the "double exponential" model for Lake Huron is realistic, what should the current PCB level in the eggs have now risen to? [Hint: Recall that for first-order reactions, $t_{1/2} = 0.69/k$]

8. Consider the PCDF shown below. Deduce which PCBs could produce this furan if they are moderately heated in air, given that PCDFs can result from HCl elimination with or without 2,3 interchange, or from Cl_2 elimination. [Hint: See Box 8-2]

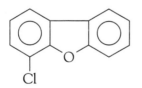

9. By comparing the average dioxin levels in humans to those in our food, decide whether or not dioxin is biomagnified in the transition. Given the food typically eaten by domestic animals such as cattle and chickens compared to that of the freshwater fish we eat, can you explain why dioxin TEQ levels in Figure 8-8 for the fish exceed those for the animals? Why is the TEQ value for butter so much higher than that for milk, and that for hot dogs greater than the general levels for meat? Why is the

TEQ value for a vegan diet so very low? Given the vegan diet TEQ and that for the domestic animals, can you predict whether biomagnification occurs for the latter relative to its diet?

10. Predict the order of toxicity of the following three PCB isomers: 2,4,3',4'-tetrachlorobiphenyl; 3,4,5,4'-tetrachlorobiphenyl; 2,4,2',6'-tetrachlorobiphenyl. [Hint: Will these molecules be coplanar?]

11. Fly ash from municipal waste incinerators can contain high levels of PDDDs and PDDFs, which are formed in the incineration process from various chlorinated starting materials and by various mechanisms. As described in this chapter, one significant mechanism is the condensation of chlorinated phenols, which was illustrated for the wood preservative pentachlorophenol (PCP). What dioxin(s) could be formed from **(a)** 2,6-dichlorophenol and **(b)** 2,4,6-trichlorophenol, assuming condensation with loss of two HCl molecules (see Box 8-1)? Name these dioxins, and predict which one would have the highest toxicity.

12. In a recent study by N. Kazerouni et al. [*Food Chemistry and Toxicology* 39 (2001): 423], the levels of benzo[*a*]pyrene found in cooked hamburgers was found to depend significantly on the cooking method and the cooking time. In the case of oven-broiled burgers, levels of 0.01 ng/g were found for both medium and very well done, whereas in the case of barbecued burgers, levels of 0.09 and 1.52 ng/g were found for medium and very well done burgers, respectively. Explain the observed difference in benzo[*a*]pyrene formation in oven-broiled as compared to barbecued burgers, and the difference in medium versus well done in the case of barbecued burgers. How many micrograms of benzo[*a*]pyrene would be ingested in the consumption of a typical "quarter-pounder" hamburger, if it was very well barbecued (ignore loss of mass during cooking)?

Further Reading

1. O. I. Kalantzi et al., "The Global Distribution of PCBs and Organochlorine Pesticides in Butter," *Environmental Science and Technology* 35 (2001): 1013.

2. J. I. Baker and R. A. Hites, "Is Combustion the Major Source of Polychlortinated Dibenzo-*p*-Dioxins and Dibenzofurans to the Environment?" *Environmental Science and Technology* 34 (2000): 2879.

3. M. Warner et al., "Serum Dioxin Concentrations and Breast Cancer Risk in the Seveso Women's Health Study," *Environmental Health Perspectives* 110 (2002): 625; P. H. Jongbloet et al., "Where the Boys Aren't: Dioxin and the Sex Ratio" *Environmental Health Perspectives* 110 (2002): 1.

4. S. Patandin et al., "Effects of Environmental Exposure to Polychlorinated Biphenyls and Dioxins on Cognitive Abilities in Dutch Children at 42 Months of Age," *Journal of Pediatrics* (January 1999): 33.

5. J. M. Zimpleman, "Dioxin, Not Doomsday," *Journal of Chemical Education* 76 (1999): 1662.

6. J. M. Stellman et al., "The Extent and Pattern of Usage of Agent Orange and Other Herbicides in Vietnam," *Nature* 422 (2003): 681.

7. C-E. Bostrom et al., "Cancer Risk Assessment, Indicators, and Guidelines for Polycyclic Aromatic Hydrocarbons in the Ambient Air," *Environmental Health Perspectives* 110 (suppl. 3) (2002): 451.

8. R. Renner, "What Fate Brominated Fire Retardants?" *Environmental Science and Technology* (1 May 2000): 222A; P. A. Darnerud et al., "Polybrominated Diphenyl Ethers: Occurrence, Dietary Exposure, and Toxicology," *Environmental Health Perspectives* 109 (suppl. 1) (2001): 49.

9. C. Maczka et al., "Evaluating Impacts of Hormonally Active Agents in the Environment," *Environmental Science and Technology* (1 March 2000): 136A; G. M. Solomon and T. Schletter, "Environment and Health. 6. Endocrine Disruption and Potential Human Health Implications," *Canadian Medical Association Journal* 163 (2000): 1471; S. H. Safe, "Endocrine Disruptors and Human Health: Is There a Problem? An Update," *Environmental Health Perspectives* 108 (2000): 487.

Websites of Interest

Log on to www.whfreeman.com/envchem3e/ and click on Chapter 8.

Electron Capture Detection of Pesticides

Chlorine-containing compounds of the type that have been discussed in the preceding two chapters usually occur in the environment at very small concentrations, but can be detected and quantified by techniques such as the one discussed in this box.

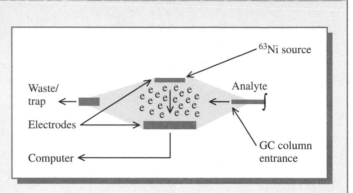

The widespread nature of pesticides in the environment makes their detection an important task, but their often low concentration makes this job difficult. One means of detecting very small amounts of environmentally important chemicals is to use very sensitive chromatographic detectors. In the case of methane this is accomplished with the flame ionization detector (see Environmental Instrumental Analysis Box 2), whose detection limit for CH_4 is in the picogram (10^{-12} g) range.

The most common gas chromatographic detector for halogen-containing pesticides is the **electron capture detector** (ECD). Since many important pesticides contain chlorine, a detector system that responds to molecules that contain this element is the key to this kind of sensitive biospheric analysis. Examples of chlorinated target compounds include DDT (and its breakdown product DDE), lindane, and chlordane. The only chlorine-containing compounds whose detection does not lend itself to this technique are those whose high boiling points preclude GC analysis.

The electron capture detector, like all GC detectors, is located at the end of the chromatographic column (see Environmental

Instrumental Analysis Box 2) in a temperature-controlled (and programmable) oven. When analytes (compounds that have been separated by the chromatographic process) exit the column, they enter the ECD and are detected.

The principle upon which the ECD works involves the disruption of a detector's electronic standing current by the arrival of an analyte containing *electron-loving* (electrophilic) atoms (such as halogens). That disruption is the basis for the ECD signal. The standing current is generated in the following way. Most electron capture detectors have a piece of radioactive nickel-63 fixed on the wall of the detection chamber. This unstable element (half-life 92 years) continuously emits β particles (high-energy electrons from nuclear decay) at a relatively constant rate. The GC carrier gas used in this analysis is usually a mixture of helium and a small amount ($\sim$5%) of another volatile compound such as methane at a constant concentration. In this way a constant ratio of carrier gas and CH_4 flows into the ECD. The β particles from the ^{63}Ni collide with some of the methane molecules in the carrier gas and create a "cloud" of slow-moving electrons in the detection chamber. This cloud creates an electrical potential between two electrodes in the detection chamber, and the resultant current is amplified and sent to the computer (or integrator). Since this standing current is present whenever the detector is on and the carrier gas is flowing, the computer receives a constant detector signal. The figure on the facing page shows the major components of the ECD.

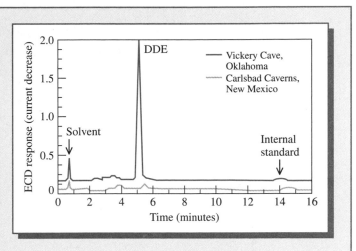

The ECD's standing current changes when an electron-rich analyte arrives in the ECD from the end of the GC column after chromatographic separation: target compounds decrease the standing current because some of the electrons are captured by the electrophilic atoms that are present in the analyte. The more of the compound that arrives, the larger is the decrease in the current.

The computer measures the amount of this decrease and correlates the detector signal with the analyte concentration; however, unlike the positive FID signal (in which more analyte means more signal), the ECD signal is "upside down"—its information is a measure of *missing* signal. The result is, however, the same; the amount of each target analyte can be sensitively and reproducibly determined by the ECD. Furthermore, like other chromatographic systems, the *time* that each compound exits the column and generates its detector signal can be used as a means of identification if other analyses and chemical standards are used.

Among many other applications, the ECD has been used to measure the presence and amount of DDT and the related substance DDE in the tissue of Mexican free-tailed bats (*Tadarida brasiliensis*). These animals absorb these compounds from their diet of insects that have been exposed to DDT in the environment. Although the DDT content of this organism is very low, the breakdown product, p,p'-DDE, remains detectable. The figure above shows two (superimposed) chromatograms resulting from ECD analysis of carcasses of female bats collected from two southwestern U.S. caves, Carlsbad Caverns, New Mexico, and Vickery Cave, Oklahoma. Although no DDT was detected in either animal, the DDE content of the bat carcass (all of the tissue but the brain and intestines) from Vickery Cave was approximately 41.9 μg DDE/g total fat.

Reference: M. L. Thies and K. McBee, "Cross-Placental Transfer of Organochloride Pesticides in Mexican Free-Tailed Bats from Oklahoma and New Mexico," *Archives of Environmental Contamination and Toxicology* 27(1994): 239–242.

PART

III

WATER

CHAPTER 9

The Chemistry of Natural Waters

All life forms on Earth depend on water. Each human being needs to consume several liters of fresh water daily to sustain life. However, fresh water is at a premium. Over 97% of the world's water is seawater, unsuitable for drinking and for most agricultural purposes. Three-quarters of the fresh water is trapped in glaciers and icecaps. Lakes and rivers are one of the main sources of drinking water, even though taken together they constitute less than 0.1% of the total water supply.

Recently it was estimated that humanity currently consumes, mostly for agriculture, about one-fifth of the accessible runoff water destined for the seas; this fraction is predicted to rise to about three-quarters by 2025. Although only 10% of the world's population in 2000 lived under conditions of water stress or scarcity, this figure is expected to rise to 38% by 2025 (Figure 9-1).

It is important to understand the types of chemical activity that prevail in natural waters and how the science and application of chemistry can be employed to purify water intended for drinking purposes. Although some discussion of pollution problems is contained in this chapter, the remediation of contaminated water is considered in detail in Chapter 10. It will be convenient to divide our considerations of water chemistry in this chapter into the two common reaction categories: acid–base reactions and oxidation–reduction (redox) reactions. Acid–base and solubility phenomena predominantly control the concentrations of dissolved inorganic ions such as carbonate in waters, whereas the organic content of water is dominated by redox reactions.

The phenomenon of acid mine drainage can produce extensive pollution of natural waterways and their natural surroundings by heavy metals and sulfuric acid, as illustrated by this river in Tennessee. (William Campbell, Livingston, MT)

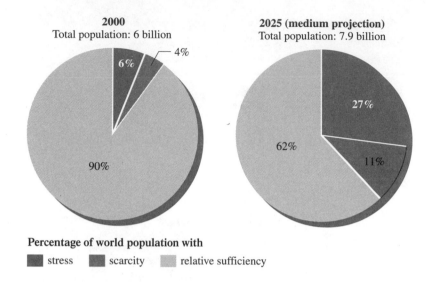

FIGURE 9-1 Global water supply in 2000 and projected for 2025. [Source: R. Engelman et al., *People in the Balance* (Washington, DC: Population Action International, 2000), as reproduced in *Nature* 422 (2003):252.]

The pH and principal ion concentrations in most natural water systems are controlled by the dissolution of atmospheric carbon dioxide and rock- and soil-bound carbonate ions; such reactions are considered in detail later in the chapter. Before considering these acid–base processes, we consider some important redox processes, mainly those involving dissolved oxygen.

Oxidation–Reduction Chemistry in Natural Waters

Dissolved Oxygen

By far the most important oxidizing agent in natural waters is dissolved **molecular oxygen,** O_2. Upon reaction, each of the oxygen atoms in O_2 is reduced from the zero oxidation state to the -2 state in H_2O or OH^-. The half-reaction that occurs in acidic solution is

$$O_2 + 4\ H^+ + 4\ e^- \longrightarrow 2\ H_2O$$

whereas that in basic aqueous solution is

$$O_2 + 2\ H_2O + 4\ e^- \longrightarrow 4\ OH^-$$

The concentration of dissolved oxygen in water is small and therefore precarious from the ecological point of view. For the dissolution process

$$O_2(g) \rightleftharpoons O_2(aq)$$

the appropriate equilibrium constant is the Henry's law constant, K_H (see Chapter 4), which for oxygen at 25°C has the value 1.3×10^{-3} mol/L atm:

$$K_H = [O_2(aq)]/P_{O_2} = 1.3 \times 10^{-3} \text{ mol/L atm} \quad \text{at 25°C}$$

Since in dry air at sea level the partial pressure, P_{O_2}, of oxygen is 0.21 atm, it follows that the solubility of O_2 is 8.7 mg/L H_2O (see Problem 9-1). This value can also be stated as 8.7 ppm, since, as discussed in Chapter 7, for condensed phases ppm concentrations are based on mass rather than moles.

Because the solubilities of gases increase with decreasing temperature, the amount of O_2 that dissolves at 0°C (14.7 ppm) is greater than the amount that dissolves at 35°C (7.0 ppm). The median concentration of oxygen found in natural, unpolluted surface waters in the United States is about 10 ppm.

PROBLEM 9-1

Confirm by calculation the value of 8.7 mg/L for the solubility of oxygen in water at 25°C.

PROBLEM 9-2

Given the solubility cited above for O_2 at 0°C, calculate the value of K_H at this temperature.

River or lake water that has been artificially warmed can be considered to have undergone **thermal pollution** in the sense that, at equilibrium, it will contain less oxygen than colder water because of the decrease in gas solubility with increasing temperature. To sustain life, most fish species require water containing at least 5 ppm of dissolved oxygen; consequently, their survival in warmed water can be problematic. Thermal pollution often occurs in the region of electric power plants (whether fossil-fuel, nuclear, or solar), since they draw cold water from a river or lake, use it for cooling purposes, and then return the warmed water to its source.

Oxygen Demand

The most common substance oxidized by dissolved oxygen in water is organic matter of biological origin, such as dead plant matter and animal wastes. If, for the sake of simplicity, the organic matter is assumed to be entirely polymerized carbohydrate (e.g., plant fiber) with an approximate empirical formula of CH_2O, the oxidation reaction would be

$$CH_2O(aq) + O_2(aq) \longrightarrow CO_2(g) + H_2O(aq)$$
carbohydrate

(The oxidation state of carbon increases from 0 to +4 in the process; thus it is oxidized by the oxygen, which decreases its state from 0 in O_2 to −2 in the product.) Similarly, dissolved oxygen in water is consumed by the oxidation of dissolved *ammonia*, NH_3, and *ammonium ion*, NH_4^+ (substances that, like organic matter, are present in water as a result of biological activity), eventually to *nitrate ion*, NO_3^- (see Problem 9-4).

PROBLEM 9-3

Show that 1 L of water saturated with oxygen at 25°C is capable of completely oxidizing 8.2 mg of polymeric CH_2O. [Hint: Use the balanced chemical reaction for the process given above.]

PROBLEM 9-4

Determine the balanced redox reaction for the oxidation of ammonia to nitrate ion by O_2 in alkaline (basic) solution. Does this reaction make the water more alkaline (i.e., increase the pH) or less? [Hint: Recall the redox balancing procedure you used in introductory chemistry.]

Water that is aerated by flowing in shallow streams and rivers is constantly replenished with oxygen. However, stagnant water or that near the bottom of a deep lake is usually almost completely depleted of oxygen because of its reaction with organic matter and the lack of any mechanism to replenish it quickly, diffusion being a slow process and turbulent mixing being absent.

The capacity of the organic and biological matter in a sample of natural water to consume oxygen, a process catalyzed by bacteria present, is called its **biochemical oxygen demand,** BOD. It is evaluated experimentally by determining the concentration of dissolved O_2 at the beginning and at the end of a period (usually five days) in which a sealed water sample seeded with bacteria is maintained in the dark at a constant temperature, usually either 20 or 25°C. A neutral pH is maintained by use of a buffer consisting of two of the ions of phosphoric acid, $H_2PO_4^-$ and HPO_4^{2-}:

$$H_2PO_4^- \rightleftharpoons HPO_4^{2-} + H^+$$

The BOD equals the amount of oxygen consumed as a result of the oxidation of dissolved organic matter in the sample. The oxidation reactions are catalyzed in the sample by the action of microorganisms present in the natural water. If it is suspected that the sample will have a high BOD, it first is diluted with pure, oxygen-saturated water so that sufficient O_2 will be available to oxidize all the organic matter; the results are corrected for this dilution. Usually the reaction is allowed to proceed for five days before the residual oxygen is determined. The oxygen demand determined from such a test, often designated BOD_5, corresponds to about 80% of that which would be determined if the experiment were allowed to proceed for a very long time—which, of course, is not very practical. The median BOD for unpolluted surface water in the United States is about 0.7 mg O_2/L, which is considerably less than the maximum solubility of O_2 in water (of 8.7 mg/L at 25°C). In contrast, the BOD values for sewage are typically several hundreds of milligrams of oxygen per liter.

A faster determination of oxygen demand can be made by evaluating the **chemical oxygen demand,** COD, of a water sample. **Dichromate ion,** $Cr_2O_7^{2-}$, can be dissolved as one of its salts, such as $Na_2Cr_2O_7$, in sulfuric acid: the result is a powerful oxidizing agent. It is this mixture, rather than O_2, that is used to ascertain COD values. The reduction half-reaction for dichromate when it oxidizes the organic matter is

$$Cr_2O_7^{2-} + 14\,H^+ + 6\,e^- \longrightarrow 2\,Cr^{3+} + 7\,H_2O$$

(In practice, excess dichromate is added to the sample and the resulting solution is back-titrated with Fe^{2+} to the endpoint.) The number of moles of O_2 that the sample would have consumed in accomplishing the oxidation of the same material equals 6/4 (= 1.5) times the number of moles of dichromate, since the latter accepts 6 electrons per ion whereas O_2 accepts only 4:

$$O_2 + 4\,H^+ + 4\,e^- \longrightarrow 2\,H_2O$$

Thus the number of moles of O_2 required for the oxidation is 1.5 times the number of moles of dichromate actually used.

PROBLEM 9-5

A 25-mL sample of river water was titrated with 0.0010 M $Na_2Cr_2O_7$ and required 8.3 mL to reach the endpoint. What is the chemical oxygen demand, in milligrams of O_2 per liter, of the sample?

PROBLEM 9-6

The COD of a water sample is found to be 30 mg O_2/L. What volume of 0.0020 M $Na_2Cr_2O_7$ will be required to titrate a 50-mL sample of the water?

The difficulty with the COD index as a measure of oxygen demand is that acidified dichromate is such a strong oxidant that it oxidizes substances that are very slow to consume oxygen in natural waters and that therefore pose no real threat to their dissolved oxygen content. In other words, dichromate oxidizes substances that would not be oxidized by O_2 in the determination of the BOD. Because of this excess oxidation, namely of stable organic matter such as cellulose to CO_2 and of Cl^- to Cl_2, the COD value for a water sample as a rule is slightly greater than its BOD value. Neither method of analysis oxidizes aromatic hydrocarbons or many alkanes, which, in any event, resist degradation in natural waters.

It is not uncommon for water polluted by organic substances associated with animal or food waste or sewage to have an oxygen demand that exceeds the maximum equilibrium solubility of dissolved oxygen. Under such circumstances, unless the water is continuously aerated, it will soon be depleted of its oxygen, and fish living in the water will die. The treatment of wastewater to reduce its BOD is discussed in Chapter 10.

Finally, we note that there are two other measures of the amount of organic substances present in natural waters. The concentration of **total organic carbon,** TOC, is used to characterize the dissolved *and* suspended organic matter in raw water. For example, the TOC concentration usually has a value of approximately 1 mg/L, i.e., 1 ppm carbon, for groundwater. The parameter **dissolved organic carbon,** DOC, is used to characterize only organic material that is actually dissolved, not suspended. For surface waters, the DOC averages about 5 ppm, though bogs and swamps can have DOC values that are 10 times this amount, and untreated sewage typically has a DOC value of hundreds of ppm. The largest component of organic carbon in natural waters is usually carbohydrates, but many other compounds, including proteins and low-molecular-weight aldehydes, ketones, and carboxylic acids are also present. The humic materials in water are discussed in Chapter 12.

Green Chemistry: Enzymatic Preparation of Cotton Textiles

Globally, over 40 billion pounds (20 million kilograms) of cotton are produced each year. Even with the invasion of synthetic fibers such as nylon, polyester, and acrylics, cotton still holds over 50% of the market share for apparel and home furnishings that are sold in the United States. In preparing raw cotton for use as a fiber, several steps, including desizing, scouring, and bleaching, are required, leaving a fiber that is 99% cellulose. These steps use copious amounts of chemicals, water, and energy and produce millions of pounds of aqueous waste that is high in BOD and COD.

Raw cotton fiber is composed of several concentric layers. The outermost layer is composed of fats, waxes, and pectin, while the inner layers consist primarily of cellulose. The fats and waxes make the raw cotton fiber waterproof, and the pectin acts as a powerful glue to hold the layers together. In order to prepare cotton for use as a fiber for bleaching and dyeing, the outer layer must be removed. This process, which is known as scouring, has traditionally been done by immersing the cotton in 18–25% aqueous **sodium hydroxide** solution at elevated temperatures. This results in hydrolysis of the fats (saponification) and waxes, which solubilizes the components of the outer layer so that they can simply be rinsed away. Scouring results in fibers of even and high wetability. In addition to sodium hydroxide, chelating agents and emulsifiers are added during the scouring process. To end the process, the mixture is neutralized with acetic acid and rinsed several times.

The scouring process is estimated by the U.S. EPA to account for about half of the total BOD produced in the preparation of cotton fibers. The BOD in wastewater from cotton production generally exceeds 1100 mg/L, which is several times that of raw sewage. In addition to the large requirement of chemicals, energy, and water that are used, and the concomitant pollution that is produced, another disadvantage of this process is the unintended weakening and loss of some of the cellulose fibers.

An alternative to the traditional scouring process, known as *Biopreparation*, was developed by Novozymes–North America Inc. and won a Presidential Green Chemistry Challenge Award in 2001. Biopreparation employs an enzyme (a pectin *lyase*), which selectively degrades pectin at ambient temperatures. Since pectin acts as a glue to hold the outer layer of the cotton fiber together, destruction of the pectin results in disintegration of this layer. Because the *lyase* is selective for only pectin, its use is a much less aggressive process than the typical scouring process and removes much less organic material (including cellulose) from the cotton. Since the dissolved organic materials are what contribute to the high BOD and COD of the wastewater, this enzymatic treatment lowers the BOD by 20% and the COD by 50%.

In addition to these environmental advantages, Biopreparation eliminates the use of sodium hydroxide solutions at elevated temperatures, which in turn

- lowers the pH of the wastewater,

- eliminates the need for neutralization with acetic acid and the concomitant wastes,

- lowers energy requirements, and

- lowers rinsing requirements, reducing water consumption by 30–50%.

Elimination of the use of sodium hydroxide also provides for a reduction in risk to workers, and the reduced degradation of cellulose provides more robust fibers and higher yield.

Decomposition of Organic Matter in Water

Dissolved organic matter will decompose in water under anaerobic (oxygen-free) conditions if appropriate bacteria are present. Anaerobic conditions occur naturally in stagnant water such as swamps and at the bottom of deep lakes. The bacteria act on organic carbon by disproportionating it; in other words, some carbon is oxidized to CO_2 and the rest is reduced to CH_4:

$$\underset{\text{organic matter}}{2\ CH_2O} \xrightarrow{\text{bacteria}} CH_4 + CO_2$$

This is an example of a **fermentation** reaction, which is defined as one in which both the oxidizing and the reducing agents are organic materials. Since methane is almost insoluble in water, it forms bubbles that can be seen rising to the surface in swamps and it sometimes catches fire; indeed, methane was originally called *marsh* or *swamp* gas. The same chemical reaction occurs in *digestor* units used by rural inhabitants in semitropical developing countries (India, for instance) to convert animal wastes into methane gas that can be used as a fuel. The reaction also occurs in landfills, as discussed in Chapter 12.

Since anaerobic conditions are reducing conditions in the chemical sense, insoluble Fe^{3+} compounds that are present in sediments at the bottom of lakes are converted by reduction into soluble Fe^{2+} compounds by the electrons available. The Fe^{2+} compounds then dissolve in the lake water:

$$\underset{\text{insoluble Fe(III)}}{Fe^{3+}} + e^- \longrightarrow \underset{\text{soluble Fe(II)}}{Fe^{2+}}$$

It is not uncommon to find aerobic and anaerobic conditions in different parts of the same lake at the same time, particularly in the summer, when a stable stratification of distinct layers often occurs (see Figure 9-2). Water at the top of the lake is warmed by the absorption of sunshine, while that below the level of penetration of sunlight remains cold. Since warm water is less dense than cold (at temperatures above 4°C), the warm upper layer floats on the cold layer below, and little thermal transfer between them occurs. The top layer, called the *epilimnion*, usually contains near-saturation levels of dissolved oxygen, due both to its contact with air and to the O_2 produced during photosynthesis by algae. Since conditions in the top layer are aerobic, elements exist there in their *most oxidized* forms: carbon as CO_2 or H_2CO_3 or HCO_3^-, sulfur as SO_4^{2-}, nitrogen as NO_3^-, and iron as insoluble $Fe(OH)_3$. Near the bottom, in the *hypolimnion*, the water is oxygen-depleted since it has no contact with air and since O_2 is consumed when biological material, such as dead algae that has sunk to these depths, decomposes. Under such anaerobic conditions, elements exist in their *most reduced* forms: carbon as CH_4, sulfur as H_2S, nitrogen as NH_3 and NH_4^+, and iron as soluble Fe^{2+}. Anaerobic conditions usually do not last indefinitely. In the fall and winter the top layer of water is cooled by cold air passing above it, so that eventually the oxygen-rich water at the top becomes more dense than the water below it and gravity

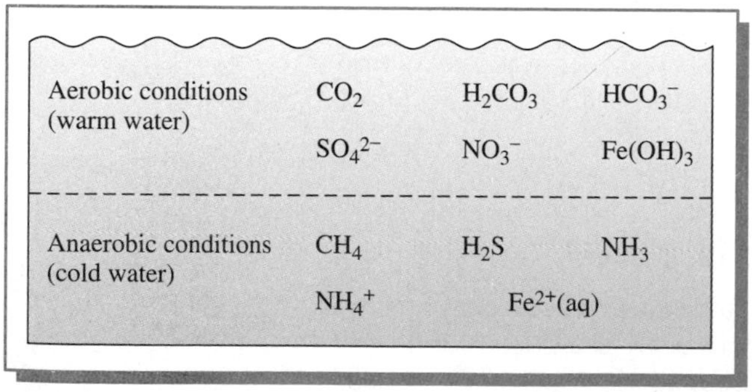

Aerobic conditions (warm water)	CO_2	H_2CO_3	HCO_3^-
	SO_4^{2-}	NO_3^-	$Fe(OH)_3$
Anaerobic conditions (cold water)	CH_4	H_2S	NH_3
	NH_4^+	$Fe^{2+}(aq)$	

FIGURE 9-2 The stratification of a lake in the summer, showing the typical forms of the main elements it contains at different levels.

induces mixing between the layers. Thus in the winter and early spring the environment near the bottom of a lake usually is aerobic.

The pE Scale

Environmental scientists sometimes use the concept of pE to characterize the extent to which natural waters are chemically reducing in nature, in analogy to the way in which pH is used to characterize their acidity. In particular, **pE** is defined as the negative base-10 logarithm of the *effective* concentration— i.e., of the activity—of electrons in water, notwithstanding the fact that free electrons do not exist in solution (any more than do bare protons, H^+ ions). pE values are dimensionless numbers, i.e., they have no units.

• *Low* pE values signify that electrons are readily available from substances dissolved in the water, so the medium is very reducing in nature.

• *High* pE values signify that the dominant dissolved substances are oxidizing agents, so few electrons are available for reduction processes.

When several acids or bases are present in a water sample, usually one of them makes a dominant contribution to the hydrogen or hydroxide ion concentration. In such situations, the position of equilibrium for the other, less dominant weak acids or bases is determined by the H^+ or OH^- level set by the dominant process. In a similar way, in natural waters one or another redox equilibrium reaction is dominant and it determines the electron availability for the other redox reactions that occur simultaneously. If we know the position of the equilibrium for the dominant process, we can calculate the pE of the water, and from it the position of the equilibrium—and hence the dominant species—in the other reactions.

When a significant amount of O_2 is dissolved in water, the reduction of the oxygen to water is the dominant reaction determining overall electron availability:

$$\tfrac{1}{4} O_2 + H^+ + e^- \rightleftharpoons \tfrac{1}{2} H_2O$$

In such circumstances, the pE of the water is related to its acidity and to the partial pressure of oxygen by the following equation, which we will discuss later:

$$pE = 20.75 + \log([H^+] P_{O_2}^{1/4})$$
$$= 20.75 - pH + \tfrac{1}{4} \log (P_{O_2})$$

For a neutral sample of water that is saturated with oxygen from air, i.e., when $P_{O_2} = 0.21$ atm, and is free of carbon dioxide so that its pH = 7, the pE value is calculated from this equation to be 13.9. If the concentration of dissolved oxygen is less than the equilibrium amount, then the equivalent partial pressure of atmospheric oxygen is less than 0.21 atm, so the pE value is less than 13.9 and in some cases even negative.

The pE expression given looks very much like the Nernst equations encountered in the study of electrochemistry. Indeed, for a one-electron process, the pE value for a water sample is simply the electrode potential, E, for whatever process determines electron availability, but divided by RT/F, where R is the gas constant, T the absolute temperature, and F the Faraday constant. At 25°C, RT/F is 0.0591, so

$$pE = E/0.0591$$

Thus the pE expression for any half-reaction in water can be obtained from its standard electrode potential, E^0, corrected by the usual concentration and/or pressure terms and evaluated for a *one-electron* reduction. For example, for the half-reaction linking the reduction of nitrate ion to ammonium ion, we first write the process as a (balanced) one-electron reduction:

$$\tfrac{1}{8}\,NO_3^- + \tfrac{5}{4}\,H^+ + e^- \rightleftharpoons \tfrac{1}{8}\,NH_4^+ + \tfrac{3}{8}\,H_2O$$

For this reaction, $E^0 = +0.836$ V (from standard tables), so $pE^0 = E^0/0.0591 = +14.15$. The equation for pE involves the subtraction from the standard pE^0 of the logarithm of the ratio of concentrations of products to reactants, each raised to its coefficient in the one-electron half-reaction:

$$pE = pE^0 - \log([NH_4^+]^{1/8}/[NO_3^-]^{1/8}[H^+]^{5/4})$$
$$= 14.15 - \tfrac{5}{4}pH - \tfrac{1}{8}\log([NH_4^+]/[NO_3^-])$$

(Here we have used the properties of logarithms that $\log a^x = x \log a$ and that $\log(1/b) = -\log b$.) As usual, the concentration of water does not appear in the expression, because its effect is already included in pE^0.

PROBLEM 9-7

Deduce the equilibrium ratio of concentrations of NH_4^+ to NO_3^- at a pH of 6.0 (a) for aerobic water having a pE value of $+11$ and (b) for anaerobic water for which the pE value is -3.

Returning to the subject of the dominant reaction that determines pE in natural waters, we note that low values of dissolved oxygen are usually caused by the operation of microorganism-catalyzed organic decomposition reactions, and their dissolved products, rather than O_2, can determine electron availability. For example, in cases of low oxygen availability, the pE of water can be determined by ions such as nitrate or sulfate. In the extreme case of the anaerobic conditions found at the bottoms of lakes in the summer and in swamps and rice paddies, the electron availability is determined by the ratio of dissolved methane (a reducing agent) to dissolved carbon dioxide (an oxidizing agent), both of which are produced by the fermentation of organic matter discussed above. They are connected in the redox sense by the half-reaction

$$\tfrac{1}{8} CO_2 + H^+ + e^- \rightleftharpoons \tfrac{1}{8} CH_4 + H_2O$$

The pE value for water controlled by this half-reaction is

$$pE = 2.87 - pH + \tfrac{1}{8} \log(P_{CO_2}/P_{CH_4})$$

For example, if the partial pressures of the two gases are equal and the water is neutral, the pE value is -4.1. Thus the lower levels of a stratified lake are characterized by negative pE values, whereas the oxygenated top layer has a substantially positive pE.

The pE concept is useful in predicting the ratio of oxidized to reduced forms of an element in a water body when we know how the electron availability is controlled by another species. Consider, e.g., the equilibrium between the two common ions of iron:

$$Fe^{3+} + e^- \rightleftharpoons Fe^{2+}$$

For this reaction, since $pE^0 = 13.2$,

$$pE = 13.2 + \log([Fe^{3+}]/[Fe^{2+}])$$

If the pE is determined by another redox process and its value is known, the ratio of Fe^{3+} to Fe^{2+} can be deduced. Thus for the oxygen-depleted water discussed that has a pE value of -4.1,

$$-4.1 = 13.2 + \log([Fe^{3+}]/[Fe^{2+}])$$

so

$$\log([Fe^{3+}]/[Fe^{2+}]) = -17.3$$

and hence

$$[Fe^{3+}]/[Fe^{2+}] = 5 \times 10^{-18}$$

In contrast, for a sample of aerobic water that has a pE of 13.9, the calculated ratio is 5:1 in favor of the Fe^{3+} ion. The transition between the dominance of one form over the other occurs at the pE value at which their concentrations are equal:

$$pE = 13.2 + \log(1) = 13.2 + 0 = 13.2$$

PROBLEM 9-8

Find the pE value for acidic water at which the ratio of concentrations of Fe^{3+} to Fe^{2+} is 100:1.

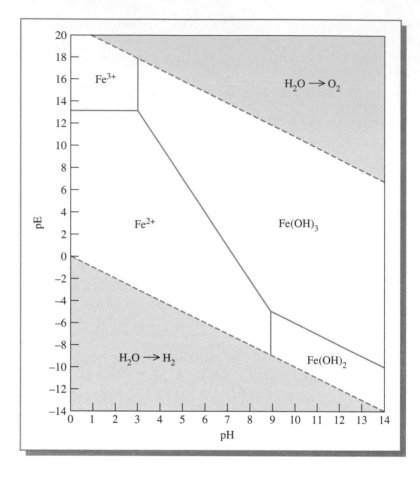

FIGURE 9-3 The pE–pH diagram for the iron system at 10^{-5} M concentration. [Source: Adapted from S. E. Manahan, *Environmental Chemistry*, 4th ed. (Boston, MA: Willard Grant Press / PWS Publishers, 1984).]

pE–pH Diagrams

A visual representation of the zones of dominance for the various oxidation states of an element in water can be displayed in a **pE–pH diagram,** as shown in Figure 9-3 for iron. It is clear from the diagram that the situation is more complicated than that indicated in our calculations, since, in moderately acidic or alkaline environments, the solid hydroxides $Fe(OH)_2$ and $Fe(OH)_3$ also come into play in the equilibria. The solid lines in the diagram indicate combinations of pE and pH values where the concentrations of the two species on either side of the line are equal. Thus we see from the top-left side of the figure that equilibrium between dissolved Fe^{2+} and Fe^{3+} is important only for pH < 3. Equality of the concentrations of these two dissolved forms corresponds to the short horizontal line in Figure 9-3. As expected from our calculations, the transition occurs at pE = 13.2 regardless of pH; hence this line is horizontal.

If iron is in the 3+ state at higher pH, it exists predominantly as the solid $Fe(OH)_3$, whereas solutions containing iron in the 2+ state do not precipitate as $Fe(OH)_2$ until the solution becomes basic.

The shaded regions at the top-right and bottom-left sides of the pE–pH diagram represent extreme conditions under which water itself is oxidized or reduced, yielding O_2 or H_2, respectively:

$$2\,H_2O \longrightarrow O_2 + 4\,H^+ + 4\,e^-$$

and

$$2\,H_2O + 2\,e^- \longrightarrow H_2 + 2\,OH^-$$

The shaded regions correspond to pE–pH combinations in which water itself is unstable with respect to decomposition.

PROBLEM 9-9

Assuming that oxygen determines the electron availability in an aqueous solution and that the partial pressure equivalent to the amount dissolved is 0.10 atm, deduce the ratio of dissolved CO_2 to dissolved CH_4 at a pH of 4.

Sulfur Compounds in Natural Waters

The common inorganic oxidation states in which sulfur is encountered in the environment are illustrated in Table 9-1; they range from the highly reduced -2 state found in hydrogen sulfide gas, H_2S, and insoluble minerals containing the **sulfide ion,** S^{2-}, to the highly oxidized $+6$ state encountered in sulfuric acid, H_2SO_4, and in salts containing the **sulfate ion,** SO_4^{2-}. In organic and bioorganic molecules such as amino acids, intermediate levels of sulfur oxidation are present. When such substances decompose anaerobically, **hydrogen sulfide,** H_2S, and other gases such as CH_3SH and CH_3SSCH_3 containing sulfur in highly reduced forms are released, giving

TABLE 9-1	Common Oxidation States for Sulfur				
	Increasing levels of sulfur oxidation $\longrightarrow$				
Oxidation state of S	-2	-1	0	$+4$	$+6$
Aqueous solution and salts	H_2S HS^- S^{2-}	 S_2^{2-}		H_2SO_3 HSO_3^- SO_3^{2-}	H_2SO_4 HSO_4^- SO_4^{2-}
Gas phase	H_2S			SO_2	SO_3
Molecular solids			S_8		

swamps their unpleasant odor. The occurrence of such gases as air pollutants was discussed in Chapter 2.

As discussed in Chapter 2, hydrogen sulfide is oxidized in air first to *sulfur dioxide*, SO_2, and then fully to sulfuric acid or a salt containing the sulfate ion. Similarly, hydrogen sulfide dissolved in water can be oxidized by certain bacteria to elemental sulfur or more completely to sulfate. Overall, the complete oxidation reactions correspond to

$$H_2S + 2\,O_2 \longrightarrow H_2SO_4$$

Some anaerobic bacteria are able to use sulfate ion as the oxidizing agent to convert organic matter, such as polymeric CH_2O, to carbon dioxide when the concentration of oxygen in the water is very low; the SO_4^{2-} ions are reduced in the process to elemental sulfur or even to hydrogen sulfide:

$$2\,SO_4^{2-} + 3\,CH_2O + 4\,H^+ \longrightarrow 2\,S + 3\,CO_2 + 5\,H_2O$$

Such reactions are especially important in seawater, for which the sulfate ion concentration is much higher than the average for fresh-water systems.

PROBLEM 9-10

(a)   Balance the reduction half-reaction that converts SO_4^{2-} to H_2S under acidic conditions.
(b)   Deduce the expression relating pE to pH, the concentration of sulfate ion, and the partial pressure of hydrogen sulfide gas, given that for the half-reaction, $pE^0 = +5.75$.
(c)   Deduce the partial pressure of hydrogen sulfide when the sulfate ion concentration is 10^{-5} M and the pH is 6.0 for water that is in equilibrium with atmospheric oxygen.

Acid Mine Drainage

One characteristic reaction of groundwater, which by definition is not well-aerated since it has spent much time not exposed to air, is that when it reaches the surface and O_2 has an opportunity to dissolve in it, its rather high level of soluble Fe^{2+} is converted to insoluble Fe^{3+} and an orange-brown deposit of $Fe(OH)_3$ is formed (see also Figure 9-3):

$$4\,Fe^{2+} + O_2 + 2\,H_2O \longrightarrow 4\,Fe^{3+} + 4\,OH^-$$

$$4\,[Fe^{3+} + 3\,OH^- \longrightarrow Fe(OH)_3(s)]$$

The overall reaction is

$$4\,Fe^{2+} + O_2 + 2\,H_2O + 8\,OH^- \longrightarrow 4\,Fe(OH)_3(s)$$

An analogous reaction occurs in some underground mines. Normally FeS_2, called *iron pyrite* or *fool's gold*, is a stable, insoluble component of underground rocks as long as it does not come into contact with air. However, as a result of the mining of coal and other substances, some of it is exposed to oxygen and becomes partially solubilized upon oxidation. The *disulfide ion*, S_2^{2-}, which contains sulfur in the -1 oxidation state, is oxidized to sulfate ion, SO_4^{2-}, which contains sulfur in the $+6$ state:

$$S_2^{2-} + 8\,H_2O \longrightarrow 2\,SO_4^{2-} + 16\,H^+ + 14\,e^-$$

The oxidation of the sulfur is accomplished mainly by O_2:

$$7\,[O_2 + 4\,H^+ + 4\,e^- \longrightarrow 2\,H_2O]$$

When this half-reaction is added to twice the oxidation half-reaction, the net redox reaction is obtained:

$$2\,S_2^{2-} + 7\,O_2 + 2\,H_2O \longrightarrow 4\,SO_4^{2-} + 4\,H^+$$

Since the sulfate of the ferrous ion, Fe^{2+}, is soluble in water, the iron pyrite is effectively solubilized by the reaction. More importantly, the reaction produces a large amount of concentrated acid (note the H^+ product), only some of which is consumed by the air oxidation of Fe^{2+} to Fe^{3+} that accompanies the process:

$$4\,Fe^{2+} + O_2 + 4\,H^+ \longrightarrow 4\,Fe^{3+} + 2\,H_2O$$

This reaction is catalyzed by bacteria.

Combining the last two reactions in the correct ratio, i.e., 2:1, we obtain the overall reaction for the oxidation of both the iron and the sulfur:

$$4\,FeS_2 + 15\,O_2 + 2\,H_2O \longrightarrow 4\,Fe^{3+} + 8\,SO_4^{2-} + 4\,H^+$$
$$\text{i.e., } 2\,Fe_2(SO_4)_3 + 2\,H_2SO_4$$

In other words, the oxidation of the fool's gold produces soluble *iron(III) sulfate*, $Fe_2(SO_4)_3$, and *sulfuric acid*, H_2SO_4. The Fe^{3+} ion is soluble in the highly acidic water that is first produced (see Figure 9-3), the pH of which can be as low as zero, with the usual range being $0-2$. However, once the drainage from the highly acidic mine water becomes diluted and its pH consequently rises, a yellowish-brown precipitate of $Fe(OH)_3$ forms from Fe^{3+}, discoloring the water and the waterway (see Problem 9-11 and Figure 9-3). Thus the pollution associated with acid mine drainage is characterized first by the seeping from the mine of copious amounts of both acidified water and a rust-colored solid. Unfortunately, the concentrated acid can liberate toxic heavy metals from their ores in the mine, further adding to the pollution of the waterway.

Interestingly, the oxidation of disulfide ion to sulfate ion in this process is accomplished to some extent by the action of Fe^{3+} as an oxidizing agent, rather than by O_2:

$$S_2^{2-} + 14\,Fe^{3+} + 8\,H_2O \longrightarrow 2\,SO_4^{2-} + 14\,Fe^{2+} + 16\,H^+$$

The phenomenon of acid drainage is currently of particular importance in the many abandoned mines in the mountains of Colorado. However, the most acidic water in the world comes from Richmond Mine at Iron Mountain, California. There the pH can be as low as -3.6, because the high temperatures (up to 47°C) of the mine water cause much of it to evaporate, thus concentrating the acid. By comparison, the most acidic natural waters occur near the Ebeko volcano in Russia, with a pH as low as -1.7; the acidity is due to hydrochloric and sulfuric acids in the hot spring water.

PROBLEM 9-11

The K_{sp} values for $Fe(OH)_2$ and $Fe(OH)_3$ are 7.9×10^{-15} and 6.3×10^{-38}, respectively. Calculate the solubilities of Fe^{2+} and Fe^{3+} at a pH of 8, assuming they are controlled by their hydroxides. Also calculate the pH values at which the ion solubilities reach 100 ppm.

Nitrogen Compounds in Natural Waters

In some natural waters nitrogen occurs in inorganic and organic forms that are of concern with respect to human health. As discussed in Chapter 4, there are several environmentally important forms of nitrogen that differ in the extent of oxidation of the nitrogen atom.

The most reduced forms are all in the -3 oxidation state, as occurs in ammonia, NH_3, and its conjugate acid, the ammonium ion, NH_4^+. The most oxidized form, the $+5$ state, occurs as the nitrate ion, NO_3^-, which exists in salts, in aqueous solutions, and in **nitric acid,** HNO_3. In solution, the most important intermediates between these extremes are the **nitrite ion,** NO_2^- ($+3$ state), and **molecular nitrogen,** N_2 (0 state). The common oxidation states of nitrogen, along with the most important examples for each, are illustrated in Table 9-2. The pE–pH diagram for the existence of these forms in aqueous solution is shown in Figure 9-4. Notice the relatively

TABLE 9-2	Common Oxidation States for Nitrogen						
	Increasing levels of nitrogen oxidation $\longrightarrow$						
Oxidation state of N	-3	0	$+1$	$+2$	$+3$	$+4$	$+5$
Aqueous solution and salts	NH_4^+ NH_3				NO_2^-		NO_3^-
Gas phase	NH_3	N_2	N_2O	NO		NO_2	

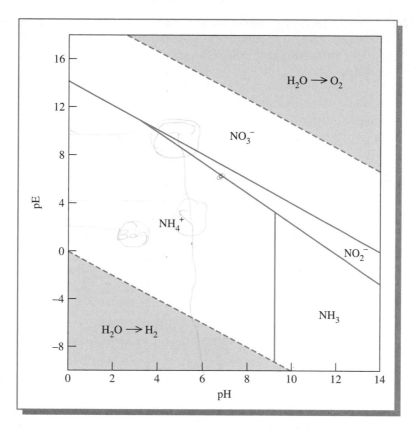

FIGURE 9-4 pE–pH diagram for inorganic nitrogen in an aqueous system. [Source: Adapted from C. N. Sawyer, P. L. McCarty, and G. F. Parkin, *Chemistry for Environmental Engineering*, 4th ed. (New York: McGraw-Hill, 1994).]

small field of dominance for the nitrite ion, NO_2^-, in which the nitrogen oxidation state has the intermediate value of $+3$.

The equilibrium between the most highly reduced and the most highly oxidized forms of nitrogen is given by the half-reaction

$$NH_4^+ + 3\,H_2O \rightleftharpoons NO_3^- + 10\,H^+ + 8\,e^-$$

Thus oxidation is highly pH dependent, being disfavored in highly acidic environments. Previously we derived the equation relating pE to pH for this reaction written as a reduction; this equation defines the diagonal line in Figure 9-4 that separates these two ions when their concentrations are equal; so since $pE^0 = 14.15$,

$$pE = 14.15 - \tfrac{5}{4}\,pH - \tfrac{1}{8}\log(1) = 14.15 - \tfrac{5}{4}\,pH$$

Recall from Chapter 4 that in the microorganism-catalyzed process of nitrification, ammonia and ammonium ion are oxidized to nitrate, whereas in the corresponding denitrification process, nitrate and nitrite are reduced to molecular nitrogen. Both processes are important in soils and in natural waters.

In aerobic environments such as the surface of lakes, nitrogen exists as the fully oxidized nitrate, whereas in anaerobic environments such as the bottom of stratified lakes, nitrogen exists as the fully reduced forms ammonia and ammonium ion (Figure 9-2). Nitrite ion occurs in anaerobic environments such as waterlogged soils that are not sufficiently reducing to convert the nitrogen all the way to ammonia. Most plants can absorb nitrogen only in the form of nitrate ion, so any ammonia or ammonium ion used as fertilizer must first be oxidized by microorganisms before it is useful to plant life.

Acid–Base Chemistry in Natural Waters: The Carbonate System

Natural waters, even when "pure," contain significant quantities of dissolved carbon dioxide and the anions it produces, as well as calcium and magnesium cations. In addition, the pH of such natural water is rarely exactly 7.0, the value expected for pure water. In this section the natural processes that involve these substances in natural waters are analyzed.

The CO_2–Carbonate System

The acid–base chemistry of many natural water systems, including both rivers and lakes, is dominated by the interaction of the **carbonate ion,** CO_3^{2-}, a moderately strong base, with the weak acid **carbonic acid,** H_2CO_3. The carbonic acid results from the dissolution of atmospheric carbon dioxide gas in water and from the decomposition of organic matter in the water. There is usually an equilibrium between the gas in the air and the aqueous acid:

$$CO_2(g) + H_2O(aq) \rightleftharpoons H_2CO_3(aq) \qquad (1)$$

[In fact, much of the dissolved carbon dioxide exists as $CO_2(aq)$ rather than as $H_2CO_3(aq)$, but following conventional practice we combine the two forms and represent it all as $H_2CO_3(aq)$.] The carbonic acid is also in equilibrium in the aqueous medium with **bicarbonate ion,** HCO_3^- (also called *hydrogen carbonate ion*), and hydrogen ion:

$$H_2CO_3 \rightleftharpoons H^+ + HCO_3^- \qquad (2)$$

The predominant source of the carbonate ion is limestone rocks, which are largely made up of **calcium carbonate,** $CaCO_3$. Although this salt is almost insoluble, a small amount of it dissolves when water passes over it:

$$CaCO_3(s) \rightleftharpoons Ca^{2+} + CO_3^{2-} \qquad (3)$$

Natural waters that are exposed to limestone are called **calcareous waters.** The dissolved carbonate ion acts as a base, producing bicarbonate ion and **hydroxide ion,** OH^-, in the water:

$$CO_3^{2-} + H_2O \rightleftharpoons HCO_3^- + OH^- \qquad (4)$$

These reactions occurring in the natural three-phase air, water, and rock system are summarized in Figure 9-5.

In the discussions that follow, we analyze the effects on the composition of a body of water of the simultaneous presence of both carbonic acid and calcium carbonate. However, to obtain a qualitative understanding of this rather complicated system, the effect of the carbonate ion alone is first considered.

Water in Equilibrium with Solid Calcium Carbonate

For simplicity, we first consider a (hypothetical) body of water that is in equilibrium with excess solid calcium carbonate and in which all other reactions are of negligible importance. The only process of interest in this case is reaction (3). Recall from introductory chemistry that the appropriate equilibrium constant for processes that involve the dissolution of slightly soluble salts in water is the **solubility product,** K_{sp}, which equals the product of the concentrations of the ions, each raised to its coefficient in the balanced equation. Thus for reaction (3), K_{sp} is related to the equilibrium concentrations of the ions by

$$K_{sp} = [Ca^{2+}][CO_3^{2-}]$$

For $CaCO_3$ at 25°C, the numerical value of K_{sp} is 4.6×10^{-9}, where [ ] indicates molar concentrations. (Following common practice, we will henceforth omit the units for equilibrium constants and assume that concentrations are in moles per liter and pressures are in atmospheres.) It follows from the stoichiometry of reaction (3) that as many calcium ions are produced as carbonate ions, and that in this simplified system both ion concentrations are equal to S, the solubility of the salt:

$$S = \text{solubility of } CaCO_3$$
$$= [Ca^{2+}] = [CO_3^{2-}]$$

After inserting S for the ion concentrations in the K_{sp} equation, we obtain

$$S^2 = 4.6 \times 10^{-9}$$

so, by taking the square root of each side of the equation, a value for S can be extracted:

$$S = 6.8 \times 10^{-5}\,\text{M}$$

Thus the solubility of calcium carbonate is estimated to be 6.8×10^{-5} mol/L H_2O, assuming that all other reactions are negligible.

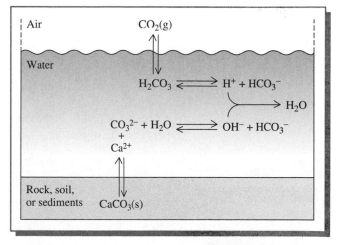

FIGURE 9-5 Reactions among the three phases (air, water, rocks) of the carbon dioxide–carbonate system

According to reaction (4), dissolved carbonate ion acts as a base in water. The relevant equilibrium constant for this process is the base ionization constant, K_b, where

$$K_b(CO_3^{2-}) = [HCO_3^-] [OH^-]/[CO_3^{2-}]$$

Since the equilibrium in this reaction lies to the right in solutions that are not very alkaline, an approximation of the overall effect of the simultaneous occurrence of reactions (3) and (4) can be obtained by adding the equations for the two reactions. The overall reaction is then

$$CaCO_3(s) + H_2O(aq) \rightleftharpoons Ca^{2+} + HCO_3^- + OH^- \qquad (5)$$

Thus the dissolution of calcium carbonate in neutral water results in the production of calcium ion, bicarbonate ion, and hydroxide ion.

It is a principle in equilibrium calculations that, if several reactions are added together, the equilibrium constant K for the *combined* reaction is the *product* of the equilibrium constants for the individual processes. Thus, since reaction (5) is the sum of reactions (3) and (4), its equilibrium constant K_5 must equal $K_{sp}K_b$, the product of the equilibrium constants for reactions (3) and (4). Since $K_a = 4.7 \times 10^{-11}$ for HCO_3^-, and since the acid and base ionization constants for any acid–base conjugate pair such as HCO_3^- and CO_3^{2-} are simply related by the equation

$$K_a K_b = K_w = 1.0 \times 10^{-14}$$

it follows that for the conjugate base CO_3^{2-}

$$K_b(CO_3^{2-}) = K_w/K_a(HCO_3^-)$$

so

$$K_b = 1.0 \times 10^{-14}/4.7 \times 10^{-11} = 2.1 \times 10^{-4}$$

Thus, since K_5 for the overall reaction (5) is $K_{sp}K_b$, its value is $(4.6 \times 10^{-9}) \times (2.1 \times 10^{-4}) = 9.7 \times 10^{-13}$.

The equilibrium constant for reaction (5) is related to the ion concentrations by the equation

$$K_5 = [Ca^{2+}] [HCO_3^-] [OH^-]$$

If we make the approximation that reaction (5) is the *only* process of relevance in the system, then from its stoichiometry we have a new expression for the solubility of $CaCO_3$, namely

$$S = [Ca^{2+}] = [HCO_3^-] = [OH^-]$$

Upon substitution of S for the concentrations, we obtain

$$S^3 = 9.7 \times 10^{-13}$$

Taking the cube root of both sides of this equation, we find

$$S = 9.9 \times 10^{-5} \, M$$

Thus the estimated solubility for $CaCO_3$ is 9.9×10^{-5} M, in contrast to the lesser value of 6.8×10^{-5} M that we obtained when the reaction of carbonate ion was ignored. The $CaCO_3$ solubility here is greater than that estimated from reaction (3) alone, since so much of the carbonate ion produced subsequently disappears (as an ion) by reacting with water molecules. In other words, the equilibrium in reaction (3) is shifted to the right since a large fraction of its product reacts further [reaction (4)].

PROBLEM 9-12

Repeat the calculation of the solubility of calcium carbonate by the approximate single equilibrium method using a realistic wintertime water temperature of 5°C; at that temperature, $K_{sp} = 8.1 \times 10^{-9}$ for $CaCO_3$, $K_a = 2.8 \times 10^{-11}$ for HCO_3^-, and $K_w = 2.0 \times 10^{-15}$. Does the solubility of calcium carbonate increase or decrease with increasing temperature?

PROBLEM 9-13

Consider a body of water in equilibrium with solid calcium sulfate, $CaSO_4$, for which $K_{sp} = 3.0 \times 10^{-5}$ at 25°C. Calculate the solubility of calcium sulfate in water assuming that other reactions are negligible. From your answer, calculate the fraction of the sulfate ion that reacts with water to form *bisulfate* (hydrogen sulfate) ion:

$$SO_4^{2-} + H_2O \rightleftharpoons HSO_4^- + OH^-$$

For HSO_4^-, $K_a = 1.2 \times 10^{-2}$. Note that a strategy that considers the two reactions separately should be appropriate in this case. Does the basicity of the sulfate ion substantially increase the solubility of the salt, as it did for calcium carbonate?

In these calculations we have used the approximation of a single reaction, (5), to represent the two-reaction system. We could improve on our estimates of concentrations now by considering the reaction of the bicarbonate ion with water (to produce hydrogen ion and carbonate ion, having already implicitly considered its equilibrium with carbonate). Alternatively, we could arrive at the same end result by applying an iterative procedure (not shown) to the solution of the algebraic equations involved in the original reactions (3) and (4) themselves rather than combining them; the final

TABLE 9-3	Calculated Ion Concentrations for Aqueous Equilibrium Systems		
Ion	CaCO$_3$ single equation	CaCO$_3$ iterative	CO$_2$ and CaCO$_3$
[HCO$_3^-$]	9.9×10^{-5} M	8.8×10^{-5} M	1.0×10^{-3} M
[CO$_3^{2-}$]	—	3.7×10^{-5} M	8.8×10^{-6} M
[Ca^{2+}]	9.9×10^{-5} M	1.25×10^{-4} M	5.2×10^{-4} M
[OH$^-$]	9.9×10^{-5} M	8.8×10^{-5} M	1.8×10^{-6} M
[H$^+$]	1.0×10^{-10} M	1.1×10^{-10} M	5.6×10^{-9} M
pH	10.0	9.9	8.3

results are listed in Table 9-3. They differ only slightly from those we obtained using the combined equation (5).

From these results it is clear that the saturated aqueous solution of calcium carbonate is moderately alkaline; its pH can be obtained by standard procedures from the hydroxide ion concentration of 8.7×10^{-5} M:

$$pH = 14 - pOH = 14 - \log_{10}[OH^-] = 14 - \log_{10}(8.7 \times 10^{-5}) = 9.9$$

That the solution is alkaline is not surprising, given that the carbonate ion, as weak bases go, is a moderately strong one.

Water in Equilibrium with Both CaCO$_3$ and Atmospheric CO$_2$

The system discussed in the previous section is somewhat unrealistic since it fails to consider the other important carbon species in water, namely, carbon dioxide and carbonic acid, and reactions (1) and (2) that involve them. Recall that pure water in equilibrium with atmospheric carbon dioxide is slightly acidic due to the operation of these two processes (Problem 3-14). These reactions will now be considered in the context of a body of water that is also in equilibrium with solid calcium carbonate, i.e., the three-phase system illustrated in Figure 9-5.

At first sight, it might seem that since reaction (2) provides another source of bicarbonate ion, then by Le Châtelier's principle the production of bicarbonate from reaction (4) of carbonate with water should be suppressed. However, a more important consideration is that reaction (2) produces hydrogen ion, which combines with the hydroxide ion that is produced in reaction (4) by the interaction of carbonate ion with water:

$$H^+ + OH^- \rightleftharpoons H_2O \tag{6}$$

Consequently, the equilibrium positions of both reactions that produce bicarbonate ion are shifted to the right due to the disappearance of one of their products by reaction (6).

If reactions (1) through (4), plus reaction (6), are all added together to deduce the net process, after canceling common terms the net result is

$$CaCO_3(s) + CO_2(g) + H_2O(aq) \rightleftharpoons Ca^{2+} + 2\,HCO_3^- \qquad (7)$$

In other words, combining equimolar amounts of solid calcium carbonate and atmospheric carbon dioxide yields aqueous *calcium bicarbonate*, without any apparent production or consumption of acidity or alkalinity:

calcium carbonate(rock) + carbon dioxide(air) = calcium bicarbonate
(in solution)

Natural waters in which this overall process occurs can be viewed as the site of a giant titration of an acid that originates with CO_2 from air with a base that originates with carbonate ion from rocks.

It should be noted that each of the individual reactions that were added is itself an equilibrium that does not lie entirely to the right. Since the reactions differ in their extent of completion, it is an approximation to state that the overall reaction (7) is the only resulting reaction. Nevertheless, it is the dominant process, and it is mathematically convenient to first consider this process alone in estimating the extent to which $CaCO_3$ and CO_2 dissolve in water when both are present.

Since reaction (7) equals the sum of reactions (1) to (4), plus (6), its equilibrium constant K_7 is the product of their equilibrium constants:

$$K_7 = K_{sp}K_bK_HK_a/K_w$$

Here K_a is the acid dissociation constant 4.5×10^{-7} for carbonic acid [i.e., for reaction (2), which has been corrected for the fact that much of the dissolved carbon dioxide does not in fact exist in the diprotonated form, as explained previously]. K_H is the Henry's law constant (see Chapter 3) for reaction (1). K_w is the ion product for water, and consequently the equilibrium constant for reaction (6) is $1/K_w$. The other constants in the equation for K_7 have been defined previously. Thus at 25 °C, for the overall reaction (7)

$$K_7 = 1.5 \times 10^{-6}$$

From the balanced equation for the reaction, the expression for K_7 is

$$K_7 = [Ca^{2+}][HCO_3^-]^2/P_{CO_2}$$

Here P_{CO_2}, the partial pressure of carbon dioxide in the atmosphere, is 0.00037 atm, since the current atmospheric concentration of CO_2 is 373 ppm (Chapter 4).

If the calcium concentration again is called S, then from the stoichiometry of reaction (7), the bicarbonate concentration must be twice as large, or $2S$; after substituting the concentrations into the equation for K_7 we obtain

$$S\,(2S)^2/0.00037 = 1.5 \times 10^{-6}$$

or

$$S^3 = 1.4 \times 10^{-10}$$

Taking the cube root of both sides, we can evaluate S:

$$S = 5.2 \times 10^{-4}\,\text{mol/L} = [\text{Ca}^{2+}]$$

and thus

$$[\text{HCO}_3{}^-] = 2S = 1.0 \times 10^{-3}\,\text{M}$$

The amount of CO_2 dissolved is also equal to S and is 34 times the amount that dissolves without the presence of calcium carbonate (see results of Problem 3-14). Furthermore, the calculated calcium concentration is four times that calculated without the presence of carbon dioxide. Thus the acid reaction of dissolved CO_2 and the base reaction of dissolved carbonate have a synergistic effect on each other that increases the solubility of both the gas and the solid. In other words, water that contains carbon dioxide more readily dissolves calcium carbonate. In fact, groundwater may become supersaturated with carbon dioxide as a result of biological decomposition processes, and in that case, the calcium carbonate solubility increases even more—at least until the water reaches the surface, when degassing of the CO_2 occurs.

PROBLEM 9-14

Repeat the calculation for the solubility of $CaCO_3$ in water that is also in equilibrium with atmospheric CO_2 for a water temperature of 5°C. At this temperature, $K_H = 0.065$ for CO_2 and K_a for H_2CO_3 is 3.0×10^{-7}; see Problem 9-12 for other necessary data.

Finally, the residual concentrations of $CO_3{}^{2-}$, of H^+, and of OH^- in the system can be deduced from equilibrium constants for reactions in the (1)–(4) and (6) set, since equilibria in these processes are still in effect, notwithstanding the overall reaction (7). Thus from reaction (3),

$$[\text{CO}_3{}^{2-}] = K_{sp}/[\text{Ca}^{2+}] = 4.6 \times 10^{-9}/5.1 \times 10^{-4} = 8.8 \times 10^{-6}\,\text{M}$$

$$[\text{OH}^-] = K_b\,[\text{CO}_3{}^{2-}]/[\text{HCO}_3{}^-] = (2.1 \times 10^{-4}) \times (8.8 \times 10^{-6})/(1.0 \times 10^{-3})$$
$$= 1.8 \times 10^{-6}\,\text{M}$$

and finally from reaction (6)

$$[\text{H}^+] = K_w/[\text{OH}^-] = 1.0 \times 10^{-14}/1.8 \times 10^{-6} = 5.6 \times 10^{-9}\,\text{M}$$

(Since both [OH$^-$] and [H$^+$] represent negligible fractions of [HCO$_3$$^-$] here, it is clear that we need not consider the reverse of reactions (2) and (4) in an iterative process to obtain accurate final results in this case.) From this calculated value for the hydrogen ion concentration, we conclude that river and lake water at 25°C, with a pH determined by saturation with CO$_2$ and CaCO$_3$, should be slightly alkaline, with a pH of about 8.3.

Typically, the pH of such calcareous waters lies in the range of 7 to 9, in reasonable agreement with our calculations. Calcareous waters are much more capable of resisting acidification from acid rain than noncalcareous ones, due to the presence of bicarbonate and carbonate ions, which will react to add H$^+$.

Due to the smaller amount of bicarbonate in noncalcareous waters, the pH values are usually close to 7. If they are subject to acid rain, the pH can become substantially lower since there is little HCO$_3$$^-$ or CO$_3$$^{2-}$ readily available with which to neutralize the acid.

About 80% of natural surface waters in the United States have pH values between 6.0 and 8.4. Lakes and rivers into which acid rain falls will have elevated levels of sulfate ion and perhaps of nitrate ion since the principal acids in the precipitation are H$_2$SO$_4$ and HNO$_3$ (see Chapter 2).

PROBLEM 9-15

In waters subject to acid rain the pH is not determined by the CO$_2$–carbonate system but rather by the strong acid from the precipitation. Assuming that equilibrium with atmospheric carbon dioxide is in effect, calculate the concentration of HCO$_3$$^-$ in natural waters with pH = 6, 5, and 4 at 25°C, for which K_H for CO$_2$ is 3.4×10^{-2} and K_a for H$_2$CO$_3$ is 4.5×10^{-7}.

PROBLEM 9-16

Using algebraic expressions and numerical values for the K_a of both H$_2$CO$_3$ and HCO$_3$$^-$, calculate the pH values for which [H$_2$CO$_3$] = [HCO$_3$$^-$] and for which [HCO$_3$$^-$] = [CO$_3$$^{2-}$]. From your answers, decide the pH domains in which the various carbon-containing species are dominant.

Ion Concentrations in Natural Waters and Drinking Water

The Abundant Ions in Fresh Water

As is evident from Table 9-4, the most abundant ions found in samples of unpolluted fresh calcareous water are usually calcium and bicarbonate, as expected from our previous analysis. Commonly, such water also contains **magnesium ion,** Mg^{2+}, principally from the dissolution of MgCO$_3$; plus some sulfate ion, SO$_4$$^{2-}$; smaller amounts of *chloride ion*, Cl$^-$, *sodium ion*, Na$^+$; and even smaller amounts of **fluoride ion,** F$^-$, and *potassium ion*, K$^+$.

TABLE 9-4	River Water Concentrations and Drinking Water Standards for Ions				
	River water molar concentration		Drinking water concentration in ppm		
				Maximum recommended concentration	
Ion	Average for world	Average for U.S.	Average U.S.	U.S.	Canada
^{a}HCO$_3^-$	9.2×10^{-4}	9.6×10^{-4}	60		
Ca^{2+}	3.8×10^{-4}	3.8×10^{-4}	15		
Mg^{2+}	1.6×10^{-4}	3.4×10^{-4}	8		
Na$^+$	3.0×10^{-4}	2.7×10^{-4}	6		200
Cl$^-$	2.3×10^{-4}	2.2×10^{-4}	8	250	250
SO$_4^{2-}$	1.1×10^{-4}	1.2×10^{-4}	12	250	500
K$^+$	5.4×10^{-5}	5.9×10^{-5}	2		
F$^-$	—	5.3×10^{-6}	0.1	0.8–2.4	1.5
NO$_3^-$	1.4×10^{-5}	—			
Fe^{3+}	7.3×10^{-6}	—			

^{a}Note: The value for bicarbonate is actually the total alkalinity.
Sources: World data from R. A. Larson and E. J. Weber, *Reaction Mechanisms in Environmental Organic Chemistry* (Boca Raton, FL: Lewis Publishers).

The overall reaction (7) of carbon dioxide and calcium carbonate implies that the molar ratio of bicarbonate ion to calcium ion should be 2:1, and this is indeed a rule that is closely obeyed on average in river water in North America and Europe. The calculated calcium ion concentration, 5.1×10^{-4} M, agrees well with the North American river-water average value of 5.3×10^{-4} M, and similarly for the bicarbonate ion data. The close agreement between the calculated and the experimental results is somewhat fortuitous, because river-water temperatures on average lie below 25°C—which results in a higher CO_2 solubility than has been assumed—and because several minor factors were oversimplified in the calculation. In fact, even calcareous river water is usually unsaturated with respect to $CaCO_3$ rather than saturated, as was implicitly assumed.

Water in rivers and lakes that is not in contact with carbonate salts contains substantially fewer dissolved ions than are present in calcareous waters. The concentration of sodium and potassium ions may be as high as those of calcium, magnesium, and bicarbonate ions in these fresh waters. Even in areas with no limestone in the soil, the waters contain some bicarbonate ion due to the weathering of *aluminosilicates* in soil and rock in the presence of atmospheric carbon dioxide. The weathering reaction can be written in general terms as

$$M^+(Al\text{-}silicate^-)(s) + CO_2(g) + H_2O \longrightarrow M^+ + HCO_3^- + H_4SiO_4$$

Here M is a metal such as potassium, and the anion (shown in parentheses) is one of the many aluminosilicate ions found in rocks (see Chapter 12). The weathering of *potassium feldspar* is an example of one of the most important sources of potassium ion in natural waters:

$$3\ KAlSi_3O_8(s) + 2\ CO_2(g) + 14\ H_2O(aq) \longrightarrow$$
$$2\ K^+(aq) + 2\ HCO_3^-(aq) + 6\ H_4SiO_4(aq) + KAl_3Si_3O_{10}(OH)_2(s)$$
$$\text{silicic acid} \qquad\qquad \text{a clay mineral}$$

Thus bicarbonate normally is the predominant anion in both calcareous and noncalcareous waters since it is produced by the dissolution of limestone and aluminosilicates, respectively.

The average compositions of river water in the United States and in the world as a whole are given in Table 9-4. As discussed, the values for the calcium and magnesium ion concentrations vary significantly from place to place, depending on whether or not the underlying soil is calcareous.

Fluoride Ion in Water

The level of fluoride ion, F^-, in water also displays substantial variations, from less than 0.01 ppm to more than 20 ppm, in different regions of the world. The source of most F^- is weathering of the mineral *fluorapatite*, $Ca_5(PO_4)_3F$.

In Mexico and some European countries sodium fluoride is added to table salt. In many communities of English-speaking countries [including the United States (about half the population), Canada, Australia, and New Zealand] in which the F^- concentration in the drinking water source is low, a soluble fluoride source such as *fluorosilicic acid*, H_2SiF_6, or its sodium salt, which react with water to release fluoride ion, is often added to bring the fluoride level up to about 1 ppm, i.e., 5×10^{-5} M. This value, at least in the past, was considered to be optimum in strengthening children's teeth against decay while providing a margin of safety. If the fluoride level is in excess of this value, as it is in some natural waters, deleterious effects on teeth, such as mottling, can occur. The maximum contaminant level (MCL) for fluoride in U.S. drinking water is 4 ppm. Almost all brands of toothpaste available in developed countries contain added fluoride in the form of *sodium fluoride*, NaF, *stannous fluoride*, SnF_2, or *sodium monofluorophosphate*. Most children in North America also receive topical fluoride not only from their toothpaste, but in some cases from applications by their dentists.

The addition of fluoride ion to public supplies of drinking water continues to be a controversial subject because at high concentrations fluoride is known to be poisonous and perhaps carcinogenic, and because some people feel that it is unethical to force everyone to drink water to which a substance has been added. In fact, for many people, the total amount of fluoride ion ingested from food and beverages (especially tea) exceeds that from water.

Bottled Drinking Waters

The maximum concentration of ions recommended for drinking water in the United States and in Canada is also listed in Table 9-4. The concentration of sodium ion, Na^+, in water is of interest since high consumption of it from water and salted food is believed to increase blood pressure, which may lead to cardiovascular disease. Excessive sulfate, beyond 500 ppm, may cause a laxative effect in some people. It is interesting to note that some varieties of bottled drinking water, which people presumably drink in preference to tap water due to health concerns about the latter, exceed the recommended values for some ions. Several of the well-known bottled waters exceeded the drinking water standards for arsenic and/or fluoride in a 1999 survey by the U.S. National Resources Defense Council. On the other hand, they were remarkably free of chloroform, a substance that plagues water supplies in many municipalities, as discussed in Chapter 10. A survey in 2000 found that several brands would not meet the new 10-ppb standard for arsenic (Chapter 11) and that bisphenol-A leached from the plastic into the water contained in most large polycarbonate jugs.

Suppliers of bottled water sometimes advertise their products as having "zero" concentrations of fluoride and/or sodium ions or as being *sodium-free* or *fluoride-free*. These are misleading statements since the actual concentrations are not zero but below the level of detection in the analytical method used by the bottler or below a threshold specified by the government. Zero is not a meaningful answer to the question of "how much" of a chemical is present in a sample.

Seawater

The total concentration of ions in seawater is much higher than that in fresh water since it contains large quantities of dissolved salts. The predominant species in seawater are sodium and chloride ions, which occur at about 1000 times their average concentration in fresh water. Seawater also contains some Mg^{2+} and SO_4^{2-}, and lesser amounts of many other ions. If seawater is gradually evaporated, the first salt to precipitate is $CaCO_3$ (present to the extent of 0.12 g/L), followed by $CaSO_4 \cdot H_2O$ (1.75 g/L), then NaCl (29.7 g/L), $MgSO_4$ (2.48 g/L), $MgCl_2$ (3.32 g/L), NaBr (0.55 g/L), and finally KCl (0.53 g/L). Thus "sea salt" is a mixture of all these salts, which taken together constitute about 3.5% of the mass of seawater. Due primarily to the CO_2–bicarbonate–carbonate equilibrium system discussed previously for fresh water, the average pH of surface ocean water is about 8.1. Seawater has a low organic content, its DOC value being about 1 mg/L.

Alkalinity Indices for Natural Waters

The actual concentrations of the cations and anions in a real water sample cannot simply be assumed to be equal to the theoretical values calculated above for calcium, carbonate, and bicarbonate for two reasons:

- the water may not be in equilibrium with either solid calcium carbonate or with atmospheric CO_2, and

- other acids or bases may also be present.

The index devised by analytical chemists to represent the actual concentration in water of the basic anions is provided by the alkalinity value for the sample. Alkalinity is a measure of the ability of a water sample to act as a base by reacting with protons. In practical use, the alkalinity of a body of water is a convenient measure of the capacity of the water body to resist acidification by neutralization when acid rain falls into it. From an operational viewpoint, **alkalinity** (more properly termed *total alkalinity*) is the number of moles of H^+ required to titrate 1 L of a water sample to the endpoint. For a solution containing carbonate and bicarbonate ions, as well as OH^- and H^+, by definition

$$\text{(total) alkalinity} = 2\,[CO_3^{2-}] + [HCO_3^{-}] + [OH^-] - [H^+]$$

The factor of 2 appears in front of carbonate ion concentration since, in the presence of H^+, it is first converted to bicarbonate ion, which is then converted by a second hydrogen ion to carbonic acid:

$$CO_3^{2-} + H^+ \rightleftharpoons HCO_3^{-}$$

$$HCO_3^{-} + H^+ \rightleftharpoons H_2CO_3$$

Minor contributors to the alkalinity of fresh-water systems can include dissolved ammonia, the anions of *phosphoric, boric,* and *silicic* acids, and H_2S, as well as natural organic matter.

By convention in analytical chemistry, *methyl orange* is used as the indicator in titrations by which total alkalinity is determined. Methyl orange is chosen because it does not change color until the solution is slightly acidic (pH = 4); under such conditions, not only has all the carbonate ion in the sample been transformed to bicarbonate, but virtually all the bicarbonate ion has been transformed to carbonic acid (see Problem 9-17).

Typical values of alkalinity for natural water samples range from 0.05 to 2 mmol/L. High values in this range correspond to waters much more able to resist acidification from precipitation that those with lower values. Alkalinity is a better measure of water's potential to resist acidification than is its pH, or rather its pOH, since it incorporates the concentrations of the weak bases such as carbonate and bicarbonate that will come into play when acid is added, rather than reflecting only the concentration of hydroxide ion.

Another index frequently encountered in the analysis of natural waters is the **phenolphthalein alkalinity,** which is a measure of the concentration of the carbonate ion and of other similarly basic anions. In order to titrate only CO_3^{2-} and not HCO_3^- as well, the indicator *phenolphthalein* or one with similar characteristics is used. Phenolphthalein changes color at a pH between 8 and 9, so it provides a fairly alkaline endpoint. At such pH values only a negligible amount of the bicarbonate ion has been converted to carbonic acid, but the majority of CO_3^{2-} has been converted to HCO_3^-. Thus

$$\text{phenolphthalein alkalinity} = [CO_3^{2-}]$$

PROBLEM 9-17

Calculate the value of the ratios $[HCO_3^-]/[CO_3^{2-}]$ and $[H_2CO_3]/[HCO_3^-]$ at pH values of 4 and 8.5 to confirm the statements made concerning the nature of the species present at the methyl orange and phenolphthalein endpoints of the titrations. [Hint: Use the equilibrium constant expressions and K values for reactions (2) and (4).]

PROBLEM 9-18

Calculate the value expected for the total alkalinity and for the phenolphthalein alkalinity of a 25°C saturated solution of calcium carbonate in water, and compare them to the values for a solution that is also in equilibrium with atmospheric carbon dioxide. Use the concentrations listed in the last two columns of Table 9-3.

PROBLEM 9-19

Calculate the total alkalinity for a sample of river water whose phenolphthalein alkalinity is known to be 3.0×10^{-5} M, whose pH is 10.0, and whose bicarbonate ion concentration is 1.0×10^{-4} M.

The alkalinity value for a lake is sometimes used by biologists as a measure of its ability to support aquatic plant life, a high value indicating a high potential fertility. The reasons for such a situation are often as follows: Algae extract the carbon dioxide they need for photosynthesis from bicarbonate ion, which is plentiful in calcareous waters, by a reversal of the $CO_2 - CaCO_3$ reaction discussed previously:

$$2\,HCO_3^-(aq) \longrightarrow CO_2 + CaCO_3(s)$$

Indeed, small crystals of calcium carbonate are sometimes observed in lakes undergoing active photosynthesis.

$$CO_2 + H_2O + \text{sunlight} \longrightarrow \underset{\text{(as algae)}}{CH_2O \text{ polymer}} + O_2$$

In noncalcareous waters, which have low alkalinity and low calcium content, dissociation of the bicarbonate ion in the water forms not only carbon dioxide but also hydroxide ion:

$$HCO_3^- \rightleftharpoons CO_2 + OH^-$$

The algae readily exploit this CO_2 for their photosynthetic needs, at the cost of allowing a buildup of hydroxide ion to such an extent that the lake water becomes quite basic, with a pH as high as 12.3 measured in some cases.

The Hardness Index for Natural Waters

As a measure of certain important cations present in samples of natural waters, analytical chemists often use the **hardness index,** since it measures the total concentration of the ions Ca^{2+} and Mg^{2+}, the two species that are principally responsible for hardness in water supplies. Chemically, the hardness index is defined as

$$\text{hardness} = [Ca^{2+}] + [Mg^{2+}]$$

Experimentally, hardness can be determined by titrating a water sample with *ethylenediaminetetraacetic acid* (EDTA), a substance that forms very strong complexes with metal ions other than those of the alkali metals (see Chapter 11 for details). Traditionally, hardness is expressed not as a molar concentration of ions but as *the mass in milligrams per liter of calcium carbonate that contains the same total number of dipositive (2+) ions*. Thus, e.g., a water sample that contains a total of 0.0010 mol of $Ca^{2+} + Mg^{2+}$ per liter would possess a hardness value of 100 mg of $CaCO_3$, since the molar mass of $CaCO_3$ is 100 g and thus 0.0010 mol weighs 0.1 g, or 100 mg.

Most calcium enters water from either $CaCO_3$ in the form of limestone or from mineral deposits of $CaSO_4$. The source of much of the magnesium is *dolmitic limestone*, $CaMg(CO_3)_2$. Hardness is an important characteristic of natural waters, since calcium and magnesium ions form insoluble salts with the anions in soaps, thereby forming a scum in washwater. Water is termed hard if it contains substantial concentrations of calcium and/or magnesium ions; thus calcareous water is hard. Some scientists define water as being hard if its hardness index exceeds 150 mg/L.

Many areas possess soils that contain little or no carbonate ion, and its dissolution and reaction with CO_2 to produce bicarbonate do not occur. Such soft water typically has a pH much closer to 7 than does hard water, since it contains few basic anions. However, there are lakes with little dissolved calcium or magnesium but relatively high concentrations of dissolved *sodium carbonate*, Na_2CO_3; such lakes have a very low degree of hardness but are high in alkalinity.

Interestingly, people who live in hard-water areas are found to have a lower average death rate from ischemic heart disease than people living in areas with very soft water. It is not clear whether the advantage of drinking hard water stems from its supply of magnesium ion to the body or from the protection that hard water provides from the presence of other ions such as sodium and those of the heavy metals (see Chapter 11).

PROBLEM 9-20

What is the value of the hardness index for a 500-mL sample of water that contains 0.0040 g of calcium ion and 0.0012 g of magnesium ion?

PROBLEM 9-21

Calculate the hardness, in milligrams of $CaCO_3$ per liter, of water that is in equilibrium at 25°C with carbon dioxide and calcium carbonate, using results in the final column of Table 9-3. Is the calculated value greater or less than the median hardness value of 37 mg/L found for surface waters in the United States?

Aluminum in Natural Waters

The concentration of **aluminum** ions in natural waters normally is quite small, typically about 10^{-6} M. This low value is due to the fact that in the typical pH range for natural waters (6 to 9), the solubility of the aluminum contained in rocks and soils to which the water is exposed is very small. The solubility of aluminum compounds in water is controlled by the insolubility of $Al(OH)_3$. Given that the K_{sp} of the hydroxide is about 10^{-33} at usual water temperatures, then for the reaction

$$Al(OH)_3 \rightleftharpoons Al^{3+} + 3\,OH^-$$

it follows that

$$[Al^{3+}]\,[OH^-]^3 = 10^{-33}$$

Take, for instance, a sample of water whose pH is 6. Since the hydroxide concentration in such water is 10^{-8} M, it follows that

$$[Al^{3+}] = 10^{-33}/(10^{-8})^3 = 10^{-9}\,M$$

Although this value is very small, for every one-unit decrease in the pH, the concentration of aluminum ion increases by a factor of 10^3, so it reaches 10^{-6} M at a pH of 5 and 10^{-3} M at a pH of 4. Thus aluminum is much more soluble in highly acidified rivers and lakes than in those where pH values do not fall below 6 or 7. Indeed, Al^{3+} is usually the principal cation in waters whose pH is less than 4.5, exceeding even the concentrations of Ca^{2+} and Mg^{2+}, which are the dominant cations at pH values greater than 4.5.

In the recent past, fears arose that human ingestion of aluminum from drinking water and from the use of aluminum cooking pots was a major cause of Alzheimer's disease; however, the research on which this conclusion was reached could not be reproduced. Today many neuroscientists do not believe that there is a strong connection between the disease and intake of the metal, since past epidemiological studies on this matter have not been definitive or consistent. However, Canadian and Australian research reported in the mid-1990s indicates that consumption of drinking water with more than 100 ppb aluminum—not an uncommon level in drinking water purified by aluminum sulfate (see Chapter 10)—can lead to neurological damage such as memory loss and perhaps to a small increase in the incidence of Alzheimer's disease.

It is thought that the principal deleterious effect of acid waters on fish arises from the solubilization of aluminum from soil and its subsequent existence as a free ion in the acidic water, as discussed in Chapter 2. Unfortunately, the $Al(OH)_3$ precipitates as a gel on contact with the less acidic gills of the fish, and the gel prevents the normal intake of oxygen from water, thus suffocating the fish.

It is also believed that aluminum mobilization in soils is one of the stresses that acid rain places on trees, resulting in the dieback of forests. Soils that contain limestone are usually considered to be buffered against much change in pH due to the ability of carbonate and bicarbonate ion to neutralize H^+, but over a period of decades, surface soil may gradually lose its carbonate content by a continual bombardment by acid rain. Thus soils receiving acid rain eventually become acidified. When the pH of the soil drops below about 4.2, aluminum leaching from soil and rocks becomes particularly appreciable. Such acidification has occurred in some regions in central Europe, including Poland, the former Czechoslovakia, and eastern Germany, and the resulting solubilization of aluminum may have contributed to the forest diebacks observed there in the 1980s.

PROBLEM 9-22

What is the concentration, in grams per liter, of dissolved aluminum in water with a pH of 5.5?

PROBLEM 9-23

Calculate the pH value at which the aluminum ion concentration dissolved in water is 0.020 M, assuming that it is controlled by the equilibrium with solid aluminum hydroxide.

Perchlorates

Perchlorate ion, ClO_4^-, is a pollutant in the drinking water supply of about 15 million Americans. Large quantities of *ammonium perchlorate,* NH_4ClO_4, are manufactured for use as oxidizing agents in solid rocket propellants, fireworks, batteries, and automobile air bags. Because rocket fuel has a limited shelf life, it must be replaced regularly. Large amounts of perchlorates were washed out of missiles and rocket boosters onto the ground or into holding lagoons in the second half of the twentieth century. The map of perchlorate releases in the United States indicates that most occur in the south-central and western states, as shown in Figure 9-6. Concentrations of perchlorate in drinking water in the U.S. Southwest range from 5 to 20 ppb. Perchlorate concentrations of about 70 ppb have been found in lettuce crops grown in areas irrigated with water from the Colorado River. The ion has now also been detected in milk samples in Texas, at levels of 2–6 ppb.

At high doses, perchlorate affects human health by reducing hormone production in the thyroid, where it competes with iodide ion. The hazard at low concentrations, if any, is not known. However, research reported in 2002 on volunteers indicated that the no-effect level for the inhibition of iodine uptake corresponds to a drinking water concentration of at least 180 ppb.

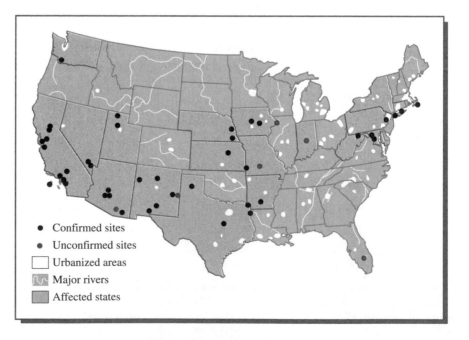

FIGURE 9-6 Regions of perchlorate use and contamination in the United States. [Source: B. E. Logan, "Assessing the Outlook for Perchlorate Remediation," *Environmental Science and Technology* (1 December 2001): 484A.]

● Confirmed sites
• Unconfirmed sites
☐ Urbanized areas
〰 Major rivers
▨ Affected states

No federal U.S. standard for the ion has yet been set, although several states have set their own limits. From the *oral reference dose* (RfD, discussed in Chapter 8) of 0.5 μg/kg/day, California has set a drinking water limit of 18 ppb. The limit in Arizona is 14 ppb and in Texas, 22 ppb. In 2002 a draft report by the U.S. EPA proposed a drinking water standard of 1 ppb as safe for human health, but this value has been criticized as too low by the Department of Defense and companies that make or use perchlorates. The EPA based its recommendation on studies that indicate that mothers who drink perchlorate-contaminated water could give birth to children whose IQs could be affected negatively because correct maternal thyroid hormone levels are vital to fetal brain development. The National Academy of Sciences has been asked to determine the appropriate MCL (as defined in Chapter 7).

Perchlorate is a difficult ion to remove from water supplies since it is a highly water-soluble anion that is very inert and does not adsorb readily either to mineral surfaces or to activated carbon. Some anion exchange resins successfully remove perchlorate, though it tends to remain in solution until all other anions have been absorbed. Barrier precipitation methods, e.g., using elemental iron, are not successful because the perchlorate anion is so unreactive. Certain bacteria biodegrade perchlorate to chloride ion, so biodegradation in engineered systems shows great potential for the future control of perchlorate.

Review Questions

1. Write the balanced half-reaction for O_2 that occurs in acidic waters when it oxidizes organic matter.

2. How does temperature affect the solubility of O_2 in water? Explain what is meant by *thermal pollution*.

3. Define *BOD* and *COD* and explain why their values for the same water sample can differ slightly. Explain why natural waters can have a high BOD.

4. What do the acronyms *TOC* and *DOC* stand for, and how do they differ in terms of what they measure?

5. Write the half-reaction used in the COD titration that converts dichromate ion to Cr^{3+} ion and balance it.

6. Write the balanced chemical reaction by which organic carbon, represented as CH_2O, is disproportionated by bacteria under anaerobic conditions.

7. Draw a labeled diagram classifying the top and bottom layers of a lake in summer as either oxidizing or reducing in character and showing the stable forms of carbon, sulfur, nitrogen, and iron in the two layers.

8. What is meant by the *pE* of an aqueous solution? What does a low (negative) pE value imply about the solution? What species determines the pE value in aerated water?

9. What are some examples of highly reduced and of highly oxidized sulfur in environmentally important compounds? Write the balanced reaction by which sulfate can oxidize organic matter.

10. Explain the phenomenon of *acid mine drainage*, writing balanced chemical equations as appropriate. How does Fe^{3+} also act as an oxidizing agent here?

11. What is the acid and what is the base that dominate the chemistry of most natural water systems and whose interaction produces bicarbonate ion?

12. What is the source of most of the carbonate ion in natural waters? What name is given to waters that are exposed to this source?

13. Write the approximate net reaction between carbonate ion and water in a system that is *not* also exposed to atmospheric carbon dioxide. Is the resulting water acidic, alkaline, or neutral?

14. Write the approximate net reaction between carbonate ion and water in a system that *is* exposed to atmospheric carbon dioxide. Is the resulting water acidic, alkaline, or neutral? Explain why the production of bicarbonate ion from carbonate ion does not inhibit its production from carbon dioxide, and vice versa.

15. If two equilibrium reactions are added together, what is the relationship between the equilibrium constants for the individual reactions and that for the overall reaction?

16. What is the natural source of fluoride ion in water? How and why is the fluoride level in drinking water artificially increased to about 1 ppm in many municipalities?

17. Define the *total alkalinity index* and the *phenolphthalein alkalinity index* for water.

18. Define the *hardness index* for water.

19. Which are the most abundant ions in clean, fresh water?

20. Explain why aluminum ion concentrations in acidified waters are much greater than those in neutral water. How does the increased aluminum ion level affect fish and trees?

21. What is the formula for the perchlorate ion? What is the origin of perchlorate ion in U.S. drinking water?

Green Chemistry Questions

1. What takes place during the scouring of cotton, and why is this process necessary for the production of finished cotton fibers?

2. Biopreparation (an enzymatic process) has replaced the use of large amounts of sodium hydroxide in the scouring of cotton.

(a) Describe any environmental problems or worker hazards associated with the use of sodium hydroxide solutions in the scouring of cotton.

(b) Would the same environmental problems or worker hazards be eliminated by the use of Biopreparation?

3. The development of Biopreparation by Novozymes–North America Inc. won a Presidential Green Chemistry Challenge Award.

(a) Which of the three focus areas (see the Introduction to Green Chemistry) for these awards does this award best fit into?

(b) List at least three of the twelve principles of green chemistry (see the Introduction to Green Chemistry) that are addressed by the chemistry developed by Novozymes–North America Inc.

Additional Problems

1. The TOC parameter for water samples is measured by oxidizing the organic material to carbon dioxide and then measuring the amount of this gas evolved from the solution. If a 5.0-L sample of wastewater produced 0.25 mL of carbon dioxide gas, measured at a pressure of 0.96 atm and a temperature of 22°C, calculate the TOC value for the sample. Assuming the average composition of the organic matter to be CH_2O, calculate the COD value for the water sample due to its organic content. [The gas constant $R = 0.0821$ L atm/mol K.]

2. Consider the reduction of nitrate ion to nitrite ion in a natural water system.

(a) Write the balanced one-electron half-reaction for the process if it occurs in acidic media.

(b) Given that for this reaction, $E^0 = +0.881$ V, calculate pE^0.

(c) From your answer to part (a), deduce the expression relating pE to pE^0 and ion concentrations.

(d) From your result in part (c), obtain an equation relating the pE and pH conditions under which the ratio of nitrate to nitrite is 100:1.

(e) From your result in part (c), deduce the ratio of nitrite to nitrate under conditions of pE = 12, pH = 5.

3. Calculate the solubility of lead(II) carbonate, $PbCO_3$ ($K_{sp} = 1.5 \times 10^{-13}$), in water, given that most of the carbonate ion it produces subsequently reacts with water to form bicarbonate ion. Recalculate the solubility, assuming that none of the carbonate ion reacts to form bicarbonate ion. Is your result significantly different from that calculated assuming complete reaction of carbonate with water?

4. The bicarbonate ion, HCO_3^-, can potentially act as an acid or as a base in water. Write the chemical equations for these two processes and, from the information given in this chapter, determine the corresponding acid and base dissociation constants. Given the relative magnitudes of the dissociation constants, decide whether the dominant reaction of bicarbonate in water will be as an acid or as a base. Calculate the pH of an aqueous 0.010 M solution of sodium bicarbonate in water using the dominant reaction alone and assuming that the amounts of carbonate ion and carbonic acid from other sources are negligible in this case.

5. A sample of lake water at 25°C is analyzed and the following parameters are found:

total alkalinity $= 6.2 \times 10^{-4}$ M
phenolphthalein alkalinity $= 1.0 \times 10^{-5}$ M
pH = 7.6
hardness = 30.0 mg/L
$[Mg^{2+}] = 1.0 \times 10^{-4}$ M

Extract all possible single-ion concentrations that you can by using these data. Also determine whether or not the water is at equilibrium with respect to the carbonate–bicarbonate system and whether or not it is saturated with calcium carbonate.

6. How many years would it take for 1 L of water of 0.030 mmol/L total alkalinity in a lake to exhaust its capacity to absorb H^+ if 0.050 L of rain with pH = 4.0 was added to it annually?

7. Using the information you gathered for Problems 9-16 and 9-17, calculate for 0.5-pH intervals from pH 3 to 11, the fraction of carbon that exists as carbonic acid, as bicarbonate ion, and as carbonate ion, and plot curves for these fractions on a single graph. [Hint: Calculate the ratios of carbonic acid and of carbonate ion to bicarbonate ion and deduce the fractions from these results. The use of a computer spreadsheet here could minimize your computations.]

8. Table 9-4 gives the concentration of various ions in river water. The corresponding values for seawater are significantly higher. For example, the concentrations of Na^+ and Ca^{2+} in seawater (based on 35% total salinity; density = 1.02478 g/cm) are 10.752 and 0.416 g/kg, respectively [G. Neuman and W. J. Pierson, Jr., *Principles of Physical Oceanography* (Englewood Cliffs, NJ: Prentice-Hall, 1966)].

Determine the ratio of the concentration of each of these ions in seawater versus fresh water (use the world average). Why is the ratio significantly smaller for Ca^{2+} than for Na^+?

9. To determine the BOD of a water sample, the O_2 concentration must be measured before and after the five-day incubation process. This determination can be done in a number of ways, including titration and oxygen electrodes. The standard titration procedure used is the Winkler method, in which the oxygen in a small aliquot of the water sample is reacted with $MnSO_4$ in basic solution. This precipitates MnO_2, which converts added I^- to I_2; the latter can be quantitatively determined by titration with standardized sodium thiosulfate. The overall set of equations is

$$2\,Mn^{2+}(aq) + 4\,OH^-(aq) + O_2(aq) \longrightarrow$$
$$2\,MnO_2(s) + 2\,H_2O(l)$$
$$MnO_2(s) + 4\,H^+(aq) + 2\,I^-(aq) \longrightarrow$$
$$Mn^{2+}(aq) + I_2(aq) + 2\,H_2O(l)$$
$$I_2(aq) + 2\,S_2O_3^{2-}(aq) \longrightarrow$$
$$S_4O_6^{2-}(aq) + 2\,I^-(aq)$$

Two 10.00-mL aliquots of a natural water sample were analyzed using the Winkler method, one before and the other after incubation. They required 10.15 and 2.40 mL, respectively, of a 0.00100 M standard $Na_2S_2O_3$ solution to titrate the I_2 produced. Calculate the BOD of this natural water sample in milligrams per liter. Would this be considered to be a polluted water sample?

10. Hardness can be quantitatively measured by atomic absorption (AA) spectroscopy. In this technique, the water to be tested is sprayed into an acetylene flame to atomize the compounds. The atoms in the flame are then analyzed by their absorption of light. Particular metal ions can be determined using specific wavelengths of light. In order to quantitatively determine the concentration of a particular metal ion, a calibration curve of absorbance versus concentration must be constructed using standard solutions; the concentrations of test solutions are then obtained from this curve based on the measured absorbance. In an experiment set up for Ca^{2+} analysis, the following absorbances were obtained for a set of four standard solutions: 2.0 ppm, 0.121; 4.0 ppm, 0.231; 6 ppm, 0.334; and 8 ppm, 0.428. The absorbances measured for test solutions were: tap water, 0.299; seawater, 0.218; and rainwater, 0.055. Note: the tap water sample was diluted by a factor of 5 and the seawater by a factor of 100 before the measurement. Plot a calibration curve (do not assume it to be linear) and determine the concentration of Ca^{2+} in each of these samples. Comment on the relative values.

Further Reading

1. W. Stumm and J. J. Morgan, *Aquatic Chemistry: Chemical Equilibria and Rates in Natural Waters* 3rd ed. (New York: Wiley-Interscience, 1996).

2. E. Rubenowitz et al., "Magnesium and Calcium in Drinking Water and Death from Acute Myocardial Infarction in Women," *Epidemiology* 10 (1999): 31.

3. "What's in That Bottle?" *Consumer Reports* (January 2003): 38.

4. B. Hileman, "Fluoridation of Water," *Chemical and Engineering News* (1 August 1988) 26.

5. B. J. Logan, "Assessing the Outlook for Perchlorate Remediation," *Environmental Science and Technology* (1 December 2001): 483A.

Websites of Interest

Log on to www.whfreeman.com/envchem3e/ and click on Chapter 9.

The Pollution and Purification of Water

The pollution of natural waters by both biological and chemical contaminants is a worldwide problem. There are few populated areas, whether in developed or undeveloped countries, that do not suffer from one form of water pollution or another. In this chapter we shall survey the various methods—both traditional and innovative—by which water can be purified.

Aerial view of the settling and treatment ponds at the sewage treatment plant in Kings County, Washington State. (Brand X Pictures)

Water Disinfection

The quality of "raw" (untreated) water, whether drawn from surface water or from groundwater, that is intended eventually for drinking varies widely, from almost pristine to highly polluted. Because both the type and quantity of pollutants in raw water vary, the processes used in purification also vary from place to place. The most commonly used procedures are shown in schematic form in Figure 10-1. Before discussing the major topic of disinfection, we shall discuss the various nondisinfection steps that are often taken in the overall purification process.

Aeration of Water

Aeration is commonly used in the improvement of water quality. Municipalities aerate drinking water that is drawn from underground aquifers in order to remove dissolved gases such as the foul-smelling H_2S and *organosulfur* compounds, as well as volatile organic

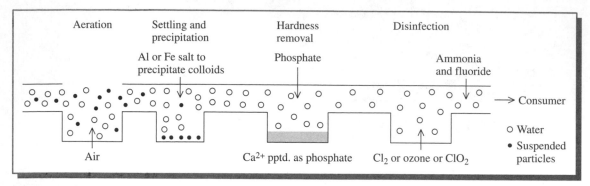

FIGURE 10-1 The common stages of purification of drinking water.

compounds, some of which may have a detectable odor. Aeration of drinking water also results in reactions that produce CO_2 from the most easily oxidized organic material. If necessary for reasons of odor, taste, or health, most of the remaining organics can be removed by then passing the water over activated carbon, although this process is relatively expensive, so rather few communities use it (see Box 10-1). Another advantage of aeration is that the increased oxygen content of water oxidizes water-soluble Fe^{2+} to Fe^{3+}, which then forms insoluble hydroxides (and related species) that can be removed as solids:

$$Fe^{3+} + 3\,OH^- \longrightarrow Fe(OH)_3(s)$$

(Recall that ions in equations without a state specified are assumed to be in aqueous solution.)

After aeration, colloidal particles in the water are removed. If the water is excessively hard, **calcium** and **magnesium** are removed before the final stages of disinfection and the addition of fluoride (see Figure 10-1).

Removal of Calcium and Magnesium

If the water comes from wells in areas with limestone bedrock, it will contain significant levels of Ca^{2+} and Mg^{2+} ions, which are usually removed during processing. Calcium can be removed from water by addition of phosphate ion in a process analogous to that discussed later for phosphate removal; here, however, phosphate is *added* in order to precipitate the calcium ion. More commonly, calcium ion is removed by precipitation and filtering of the insoluble salt $CaCO_3$. The carbonate ion is either added as *sodium carbonate*, Na_2CO_3, or if sufficient HCO_3^- is naturally present in the water, *hydroxide ion*, OH^-, is added to convert dissolved *bicarbonate ion* to *carbonate*, CO_3^{2-}:

$$OH^- + HCO_3^- \longrightarrow CO_3^{2-} + H_2O$$
$$Ca^{2+} + CO_3^{2-} \longrightarrow CaCO_3(s)$$

BOX 10-1 | Activated Carbon

Activated carbon (activated charcoal) is a very useful solid for purifying water of small organic molecules present in low concentrations. The ability of this material to remove contaminants from water and to improve its taste, color, and odor has been known for a long time; indeed, the ancient Egyptians used charcoal-lined vessels to store water for drinking purposes.

Activated carbon is produced by anaerobically charring a high-carbon-content material such as peat, wood, or lignite (a soft brown coal) at temperatures below 600°C, followed by a partial oxidation process using carbon dioxide or steam at a slightly higher temperature.

The removal of contaminants by activated carbon is a physical adsorption process and therefore is reversible if sufficient energy is applied. The characteristic that makes activated carbon such an excellent adsorber is its huge surface area, about 1400 m^2/g. This surface is internal to the individual carbon particles, so that crushing the material neither increases nor decreases the area. The internal structure of the solid involves series of channels (pores) of progressively decreasing size that are produced by the charring and partial oxidation processes. The internal sites where adsorption occurs are large enough only for small molecules, including chlorinated solvents. At the typical ppm concentrations found for organic contaminants in water, each gram of activated carbon can adsorb a few percent of its mass in contaminants such as chloroform and the dichloroethenes, and much higher masses of TCE, PCE, and pesticides such as dieldrin, heptachlor, and DDT.

Once a sample of activated carbon has reached near-saturation in terms of adsorbed organics, three alternatives are available. It can be simply disposed of in a landfill, it can be incinerated to destroy it and the adsorbed contaminants, or it can be heated to rejuvenate the surface by driving off the organic pollutants, which can then be incinerated or catalytically oxidized.

Magnesium ion precipitates as *magnesium hydroxide*, $Mg(OH)_2$, when the water is made sufficiently alkaline, i.e., when the OH^- content is increased. After removal of the solid $CaCO_3$ and $Mg(OH)_2$, the pH of the water is readjusted to near-neutrality by bubbling carbon dioxide into it.

PROBLEM 10-1

Ironically, calcium ion is often removed from water by adding hydroxide ion in the form of $Ca(OH)_2$. Deduce a balanced chemical equation for the reaction of calcium hydroxide with dissolved calcium bicarbonate, $Ca(HCO_3)_2$, to produce insoluble calcium carbonate. What molar ratio of $Ca(OH)_2$ to dissolved calcium should be added to ensure that almost all the calcium is precipitated?

Disinfection to Reduce Illnesses

In terms of causing immediate sickness and even death, biological contaminants of water are almost always much more important than chemical ones. For that reason, we begin our discussion of the purification of water by extensively discussing its disinfection, i.e., the elimination of microorganisms that can cause illness.

Many of the microorganisms in raw water are present as a result of contamination by human and animal feces. The microorganisms are principally

- **bacteria,** including those of the *Salmonella* genus, one species of which causes typhoid (this category also includes *E. coli O157:H7*, whose transmission in water caused a number of deaths in recent years, including an outbreak in Walkerton, Ontario, in 2000);

- **viruses,** including polio viruses, the hepatitus-A virus, and the Norwalk virus; and

- **protozoans** (single-celled animals), including *Cryptosporidium* and *Giardia lamblia.*

Because many microorganisms of these types are pathogenic, causing mild to serious and sometimes fatal illnesses, they must be removed from water before it is suitable for drinking.

Filtering of Water

In addition to dissolved chemicals, the raw water that is obtained from rivers, lakes, or streams contains a multitude of tiny particles, some of which consist of or contain microorganisms. The larger of these particles are often removed from the water simply by filtering it. Recently it was realized that forcing raw water through filters with especially small openings can be used instead of chemical or ultraviolet radiation methods to disinfect water by removing viruses and bacteria, and even some dissolved chemicals.

Filtering of water by passing it through a bed of sand is the oldest form of water purification known, dating back to ancient times. The sand retains suspended solids of all types, including microorganisms, down to about 10 μm in size.

Removal of Colloidal Particles by Precipitation

Most municipalities allow raw water to settle, since this permits large particles to settle out or to be readily separated. However, much of the insoluble matter—which originates from rocks and soil and from the disintegration and decomposition of water-based plants and animals—will not precipitate spontaneously since it is suspended in water in the form of **colloidal**

particles. These are particles that have diameters ranging from 0.001 to 1 μm and consist of *groups* of molecules or ions that are weakly bound together. These groups dissolve as a unit, rather than breaking up and dissolving as individual ions or molecules. In many cases the individual units within a colloidal particle are spatially organized so that the surface of the particle contains ionic groups. The ionic charges on the surface of one particle repel those on neighboring particles, preventing their aggregation and subsequent precipitation.

Colloidal particles must be removed from drinking water for both aesthetic and health-safety reasons. To capture the colloidal particles, a small amount of either *iron(III) sulfate*, $Fe_2(SO_4)_3$, or *aluminum sulfate*, $Al_2(SO_4)_3$ (alum), is deliberately added to the water. At neutral or alkaline pH values (7 and up) both the Fe^{3+} and Al^{3+} ions produced from the salts form gelatinous hydroxides that physically incorporate the colloidal particles and form a removable precipitate. The water is greatly clarified once this precipitate has been removed. Commonly, after the removal of the colloidal particles, the water is filtered through sand and/or some other granular material.

Although the approximate formulas of the precipitates are $Fe(OH)_3$ and $Al(OH)_3$, the actual situation is much more complex. For example, aluminum forms a polymeric cation, $Al_{13}O_4(OH)_{24}{}^{7+}$, that forms a loose network structure held together by hydrogen bonds. This network entraps the colloidal particles and creates the precipitate. Only if the pH rises to a high value does the aluminum in solution form the expected hydroxide $Al(OH)_3$. Since the concentration of aluminum sulfate added to the water is only about 10 μmol/L, very little aluminum ion is left in the treated water.

PROBLEM 10-2

Calculate the approximate number of atoms contained in colloidal particles of (a) 1 μm and (b) 0.01 μm diameters, assuming that their densities are similar to that of water and that the atomic mass of the atoms averages 10 g/mol.

Disinfection by Membrane Technology

Water can be purified of most contaminant ions, molecules, and small particles including viruses and bacteria by passing it through a membrane in which the individual holes, called *pores*, are of uniform and microscopic size. The range of sizes of the various contaminants in raw water are summarized in Figure 10-2. Clearly, for a technique to be effective in providing a barrier, the pore size of the membrane must be smaller than the contaminant size.

In the processes of **microfiltration** and **ultrafiltration,** a membrane or some other analogous barrier containing pores of 0.002 to 10 μm diameter (2–10,000 nm) is employed to remove constituents larger than these sizes

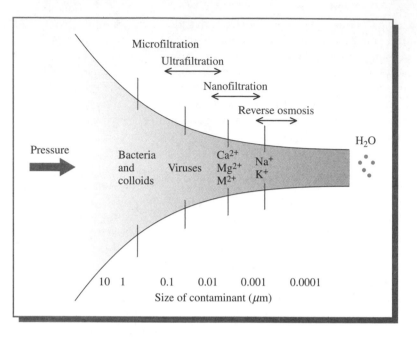

FIGURE 10-2 Filtration of contaminants by various methods.

from water. The water can be forced through the barrier by pressure or can be drawn through it by suction, leaving behind the larger impurities. In one modern version of this technology, the barrier is composed of thousands of strands of plastic tubing with walls that are pierced with thousands of tiny pores of similar size.

Some bacteria and colloid particles are as small as 0.1 μm and can pass through conventional filters and even some microfilters (Figure 10-2). Viruses can be as small as 0.01 μm and therefore require at least the ultrafiltration level to eliminate them all. However, filtration using membranes can be successfully used to disinfect water if a sufficiently small pore size is used.

Neither microfiltering nor ultrafiltering removes dissolved ions or small organic molecules. Generally speaking, before water is treated using membranes with even smaller pores (see below), it must be pretreated to remove the larger particles—especially colloids—which would otherwise foul the finer membrane by leaving deposits.

Membrane systems have been developed recently that purify water of virtually all contaminants by **nanofiltration.** Water is pumped under pressure through fine membranes that have pores only about 1 nm wide, which therefore remove not only bacteria and viruses, but also any larger organic molecules that would nourish the regrowth of bacteria. These **nanofilters** still allow water molecules to pass through the filter, since the molecules are only a few tenths of a nanometer in size. Unlike ultrafiltration, nanofiltration can be used to soften water, since divalent ions such as Ca^{2+} and Mg^{2+} are larger than the pores and do not pass through. Monovalent ions such as sodium and chloride also pass through some nanofilters, but not through those with subnanometer pore sizes. As a consequence, some nanofilter membrane systems can be used to desalinate seawater and to help purify wastewater, as discussed later in this chapter.

The ultimate in membrane filtration occurs in the widely used technique **reverse osmosis,** sometimes called *hyperfiltration*. Here water is forced under high pressure to pass through a *semipermeable membrane* composed of an organic polymeric material such as *cellulose acetate* or *triacetate*. Since only water can pass through the pores, the liquid on the other side of the membrane is pure water. The solution on the impact side of the membrane becomes more and more concentrated in contaminants as time goes on and is discarded.

All particles, molecules (including even small organic molecules), and ions down to less than 1 nm (0.001 μm) in size are removed by reverse osmosis. It is particularly useful for removing alkali and alkaline earth metal ions, as well as salts of heavy metals. Thus it is useful in hospitals and renal units for producing water that is exceptionally free of ions. Indeed, reverse osmosis is widely used in the Middle East and elsewhere to desalinate seawater.

As mentioned, it is common to pretreat polluted water by other methods to remove the larger particles such as bacteria before subjecting it to reverse osmosis, because of membrane fouling problems. Reverse osmosis does tend to be wasteful of water, since so much of it is discarded. Because of the high pressures needed, it also tends to be energy intensive. Small reverse osmosis units are available for under-the-sink installation in homes to remove unwanted contaminants such as lead and nitrate ions and organic molecules from water obtained from domestic supplies.

Disinfection by Ultraviolet Irradiation

Ultraviolet light can also be used to disinfect and purify water. Powerful lamps containing mercury vapor whose excited atoms emit UV-C light centered at 254 nm are immersed in the water flow. About 10 seconds of irradiation are usually sufficient to eliminate the toxic microorganisms, including *Cryptosporidium*. The germicidal action of the light disrupts the DNA in microorganisms, preventing their subsequent replication and thereby inactivating the cells. At the molecular level, absorption of UV-C light results in the formation of new covalent bonds between nearby thymine units on the same strand of DNA. If sufficient thymine dimers are formed, the DNA molecule becomes so distorted that subsequent replication of the organism is prevented.

The use of ultraviolet light to purify water is complicated by the presence of dissolved iron and humic substances, both of which absorb the UV light and thus reduce the amount available for disinfection. Small solid particles suspended in the water also inhibit the action of the UV light since they can shade or absorb bacteria and also scatter or absorb the light. An advantage of UV disinfection technology is that small units can be employed to serve small population bases, whether in the developed or developing world, so the continuous monitoring activity of chemical systems is avoided. As discussed later, UV light can also be used to free water of dissolved organic compounds, but via a different mechanism.

Disinfection by Chemical Methods: Ozone and Chlorine Dioxide

To rid drinking water of harmful bacteria and viruses, especially those arising from fecal matter from both humans and animals, by chemical means requires an oxidizing agent more powerful than O_2. In some localities, particularly in France and other parts of western Europe but also in some North American cities—e.g., Montreal and Los Angeles—**ozone** is used for this purpose. Since O_3 cannot be stored or shipped because of its very short lifetime, it must be generated on-site by a relatively expensive process involving electrical discharge (20,000 V) in dry air. The resulting ozone-laden air is bubbled through the water; about 10 min of contact is usually sufficient for disinfection. Since the lifetime of ozone molecules is short, there is no residual protection in the purified water to protect it from future contamination. Some contaminants in water react with the ozone itself and others with free radicals such as hydroxyl and hydroperoxy (Chapters 1–3) that are produced when the ozone reacts with water.

Unfortunately, the reaction of ozone with bromine in water leads to the formation of oxygen-containing organic compounds, particularly those containing the *carbonyl group*, $\diagdown\!\!C\!\!=\!\!O$, such as formaldehyde and other low-molecular-weight aldehydes and various other compounds, some of which are toxic. In addition, ozone reacts with *bromide ion*, Br^-, present in the water to produce the **bromate ion,** BrO_3^-, a carcinogen in test animals, which is also probably carcinogenic in humans. The reaction of ozone with bromide, a natural constituent of water that often is present at ppm concentrations, occurs in several steps; the overall reaction is

$$Br^- + 3\,O_3 \longrightarrow BrO_3^- + 3\,O_2$$

The bromate ion produced by ozonation may subsequently react with organic matter in the water to produce toxic organobromine compounds, although experiments have shown that the only brominated product is *dibromoacetonitrile*, $CHBr_2CN$. The MCL (maximum contaminant level) of bromate ion in drinking water is set at 10 ppb (0.010 ppm) by the U.S. EPA. Substances such as bromate ion that are produced during water purification are called

disinfection by-products or DBPs. *All* known chemical methods of disinfecting water produce DBPs.

Similarly, **chlorine dioxide** gas, ClO_2, is used in more than 300 North American and in several thousand European communities to disinfect water. The ClO_2 molecules, themselves free radicals, oxidize organic molecules by extracting electrons from them:

$$ClO_2 + 4\,H^+ + 5\,e^- \longrightarrow Cl^- + 2\,H_2O$$

The organic cations created in the accompanying oxidation half-reaction subsequently react further and eventually become more fully oxidized.

Since chlorine dioxide is *not* a chlorinating agent—it does not generally introduce chlorine atoms into the substances with which it reacts—and since it oxidizes the dissolved organic matter, much smaller amounts of toxic organic chemical by-products are formed than if molecular chlorine were used. As is the case with ozone, ClO_2 cannot be stored since it is explosive at the high concentrations that its practical use calls for, so it must be generated on-site. This is accomplished by oxidizing its reduced form, the *chlorite ion*, ClO_2^-, from the salt *sodium chlorite*, $NaClO_2$:

$$ClO_2^- \longrightarrow ClO_2 + e^-$$

Some of the chlorine dioxide in these processes is converted to ClO_2^- and ClO_3^- (chlorate) ions. The presence of chlorite and chlorate ions as residuals in the final water has raised health concerns due to their potential toxicity. The U.S. EPA has set an MCL of 1.0 ppm for chlorite ion and an MRDL (maximum residual disinfectant level) of 0.8 ppm for chlorine dioxide in drinking water.

Disinfection by Chlorination: History

The most common water purification agent used in North America is **hypochlorous acid,** HOCl; about half the U.S. population uses surface water, and one-quarter of the population uses groundwater, that is disinfected by HOCl. This neutral, covalent compound kills microorganisms, as it readily passes through their cell membranes. In addition to being effective, disinfection by **chlorination** is relatively inexpensive. Incorporating a small excess of the chemical in the treated water provides it with residual disinfection power during subsequent storage and transmission to the consumer. Chlorination is more common than ozonation in North America because generally the raw water is less polluted. Chlorination of public water supplies in the United States, Canada, and Great Britain began in the early years of the twentieth century. For the previous 50 years, chlorination had been practiced on an emergency basis during epidemics.

Disinfection by Chlorination: Production of Hypochlorous Acid

Like ozone, HOCl is not stable in concentrated form, so it cannot be stored. For large-scale installations, e.g., municipal water treatment plants, it is generated by dissolving **molecular chlorine** gas, Cl_2, in water. At moderate pH values the equilibrium in the reaction of chlorine with water lies far to the right and is achieved in a few seconds:

$$Cl_2(g) + H_2O(aq) \rightleftharpoons HOCl(aq) + H^+ + Cl^-$$

Thus a dilute aqueous solution of chlorine in water contains very little aqueous Cl_2 itself. If the pH of the reaction water were allowed to become too high, the result would be the ionization of the weak acid HOCl to the **hypochlorite ion,** OCl^-, which is less able to penetrate bacteria because of its electrical charge. Once chlorination is complete, the pH is adjusted upward, if necessary, by the addition of lime.

In small-scale applications of chlorination, as in swimming pools, the handling of cylinders of Cl_2 is inconvenient and dangerous. The chlorine can be produced as needed on the spot by the electrolysis of salty water. More commonly, hypochlorous acid is generated instead from the salt *calcium hypochlorite*, $Ca(OCl)_2$, or is supplied as an aqueous solution of *sodium hypochlorite*, NaOCl. In water, an acid–base reaction occurs to convert most of the OCl^- in these substances to HOCl:

$$OCl^- + H_2O \rightleftharpoons HOCl + OH^-$$

Close control of the pH in an environment like a swimming pool is necessary to avoid a shift to the left of the equilibrium for this reaction, which can occur under very alkaline conditions. On the other hand, corrosion of pool construction materials can occur in acidic water, so the pH is usually maintained above 7 to prevent such deterioration. Maintenance of an alkaline pH also prevents the conversion of dissolved **ammonia,** NH_3, to the **chloramines,** NH_2Cl, $NHCl_2$, and especially NCl_3, which is a powerful eye irritant:

$$NH_3 + 3\ HOCl \longrightarrow NCl_3 + 3\ H_2O$$

It is desirable to adjust the position of the equilibrium in the $OCl^- \rightarrow$ HOCl reaction to favor the predominance of the disinfectant molecular species, HOCl. Since the equilibrium between HOCl and OCl^- shifts rapidly in favor of the ion between pH values of 7 and 9, however, the acidity level must be meticulously controlled. Swimming pool acidity can be adjusted by the addition of acid (in the form of *sodium bisulfate*, $NaHSO_4$, which contains the acid HSO_4^-) or a base (Na_2CO_3) or a buffer ($NaHCO_3$, which contains the amphoteric anion HCO_3^-). Chlorine must be constantly replenished in outdoor pools since UV-B and the short-wavelength components of UV-A light in sunshine are absorbed by and decompose the hypochlorite ion:

$$2 \, ClO^- \xrightarrow{\text{UV}} 2 \, Cl^- + O_2$$

Hypochlorous acid can also be generated by the reaction with water of a chlorine-containing derivative of *isocyanuric acid*, $C_3N_3O_3H_3$:

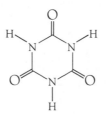

Either the trichloro derivative, in which each hydrogen is replaced by Cl to give $C_3N_3O_3Cl_3$, or the sodium dichloro derivative, $C_3N_3O_3Cl_2Na$, is used. In either case, the OH group of water combines with the chlorine to produce HOCl and the hydrogen of H_2O is bonded to the nitrogen, giving isocyanuric acid:

$$C_3N_3O_3Cl_3 + 3 \, H_2O \rightleftharpoons C_3N_3O_3H_3 + 3 \, HOCl$$

Since this process is an equilibrium, not all the material added is immediately converted to hypochlorous acid. As HOCl is used up, both by its use as a disinfectant and by dissociation in sunlight of its ionic form, the equilibrium shifts to the right and more HOCl is produced. None of the various forms of isocyanuric acid absorb UV light, so its chlorine is "protected" against decomposition by sunlight. Since the bulk forms of chlorinated isocyanic acid are expensive, it is common to supply hypochlorite from a cheaper source and to add isocyanuric acid as a stabilizer, temporarily reversing the above reaction to "store" the chlorine until it is needed.

Disinfection by Chlorination: By-Products and Their Health Effects

An important drawback to the use of chlorination for disinfecting water is the concomitant production of chlorinated organic substances, some of which are toxic, since HOCl is not only an oxidizing agent but also a chlorinating agent. An example of these important by-products is *halogenated acetic acids* (haloacetic acids), such as $CH_2Cl—COOH$, which the U.S. EPA restricts to 60 ppb as an MCL annual average for drinking water.

If the water to be disinfected contains *phenol*, C_6H_5OH, or one of its derivatives, chlorine readily replaces some of the hydrogen atoms on the ring to give rise to chlorinated phenols: these compounds have an offensive odor and taste and are toxic. Some communities switch from chlorine to chlorine dioxide when their raw-water supply is temporarily contaminated with phenols.

A more general problem with chlorination of water lies in the production of **trihalomethanes** (THMs). Their general formula is CHX_3, where the three X atoms can be chlorine or bromine or a combination of the two. The THM of principal concern is **chloroform,** $CHCl_3$, which is produced when hypochlorous acid reacts with organic matter dissolved in the water (see Box 10-2). Chloroform is a suspected liver carcinogen in humans, and it may also have negative reproductive and developmental effects. Its presence, even at very low levels of approximately 30 ppb, raises the specter that chlorinated drinking water may pose a health hazard, although one that pales by comparison with the benefits that it confers in the elimination of fatal waterborne diseases. Recently the annual average limit of total THMs in drinking water in the United States and the European Union was reduced to 80 ppb. The

---

BOX 10-2 | The Mechanism of Chloroform Production in Drinking Water

Humic acids, with which HOCl reacts to form chloroform, are water-soluble, nonbiodegradable components of decayed plant matter. Of particular importance are humic acids that contain 1,3-dihydroxybenzene rings. The carbon atom (#2) located between those carrying the —OH groups is readily chlorinated by HOCl, as in this elementary case:

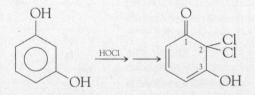

Subsequently the ring cleaves between C-2 and C-3 to yield a chain:

$$R—\underset{\underset{O}{\|}}{C}—CHCl_2$$

In the presence of the HOCl, the terminal carbon becomes trichlorinated, and the —CCl₃ group is readily displaced by the OH⁻ in water to yield chloroform:

$$R—\underset{\underset{O}{\|}}{C}—CHCl_2 \xrightarrow{HOCl} R—\underset{\underset{O}{\|}}{C}—CCl_3$$

$$\xrightarrow[H^+]{OH^-} R—\underset{\underset{O}{\|}}{C}—OH + CHCl_3$$

Analogous sequences of reactions produce bromoform, $CHBr_3$, and mixed chlorine-bromine trihalomethanes from the action on humic materials of hypobromous acid, HOBr, which is formed when bromide ion in water displaces chlorine from HOCl:

$$HOCl + Br^- \rightleftharpoons HOBr + Cl^-$$

previous limit of 100 ppb is still used in Canada. In fact, these 80–100 ppb limits are set not only to regulate the THM chemicals themselves but also as a limit on the production of *other* chlorinated organic DBPs.

The U.S. EPA attempted in the late 1990s to set a **maximum contaminant level goal** (MCLG) of 300 ppb for chloroform itself in drinking water. An MCLG is the maximum level at which the contaminant is believed to be safe, allowing for adequate margins of safety, but unlike the MCL (Chapter 7), it is not an enforceable standard. A nonzero level was set because chloroform is believed to operate indirectly as a carcinogen, not by damaging DNA, but by causing tissue damage that leads to rapid cell proliferation, which in turn increases the likelihood that cancer will form in the damaged tissue. A threshold below which no effects are likely to be observed is expected for carcinogens that operate in this manner. However, some environmental groups objected and convinced the EPA in 1998 to reset the goal to zero, as had been the EPA policy for any carcinogenic compounds in water. In 2000, however, in accordance with a ruling in a court case initiated by the chlorine industry, the zero MCLG was removed.

The level of trihalomethanes formed in water depends strongly on the organic content of the raw water, since they are formed from reaction of organics with HOCl (see Box 10-2). THM levels can reach 250 ppb in areas of Scotland and Northern Ireland that have peat moorlands. Water exposed to bogs in Newfoundland, Canada, has generated THM levels in excess of 400 ppb. As of the early 1990s, about 1% of the larger U.S. drinking water utilities that used surface waters, and none that used groundwater, had average THM levels exceeding 100 ppb. The THM content of chlorinated water could be decreased by using activated carbon either to remove dissolved organic compounds before the water is chlorinated or to remove THMs and other chlorinated organics after the process, although THMs are not very efficiently adsorbed by the carbon and it is an expensive process.

An analysis has been reported of all the recent epidemiological studies relating the chlorination of water to cancer rates in various communities in the United States. The conclusion was that the risk of bladder cancer in humans increased by 21% and that of rectal cancer by 38% for Americans who drank chlorinated surface water in the past. A similar recent study in Ontario of people who drank water for 35 years or more found even higher risk factors for bladder cancer at THM levels greater than 50 ppb, and for colon cancer at levels exceeding 75 ppb, but found no correlations of THMs with rectal cancer rates.

Given that slightly more than half the population of the United States drinks surface water, one effect of chlorination is to have increased bladder cancer incidence by about 4200 cases per year and rectal cancer incidence by about 6500 cases annually. Because of these risks, some communities are considering a switch, or have already switched, to water disinfection by ozone or chlorine dioxide, since these agents produce little or no chloroform. The extent of chlorination has already been reduced in most American communities relative to the levels that produced these statistics.

Several other mutagenic chlorinated organic DBPs formed during chlorination have been detected in water, in addition to chloroform. It is not clear whether the main carcinogen in the chlorinated drinking water is THM itself or some nonvolatile, higher-molecular-weight, mutagenic by-product present at still lower concentrations that would presumably be proportional to THM concentrations. The same risks do not usually apply to chlorinated well water, since its organochlorine content is much lower (only 0.8 ppb on average, versus 51 ppb for surface water) because it contains much smaller amounts of organic matter in the first place. Brands of carbonated water that use municipal drinking water as their source also contain dissolved chloroform. A recent study showed that exposure to chloroform by dermal contact and inhalation of the gases deabsorbed from hot water during showers and baths contribute about as much to one's intake of THMs as does drinking the water itself. Swimming in pools in which the water is chlorinated for disinfection also contributes significantly to dermal exposure.

Recently public health officials have expressed concern about the possible link between THMs and adverse human reproductive outcomes, including first-term miscarriages, stillbirths, impaired fetal growth, and certain birth defects. Even though the existing research in this area is not yet definitive, some officials suggest that women drink bottled water rather than chlorinated tap water during their first three months of pregnancy.

Disinfection by Chlorination: Advantages over Other Methods

Notwithstanding the preceding discussion of chlorination by-products, it is important to point out that the disinfection of water is extremely important in protecting public health and saves many more lives—by a very large factor—than are affected negatively. For example, both typhoid and cholera were widespread in both Europe and North America a century ago but have been almost completely eradicated in the developed world, thanks to chlorination and other disinfection methods for drinking water and to improved sanitation in general. The same is not true in many developing countries; e.g., there were more than half a million cases of cholera in Peru in the early 1990s. Overall, about 20 million people, most of them infants, die from waterborne diseases annually worldwide in underdeveloped countries, where water purification is often erratic or even nonexistent. Under no circumstances should effective disinfection of water be abandoned because of concern for the by-products of chlorination!

An advantage chlorination has over disinfection by chlorine dioxide or ozone or UV is that some chlorine remains dissolved in water after it has left the purification plant, so the water is protected from subsequent bacterial contamination before it is consumed. Indeed, some chlorine is usually added to water purified by the other methods to provide this protection. There is very little danger of significant chloroform production in the purified water since its organic content has been virtually eliminated before the chlorine is introduced. If the chlorine level in water purified by chlorination is too high, it can be lowered by the addition of sulfur dioxide.

The residual chlorine in water often exists in the form of the chloramines NH_2Cl, $NHCl_2$, and NCl_3, which are produced from reaction with dissolved ammonia gas. Although not as fast as HOCl in disinfecting water, the mono- and dichloroamines especially are good disinfectants. The mixture of chloramines, called **combined chlorine,** is longer-lived than hypochlorous acid and thus provides longer residual protection. Indeed, ammonia is often added to purified drinking water in order to convert the residual chlorine to the combined form (Figure 10-1). Chloramines are sometimes used, rather than chlorine or ozone or chlorine dioxide, as the main disinfectant in the purification of drinking water. They have the advantage over chlorine of producing small (though not zero) amounts of THMs and haloacetic acids. The EPA has set MRDLs of 4.0 ppm for both chlorine and chloramine in drinking water.

Bromine rather than chlorine is sometimes used as the disinfectant in swimming pools. The main disinfecting agent in bromination is *hypobromous acid*, HOBr, analogous to the role of hypochlorous acid in chlorination. HOBr reacts more rapidly with dissolved ammonia than does HOCl, producing mainly NH_2Br, which is also a good disinfectant.

To disinfect water for drinking purposes, hikers either boil raw water or treat it chemically with either chlorine, in the form of bleach, which provides HOCl, or iodine, as *elemental I_2* or *hypoiodous acid*, HIO. Concerns have been expressed about chronic health problems such as thyroid disfunction associated with long-term use of iodine, however. Treating the water with elemental iodine tends to make it unpalatable as well.

PROBLEM 10-3

Assuming that the nitrogen atom in monochloramine, NH_2Cl, has an oxidation number of -3, calculate that of the chlorine. Using the principle that unlike charges attract, predict whether it will be the hydrogen ion or the hydroxide ion from dissociated water molecules that will extract the Cl from NH_2Cl; from your result, predict the products of the decomposition reaction of chloramine in water.

A drinking water quality issue of current concern involves the pathogenic protozoa called *Cryptosporidium*, which was responsible for the death of 100 people and for about 400,000 cases of watery diarrhea in Milwaukee in 1993.

Less serious *Cryptosporidium* outbreaks occurred in Oxford, England, in 1989 and in Saskatchewan, Canada, in 2001. This deadly parasite is resistant to standard methods of disinfection such as chlorination at normal levels and is so small (about 5 μm in diameter) that it easily passes through the standard filters used to separate sediments. Several possible solutions have been advanced, including ozonation or the use of ultrafiltration or UV irradiation or the application of monochloramine following chlorination. A longer-than-usual exposure of water containing *Cryptosporidium* is necessary with ozonation, since the activation energy for destruction of protozoa by ozone is about twice as large as that for bacteria (80 versus about 40 kJ/mol).

Another protozoa, *Giardia lamblia*, also causes many instances of water-borne disease. Like *Cryptosporidium*, it is also somewhat resistant to chlorination, but since it is larger (about 10 μm in diameter), it is more easily removed by filtration through sand.

Groundwater: Its Supply, Chemical Contamination, and Remediation

The Nature and Supply of Groundwater

The greater part of the available fresh water on Earth lies underground, half of it at depths exceeding a kilometer. As one digs into the ground below the initial belt of soil moisture, the **aeration** or **unsaturated zone,** where the particles of soil are covered with a film of water but where air is present between the particles, is encountered next. At lower depths is the **saturated zone,** in which water has displaced all the air from these **pore spaces. Groundwater** is the name given to the fresh water in the saturated zone (see Figure 10-3); it makes up 0.6% of the world's total water supply. The ultimate source of groundwater is precipitation that falls onto the surface; a small fraction of it eventually filters down to the saturated zone. Underground water ranges in "age" from a few years to millions of years. For example, in zones that are currently arid, much of the groundwater currently being accessed has been there since the wetter conditions of the last ice age and will not be quickly replaced. The distribution of water among its major pools and the annual fluxes between them are summarized in Figure 10-4.

The top of the groundwater (saturated) region is called the **water table.** In some places it occurs right at the surface of the soil, a phenomenon that gives rise to swamps. Where the water table lies above the soil, we encounter lakes and streams.

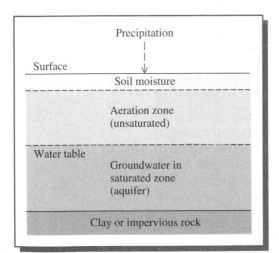

FIGURE 10-3 Groundwater location in relation to regions in the soil.

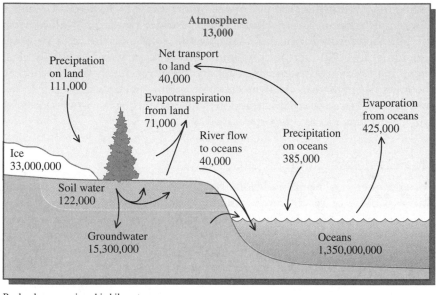

Pool volumes are in cubic kilometers.
Fluxes are in cubic kilometers per year.

FIGURE 10-4 Global pools and fluxes of water on Earth, showing the size of groundwater storage relative to other major water stores and fluxes. All pool volumes are in cubic kilometers, and all fluxes are in cubic kilometers per year. [Source: W.H. Schlesinger, *Biogeochemistry—An Analysis of Global Change*, 2nd ed, (San Diego: Academic Press, 1997, Chapter 10).]

If groundwater is contained in soil composed of porous rocks such as sandstone, or in highly fractured rock such as gravel or sand, and if the water is bounded at its lower depths by a layer of clay or impervious rocks, then it constitutes a permanent reservoir—a sort of underground lake—called an **aquifer.** This groundwater can be extracted by wells and is the main supply of drinking water for almost half the population of North America and over 1.5 billion people worldwide. In the United States in 1990 groundwater supplied 39% of the water used for public supplies and 96% of that withdrawn for individual domestic systems, the latter being very common in rural homes. In Europe the proportion of public drinking water extracted from aquifers ranges from nearly 100% for Denmark, Austria, and Italy, to about two-thirds in Germany, Switzerland, and the Netherlands, to less than one-third in Great Britain and Spain. Some aquifers lie below several layers of impermeable rock or soil; these are called *confined* or *artesian* aquifers.

In the United States the majority of groundwater use is for irrigation purposes, almost all of it in the western states. The massive extraction of water from American aquifers has given rise to fears about future supplies of fresh water (and about the sinking of land above the aquifers), since such aquifers are replenished only very slowly. In the High Plains of the central United States more than half the groundwater in storage has been depleted in some areas. In northern China the depletion of shallow aquifers is forcing the sinking of wells more than 1 km deep to reach a new supply of groundwater. Indeed, groundwater depletion—along with the buildup of salts in the soil—is now the dominant threat to irrigated agriculture. In addition, the contamination of groundwater by chemicals is becoming a serious concern in many areas. Currently, about one-third of the world's population lives in countries that already experience some shortage of fresh water; this fraction will probably rise to two-thirds by 2025, according to U.N. reports.

The Contamination of Groundwater

Groundwater has been traditionally considered to be a pure form of water. Because of the filtration through soil and its long residence time underground, it contains much less natural organic matter and many fewer disease-causing microorganisms than water from lakes or rivers, although the latter point may be a misconception, according to recent evidence. Some groundwater is naturally too salty or too acidic for either drinking or irrigation purposes and may contain too much sodium, sulfide, or iron ion for many uses.

Although humans have been concerned about the pollution of surface water in rivers and lakes for a long time, the contamination of groundwater by chemicals was not recognized as a serious environmental problem until the 1980s, notwithstanding the fact that it had been occurring for half a century. To a large extent, groundwater contamination was neglected because it was not immediately visible—it was "out-of-sight, out-of-mind"—even though groundwater is a major source of drinking water. We were ignorant of the long-range consequences of our waste disposal practices. Ironically, surface water can be cleaned up relatively easily and quickly, whereas groundwater pollution is a much harder, much more expensive, long-range problem to solve.

Because we are now aware of the consequences—including high remediation costs—of the uncontrolled disposal of organic wastes, most large corporations in developed countries have become much more responsible in their disposal of chemicals. Unfortunately, the collective discharges from smaller sources, including many municipalities, small industries, and farms, have not yet been controlled. Similarly, the huge number of septic tanks that exist are collectively a major source of nitrate, bacteria, viruses, detergents, and household cleaners to groundwater.

Nitrate Contamination of Groundwater

The inorganic contaminant of greatest concern in groundwater is the **nitrate ion,** NO_3^-, which commonly occurs in both rural and suburban aquifers. Although uncontaminated groundwater generally has nitrate nitrogen levels of less than

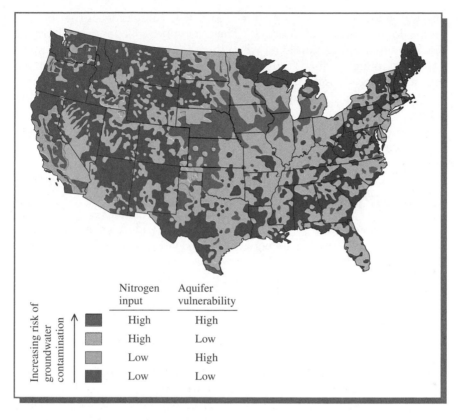

Increasing risk of groundwater contamination

	Nitrogen input	Aquifer vulnerability
	High	High
	High	Low
	Low	High
	Low	Low

FIGURE 10-5 The risk of nitrate contamination of the groundwaters in the United States. [Source: B. T. Nolan et al., "Risk of Nitrate in Groundwaters of the United States— A National Perspective," *Environmental Science and Technology* 31 (1997): 2229.]

2 ppm, about 9% of shallow aquifers—from which well water often is extracted—in the United States now have nitrate levels that exceed the 10 ppm nitrogen MCL value (see Figure 10-5). Exceeding this limit is much rarer (1%) for public U.S. water supplies, partly because they are drawn from deeper aquifers; these are generally less contaminated because of their depth, because their location is remote from large sources of contamination, and because natural remediation by denitrification in the low-oxygen conditions can occur.

Nitrate in groundwater originates mainly from four sources:

• application of nitrogen fertilizers, both inorganic and animal manure, to cropland,

• atmospheric deposition,

• human sewage deposited in septic systems, and

• cultivation of the soil.

For example, almost 12 million tons of nitrogen are applied annually as fertilizer for agriculture in the United States, and manure production contributes almost 7 million tons more. As discussed below, cultivation of the soil also results in the release of nitrogen. The atmospheric deposition of nitrate results from its production in air when the NO_X emissions from vehicles and power plants are oxidized to nitric acid and then neutralized to ammonium nitrate (Chapters 2 and 3).

In most cases, the reduced forms of nitrogen are oxidized in the soil to nitrate, which then migrates down to the groundwater, where it dissolves in water and is diluted. Because nitrate removal from well water is very expensive, water contaminated with high levels normally is not used for human consumption, at least in public supplies.

PROBLEM 10-4

The nitrate concentration in an aquifer is 20 ppm and its volume is 10 million liters. What mass of ammonia upon oxidation would have produced this mass of nitrate?

Nitrates in Water

Concern has been expressed recently about the increasing levels of nitrate ion in drinking water, particularly in well water in rural locations; the main source of this NO_3^- is runoff from agricultural lands into rivers and streams. Initially, oxidized animal wastes (manure) and unabsorbed *ammonium nitrate* and other nitrogen fertilizers were thought to be the culprits. Nitrogen unused by plants is often converted naturally to nitrate, which is highly soluble in water and can easily leach down into groundwater. It now *also* appears that intensive cultivation of land, even without the application of fertilizer or manure, facilitates the oxidation of reduced nitrogen to nitrate in decomposed organic matter in the soil by providing aeration and moisture.

Rural areas with high nitrogen input, well-drained soil, and little woodland are at particular risk for nitrate contamination of groundwater. Denitrification of nitrate to nitrogen gas and uptake of nitrate by plants can occur in forested areas that separate agricultural farms from streams, thereby lowering the risk of contamination in areas with significant woodland.

In urban areas, the use of nitrogen fertilizers on domestic lawns and golf courses, parks, etc. contributes nitrate to groundwater. Septic tanks and cesspools also are significant contributors where they exist.

Excess nitrate ion in wastewater flowing into seawater, e.g., in the Baltic Sea, has resulted in algal blooms that pollute the water after they die. Nitrate ion normally does not cause this effect in bodies of fresh water, where phosphorus rather than nitrogen is usually the limiting nutrient; increasing the nitrate concentration there without an increase in phosphate levels does not lead to an increased amount of plant growth. There are, however, instances where nitrogen rather than phosphorus temporarily becomes the limiting nutrient even in fresh waters.

Health Hazards of Nitrates in Drinking Water

Excess nitrate ion in drinking water is a potential health hazard since it can result in **methemoglobinemia** in newborn infants as well as in adults with a particular enzyme deficiency. The pathological process, in brief, is as follows.

Bacteria, e.g., in unsterilized feeding bottles or in the baby's stomach, reduce some of the nitrate to **nitrite ion,** NO_2^-:

$$NO_3^- + 2\,H^+ + 2\,e^- \longrightarrow NO_2^- + H_2O$$

The nitrite combines with and oxidizes the hemoglobin in blood, thereby preventing the proper absorption and transfer of oxygen to cells. The baby turns blue and suffers respiratory failure. (In almost all adults, the oxidized hemoglobin is readily reduced back to its oxygen-carrying form, and the nitrite is readily oxidized back to nitrate; also, nitrate is mainly absorbed in the digestive tract of adults before reduction to nitrite can occur.) The occurrence of methemoglobinemia, or "blue-baby syndrome," is now relatively rare in industrialized countries but it is still of concern in some developing countries.

Recently an increase in the risk of acquiring non-Hodgkin's lymphoma has been found for persons consuming drinking water having the highest levels (long-term average of 4 ppm or more of nitrogen as nitrate) of nitrate in drinking water for some communities in Nebraska. As discussed in the next section, excess nitrate ion in drinking water is also of concern because of its potential link with stomach cancer. Recent epidemiological investigations have, however, failed to establish any positive, statistically significant relationship between nitrate levels in drinking water and the incidence of stomach cancer. However, a study reported in 2001 found that women in Iowa who drank water from municipal supplies with elevated nitrate levels (>2.46 ppm) were almost three times as likely to be diagnosed with bladder cancer than those least exposed (<0.36 ppm in their drinking water).

The expenditure of public money on nitrate-level reductions in drinking water has become a controversial subject. In Great Britain, in particular, hundreds of millions of dollars have been spent on achieving the 50-ppm maximum level of nitrate ion set by the European Union. The location of areas in the United States that have a high risk of nitrate contamination of groundwater is shown in Figure 10-5. The natural concentration of nitrate in U.S. groundwater generally is less than 2 ppm. Groundwater from domestic wells is more likely to be contaminated with nitrate than public supply wells, since the latter are usually deeper and located farther away from nitrate sources. In addition, the low-oxygen conditions in deep aquifers promote the denitrification of nitrate. The U.S. EPA MCL of 10 ppm of nitrate nitrogen was set to avoid blue-baby syndrome. Since this is now almost nonexistent in the United States (only two cases since the mid-1960s), some policy analysts think the 10-ppm value is too stringent.

PROBLEM 10-5

Convert the EU nitrate standard of 50 ppm to its nitrogen content alone. Is the EU standard more or less stringent than the U.S. regulatory limit of 10 ppm (mg) of nitrate nitrogen per liter (i.e., 10 ppm N)?

Nitrosamines in Food and Water

Some scientists have warned that excess nitrate ion in drinking water and foods could lead to an increase in the incidence of stomach cancer in humans, since some of it is converted in the stomach to nitrite ion. The problem is that the nitrites could subsequently react with amines to produce **N-nitrosamines,** compounds that are known to be carcinogenic in animals. N-nitrosamines are amines in which two organic groups and an —N=O unit are bonded to the central nitrogen:

$$
R_2NNO \quad \text{or} \quad
\begin{array}{c}
R \\
\diagdown \\
N-N=O \\
\diagup \\
R
\end{array}
$$

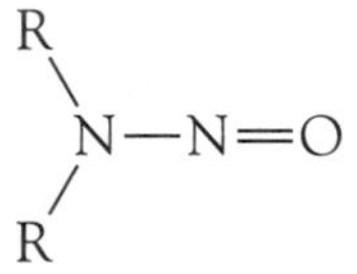

N-nitrosamines

Of concern not only with respect to its production in the stomach and its occurrence in foods and beverages (e.g., cheeses, fried bacon, smoked and/or cured meat and fish, and beer), but also as an environmental pollutant in drinking water, is the compound in which R is the methyl group, CH_3; it is called **N-nitrosodimethylamine** or NDMA:

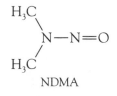

$$
\begin{array}{c}
H_3C \\
\diagdown \\
N-N=O \\
\diagup \\
H_3C
\end{array}
$$

NDMA

This organic liquid is somewhat soluble in water (about 4 g/L) and somewhat soluble in organic liquids. It is a probable human carcinogen, and a potent one if extrapolation from animal studies is a reliable guide. It can transfer a methyl group to a nitrogen or oxygen of a DNA base and thereby alter the instructional code for protein synthesis in the cell.

In the early 1980s it was found that NDMA was present in beer to the extent of about 3,000 ppt. Since that time, commercial brewers have modified the drying of malt so that the current levels of NDMA in U.S. and Canadian beers are only about 70 ppt.

Large quantities of nitrate are used to "cure" pork products such as bacon and hot dogs. In these foods, some of the nitrate ion is biochemically reduced to nitrite ion, which prevents the growth of the organism responsible for botulism. Nitrite ion also gives these meats their characteristic taste and color by combining with hemoproteins in blood. Nitrosamines are produced from excess nitrite during frying (e.g., of bacon) and in the stomach, as discussed. Government agencies have instituted programs to decrease the residual nitrite levels in cured meats. Some manufacturers of these foods now add vitamin C or E to the meat to block the formation of nitrosamines. Based on average levels of NDMA in various foods and the average daily intake for each of them, most of us now ingest more NDMA from consumption of cheese (which is often treated with nitrates) than from any other source.

Even though the commercial production of NDMA has been phased out, it can be formed as a by-product of the use of amines in industrial processes such as rubber tire manufacturing, leather tanning, and pesticide production.

The levels of NDMA in drinking water drawn from groundwater is of concern in some localities that have industrial point sources of the compound. For example, following the discovery that the water supply of one town had been contaminated by up to 100 ppt of NDMA from a tire factory, the province of Ontario, Canada, adopted a guideline maximum of 9 ppt of NDMA in drinking water, which corresponds to a lifetime cancer risk of 1 in 100,000. By contrast, the guideline for water in the United States is set at 0.68 ppt, which corresponds to a cancer risk of 1 in a million but which actually lies considerably below the detection limit (about 5 ppt) for the compound.

PROBLEM 10-6

Write balanced redox half-reactions (assuming acidic conditions) for the conversion of NH_4^+ to NO_3^- and of NO_2^- to N_2.

Groundwater Contamination by Organic Chemicals

The contamination of groundwater by organic chemicals is a major concern. The compounds that are most often detected in groundwater-based U.S. community public water supplies, including those near hazardous waste sites, are summarized in Table 10-1. Municipal landfills as well as industrial waste disposal sites are often the sources of the contaminants. Liquid that contains dissolved matter that drains from a terrestrial source, such as a landfill, is called

TABLE 10-1	Organic Compounds Commonly Found in U.S. Groundwater-Based Community Water Supplies and Their Properties		
Chemical		**Density (g/mL)**	**Water solubility (g/L)**
Present at 25–50% of sites:			
Chloroform (trichloromethane)		1.48	8.2
Bromodichloromethane		1.98	4.4
Dibromochloromethane		2.45	2.7
Bromoform (tribromomethane)		2.89	3.0
Present at a smaller fraction of sites:			
Trichloroethene		1.46	1.1
Tetrachloroethene (perchloroethene)		1.62	0.15
1,1,1-Trichloroethane		1.34	1.5
1,2-Dichloroethenes		1.26, 1.28	3.5, 6.3
1,1-Dichloroethane		1.18	5.5
Carbon tetrachloride		1.46	0.76
Dichloroiodomethane		1.58	
Xylenes		0.86–0.88	0.18 (o)
1,2-Dichloropropane		1.16	2.8
Benzene		0.88	1.8
Toluene		0.87	0.54
Also commonly present at wells close to hazardous waste sites:			
Methylene chloride		1.33	20
Ethylbenzene		0.87	0.15
Acetone		0.79	sol
1,1-Dichloroethene		1.22	2.3
1,2-Dichloroethane		1.24	8.5
Vinyl chloride (chloroethene)		gas	8.8
Methyl ethyl ketone		0.80	268
Chlorobenzene		1.11	0.47
1,1,2-Trichloroethane		1.44	4.5
Chloroethane		0.90	5.7
Fluorotrichloromethane		gas	1.1
1,1,2,2-Tetrachloroethane		1.60	2.7
Methyl isobutyl ketone		0.80	19

Source: Based on U.S. EPA surveys of about 2% of U.S. water supplies.

a **leachate.** In rural areas, the contamination of shallow aquifers by organic pesticides, such as atrazine leached from the surface, has become a concern.

The typical organic contaminants in most major groundwater supplies are

- chlorinated solvents, especially **trichloroethene** (TCE, also called trichloroethylene), C_2HCl_3, and **perchloroethene** (PCE, also called perchloroethylene or tetrachloroethene), C_2Cl_4. These molecules contain a C=C bond, with three or four of the four hydrogen atoms of ethene (ethylene) replaced by chlorine:

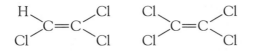

- hydrocarbons from the BTX component of gasoline and other petroleum products: *benzene*, C_6H_6, and its methylated derivatives; *toluene*, $C_6H_5(CH_3)$; and the three isomers of *xylene*, $C_6H_4(CH_3)_2$ (see Chapter 6 for structures), and

- MTBE from gasoline (see Chapter 6)

The chemicals in these groups occur commonly in groundwater at sites where manufacturing and/or waste disposal occurred, especially from 1940 to 1980. In that period little attention was paid to the ultimate fate and residence of these chemicals once they were deposited in the ground. The sources of these organic substances included leaking chemical waste dumps, leaking underground gasoline storage tanks, leaking municipal landfills, and accidental spills of chemicals on land. In fact, many substances do decay rapidly or are immobilized in the soil, so the number of compounds that are sufficiently persistent and mobile to travel to the water table and to contaminate groundwater there is relatively small.

Trichloroethene is an industrial solvent used to dissolve grease on metal, as is perchloroethene. The U.S. MCL for TCE in drinking water is 5 ppb, whereas the Canadian limit is 50 ppb. There are indications that in the near future the U.S. EPA will change the status of trichloroethene from a "probable human carcinogen" to a substance "highly likely to produce cancer in humans," since it has been linked to cancer at a variety of body sites. A link between TCE exposure and an abnormally low sperm count in males has recently been established.

PCE is not only used in metal degreasing but also finds wide application as the solvent in dry-cleaning operations, so it can be released from a large number of small sources. The most highly exposed human subset, a group of women in Cape Cod, Massachusetts, who were inadvertently exposed over several decades to high levels of PCE in their drinking water, were found to have small to moderate increases in their risk of contracting breast cancer.

Gasoline enters the soil via surface spills, leakage from underground storage tanks, and pipeline ruptures. Before 1980 underground gasoline storage tanks were made from steel; almost half of them were corroded enough to leak by the time they were 15 years old. Once they descend to groundwater, the water-soluble components of the gasoline are preferentially leached into the water and can migrate rapidly in the dissolved state. The BTX component is the most soluble of the hydrocarbons and often occurs at concentrations of 1–50 ppb in groundwater. However, the alkylated benzenes are rapidly degraded by aerobic bacteria and consequently are not long-lasting.

The MTBE component of gasoline is more water soluble than the hydrocarbons but, unlike them, it is not readily biodegraded. It is not highly toxic. The main problem is the odor and taste that it gives to water: as little as 15 ppb in water can be tasted or smelled. MTBE contamination of well water, albeit at low levels, is becoming a concern in the United States since it occurs at about a quarter of a million sites.

The Ultimate Sink for Organic Contaminants in Groundwater

The subsequent behavior of the organic compounds that do migrate to the water table depends significantly on their density relative to that of 1.0 g/mL for water. Liquids that are *less* dense ("lighter") than water form a mass that floats on the top of the water table. All hydrocarbons having a small or medium molecular mass belong to this group, including the BTX fraction of gasolines and other petroleum products (see Table 10-1). In contrast, polychlorinated solvents are *more* dense ("heavier") than water, so they tend to sink deep into aquifers; important examples are *methylene chloride*, chloroform, *carbon tetrachloride*, *1,1,1-trichloroethane*, TCE, and PCE. They all also share the properties of being persistent in soil and being slightly soluble in water (see Table 10-1 for details). Nonchlorinated but insoluble high-molecular-mass organic materials such as *creosote* and *coal tar* also belong to the heavier-than-water group.

Although the oily liquid blobs that these organic compounds form are generally found in an aquifer at a position directly below their original point of entry into the soil or close to it, the conclusion that they are horizontally immobile is misleading. Very slowly—in a process that often takes decades or centuries to complete—these low-solubility compounds gradually dissolve in the water that passes over the blob and provide a continuous supply of contaminants to the groundwater. Thus **plumes** of polluted water grow in the direction of the water's flow, thereby contaminating the bulk of the aquifer (see Figure 10-6). Because of such contamination, many drinking water wells have had to be closed.

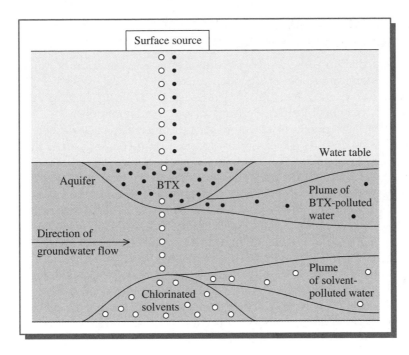

FIGURE 10-6 The contamination of groundwater by organic chemicals.

Decontamination of Groundwater: Pump-and-Treat

In the last two decades considerable energy and money have been spent in the United States on attempts to control aquifer pollution by these oily liquids. Dense organic leachates at most of the U.S. Superfund toxic waste sites (see Chapter 12) have contaminated the groundwater that lies below them. Unfortunately, no easy cure for the problem of contamination has been found. Control usually consists of **pump-and-treat** systems that pump contaminated water from the aquifer, treat it to remove its organic contaminants (using methods of the type described later in this chapter), and return the cleaned water to the aquifer or to some other water body. In one recent variation, a fine mist of the contaminated groundwater is sprayed into the air above agricultural land by a long, movable sprinkler system; the contaminant VOCs (volatile organic compounds) evaporate into the air and the cleansed water is used for irrigation.

The volume of water that must be pumped and treated in a given aquifer is huge. For organic contaminants with low water solubility, recontamination of water returned to the aquifer by additional dissolution from the blob will occur. Consequently, the treatment systems must operate in perpetuity, and there are already thousands of them spread across the United States.

Decontamination of Groundwater: Bioremediation and Natural Attenuation

Bioremediation is the term applied to the decontamination of water or soil using biochemical rather than chemical or physical processes. Recently there has been interesting progress reported, using bioremediation to cleanse water of chlorinated ethene solvent contamination.

The biodegradation of chloroethenes by *aerobic* bacteria becomes less and less efficient as the extent of chlorination increases, so it is ineffective for perchloroethene. However, under *anaerobic* conditions, the reductive biodegradation of PCE and TCE proceeds more quickly, particularly if an easily oxidized substance such as methanol is added to supply electrons for the reduction processes. Unfortunately, the stepwise dechlorination of these compounds proceeds through *vinyl chloride*, $CH_2\!=\!CHCl$, a known carcinogen. Recently a bacterium has been discovered that removes all the chlorine from organic solvents such as TCE and PCE.

Owing to the high cost and limited effectiveness of many groundwater cleanup technologies, the inexpensive process of natural attenuation—allowing natural biological, chemical, and physical processes to treat groundwater contaminants—has become popular. Indeed, it is now used at more than 25% of Superfund program sites in the United States and is the leading method for remedying the contamination of groundwater from leaking underground storage sites.

However, there is great controversy about whether or not natural attenuation is an appropriate strategy for managing groundwater contamination: many environmentalists feel that it is a cheap way for industry to avoid expensive cleanup costs. The U.S. National Research Council in 1997 appointed a committee to determine which pollutants could be treated successfully by this technique. Table 10-2 summarizes their results. Only three pollutants are highly likely to be successfully treated by natural attenuation:

- BTEX hydrocarbons (i.e., BTX hydrocarbons plus ethylbenzene),

- low-molecular-weight oxygen-containing organics, and

- methylene chloride.

In all three cases, biotransformation is the dominant process by which attenuation occurs. Notice that neither highly chlorinated organics, including TCE and PCE, nor MTBE are successfully treated in this way, nor is mercury or perchlorate ion.

Decontamination of Groundwater: In Situ Remediation

Scientists have developed a promising in situ technique for treating groundwater contaminated by volatile (mainly C_1 and C_2) chlorinated organics.

TABLE 10-2	Likelihood of Success of Groundwater Remediation by Natural Attenuation for Various Substances	
Chemical class	Dominant attenuation processes	Likelihood of success given current level of understanding
Organic Compounds		
Hydrocarbons		
BTEX	Biotransformation	High
Gasoline, fuel oil	Biotransformation	Moderate
Nonvolatile aliphatic compounds	Biotransformation, immobilization	Low
PAHs	Biotransformation, immobilization	Low
Creosote	Biotransformation, immobilization	Low
Oxygenated hydrocarbons		
Low-molecular-weight alcohols, ketones, esters	Biotransformation	High
MTBE	Biotransformation	Low
Halogenated aliphatics		
PCE, TCE, carbon tetrachloride	Biotransformation	Low
TCA	Biotransformation, abiotic transformation	Low
Methylene chloride	Biotransformation	High
Vinyl chloride	Biotransformation	Low
Dichloroethylene	Biotransformation	Low
Halogenated aromatics		
Highly chlorinated		
PCBs, tetrachlorodibenzofuran, pentachlorophenol, multichlorinated benzenes	Biotransformation, immobilization	Low
Less chlorinated		
PCBs, dioxins	Biotransformation	Low
Monochlorobenzene	Biotransformation	Moderate
Inorganic Substances		
Metals		
Ni	Immobilization	Moderate
Cu, Zn	Immobilization	Moderate
Cd	Immobilization	Low
Pb	Immobilization	Moderate

(continued on next page)

TABLE 10-2	Likelihood of Success of Groundwater Remediation by Natural Attenuation for Various Substances *(continued)*	
Chemical class	Dominant attenuation processes	Likelihood of success given current level of understanding
Cr	Biotransformation, immobilization	Low to moderate
Hg	Biotransformation, immobilization	Low
Nonmetals		
As	Biotransformation, immobilization	Low
Se	Biotransformation, immobilization	Low
Oxyanions		
Nitrate	Biotransformation	Moderate
Perchlorate	Biotransformation	Low

Source: Adapted from J. A. Macdonald, "Evaluating Natural Attenuation for Groundwater Cleanup," *Environmental Science & Technology* (1 August 2000): 346A.

They construct an underground permeable "wall" of material (mostly coarse sand) along the path of the water. The water is cleansed as it passes through the wall and never has to be pumped out of the ground (see Figure 10-7).

The ingredient placed within the sand bed that chemically cleans the water is **metallic iron,** Fe^0, in the form of small granules, a common waste product of manufacturing processes. When placed in contact with certain chlorinated

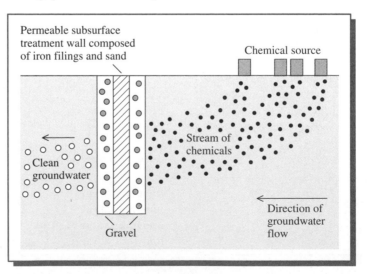

FIGURE 10-7 In situ purification of groundwater using an "iron wall."

organics dissolved in water, the iron acts as an reducing agent, giving up electrons to form the *ferrous* or Fe(II) ion, Fe^{2+}, which dissolves in the water:

$$Fe(s) \longrightarrow Fe^{2+}(aq) + 2\ e^-$$

Usually these electrons are donated to chloroorganic molecules that are temporarily adsorbed onto the metal's surface; the chlorine atoms in the molecules are consequently reduced to *chloride ions*, Cl^-, which are released into aqueous solution. This technique is an example of **reductive degradation.** For example, the reduction by electrons of trichloroethene to its completely dechlorinated form, *ethene*, can be written in unbalanced form as

$$C_2HCl_3 \longrightarrow C_2H_4 + 3\ Cl^-$$

On application of a standard redox balancing technique for alkaline solution, we obtain the balanced half-reaction

$$C_2HCl_3 + 3\ H_2O + 6\ e^- \longrightarrow C_2H_4 + 3\ Cl^- + 3\ OH^-$$

Combination of the half-reactions (after tripling the oxidation step to ensure the number of electrons lost and gained is the same) yields the overall reaction

$$3\ Fe(s) + C_2HCl_3 + 3\ H_2O \longrightarrow 3\ Fe^{2+}(aq) + C_2H_4 + 3\ Cl^- + 3\ OH^-$$

One of the by-products of the reaction is hydroxide ion, OH^-. Recall from Chapter 9 that in limestone-rich areas the groundwater contains significant concentrations of dissolved *calcium bicarbonate*, i.e., Ca^{2+} and HCO_3^-. The hydroxide ions produced in the groundwater remediation reaction react with bicarbonate to produce carbonate ion, CO_3^{2-}, which combines with dissolved calcium ions to produce insoluble calcium carbonate, $CaCO_3$, which then precipitates in the sand–iron mixture.

Field trials indicate that this new technology can work successfully for several years at least and may replace the pump-and-treat methods for many applications involving chlorinated methanes and ethanes dissolved in underground water.

Recently it has been found that coating the iron filings with *nickel* speeds up the rate of degradation of the organic compounds by a factor of ten; with this modification, the technique may be even more useful than first imagined. It has also been discovered that the elemental iron in the barriers will reduce soluble Cr^{6+} ions to insoluble Cr^{3+} oxides and can therefore remediate groundwater contaminated by Cr^{6+}. A technique for the in situ creation of elemental iron from its ions (Fe^{2+} and Fe^{3+}) by the injection of aqueous reducing agents has also been tested.

Recently an in situ technique for treating TCE and PCE by hydrogenation has been developed. The process uses dissolved H_2 gas to rapidly dechlorinate these two organics, eventually forming ethane and HCl. The reaction uses a palladium catalyst and can be done within a well bore so that the water need not be brought up to the surface.

PROBLEM 10-7

The dissolution of iron in the process described produces some molecular hydrogen gas, H_2. Show by balanced equation(s) how the hydrogen could arise from the reduction of water rather than of TCE.

PROBLEM 10-8

Deduce the overall reaction of hydroxide ion with dissolved calcium bicarbonate to produce insoluble calcium carbonate. Are there any reaction products other than $CaCO_3$?

PROBLEM 10-9

Suppose that the "iron wall" technology reduced an appreciable fraction of TCE to vinyl chloride rather than completely to ethene. Why would this be an unacceptable result environmentally? (Note that, in practice, a sufficiently thick wall of iron and sand is used to convert any vinyl chloride by-product to ethene.) [Hint: Using the index, look up the properties of vinyl chloride discussed previously.]

PROBLEM 10-10

Deduce the overall reaction by which perchloroethene is converted to ethene by metallic iron.

PROBLEM 10-11

At one test site for this remediation process, the water contained 270 ppm TCE and 53 ppm perchloroethene. Calculate the mass of iron required to remediate 1 L of this groundwater.

The Chemical Contamination and Treatment of Wastewater and Sewage

Most municipalities treat the raw sewage collected from homes, buildings, and industries (including food processing plants) through a **sanitary sewer** system before the liquid residue is deposited into a nearby source of natural waters, whether a river, lake, or ocean. Since the rainwater and melted snow that drains from streets and other paved surfaces is usually not highly contaminated, it is often collected separately by *storm sewers* and deposited directly into a body of natural water. Unfortunately, in some municipalities, storm-driven overflow occurs in the sanitary sewer system and this excess is combined with stormwater and deposited, untreated, into waterways.

The main component of sewage—other than water—is organic matter of biological origin. It occurs mainly as particles, ranging from those of macroscopic size large enough to be trapped (together with such objects as facial tissues, stones, socks, tree branches, condoms, and tampon applicators) by mesh screens to those of microscopic size, which are suspended in the water as large colloids.

Sewage Treatment

In the **primary** (or mechanical) treatment stage (see Figure 10-8), the larger particles—including sand and silt—are removed by allowing the water to flow across screens and slowly along a lagoon or settling basin. A **sludge** of insoluble particles forms at the bottom of the lagoon, while "liquid grease" (a term that here includes not only fat, oils, and waxes, but also the products formed by the reaction of soap with calcium and magnesium ions) forms a lighter-than-water layer at the top and is skimmed off. About 30% of the biochemical oxygen demand (BOD) (Chapter 9) of the wastewater is removed by the primary treatment process, even though this stage of the process is entirely mechanical in nature. The treatment and disposal of the sludge are discussed later.

After passing through conventional primary treatment, the sewage water has been much clarified but still has a very high BOD—typically several hundred milligrams per liter (ppm)—and is detrimental to fish life if released at this stage (as occurs in some jurisdictions that discharge into the ocean). The high BOD is due mainly to organic colloidal particles. In the **secondary** (biological) treatment stage, most of this suspended organic

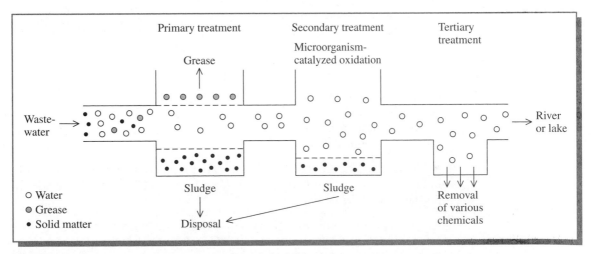

FIGURE 10-8 The common stages in the treatment of wastewater.

matter as well as that actually dissolved in the water is biologically oxidized by microorganisms to carbon dioxide and water or converted to sludge, which can readily be removed from the water. Either the water is sprinkled onto a bed of sand and gravel or plastic covered with aerobic bacteria (trickling filters) or it is agitated in an aeration reactor (activated sludge process) to effect the microorganism-driven reaction. The system is kept well-aerated to speed the oxidation. In essence, by deliberately maintaining in the system a high concentration of aerobic microorganisms, especially bacteria, the same biological degradation processes that would require weeks to occur in open waters are accomplished in several hours.

The biological oxidation processes of secondary treatment reduce the BOD of the polluted water to less than 100 mg/L, which is about 10% of the original concentration in the untreated sewage. For example, Canadian wastewater quality standards require that the BOD in treated water be reduced to 20 ppm or less. The process of nitrification also occurs to some extent, converting organic nitrogen compounds to nitrate ion and carbon dioxide. In summary, the secondary treatment of wastewater involves biochemical reactions that oxidize much of the oxidizable organic material that was not removed in the first stage. Treated water diluted with a greater amount of natural water can support aquatic life. Normally, municipalities take the water produced by secondary treatment and disinfect it by chlorination or irradiation with UV light before pumping it into a local waterway. Recent research in Japan has shown that chlorination of the effluent before its release produces mutagenic compounds, presumably by interaction of chlorine-containing substances with the organic matter that remains in the water.

A few municipalities employ **tertiary** (or **advanced** or chemical) wastewater treatment as well as primary and secondary. In the tertiary phase, specific substances are removed from the partially purified water before its final disinfection. In some cases the water produced by tertiary treatment is of sufficiently high quality to be used as drinking water. Alternatively, river water into which the effluent from sewage treatment plants has been deposited is used by municipalities downstream as drinking water. The reuse of water after it has been cleansed is particularly prevalent in Europe, where supplies of fresh water are less available than in North America and the density of the consuming population is high.

Depending on locale, tertiary treatment can include some or all of the following chemical processes:

• further reduction of BOD by removal of most of the remaining colloidal material using an aluminum salt in a process similar to that used for the purification of drinking water;

• removal of dissolved organic compounds (including chloroform) and some heavy metals by adsorption onto activated carbon, over which the water is allowed to flow (see Box 10-1);

- phosphate removal (as discussed in the next section), some phosphorus being removed in the secondary treatment stage, since microbes incorporate it as a nutrient for their growth;

- heavy-metal removal by the addition of hydroxide or *sulfide ions*, S^{2-}, to form insoluble metal hydroxides or sulfides;

- iron removal by aeration at a high pH to oxidize it to its insoluble Fe^{3+} state, possibly in combination with the use of a strong oxidizing agent to destroy organic ligands bound strongly to the Fe^{2+} ion, which would otherwise prevent its oxidation; and

- removal of excess inorganic ions.

In some wastewaters the further removal of **nitrogen** compounds—usually either ammonia or organic nitrogen compounds—is deemed necessary. Ammonia removal can be achieved by raising the pH to about 11 (with lime) to convert most ammonium ion to its molecular form, ammonia, followed by bubbling air through the water to air strip it of the dissolved ammonia gas. This process is relatively expensive, however, because it is energy-intensive. Ammonium ion can also be removed by ion exchange using certain resins that have their exchange sites initially populated by sodium or calcium ions.

Alternatively, both organic nitrogen and ammonia can be removed by first using nitrifying bacteria to oxidize all the nitrogen to nitrate ion. Then the nitrate is subjected to denitrification by bacteria to produce molecular nitrogen, which bubbles out of the water. Since this reduction step requires a substance that will be oxidized, *methanol*, CH_3OH, is added to the water if necessary and is converted to carbon dioxide in the process:

$$5\,CH_3OH + 6\,NO_3^- + 6\,H^+ \xrightarrow{\text{bacteria}} 5\,CO_2 + 3\,N_2 + 13\,H_2O$$

Of course, water contaminated by nitrate ion can also be treated by this latter step. A mathematical analysis of the kinetics of the transformations is given in Box 10-3.

PROBLEM 10-12

Given that the K_b for ammonia is 1.8×10^{-5}, deduce a formula giving the ratio of ammonia to ammonium ion as a function of the pH of water. What is the value of this ratio at pH values of 5, 7, 9, and 11?

The Origin and Removal of Excess Phosphate

One of the world's most famous cases of water pollution involves Lake Erie, which in the 1960s was said to be dying. Indeed, one of the authors of this

BOX 10-3 | Time Dependence of Concentrations in the Two-Step Oxidation of Ammonia

The bacteria-catalyzed oxidation of ammonia (or of other reduced organic nitrogen compounds) to nitrate is a reaction with two main steps, with nitrite ion, NO_2^-, an intermediate:

Step 1 $NH_3 + \frac{3}{2} O_2 \longrightarrow$
$$NO_2^- + H^+ + H_2O$$
Step 2 $NO_2^- + \frac{1}{2} O_2 \longrightarrow NO_3^-$

If sufficient oxygen is available, the rate of each reaction is first-order only in the concentration of the nitrogen reactant, so the sequence can be represented as

$$A \xrightarrow{k_1} B \xrightarrow{k_2} C$$

where A stands for ammonia, B for nitrite ion, and C for nitrate ion, and k_1 and k_2 are the pseudo-first-order rate constants. Since the rate of step 1 depends on the first power of the ammonia concentration, then the rate of disappearance of this species is

$$\frac{d[A]}{dt} = -k_1[A]$$

Since B (nitrite) is produced at this rate by step 1, but is consumed in step 2 by a process whose rate is proportional to the first power of its concentration, we can write

$$\frac{d[B]}{dt} = +k_1[A] - k_2[B]$$

These differential equations can be coupled and integrated to yield the following expressions for the evolution of [A] and [B] with time, relative to $[A]_0$, the original concentration of A:

$$[A]/[A]_0 = e^{-k_1 t}$$
$$[B]/[A]_0 = k_1(e^{-k_1 t} - e^{-k_2 t})/(k_2 - k_1)$$

As can be seen from the solution to Problem 1, the concentration of B (nitrite) rises exponentially at first, reaches a peak value, then declines slowly, if $k_1 > k_2$. Thus there occur significant concentrations of nitrite ion in water undergoing the two-step conversion of ammonia to nitrate.

PROBLEM 1

(a) Derive a general expression relating the time at which [B] reaches a peak to k_1 and k_2.
(b) Draw a graph showing the evolution of [A] and of [B] with time for the values $k_1 = 1$ and $k_2 = 2$.

book can recall visiting a once-popular beach on its north shore in the early 1970s and being repulsed by the sight and smell of dead, rotting fish on the shoreline. Lake Erie's problems stemmed primarily from an excess input of **phosphate ion,** PO_4^{3-} in the waters of its tributaries. The phosphate sources were the polyphosphates in detergents (as explained in detail later), raw sewage, and the runoff from farms that used phosphate fertilizers. Since there is commonly an excess of other dissolved nutrients in lakes, phosphate ion usually functions as the limiting (or controlling) nutrient for algal growth: the larger the supply of the ion, the more abundant the growth of algae, and its growth can be quite abundant indeed. When the vast mass of excess algae eventually dies and starts to decompose by oxidation, the water becomes depleted of dissolved oxygen, with the result that fish life is adversely affected. The lake water also becomes foul-tasting, green, and slimy, and masses of dead fish and aquatic weeds rot on the beaches. The series of changes, including the rapid degradation and aging, that occur when lakes receive excess plant nutrients from their surroundings is called **eutrophication.** When the enrichment arises from human activities, it is called **cultural eutrophication.**

To correct the problem, the United States and Canada in 1972 signed the *Great Lakes Water Quality Agreement.* Since that time, over $8 billion has been spent building sewage treatment plants to remove phosphates from wastewater before it reaches the tributaries and the lake itself. In addition, the levels of phosphates in laundry detergents were restricted in Ontario and in many of the states that border the Great Lakes. The total amount of phosphorus entering Lake Erie has now decreased by more than two-thirds. As a result, Lake Erie has sprung back to life: its once-fouled beaches are regaining popularity with tourists and its commercial fisheries have been revived.

As we have pointed out, the presence of excess phosphate ion in natural waters can have a devastating effect on an aquatic ecology because it overfertilizes plant life. Formerly, one of the largest sources of phosphate as a pollutant was detergents, and in the material that follows, the role of such phosphates is discussed.

The reaction of synthetic detergents with calcium and magnesium ions to form complex ions diminishes the cleansing potential of the detergent. *Polyphosphate ions,* which are anions containing several phosphate units linked by shared oxygens, are added to detergents as *builders* that preferentially form soluble complexes with the metal ions and thereby allow the detergent molecules to operate as cleansing agents rather than being complexed with Ca^{2+} and Mg^{2+} in the water. Another role of the builder is to make the washwater somewhat alkaline, which helps to remove dirt from certain fabrics.

In particular, great quantities of *sodium tripolyphosphate* (STP), $Na_5P_3O_{10}$, were formerly added as the builder in most synthetic detergent formulations. As shown in Figure 10-9a, STP contains a chain of alternating phosphorus and oxygen atoms, with one or two additional oxygens attached to each phosphorus. In solution, one tripolyphosphate ion can form a complex with one calcium ion by forming interactions between three of its oxygen atoms and the metal ion (Figure 10-9b).

Substances like STP that have more than one site of attachment to the metal ion, and thereby produce ring structures that incorporate the metal, are called **chelating agents** (from the Greek word for "claw"). Because several bonds are formed, the resulting chelates are very stable and do not normally

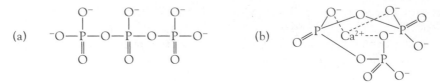

FIGURE 10-9 Structure of the polyphosphate ion: (a) uncomplexed and (b) complexed with calcium ion.

release their metal ions back into a free form. The use of chelating agents to remove metals from the body is discussed in Chapter 11.

Tripolyphosphate ion, $P_3O_{10}^{5-}$, like phosphate ion itself, is a weak base in aqueous solution and thus provides the alkaline environment that is required for effective cleaning:

$$P_3O_{10}^{5-} + H_2O \longrightarrow P_3O_{10}H^{4-} + OH^-$$

Unfortunately, when washwater containing STP is discarded, the excess tripolyphosphate enters waterways, where it slowly reacts with water and is transformed into phosphate ion, PO_4^{3-} (sometimes called ortho-phosphate):

$$P_3O_{10}^{5-} + 2 H_2O \longrightarrow 3 PO_4^{3-} + 4 H^+$$

Note that when tripolyphosphate decomposes, STP behaves as an acid rather than a base (since H^+ is formed in the reaction).

Because of environmental concerns, phosphates are now used only sparingly as builders in detergents in many areas of the world. In Canada and parts of Europe, STP was replaced largely by *sodium nitrilotriacetate* (NTA) (see Figure 10-10a). The anion of NTA acts in a similar fashion to that of STP, chelating calcium and magnesium ions using three of its oxygen atoms and the nitrogen atom (Figure 10-10b). NTA is not used as a builder in the United States because of concerns that its slow rate of degradation might lead to health hazards in drinking water. However, the early experiments with test animals that led to this concern are open to question, as are fears about its persistence and its tendency to solubilize heavy metals into water supplies.

Other builders now used include *sodium citrate*, sodium carbonate (washing soda), and *sodium silicate*. Currently, substances called **zeolites** are also employed as detergent builders. Zeolites are abundant aluminosilicate minerals (see Chapter 12) consisting of sodium, aluminum, silicon, and oxygen. The latter three elements are bonded together to form cages, which the sodium ions can enter. In the presence of calcium ion, zeolites exchange their sodium ions for Ca^{2+} (though not for Mg^{2+}), thereby sequestering it in

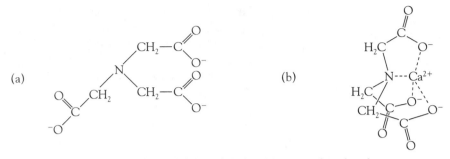

FIGURE 10-10 Structure of the nitrilotriacetate ion: (a) uncomplexed and (b) complexed with calcium ion.

a manner similar to polyphosphates. Like polyphosphates, they also control pH. One disadvantage of the use of zeolites is that they are insoluble, so their use increases the amount of sludge that must be removed at wastewater treatment plants.

Phosphate ion can be removed from municipal and industrial wastewater by the addition of sufficient calcium as the hydroxide, $Ca(OH)_2$, to form insoluble calcium phosphate precipitates such as $Ca_3(PO_4)_2$ and $Ca_5(PO_4)_3OH$ that can then be readily removed. Phosphate removal could be a standard practice in the treatment of wastewater, but it is not yet practiced in all cities. Some policymakers believe that the optimum environmental solution is to use phosphates, rather than some other builder, in detergents and then to efficiently remove them at wastewater treatment plants.

Geographically, phosphate ion enters waterways from both point and nonpoint sources. **Point sources** are specific locations such as factories, landfills, and sewage treatment plants that discharge pollutants. **Nonpoint sources** are large land areas such as farms, logged forests, septic tanks, golf courses and individual domestic lawns, stormwater runoff, and atmospheric deposition. Although each nonpoint source may provide a small amount of pollution, on account of the large number of them involved, they can generate larger *total* quantities than point sources. Therefore, now that sewage treatment plants and detergent controls have been instituted, much of the remaining phosphate arises from nonpoint agricultural sources in many areas.

Green Chemistry: Sodium Iminodisuccinate—A Biodegradable Chelating Agent

Because most chelating agents are not biodegradable or are only slowly biodegradable, they not only place a load on the environment (e.g., phosphates acts as nutrients), but it may be necessary to remove them in wastewater

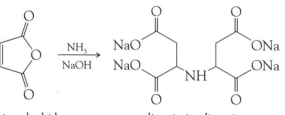

maleic anhydride   sodium iminodisuccinate

FIGURE 10-11 Synthesis and structure of sodium iminodisuccinate, a biodegradable chelating agent.

treatment plants. Unlike many chelating agents, *sodium iminodisuccinate* (IDS, see Figure 10-11) [also known as DL-aspartic-N-(1,2-dicarboxylethyl) tetrasodium salt] readily degrades in the environment. Not only is IDS biodegradable, it is also nontoxic.

IDS can be used as an effective chelating agent for absorption of agricultural nutrients, metal ion scavenging in photographic processing, and groundwater remediation, and as a builder in detergents and household and industrial cleaners. The Bayer Corporation won a Presidential Green Chemistry Challenge Award in 2001 for the development of IDS as a chelating agent and for its synthesis from maleic anhydride (Figure 10-11). This synthesis is accomplished under mild conditions, with water as the only solvent. The excess ammonia that is used is recycled back into the production of more IDS. This synthesis stands in stark contrast to typical syntheses of aminocarboxylate chelating agents, which employ hydrogen cyanide as a reagent. Bayer markets IDS as a chelating agent under the name Baypure.

Reducing the Salt Concentration in Water

The decomposition of organic and biological substances during the secondary phase of wastewater treatment usually results in the production of inorganic salts, many of which remain in the water even after the techniques discussed have been applied. Water can also become salty due to its use in irrigation or because water softener units have been recharged and their discharge disposed of as sewage. Inorganic ions can be removed from water by desalination by using one of the following techniques or by using the precipitation methods mentioned previously.

• **Reverse osmosis**   As previously mentioned, this technique is also used to produce drinking water from salt water, such as seawater (see Figure 10-12).

• **Electrodialysis:**   Here a series of membranes permeable only to either small inorganic cations or small inorganic anions are set up vertically in an alternating fashion within an electrical cell. Direct current is applied across the water, so cations migrate toward the cathode and anions toward the anode. The liquid in alternating zones becomes more concentrated (enriched) or less concentrated (purified) in ions; eventually the ion-concentrated water can be disposed of as brine and the purified water released into the environment. This technology is also used to desalinate seawater for drinking purposes.

- **Ion exchange:** Some polymeric solids contain sites that hold ions relatively weakly, so one type of ion can be exchanged for another of the same charge as it passes by it. Ion exchange resins can be formulated to possess either cationic or anionic sites that function in this manner. The exchange sites of a cationic resin of this type are initially occupied by H^+ ions and those of an anionic exchange resin are occupied by OH^- ions. When water polluted by M^+ and X^- ions is passed sequentially through the two resins, the H^+ ions on the first are replaced by M^+, and then the OH^- ions on the second are replaced by X^-. Thus the water that has passed through contains H^+ and OH^- ions rather than those of the salt, which remain behind in the resins; of course, these two ions immediately combine to form more water molecules. Ion exchange can be used to remove salts, including those of heavy metals, from wastewater.

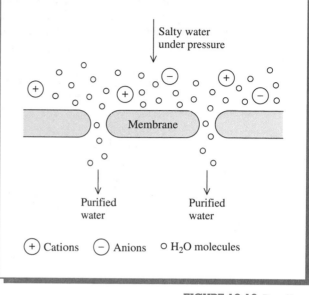

FIGURE 10-12 Desalination of water using reverse osmosis.

PROBLEM 10-13

Water polluted by inorganic ions could be purified by distilling it or by freezing it. Why do you think that such techniques are not used on a mass scale to purify water?

Transition metal cations can be removed from water using either precipitation or reduction techniques, in either case to form insoluble solids. Precipitation of sulfides or hydroxides has already been mentioned; a disadvantage of the latter is the production of a voluminous sludge that must be disposed of in an acceptable manner. Electrolytic reduction of metals leads to their deposition on the cathode. If, instead of the elemental metal, a concentrated aqueous solution of it is desired, the deposited metal can be reoxidized chemically by addition of hydrogen peroxide or electrolytically by reversing the polarity of the cell.

The Biological Treatment of Wastewater and Sewage

An alternative to the processing of wastewater through a conventional treatment plant in small communities is biological treatment in an **artificial**

marsh (also called a *constructed wetland*) that contains plants such as bull-rushes and reeds. The decontamination of the water is accomplished by the bacteria and other microbes that live among the plants' roots and rhizomes. The plants themselves take up metals through their root systems and concentrate contaminants within their cells. In facilities that have been constructed to deal with sewage, primary treatment to filter out solids, etc. in a lagoon is usually implemented before the wastewater is pumped to the marsh, where the equivalent of secondary and tertiary treatment occurs. The plant growth uses up the pollutants and increases the pH—which serves to destroy some harmful microorganisms.

One advantage of biological treatment is that great amounts of sludge are not generated, in contrast to conventional treatments. In addition, it requires neither the addition of synthetic chemicals nor the input of commercial energy. Among the problems of such facilities is decaying vegetation, which must be limited so that the BOD of the processed water does not rise too much, and the fact that the marshes require a great deal of land.

In many rural and small communities, **septic tanks** are used to decontaminate sewage, since central sewage facilities are not available. These underground concrete tanks receive the wastewater, often from only one home. Although solids settle in the tank, grease and oil rise to the top, from which they are periodically removed. The bacteria in the wastewater feed on the bottom sludge, thereby liquefying the waste. Partially purified water flows out of the tank into an underground drain, where further decontamination takes place. The system is relatively passive, compared to central facilities, and, as in the case of artificial marshes, time is required for the processes to occur. In addition, nitrogen compounds are converted to nitrate but the latter is not reduced to molecular nitrogen, so groundwater under the septic system can become contaminated by nitrate.

Drugs in Wastewater from Sewage Treatment Plants

In recent years, scientists have detected trace concentrations of various drugs—prescription, over-the-counter, illegal, and veterinary—in the water leaving sewage treatment plants and in rivers and streams into which this water then flows. The substances—commonly including *estradiol*, *ibuprofen*, the antiepileptic drug *carbamazepine*, and degradation products associated with cholesterol-reducing pharmaceuticals—are present in raw sewage after their excretion in urine or feces from humans and animals and after the disposal of unused or expired medication in toilets.

The Treatment of Cyanides in Wastewater

The **cyanide ion,** CN^-, binds strongly to many metals, especially those of the transition series, and is often used to extract them from mixtures.

Consequently, cyanide is widely used in mining, refining, and electroplating metals such as gold, cadmium, and nickel. Unfortunately, cyanide ion is very poisonous to animal life since it binds strongly to metal ions in living matter, e.g., to the iron in the proteins necessary for molecular oxygen to be utilized by cells.

Cyanide is a very stable species and does not quickly decompose on its own or in the environment. Thus it is an important water pollutant and should be destroyed chemically rather than simply disposed of in a waterway.

We can deduce the type of treatment that will be effective for cyanide by considering its acid–base and redox characteristics. Cyanide ion is the conjugate base of the weak acid HCN, *hydrocyanic acid*, which has limited solubility in water. Thus acidification of cyanide solutions will result in the release of poisonous HCN gas and therefore is not a good solution.

The redox chemistry of cyanide ion can be predicted by considering the oxidation states of the two atoms involved. If nitrogen, the more electronegative atom, is considered to be in its fully reduced -3 state, then the carbon must be in the $+2$ state. Thus one way to destroy cyanide ion is to oxidize the carbon more fully, to the $+4$ state in CO_2 or HCO_3^-. This can be accomplished by dissolved molecular oxygen if high temperatures and elevated air pressures are used:

$$\overset{+2}{2\ CN^-} + \overset{0}{O_2} + 4\ H_2O \longrightarrow \overset{+4\ -2}{2\ HCO_3^-} + 2\ NH_3$$

The use of stronger oxidizing agents, such as Cl_2 or ClO^-, not only oxidizes the carbon from $+2$ to $+4$, but can also oxidize the nitrogen from the -3 state to the zero state of molecular nitrogen:

$$\overset{+2\ -3}{2\ CN^-} + \overset{0}{5\ Cl_2} + 8\ OH^- \longrightarrow \overset{+4}{2\ CO_2} + \overset{0}{N_2} + \overset{-1}{10\ Cl^-} + 4\ H_2O$$

(Four of the ten electrons gained collectively by the chlorines are used by the carbons and six by the nitrogens in this overall process.) Other oxidizing agents that are used in cyanide treatment include **hydrogen peroxide** and/or oxygen, both with a copper salt added as a catalyst. The process can also be carried out electrochemically for high cyanide concentrations; the remaining low concentration of cyanide can be subsequently oxidized by ClO^-.

Sodium cyanide is now being used in some shallow tropical waters such as those in Indonesia to stun reef fish so that they can be captured and sold live as seafood or pets. Unfortunately, the cyanide kills smaller fish and destroys the coral.

PROBLEM 10-14

Deduce the balanced half-reaction when cyanide ion is oxidized in acid media by hypochlorite ion to bicarbonate ion and molecular nitrogen.

PROBLEM 10-15

If an oxidizing agent even more powerful than chlorine or hypochlorite were to be used in the treatment of cyanide, what other possibilities for the nitrogen-containing product would there be?

PROBLEM 10-16

For HCN in water, $K_a = 6.0 \times 10^{-10}$. Calculate the fraction of cyanide that exists as the anion rather than in the molecular form at pH values of 4, 7, and 10.

The Disposal of Sewage Sludge

The **sludge** from both the primary and secondary treatment stages of sewage is principally water and organic matter. It can be digested anaerobically, in a process that takes several weeks to complete. Bacteria in the sludge are not thereby completely eliminated, but the levels are reduced about a thousandfold. The sludge that remains after this further organic decomposition has occurred and after the supernatant water is removed is sometimes incinerated or used as landfill or simply dumped into a landfill or into a water body such as the ocean. However, sludge is high in plant nutrients, so about half the sewage sludge in North America and Europe is spread on farm fields, golf courses, and even residential lawns as low-grade fertilizer sometimes called *biosolid*.

Unfortunately, sewage sludge may contain toxic substances, which potentially could be incorporated into food grown on the land or could contaminate groundwater under the fields. In particular, heavy-metal concentrations often are higher in sewage sludge than in soil, principally because industrial wastes are sometimes released directly into sewage lines shared by households. For example, the lead level in municipal sludge can range from several hundred to several thousand parts per million, compared to an average of about 10 ppm in the Earth's crust. In a few communities an attempt is made to eliminate these materials before final disposal occurs. Some scientists have worried that food crops grown in soil fertilized by sewage sludge may incorporate some of the increased amounts of heavy metals. Control experiments indicate that plants vary greatly in the extent to which they will absorb increased amounts of the metals; e.g., the uptake of lead by lettuce is particularly large, but that by cucumbers is negligible. The concentration of arsenic in agricultural soils is greatly increased if arsenic pesticides are applied to them; crops planted on these soils subsequently absorb some of the adsorbed arsenic. Other substances of concern in using sewage sludge as fertilizer for food are *alkylphenols* from detergents, brominated fire retardants, and pharmaceuticals — especially antibiotics given to farm animals.

Modern Wastewater and Air Purification Techniques

The most important chemical (as opposed to biological) pollutants dissolved in wastewater are usually chloroorganics, phenols, cyanides, and heavy metals. We will now describe some of the high-tech methods that have recently been developed and put into practice to purify wastewater, particularly for removal of chloroorganics. Some of the same techniques are also used to cleanse the compounds from contaminated air.

The Destruction of Volatile Organic Compounds

The major stationary sources in North America of VOCs are the evaporation of organic solvents, the manufacture of chemicals, and the petroleum industry and its storage activities. Wastewater effluent that is contaminated with VOCs, e.g., the water emanating from chemical or petrochemical plants, is commonly treated by a two-step process:

1. The VOCs are removed from the wastewater by **air stripping.** In this process, air is passed upward into a downward stream of the water and the volatile materials are transferred from the liquid to the gas phase. This technique does not work well for compounds that are very water soluble.

2. The resulting VOCs, now present in low concentration in a contained mass of humid air, are destroyed by a process of **catalytic oxidation.** For example, the air, heated to 300–500°C, is passed for a short time over platinum or, depending upon the VOC, some other precious metal that is supported on alumina. The energy costs of this step are very high since it involves heating a large volume of humid air and water vapor. Note that the outlet air from such processes contains HCl if the VOCs originally contained chlorine; this compound must be removed by scrubbing with a basic substance before the air is released into the atmosphere.

The removal of VOCs from *gaseous* industrial emissions usually operates by this same catalytic oxidation process; typically the concentration of VOCs in the air stream is thereby reduced by 95%. A primary heat exchanger recovers and reuses the VOCs' heat of combustion to warm incoming gases to the operating temperature.

The **adsorption** of compounds onto activated carbon (see Box 10-1) or onto synthetic carbonaceous adsorbents is a cost-effective technology used for the removal of low-level VOC concentrations from both liquid and vapor streams, and is also useful for nonvolatile organic compounds. These adsorbents can be easily regenerated by treatment with steam or by other thermal techniques as well as by solvents; the concentrated pollutants can be subsequently destroyed by catalytic oxidation.

Advanced Oxidation Methods for Water Purification

Conventional water purification methods often do not successfully deal with synthetic organic compounds such as chloroorganics that are dissolved at low concentrations. Examples include the common groundwater pollutants trichloroethene and perchloroethene discussed earlier in this chapter. The conventional method for the treatment of water containing such pollutants is adsorption of the chloroorganics onto activated carbon; this removes the compounds but doesn't destroy them. The wastewater from pulp and paper mills also contains organochlorines that are resistant to conventional treatments.

In order to cleanse water of these extra-stable organics, so-called **advanced oxidation methods** (AOMs) have been developed and deployed. The aim of these methods is to **mineralize** the pollutants, i.e., to convert them entirely to CO_2, H_2O, and mineral acids such as HCl. Most AOMs are ambient-temperature processes that use energy to produce highly reactive intermediates of high oxidizing or reducing potential, which then attack and destroy the target compounds. The majority of the AOMs involve the generation of significant amounts of the **hydroxyl free radical,** OH, which in aqueous solution is a very effective oxidizing agent, as it is in air (see Chapters 2 and 3). The hydroxyl radical can initiate the oxidation of a molecule by extraction of a hydrogen atom or by addition to one atom of a multiple bond as it does in air (Chapter 3); in water, as an additional alternative, it can also extract an electron from an anion.

Since the generation of OH in solution is a relatively expensive process, it is economical to use AOMs to treat only the wastes that are resistant to the cheaper, conventional treatment processes. Thus integrating an AOM with pretreatment of the wastewater by biological or other processes to first dispose of the easily oxidized materials is often appropriate.

Ultraviolet (UV) light is often used to initiate the production of hydroxyl radicals and thus to begin the oxidations. Commonly hydrogen peroxide, H_2O_2, is added to the polluted water and UV light from a strong source in the 200- to 300-nm range is shone on the solution. The hydrogen peroxide absorbs the ultraviolet light (especially that closer to 200 nm than to 300 nm) and uses the energy obtained to split the O—O bond, resulting in the formation of two OH radicals:

$$H_2O_2 \xrightarrow{\text{UV}} 2\ OH$$

Alternatively and less commonly, ozone is produced and photochemically decomposed by UV. The resulting oxygen atom reacts with water to efficiently produce OH via the intermediate production of hydrogen peroxide, which is photolyzed:

$$O_3 \xrightarrow{\text{UV}} O_2 + O$$

$$O + H_2O \longrightarrow H_2O_2 \xrightarrow{\text{UV}} 2\ OH$$

A fraction of the oxygen atoms produced by ozone photolysis are electronically excited, and these react with water to directly produce hydroxyl radicals, as discussed in Chapter 1.

PROBLEM 10-17

Given that the enthalpies of formation for H_2O_2 and OH are, respectively, -136.3 and $+39.0$ kJ/mol, calculate the heat energy required to dissociate one mole of hydrogen peroxide into hydroxyl free radicals. What is the maximum wavelength of light that could bring about this transformation? [Hint: see Chapter 1.] Given that light of 254-nm wavelength is usually used, and that all the energy of each photon that is in excess of that required to dissociate one molecule is lost as waste heat, calculate the maximum percentage of the input light energy that can be used for dissociation itself.

Hydroxyl radicals for wastewater treatment can also be efficiently produced *without* the use of UV light by combining hydrogen peroxide with ozone. The chemistry of the intermediate processes here is complex, but the overall reaction between the two species is

$$H_2O_2 + 2\ O_3 \longrightarrow 2\ OH + 3\ O_2$$

This *ozone–H_2O_2 method* is more cost-effective and is easier to adapt to existing water treatment systems than is any other AOM system.

It is also possible to generate the hydroxyl radical electrolytically. In most such applications, a metal ion (such as Ag^+ or Ce^{3+}) is first oxidized to a more positively charged ion (Ag^{2+} or Ce^{4+} in our examples) that will subsequently oxidize water to H^+ and OH.

The biggest liability associated with advanced oxidation processes is that they produce toxic chemical by-products. For example, in the ozone–peroxide and peroxide–UV treatments of groundwater contaminated with trichloroethene and perchloroethene, the intermediates *trichloroacetic acid* and *dichloroacetic acid* are formed in about 1% yield.

Photocatalytic Processes

Another innovative technology for wastewater treatment involves the irradiation by UV light of solid semiconductor photocatalysts such as **titanium dioxide,** TiO_2, small particles of which are suspended in solution. Titanium dioxide is chosen as the semiconductor for such applications since it is nontoxic, is very resistant to photocorrosion, is cheap and plentiful, has a band gap in the UV-A region, and can be used at room temperature. Irradiation at wavelengths less than 385 nm produces electrons, e^-, in the conduction band and *holes,* h^+, in the valence band of the metal oxide. The holes in the valence band of the semiconductor can react with surface-bound hydroxide ions or with water molecules, thereby producing hydroxyl radicals in both cases:

$$h^+ + OH^- \longrightarrow OH$$
$$h^+ + H_2O \longrightarrow OH + H^+$$

The holes can also react directly with adsorbed pollutants, producing radical cations that readily engage in subsequent degradation reactions.

Normally, O_2 molecules dissolved in the water react with the electron produced at the semiconductor surface, a process that eventually produces more reactive free radicals, but is relatively slow. If hydrogen peroxide is added to the water instead, it will react with the electron to form the anion radical and generate reactive radicals more quickly.

The cost of the electrical energy required to generate the needed UV light is usually the major expense in operation of AOM systems. On this basis, the titanium dioxide methods are even less cost-effective than those described previously, since considerably more electricity is required per pollutant molecule destroyed. Sunlight could be used to supply the UV light, but only about 3% of its light lies in the appropriate range and is absorbed by the solid. Another problem with the TiO_2 processes is the difficulty of separating the various reactants and products from the TiO_2

particles if the metal oxide has been used in the form of a fine powder. However, there now are closed systems in which the titanium dioxide slurry is efficiently separated from the purified water and recycled back to the inlet stream.

Some scientists have experimented with immobilizing TiO_2 as a thin film (1 μm thick) on a solid surface such as glass, tile, or alumina. Indeed, TiO_2-coated tiles are now used on walls and floors in some buildings. The low-level UV light from fluorescent lighting in such rooms is sufficient to allow the destruction of gaseous and liquid-phase pollutants that touch the oxide on the tiles! For example, odors that upset people are usually present in air at concentrations of only about 10 ppm; at such levels, the UV from normal fluorescent lighting should be sufficient to remove them by TiO_2 photocatalysis. Bacterial infections such as those that cause many secondary infections in hospitals can also be eliminated by spraying walls and floors (in rooms lit by fluorescent bulbs) to give them a titanium dioxide film. Photocatalysts are quiet, unobtrusive cleansing materials.

Other Advanced Oxidation Methods

A process called **direct chemical oxidation** has been proposed for the destruction of solid and liquid organic wastes in the aqueous phase, particularly in environments such as those under buildings, where the light required for UV processes cannot conveniently be supplied. It uses one or another of the strongest known chemical oxidants, e.g., acidified **peroxydisulfate anion,** $S_2O_8^{2-}$, under ambient pressure and moderate temperatures to oxidize the wastes. Such a process needs no catalysts and produces no secondary wastes of concern. The sulfate that results from the peroxydisulfate can be recycled back to the oxidant. Other very strong oxidizing agents that have been tested are the *peroxymonosulfate anion*, HSO_5^-, and the *ferrate ion*, FeO_4^{2-}; in the latter, iron exists in the +6 oxidation state, so it is not surprising that it is a strong oxidizing agent. Unfortunately, ferrate ion suffers from the problem of instability.

PROBLEM 10-18

Deduce the balanced half-reaction (acidic media) in which the peroxydisulfate ion is converted into sulfate ion. Repeat the exercise for the conversion of oxalic acid, $C_2H_2O_4$, into carbon dioxide. Combine these half-reactions into a balanced equation and calculate the volume of 0.010 M peroxydisulfate that is required to oxidize 1 kg of oxalic acid.

Review Questions

1. Describe the function of (a) aeration and (b) addition of aluminum or iron sulfate in the purification of drinking water.

2. Describe the chemistry underlying the removal of excess calcium and magnesium ions from drinking water.

3. Describe how water can be disinfected by (a) membrane filtration and (b) ultraviolet irradiation.

4. What two other chemical methods, other than chlorination, are used to disinfect water? What are some advantages and disadvantages of these alternatives?

5. Explain the chemistry underlying the disinfection of water by chlorination. What is the active agent in the destruction of the pathogens? What are the practical sources of the active ingredient?

6. Explain why pH control of water in swimming pools is important. What compounds are formed when the chlorinated water reacts with ammonia?

7. Discuss the advantages and disadvantages of using chlorination to disinfect water, including the nature of the THM compounds.

8. What is meant by the terms *groundwater* and *aquifer*? How does the *saturated zone* of soil differ from the *unsaturated*?

9. Why did concern about groundwater pollution lag far behind that about surface water?

10. Name three important sources of nitrate ion in groundwater.

11. Construct a table that shows the common oxidation states for nitrogen. Deduce in which column the following environmentally important compounds belong: HNO_2, NO, NH_3, N_2O, N_2, HNO_3, and NO_3^-. Which of the species become prevalent in aerobic conditions in a lake? under anaerobic conditions? What is the oxidation state of nitrogen in NH_2OH?

12. Explain why excess nitrate in drinking water or food products can be a health hazard; include the relevant balanced chemical reaction showing how nitrate becomes reduced.

13. What is an *N-nitrosamine*? Write the structure and the full name for NDMA.

14. Define *leachate*.

15. Name two types of organic contaminants found in groundwater and give two examples of each type.

16. Explain the difference in vertical location in an aquifer between compounds such as chloroform and toluene.

17. Define the term *plume* and describe how it forms in an aquifer.

18. Why are the BTX and MTBE components of gasoline the ones that are most often found in groundwater? Are both components easily biodegraded?

19. What is meant by *reductive degradation*? Describe the in situ technique by which chloro-organics in aquifers can be destroyed by reductive dechlorination.

20. What procedures are involved in primary wastewater treatment? in secondary treatment?

21. List five possible water purification processes that are associated with the tertiary treatment of wastewater, including one that removes phosphate ion.

22. What polyphosphate is commonly used in detergents, and why does its use lead to environmental problems? What are the other main sources of phosphate in natural waters? What other builders are used in detergents?

23. Describe two important methods that are used to desalinate wastewater.

24. Describe the chemical processes by which cyanide ion can be removed from wastewater.

25. Describe how VOCs dissolved in wastewater are usually removed and destroyed.

26. What does AOM stand for? State the most common reactive agent in such

processes. Describe three methods by which this reactive species can be generated.

27. Describe two photocatalytic methods that can destroy organic wastes.

28. What is *direct chemical oxidation*? What are two of the strong oxidizing agents that can be used for such procedures?

Green Chemistry Questions

1. What are the environmental advantages of using iminodisuccinate compared to most chelating agents?

2. The development of iminodisuccinate by Bayer won a Presidential Green Chemistry Challenge Award.

(a) Which of the three focus areas (see the Introduction to Green Chemistry) for these awards does this award best fit into?

(b) List at least three of the twelve principles of green chemistry (see the Introduction to Green Chemistry) that are addressed by the green chemistry developed by Bayer.

Additional Problems

1. Given that $K_a = 2.7 \times 10^{-8}$ for HOCl, deduce the fraction of a sample of the acid in water that exists in the molecular form at pH values (predetermined by the presence of other species) of 7.0, 7.5, 8.0, and 8.5. [Hint: Derive an expression that relates the fraction of HOCl that is ionized to the concentration of hydrogen ions.] Would it be a good idea to allow the pool water's pH to rise to 8.5?

2. The equilibrium constant for the reaction of dissolved molecular chlorine with water to give hydrogen ions, chloride ions, and HOCl is 4.5×10^{-4}, where as usual the concentration of water is included in the K value. If the pH of the solution is determined by other processes, so that the amount of hydrogen ion contributed by the chlorine reaction is negligible, calculate the fraction of the original 50 ppm chlorine that remains as Cl_2 at pH values of 0, 1, and 2. [Notes: (1) The

dissociation of HOCl into ions is negligible at these low pH values. (2) Approximate solutions to the quadratic equation involved in these calculations will not be accurate due to the high percentage of reaction.]

3. Calculate the volume of $Ca_5(PO_4)_3OH$, the density of which is 3.1 g/mL, that is produced for each gram of sodium tripolyphosphate present in a detergent when it is removed in tertiary wastewater treatment. Estimate the annual mass of detergent used for laundry purposes for a typical household of four persons and, assuming that the phosphate levels in laundry detergents used to be about 50%, calculate the volume of $Ca_5(PO_4)_3OH$ that was required annually to dispose of its waste laundry phosphate.

4. Calculate the oxidation number of the chlorine in molecular chlorine, HOCl, chlorine dioxide, monochloramine (see Problem

10-3), and sodium chloride. Given that the last item is the stablest form of chlorine, predict whether the other substances mentioned are likely to be oxidizing agents or reducing agents, and rank them in likely order of this redox behavior. Using this analysis and the section on chlorine compounds in your introductory chemistry textbook, suggest other compounds that might be useful in disinfecting water.

5. Given their names, can you deduce the nature of the similarity in molecular structure between hydrogen peroxide and the sulfur compounds that are used in direct chemical oxidation methods? By calculating the oxidation state of the atoms in hydrogen peroxide, deduce why it can act as an oxidizing agent.

6. What could be done to dispose of the solvents that are used to extract VOCs from adsorbents?

7. Write the initial reaction step that occurs if methylchloroform, CH_3CCl_3, were to be destroyed by **(a)** reductive degradation and **(b)** hydroxyl radical attack.

8. Water samples from three wastewater streams were analyzed and the important pollutants determined to be those listed below. In each case, devise economical, practical processes (other than activated carbon treatment) for purifying the water of the pollutants: **(a)** phosphate ion, ammonium ion, and salt (in water containing bicarbonate ion); **(b)** nitrite ion, PCE, and Fe(II); and **(c)** cadmium ion, carbon tetrachloride, and glucose.

9. In the oxidation of cyanide ion, whether chemically or electrolytically, the cyanate ion, CNO^-, is an intermediate. Indeed, in some operations the oxidation is only forced this far and therefore cyanate is the final product. **(a)** Deduce the half-reactions in alkaline solution that (1) converts cyanide ion to cyanate ion and (2) oxidizes cyanate to carbonate ion and

molecular nitrogen. **(b)** Deduce suitable Lewis structures for the cyanate ion and for its isomer, the isocyanate ion, OCN^-. **(c)** By consulting handbooks containing toxicity data and solubility products, decide whether the conversion of cyanide to cyanate increaes or decreases the toxicities and water solubilities of the ions.

10. Write the reaction that occurs between hydroxyl radical and carbonate ion dissolved in water. What are two alternative substances that you could add to the water to decrease the carbonate ion concentration, one that operates by elimination of carbon dioxide and the other by precipitation of the ion, to decrease the amount of hydroxyl radical destroyed by this reaction? Given the solubility product constants quoted in this chapter, estimate the lowest practical carbonate concentration you could reach by this process, and comment on the desirability of the ions that you have introduced into the water.

11. In order for efficient disinfection to occur, treated drinking water should contain 0.3 to 0.5 mg/L of available Cl_2 (which will mostly be converted to HOCl). Given that the Henry's law constant for Cl_2 in water is 8.0×10^{-3} M/atm, what pressure of $Cl_2(g)$ would be required to obtain a concentration of 0.5 mg/L (assuming a negligible amount of available chlorine lost by reactions other than formation of HOCl)?

12. Hypochlorous acid is a much more effective disinfecting agent in its neutral, undissociated form. Given that the K_a of hypochlorous acid is 3.5×10^{-8}, determine the pH of 1.00 and 0.100 M hypochlorous acid solutions and the percentage dissociation in each case. Determine the percentage dissociation of 0.100 M hypochlorous acid in buffered solutions at pH 7.0 and pH 10.0.

13. As mentioned in the chapter, an important application of reverse osmosis relevant to the provision of safe drinking water is its use in the desalination of seawater. To extract pure water from seawater using a semipermeable membrane, pressure must be exerted on the seawater to exceed its osmotic pressure. Osmotic pressure, π, is a colligative property, which depends only on the *total* molar concentration of solutes (M) and is given by the equation $\pi = MRT$. The composition of seawater can be defined in terms of the amounts of the various salts it contains. The five most abundant salts and their amounts in kilograms per cubic meter of seawater are: $NaCl$, 28.014; $MgCl_2$, 3.812; $MgSO_4$, 1.752; $CaSO_4$, 1.283; and K_2SO_4, 0.816 [G. Neuman and W. J. Pierson, Jr., *Principles of Physical Oceanography* (Englewood Cliffs, NJ: Prentice-Hall, 1966)]. Assuming that other salts present in seawater (e.g., $CaCO_3$ and KBr), as well as molecular solutes, are present in negligible amounts relative to the top five salts, determine the pressure that must be exerted on seawater to desalinate it using reverse osmosis at 20°C.

Further Reading

1. S. L. Postel, G. C. Daily, and P. R. Ehrlich, "Human Appropriation of Renewable Fresh Water," *Science* 271 (1996): 785.

2. F. Bove et al., "Drinking Water Contaminants and Adverse Pregnancy Outcomes: A Review," *Environmental Health Perspectives* 110 (supplement 1): 61.

3. U. van Gunten, "Ozonation of Drinking Water, Parts I and II," *Water Research* 37 (2003): 1443, 1469.

4. S. D. Richardson et al., "Identification of New Ozone Disinfection Byproducts in Drinking Water," *Environmental Science and Technology* 33 (1999): 3368.

5. A. Kolch, "Disinfecting Drinking Water with UV Light," *Pollution Engineering* (October 1999): 34.

6. B. T. Nolan et al., "Risk of Nitrate in Groundwaters of the United States—A National Perspective," *Environmental Science and Technology* 31 (1997): 2229.

7. J. A. MacDonald, "Evaluating Natural Attenuation for Groundwater Cleanup," *Environmental Science and Technology* (1 August 2000): 346A.

8. L. W. Canter, R. C. Knox, and D. M. Fairchild, *Groundwater Quality Protection* (Boca Raton, FL: Lewis Publishers, 1987).

9. E. K. Nyer, *Groundwater Treatment Technology*, 2nd ed., (New York: Van Nostrand Reinhold, 1992).

10. D. M. Mackay and J. A. Cherry, "Groundwater Contamination: Pump-and-Treat Remediation," *Environmental Science and Technology* 23 (1989): 630.

11. "Environmental Processes '96: A Special Report," *Hydrocarbon Processing Magazine* (International Ed.) 75 (1996): 85 [reviews many emerging technologies that can handle water and air pollution problems].

12. D. Simonsson, "Electrochemistry for a Cleaner Environment," *Chemical Society Reviews* 26 (1997): 181.

13. N. C. Baird, "Free Radical Reactions in Aqueous Solutions: Examples from Advanced Oxidation Processes for Wastewater and from the Chemistry in Airborne Water Droplets," *Journal of Chemical Education* 74 (1997): 817.

Websites of Interest

Log on to www.whfreeman.com/envchem3e/ and click on Chapter 10.

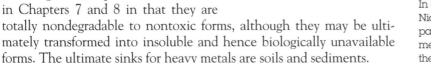

CHAPTER 11

Toxic Heavy Metals

Introduction

In chemistry, **heavy metal** refers, not to a type of rock music, but to a type of chemical element, many of which are poisonous to humans. The five main ones discussed in this chapter—**mercury** (Hg), **lead** (Pb), **cadmium** (Cd), **chromium** (Cr), and **arsenic** (As)—present the greatest environmental hazard due to their extensive use, their toxicity, and their widespread distribution. Each one has been found at toxic levels in certain locales in recent times. Metals differ from the toxic organic compounds we discussed in Chapters 7 and 8 in that they are totally nondegradable to nontoxic forms, although they may be ultimately transformed into insoluble and hence biologically unavailable forms. The ultimate sinks for heavy metals are soils and sediments.

The heavy metals occur near the middle and bottom of the periodic table. Their densities are high compared to those of other common materials. The densities of the metals of interest here are listed in Table 11-1, as are the values for water and two common "light" metals for comparison.

Although we commonly think of heavy metals as water pollutants, they are for the most part transported from place to place via the air, either as gases or as species adsorbed on, or absorbed in, suspended particulate matter. Thus, for example, over half of the heavy-metal input into the Great Lakes is due to deposition from air. Evidence from Sweden indicates that the deposition of lead into European lake sediments dates back to the time of the ancient Greeks, when silver was first mass-produced for use in coins. Apparently the rather substantial

In developing countries such as Nicaragua and Brazil, workers pan for gold by adding liquid mercury to river sediments and then boil off the mercury once it has extracted the gold. The environmental mercury pollution that results can devastate human health. (Peter Chartrand/ D. Donne Bryant Stock)

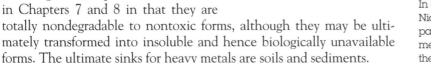

TABLE 11-1	Densities of Some Important Heavy Metals and Other Substances
Substance	Density (g/cm^3)
Hg	13.5
Pb	11.3
Cu	9.0
Cd	8.7
Cr	7.2
Sn	5.8–7.3
As	5.8
Al	2.7
Mg	1.7
H_2O	1.0

amount of lead contaminanting the crude silver escaped into the air during the refining of the metal.

Speciation and the Toxicity of Heavy Metals

Although mercury *vapor* is highly toxic, the heavy metals Hg, Pb, Cd, Cr, and As are not particularly toxic as the *condensed* free elements. However, all five are dangerous in the form of their cations, and most are also highly toxic when bonded to short chains of carbon atoms. Biochemically, the mechanism of the toxic action arises from the strong affinity of the cations for sulfur. Thus, **sulfhydryl groups,** —SH, which occur commonly in the enzymes that control the speed of critical metabolic reactions in the human body, readily attach themselves to ingested heavy-metal cations or to molecules that contain the metals. Because the resultant metal–sulfur bonding affects the entire enzyme, it cannot act normally and, as a result, human health is adversely affected, sometimes fatally. The reaction of heavy-metal cations M^{2+} (where M is Hg, Pb, or Cd) with the sulfhydryl units of enzymes, R—S—H, to produce stable systems such as R—S—M—S—R is analogous to their reaction with the simple inorganic chemical H_2S, with which they yield the insoluble solid MS.

PROBLEM 11-1

Write the balanced chemical reactions that correspond to the reaction of an M^{2+} ion (a) with H_2S and (b) with R—S—H to produce hydrogen ions and the products mentioned in the text.

A common medicinal treatment for acute heavy-metal poisoning is the administration of a compound that binds to the metal even more strongly than does the enzyme; subsequently the metal–compound combination is solubilized and excreted from the body. One compound used to treat mercury and lead poisoning is *British Anti-Lewisite* (BAL); its molecules contain two —SH groups that together capture the metal:

$$\begin{array}{ccc} CH_2 & CH & CH_2 \\ | & | & | \\ OH & SH & SH \end{array}$$

British Anti-Lewisite

Also useful for this purpose is the calcium salt of *ethylenediaminetetraacetic acid* (EDTA), a well-known compound that extracts and solubilizes most metal ions. The metal ions are complexed by the two nitrogens and the charged oxygens to form a chelate (Chapter 10). The metal-containing chelate is subsequently excreted from the body:

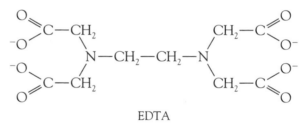

EDTA

Treatment of heavy-metal poisoning by chelation therapy is best begun early, before neurological damage has occurred. The calcium rather than the sodium salt is used so that calcium ion is not inadvertently leached from the body by the EDTA.

The toxicity of the five heavy metals depends very much on the chemical form of the element, i.e., on its **speciation.** Substances that are almost completely insoluble pass through the human body without doing much harm. The most devastating forms of the metals are

• those that cause immediate sickness or death (e.g., a sufficiently large dose of arsenic oxide) so that therapy cannot exert its effects in time, and

• those that can pass through the membrane protecting the brain—the blood–brain barrier—or the placental barrier that protects the developing fetus.

For mercury and lead, the forms that have alkyl groups attached to the metal are highly toxic. Because these compounds are covalent molecules, they are soluble in animal tissue and they can pass through biological membranes, whereas charged ions are less able to do so; e.g., the toxicities of lead as the ion Pb^{2+} and in covalent molecules differ substantially.

The toxicity of a given concentration of a heavy metal present in a natural waterway depends not only on its speciation, but also on the water's pH and on the amounts of dissolved and suspended organic matter in it, since interactions such as complexation and adsorption may well remove some of the metal ions from potential biological activity.

Bioaccumulation of Heavy Metals

Recall from Chapter 7 that some substances display the phenomenon of bio-magnification: their concentrations increase progressively along an ecological food chain. The only one of the five heavy metals under consideration that is indisputably capable of doing this is mercury. Many aquatic organisms do, however, bioconcentrate (but not biomagnify) heavy metals. For example, oysters and mussels can contain levels of mercury and cadmium that are 100,000 times greater than those in the water in which they live.

The concentrations of most heavy metals that humans encounter in drinking water are usually small and cause no direct health problems; however, exceptions do occur and will be discussed later. As is the case with toxic organic chemicals, the amounts of metals that are ingested through the food supply are usually of much greater concern than the intake attributable to drinking water. Paradoxically, the heavy metals in the fish that we ingest usually originate in fresh water.

Mercury

Mercury Vapor

Elemental mercury is employed in hundreds of applications, many of which (e.g., electrical switches) take advantage of its unusual property of being a liquid that conducts electricity well. In automobiles built before about 2000, electrical switches that operate convenience and trunk lighting contained mercury, as did instrument panels and antilock brakes; all of this mercury is lost to the environment when the cars are recycled for their steel unless the element is specifically collected, as is required in a few U.S. states.

Mercury is used in fluorescent light bulbs and in the mercury lamps employed for street lighting, since energized mercury atoms emit light in the visible wavelength region. In view of the contamination of the environment when mercury lamps are broken, there has recently been a shift toward the use of sodium vapor lamps, which present a lower toxicity hazard and are more efficient light sources. In addition, the mercury content of fluorescent lamps has been reduced by about 80% since the mid-1980s, down to about 10 mg each today.

Mercury is the most volatile of metals, and its vapor is highly toxic. Adequate ventilation is required whenever mercury is used in closed quarters, since the equilibrium vapor pressure of mercury is hundreds of times the maximum recommended exposure. Mercury vapor consists of free, neutral atoms. It diffuses from the lungs into the bloodstream, and then, because it is electrically neutral, it readily crosses the blood–brain barrier to enter the brain, where it is transformed to the +2 cation. The result is serious damage to the central nervous system, which is manifested by difficulties with coordination, eyesight, and tactile senses. Liquid mercury itself is not highly toxic, and most of that ingested is excreted. Nevertheless, children should not be allowed to play with droplets of the metal because of the danger from breathing the vapor.

The anthropogenic sources of atmospheric mercury now rival the natural input from volcanoes, formerly the predominant source of airborne mercury. Emissions of mercury from large industrial operations in developed countries have been successfully curtailed, but this represented only a small part of the problem, since the element is released from thousands of nonpoint sources. Over the 1990s the concentration of airborne mercury increased by about 1.5% per year, notwithstanding the decline in industrial emissions. Large amounts of mercury vapor are released into the air as a result of the unregulated burning of coal and fuel oil, both of which always contain trace amounts of the element (reaching several hundred ppm in some coals), and of incinerating municipal waste that contains mercury in products such as batteries. Mercury thermometers should be disposed of at a hazardous waste facility rather than being included with domestic garbage. Currently, coal-fired power plants and municipal and medical waste incinerators are the biggest sources of mercury emissions to the atmosphere in North America. The vaporized mercury is eventually oxidized and returns in rain and snow, often falling far from the site of the original emissions. The U.S. EPA proposed regulations in 2003 requiring coal-fired power plants to install pollution equipment that would remove mercury from exhaust gases before they are emitted into the air.

Recently a technique has been developed by which mercury and other gases such as nitrogen and sulfur oxides in power-plant exhaust can be oxidized in the emission stack and thereby converted to aerosols that can be collected by electrostatic precipitators. Widespread adoption of this technique could greatly reduce mercury emissions from coal-fired power plants in the future.

In air, the greater part of the elemental mercury is in the vapor (gaseous) state, with only a tiny fraction bound to airborne particles. Airborne gaseous mercury can travel long distances before being oxidized and then dissolving in rain and subsequently being deposited on land or in water bodies. This global cycling of mercury results in its being distributed even to remote parts of the planet.

Mercury Amalgams

Mercury readily forms **amalgams,** which are solutions or alloys, with almost any other metal or combination of metals. The dental amalgam that has been used to fill cavities in teeth for more than 150 years initially has a putty-like consistency and is prepared by combining approximately equal proportions of liquid mercury and a solid mixture that is mainly silver with variable amounts of copper, tin, and zinc. The slight expansion of volume that accompanies its solidification ensures that the final amalgam fills the cavity. When a filling is first placed in a tooth and whenever it is involved in the chewing of food, a tiny amount of the mercury is vaporized. Some scientists believe that mercury exposure from this source causes long-term health problems in some individuals, but an expert panel of the U.S. National Institutes of Health concluded that dental amalgams do not pose a health risk. A very recent study of adults found that no measure of exposure to mercury—either the level of the element in the urine or the number of dental fillings—correlated with any measure of mental functioning or fine-motor control.

Some countries in Europe, such as Germany, have banned the use of mercury in fillings, at least for pregnant women and small children. Mercury-free "amalgams" for use in dentistry are under development; porcelain fillings already are common, though expensive. Some fears have been expressed about the release of elemental mercury vapor into the atmosphere when human bodies that have amalgam-filled teeth are cremated, since the amalgam decomposes at high temperatures. In countries such as Sweden, crematoria are fitted with selenium filters that remove most mercury from emissions by forming mercury selenide crystals.

In some areas dentists are now required to install a separator to capture mercury from their wastewater rather than have it flow through drains and become part of municipal sewage. On average, each dentist produces about 1 kg of mercury waste per year. Dentists collectively release about the same amount of the metal as is emitted by coal-fired power plants. Because of the control of emissions by dentists, the sewage sludge used by farmers as fertilizer (Chapter 12) will have a much lower concentration of mercury in the future.

In working some ore deposits, very small amounts of elemental gold or silver are extracted from much larger amounts of the denser particles of soil or sediment by adding elemental mercury to the mixture. The mercury extracts the gold or silver by forming an amalgam, which is then roasted to distill off the mercury. From 1570 until about 1900 this process was used to extract silver from ores in Central America and South America. About 1 g of mercury was lost to the environment for every gram of silver produced, resulting in the release of almost 200,000 tonnes of mercury. The mercury was shipped to these regions from Almaden, Spain, and from Peru.

Today, the gold extraction procedure is carried out on a large scale in Brazil to obtain gold from muddy sediments, and it results in substantial mercury pollution both in the air and, because of careless handling practices, in the Amazon River itself. The health hazards to workers using processes that involve the vaporization of mercury are significant, since the element is so toxic in its gaseous form. In addition, mercury in surface sediments disturbed by deforestation in the region enters the aquatic environment, where some of it is methylated and enters the food chain. Initiatives have been undertaken by the European Union to incorporate inexpensive technology into the process to prevent the massive release of mercury into the air and to the Amazon River during the extraction of gold. Unfortunately, mercury is also being released into the river in comparable amounts as a result of the slash-and-burn farming practiced in the area.

Mercury and the Industrial Chlor-Alkali Process

An amalgam of sodium and mercury is used in some industrial chlor-alkali plants in the process that converts aqueous sodium chloride into the commercial products *chlorine* and *sodium hydroxide* (and hydrogen) by electrolysis. In order to form a concentrated, pure solution of NaOH, flowing mercury is used as the negative electrode (cathode) of the electrochemical cell. The metallic sodium that is produced by reduction in the electrolysis combines with the mercury and is removed from the NaCl solution without having reacted in the aqueous medium:

$$Na^+(aq) + e^- \xrightarrow{\text{Hg}} Na \text{ (in Na–Hg amalgam)}$$

When metals such as sodium are dissolved in amalgams, their reactivity is greatly lessened compared to that in the free state, so that the otherwise highly reactive elemental sodium in the Na–Hg amalgam does not react with the water in the original solution. Instead, the amalgam is removed and later induced, by the application of a small electrical current, to react with water in a separate chamber, producing sodium hydroxide that is free of NaCl.

The mercury is recovered after NaOH production and is recycled back to the original cell. The recycling of mercury is not complete, however, and some finds its way into the air and into the river from which the plant's cooling water is obtained and to which it is returned. Although liquid mercury is soluble neither in water nor in dilute acid, it can be oxidized to a soluble form by the intervention of bacteria that are present in natural waters. By this means the mercury becomes accessible to fish.

The mass of mercury lost to the environment from the average chlor-alkali plant has decreased enormously since the problem was identified in the 1960s.

Nevertheless, installations in North America that use mercury electrodes have largely been phased out. They were replaced by those that use a fluorocarbon membrane that separates the NaCl solution from the chloride-free solution at the negative electrode. The membrane is designed so that Na^+, but not anions, can pass through it. In both types of cells the overall reaction is

$$2\,NaCl(aq) + 2\,H_2O(l) \longrightarrow 2\,NaOH(aq) + Cl_2(g) + H_2(g)$$

In 1969 fish from Lake St. Clair (which borders Michigan and Ontario) were found to have elevated levels of mercury, presumably as a result of mercury discharges from a nearby chlor-alkali plant.

The 2+ Ion of Mercury

Like its partners zinc and cadmium in the same subgroup of the periodic table, the common ion of mercury is the 2+ species, Hg^{2+}, the **mercuric** or **mercury(II)** ion. A compound containing the mercuric ion is the ore HgS, i.e., $Hg^{2+}S^{2-}$. Like most sulfides, this salt is very insoluble in water; indeed, the wastewater at chlor-alkali plants is sometimes treated by adding a soluble salt, such as Na_2S that contains the *sulfide ion*, to precipitate the mercury as HgS:

$$Hg^{2+} + S^{2-} \longrightarrow HgS(s)$$

PROBLEM 11-2

The solubility product, K_{sp}, for HgS is 3.0×10^{-53}. Calculate the solubility of HgS in water in moles per liter and transform your answer into the number of mercuric ions per liter. According to this calculation, what volume of water in equilibrium with solid HgS contains a single Hg^{2+} ion?

In natural waters, much of the Hg^{2+} is attached to suspended particulates and is eventually deposited in sediments. This topic is considered in further detail when soil and sediment chemistry is discussed (Chapter 12).

The nitrate salt of Hg^{2+} is water soluble and was once used to treat the fur employed at that time to make felt for hats. The fur was immersed in a hot solution of *mercuric nitrate*, which made the fibers rough and twisted so they would mat together easily. As a consequence of this constant exposure to mercury, workers in the felt trade often displayed nervous disorders: muscle tremors, depression, memory loss, paralysis, and insanity. This gave rise to the expression "mad as a hatter," a phrase familiar to fans of Lewis Carroll's *Alice in Wonderland*. Mercury vapor, and to a lesser extent mercury salts, attack the central nervous system, but the main target organs for Hg^{2+} are the kidney and the liver, where it can cause extensive damage.

Mercuric oxide, HgO, is present in a paste in *mercury cell* batteries such as those used in hearing aids. If the discarded spent batteries are subsequently incinerated as garbage, the volatile mercury can be released into the air. The amount of mercury used in ordinary flashlight batteries, added as a minor constituent in the zinc electrode to prevent its corrosion and thereby extend the shelf life of the product, has been drastically curtailed—typically from about 10,000 ppm to about 300 ppm nowadays in alkaline batteries—and in many cases completely eliminated, thereby halving the mercury in domestic garbage.

The other inorganic ion of mercury, Hg_2^{2+}, is not very toxic since it combines in the stomach with chloride ion to produce insoluble Hg_2Cl_2. The U.S. MCL (maximum contaminant level) for all forms of inorganic mercury in drinking water is 2 ppb.

Methylmercury Toxicity

When in combination with anions capable of forming covalent bonds (not nitrate, oxide, or sulfate), the **mercuric ion,** Hg^{2+}, forms covalent molecules rather than ionic solids. For example, $HgCl_2$ is a molecular compound, not a salt of Hg^{2+} and Cl^-. And as chloride ion forms a covalent compound with Hg^{2+}, so does the methyl anion, CH_3^-, yielding the volatile molecular liquid **dimethylmercury,** $Hg(CH_3)_2$. The process of dimethylmercury formation occurs in the muddy sediments of rivers and lakes, especially under anaerobic conditions. The active agent in the biomethylation process is a common constituent of microorganisms; it is a derivative of vitamin B_{12} with a CH_3^- anion bound to cobalt and is called *methylcobalamin.*

The less volatile mixed compounds CH_3HgCl and CH_3HgOH, collectively called **methylmercury** (or *monomethylmercury*), are often written as CH_3HgX, or somewhat misleadingly as CH_3Hg^+. These substances, like most of those written as Hg^{2+}, consist of covalent molecules, not ionic lattices. (The methylmercury ion CH_3Hg^+ exists as such only in compounds with anions such as nitrate or sulfate.) Methylmercury compounds are even more readily formed in the same way as dimethylmercury. Methylmercury production predominates over dimethylmercury formation in acidic or neutral aqueous solutions.

Due to its volatility, dimethylmercury evaporates from water relatively quickly unless it is transformed by acidic conditions into the monomethyl form. The pathways for the production and fate of dimethylmercury and of other mercury species in a body of water are illustrated in Figure 11-1. Recent research at Trent University in Ontario has shown that methylation of inorganic mercury does occur in the anaerobic regions of lakes, especially near the interface of the epilimnion and the hypolimnion, and at the interface of the hypolimnion with the sediments, but not in aerobic water. Methylmercury in surface water is photodegraded (to as yet unknown

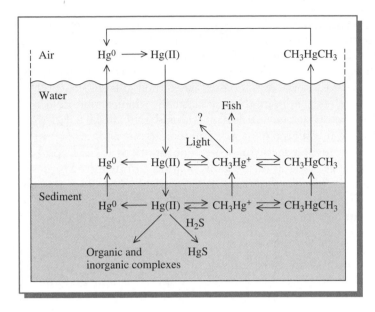

FIGURE 11-1 The cycling of mercury in fresh-water lakes. [Source: Adapted from M. R. Winfrey and J. W. M. Rudd, "Environmental Factors Affecting the Formation of Methylmercury in Low pH Lakes," *Environmental Toxicology and Chemistry* 9 (1990): 853–869.]

products) and is the most important sink for this substance in some lakes. Once elemental mercury vapor is formed in air, it undergoes the same long-range transport phenomenon discussed in Chapter 8 for organochlorines; thus even the Arctic and its wildlife are polluted by mercury.

Mercuric ion, Hg^{2+}, itself is not readily transported directly across biological membranes. Methylmercury is a more potent toxin than are salts of Hg^{2+} because it is soluble in fatty tissue in animals and bioaccumulates and biomagnifies there, and is more mobile. Once ingested, the original CH_3HgX compound is converted to compounds in which X is a sulfur-containing amino acid; in some of these forms it is soluble in biological tissue and can cross both the blood–brain barrier and the human placental barrier, presenting a twofold hazard. Methylmercury is, in fact, the most hazardous form of mercury, followed by the vapor of the element. The main toxic effects of methylmercury occur in the central nervous system. In the brain methylmercury is converted to inorganic mercury, which probably is responsible for the brain damage. Mercury vapor also oxidizes to this ion once it has entered the cell. Thus the usual barriers to Hg^{2+} are circumvented by covalent forms of mercury, Hg and CH_3HgX; by their covalency they can penetrate the defenses but can later be converted to the highly toxic 2+ ionic form.

Most of the mercury present in humans is in the form of methylmercury. Almost all of this originates from the fish in our food supply: mercury in fish is usually at least 80% methylmercury, and 95% is absorbed by the body when the fish is eaten. Mercury contamination is the reason behind about 97% of the advisories against eating fish caught in various regions of North America.

In contrast to organochlorines, which predominate in the fatty portions of fish, methylmercury can bind to the sulfhydryl group in proteins and is therefore distributed throughout the fish. Consequently, the mercury-containing part cannot be cut away before the fish is eaten. Fish absorb methylmercury that is dissolved in water as it passes across their gills (bioconcentration), and they also absorb it from their food supply (biomagnification). The ratio between methylmercury in fish muscle and that dissolved in the water in which the fish swim is often about 1 million to one and can exceed 10 million to one. The highest methylmercury concentrations (over 1 ppm) are usually found in large, long-lived predatory marine carnivores such as shark, king mackerel, tilefish, swordfish, and large tuna (sold as steaks and sushi), and in fresh-water species such as bass, trout, and pike. Indeed, the U.S. Food and Drug Administration warns women of childbearing age not to eat the first four types of these marine fish. On average, the older the fish, the more methylmercury it will have bioaccumulated. Noncarnivorous species such as whitefish do not accumulate much mercury since biomagnification in their food chain operates to a much smaller extent than in carnivorous fish.

In lakes the mercury content of fish is generally greater in acidic water, probably because both the solubility of mercury is greater and the methylation of mercury is faster at lower pH. In this way, the acidification of natural waters indirectly increases the exposure of fish-eaters to methylmercury. The relationship between mercury levels in small fish and the pH of the water is illustrated in Figure 11-2. The data in this figure are from a collection of lakes in Wisconsin, eastern Ontario, and Nova Scotia; most of the acidic lakes are in Nova Scotia.

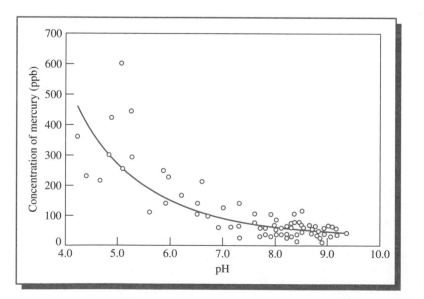

FIGURE 11-2 The relationship between pH and mercury concentration in fish for a standard fish length for 84 lakes from Ontario, Nova Scotia, and Wisconsin. [Source: D. Lean, "Mercury Pollution," *Canadian Chemical News* (January 2003): 23.]

Methylmercury Accumulation in the Environment and in the Human Body

The half-life of methylmercury compounds in humans, about 70 days, is much longer than that for Hg^{2+} salts, due in part to its greater solubility in a lipid environment. Consequently methylmercury can accumulate in the body to a much higher steady-state concentration, even if, on a daily basis, a person consumes amounts that individually would not be harmful.

PROBLEM 11-3

If the half-life in the human body of methylmercury is 70 days, what is its steady-state accumulation in a person who consumes daily 1 kg of fish containing 0.5 ppm of methylmercury? [Hint: Recall the discussion of steady-state concentrations in Chapter 4.]

Most of the well-publicized environmental problems involving mercury have arisen from the fact that the methylated form is a cumulative poison. However, at high enough concentration it can be acutely fatal. In 1997 cancer researcher Karen Wetterhahn of Dartmouth College died from mercury poisoning several months after a drop or two of pure dimethylmercury apparently seeped through latex gloves she was wearing while using the compound in experiments. Dialkylmercury compounds, including dimethylmercury, are sometimes called *supertoxic* because they are so toxic even in small amounts.

At the fishing village of Minamata, Japan, a chemical plant employing Hg^{2+} as a catalyst in polyvinyl chloride production discharged mercury-containing residues into Minamata Bay. The methylmercury compounds, mainly $CH_3Hg—SCH_3$, that subsequently formed from the inorganic mercury by biomethylation by microorganisms in the bay's sediments then bioaccumulated to concentrations as high as 100 ppm in the fish, which were the main component of the diet of many local residents. (By contrast, the current North American recommended limit for total mercury in fish to be consumed by humans is 0.5 ppm.) Thousands of people in Minamata were affected in the 1950s by mercury poisoning from this source, and hundreds of them died from it. Because the onset of symptoms in humans is delayed, the first signs of *Minamata disease* were observed in cats who ate discarded fish: they began jumping around and twitching, ran in circles, and finally threw themselves into the water and drowned. Symptoms in humans arise from dysfunctions of the central nervous system, since the target organ for methylmercury is the brain; they include numbness in arms and legs, blurring and even loss of vision, loss of hearing and muscle coordination, and lethargy and irritability.

Since methylmercury can be passed to the fetus, children born to mothers poisoned even slightly by mercury showed severe brain damage, some to a fatal extent. The infants showed symptoms similar to those of cerebral palsy: mental retardation, seizures, and motor disturbance, even paralysis. Just as in the case of high PCB levels discussed in Chapter 8, the developing fetuses were much more affected by methylmercury than were the mothers themselves. The poisonings at Minamata must surely rank as one of the major environmental disasters of modern times.

Other Sources of Methylmercury

Organic compounds of mercury have been used as fungicides in agriculture and in industry and enter the environment as a side effect of these applications. However, as a result of contact with soil, the compounds are eventually broken down and the mercury becomes trapped as insoluble compounds by attachment to sulfur present in clays and organic matter.

Hundreds of deaths in Iraq in 1956, 1960, and 1972, and a few in China and the United States, resulted from the consumption of bread made from seed grain (intended for planting) that had been treated with mercury-based fungicides to reduce seedling losses from fungus attack. The fungicides contained compounds of *ethylmercury*, $CH_3CH_2Hg^+$, the toxicity of which is presumed to be similar to that of methylmercury. In Sweden and Canada the use of mercury compounds to treat seeds led to a significant reduction in the number of birds of prey that consumed the smaller birds and mammals that fed on the scattered seed. The use of ethylmercury products in agriculture has now been curtailed in North America and western Europe.

Mercury is leached from rocks and soil into water systems by natural processes, some of which are accelerated by human activities. Flooding of vegetated areas can release mercury into water. For example, after the flooding of huge areas of northern Quebec and Manitoba to produce hydroelectric power dams, the newly submerged surface soils (and to a lesser extent the vegetation) released a considerable quantity of soluble methylmercury, formed from the "natural" mercury content of these media. The additional methylmercury resulted from contact of soil-bound Hg^{2+} with anaerobic bacteria produced by the decomposition of the immersed organic matter. In this way previously insoluble inorganic mercury was converted to methylmercury, which readily dissolved in the water. The methylmercury subsequently entered the food chain through its absorption by fish, and native persons who ate fish from these flooded areas now have substantially elevated levels of mercury in their bodies. Indeed, the methylmercury concentration in fish from these areas, 5 ppm or more, approaches that previously associated only with regions of industrial mercury pollution. Similarly, the release of mercury from soils in the Amazon jungle that have recently been cleared of trees apparently adds a sizable amount of the element to the environment, perhaps comparable to that from gold mining.

In 1999 the safety of using a mercury-based preservative in many of the vaccine preparations administered to infants was questioned by the American Academy of Pediatrics and the U.S. Public Health Service. The preservative is $CH_3CH_2-Hg-S-C_6H_4COOH$; it is said to contain the ethylmercury ion, but presumably, like methylmercury compounds, it actually is a covalent compound. This substance has now been removed by its manufacturer from all vaccines destined for the United States. The preservative was also widely used as a topical disinfectant.

The Use of Mercury in Preservatives and as Medications

Compounds of the *phenylmercury ion*, $C_6H_5Hg^+$, with acetate or nitrate as the anion, have been used to preserve paint while in the can and to prevent mildew after application of latex paint, particularly in humid areas. The phenylmercury salts are not as toxic to humans as are methylmercury compounds, since the former break down quickly into compounds of the less toxic Hg^{2+}. However, mercury compounds have been banned from indoor latex paints in the United States since 1990 because some ingestion of the element from this source is inevitable. Phenylmercury compounds were also formerly used as slimicides in the pulp-and-paper industry in order to prevent the growth of slime on wet pulp; because this practice has now been curtailed and because mercury-containing wastes are now usually treated, mercury releases from such sources have greatly decreased.

Because of their antiseptic and preservative qualities, however, some mercury compounds are still used in pharmaceuticals (especially topical antiseptics) and cosmetics. Elemental mercury was also used in some pharmaceuticals in the past. Indeed, the antidepressant pills that Abraham Lincoln took, mainly in the years before becoming president, contained the element, and some medical historians think the leaching of mercury into his bloodstream can account for his often-bizarre behavior in that period.

PROBLEM 11-4

A quantity of a mercury–chlorine compound is included in a shipment of waste to a toxic waste disposal dump. Before it can be disposed of properly, the owners of the dump need to know whether the compound is $HgCl_2$, or Hg_2Cl_2, or some other compound. They send a sample of it for analysis and find that it contains 26.1% chlorine by mass. What is the formula of the compound?

Safe Levels of Mercury in the Body

It is somewhat reassuring that both the direct effects of methylmercury on humans and the prenatal effects probably have thresholds below which no effects are observed. Currently the daily methylmercury intake of 99.9% of Americans lies below the World Health Organization's "safe limit." Nonetheless, some effects of methylmercury consumption on human vision are observed even when the concentration of (total) mercury in hair lies below the generally recognized threshold of 50 ppm.

However, if prenatal health is the main consideration and if a safety factor of 10 is applied, a substantial fraction of the population of the United States would exceed the safe limit. The World Health Organization has concluded that levels of 10–20 ppm of methylmercury in hair indicate that a pregnant woman has sufficient methylmercury in her blood to represent a threat to a developing fetus. This places at risk the developing fetuses of more than 30% of the women in some native communities in northern Canada, for example, in which fish play a large part in the diet. Although it is clear that high levels of methylmercury can result in developmental disabilities, there is continuing controversy whether methylmercury acquired through a diet high in fish and marine mammals can cause significant neurological damage to an adult or a developing fetus. The epidemiological studies of this question have, to this time, produced inconsistent results.

PROBLEM 11-5

What is the mass, in milligrams, of mercury in a 1.00-kg lake trout that just meets the Northern American standard of 0.50 ppm Hg? What mass of fish, at the 0.50-ppm Hg level, would you have to eat to ingest a total of 100 mg of mercury?

PROBLEM 11-6

The new U.S. EPA oral reference dose (RfD) for methylmercury is 0.1 μ/kg body weight/day. What mass of fish can a 60-kg woman safely eat each week if the average methylmercury level in the fish is 0.30 ppm? Approximately how many average servings of fish does this correspond to?

Lead

Although the environmental concentration of lead, Pb, is still increasing in some parts of the world, those uses that result in uncontrolled dispersion have been greatly reduced in the last two decades in many developed countries. Consequently, its concentration in soil, water, and air has decreased substantially.

Lead's relatively low melting point of 327°C allows it to be readily worked—it was the first metal to be extracted from its ores—and shaped into pipes, etc. Lead was used as a structural metal in ancient times, and for weatherproofing buildings, in water ducts, and for cooking vessels. Analysis of ice-core samples from Greenland indicates that atmospheric lead concentration reached a peak in Roman times that was not equaled again until the Renaissance. Lead is still used for roofing and flashing and for soundproofing in buildings. When combined with tin, it forms solder, the low-melting alloy used in electronics and in other applications (e.g., tin cans) to make connections between solid metals.

The boiling point of lead is 1740°C, compared to only 357°C for mercury, so the vapor pressure of lead is very much smaller than that of mercury and therefore not generally a problem. This behavior contrasts with that of mercury, for which the vapor is of environmental concern.

The effects of lead poisoning were known to the ancient Greeks, who realized that drinking acidic beverages from containers coated with lead-containing substances could result in illness. This information was not available to the Romans. Indeed, they sometimes deliberately adulterated overly acidic wine with sweet lead salts to improve the flavor. The concentration of lead in the bones of Romans is almost 100 times that found for modern North Americans. Some historians have hypothesized that chronic lead poisoning, from wine and other sources, of upper-class Romans contributed to the eventual downfall of the Roman Empire because of the metal's detrimental effects on the neurological and reproductive systems. The latter effects include dysfunctional sperm in males and an inability to bring the fetus to term in females. Due primarily to the contamination of beverages by lead from the distillation of alcohol in lead vessels, episodes of colic and gout due to lead poisoning were recorded through the Middle Ages and even until recent times.

Elemental Lead as an Environmental Risk

Lead is also found in the ammunition (lead shot) used in huge amounts by hunters of water fowl. Many ducks and geese are injured or die from chronic lead poisoning after ingestion of lead shot, which dissolves in the acidic environment inside them. Ducks consume the pellets left lying on the ground or at the bottom of ponds, since they look like food or grit. When birds prey on some ducks and other waterfowl that have been shot by hunters but not harvested, or have eaten lead shot to help grind food in their gizzards, the predators (such as bald eagles) become victims of lead poisoning. For these reasons, lead shot has been banned in the United States, Canada, the Netherlands, Norway, and Denmark. However, many loons die in North America because they swallow and are poisoned by lead sinkers and jigs still used in sport fishing.

Ionic 2+ Lead in Water and Food as an Environmental Hazard to Humans

Although the elemental form is not an environmental problem to most life forms, lead does become a real concern when it dissolves to yield an ionic form.

The stable ion of lead is the 2+ species, Pb(II) as Pb^{2+}. For example, lead forms the ionic *lead sulfide* PbS, $Pb^{2+}S^{2-}$. Lead sulfide is the metal-bearing component of the highly insoluble ore *galena*, from which almost all lead is extracted.

Lead does not react on its own with dilute acids. Indeed, elemental lead is stable as an electrode in the **lead storage battery,** even though it is in contact with fairly concentrated *sulfuric acid*. However, some lead in the solder that was used commonly in the past to seal tin cans will dissolve in the dilute acid of fruit juices and other acidic foods if air is present—i.e., once the can has been opened—since lead is oxidized by oxygen in acidic environments:

$$2\,Pb(s) + O_2 + 4\,H^+ \longrightarrow 2\,Pb^{2+}(aq) + 2\,H_2O$$

The Pb^{2+} produced by this reaction contaminates the contents of the can; for this reason, lead solder is no longer used for food containers in North America. Partly as a result of this change, the average daily lead intake for two-year-old children dropped from about 30 μg in 1982 to about 2 μg in 1991.

The 1845 Franklin Expedition to find a Northwest Passage across the Arctic is thought to have failed because the members all died from lead poisoning from the solder in the tin cans that held their food. Canadian writer Margaret Atwood has written eloquently about the incident in her short story *The Age of Lead*:

> it was the tin cans that did it, a new invention back then, a new technology, the ultimate defence against starvation and scurvy. The Franklin Expedition was excellently provisioned with tin cans, stuffed full of meat and soup and soldered together with lead. The whole expedition got lead poisoning. Nobody knew it. Nobody could taste it. It invaded their bones, their lungs, their brains, weakening them and confusing their thinking, so that at the end those that had not yet died in the ships set out in an idiotic trek across the stony, icy ground, pulling a lifeboat laden down with toothbrushes, soap, handkerchiefs and slippers, useless pieces of junk. When they were found (ten years later, skeletons in tattered coats, lying where they'd collapsed) they were headed back toward the ships. It was what they'd been eating that had killed them.
>
> [*The Age of Lead*, in *Wilderness Tips*, copyright 1991 by O. W. Toad Limited]

Some lead used to solder the joints of domestic copper water pipes, and lead used in previous decades and centuries to construct the pipes themselves, can dissolve in drinking water, particularly if the water is quite acidic or if it is particularly soft. Thus it is a good idea not to drink water that has been standing overnight in older drinking fountains or in the pipes of older dwellings; water in these plumbing systems should be allowed to run for a minute or so before being used for human consumption. On the other hand, hard water contains *carbonate ion*, CO_3^{2-}, which, together with oxygen, forms an insoluble layer containing compounds such as $PbCO_3$ on the surface of the lead. This layer prevents the metal underneath from dissolving in the water that passes over it. In some regions of England and in some cities in the northeastern United States that have soft water and networks of old lead pipes, phosphates are added to drinking water to form a similar insoluble protective coating on the inside of lead pipes and so reduce the concentration of dissolved lead.

Lead in water is more fully absorbed by the body than is lead in food. Now that many other sources of lead have been phased out, drinking water accounts for about one-fifth of the collective lead intake of Americans, the major source being food. Many domestic water treatment systems successfully remove most of the lead from drinking water.

PROBLEM 11-7

According to an informal 1992 survey, the drinking water in about one-third of the homes in Chicago had lead levels of about 10 ppb. Assuming that an adult drinks about 2 L of water a day, calculate the total lead that residents of these Chicago homes obtain daily from their drinking water.

Lead Salts as Glazes and Pigments

One form of the oxide PbO is a yellow solid that has been used at least as far back in history as ancient Egypt to glaze pottery. In glazing, the material is fused as a thin film to the surface of the pottery to make it waterproof and to give it a brilliant high gloss. Nowadays, *lead silicate* rather than the oxide or the sulfate is used for glazing since it is almost insoluble and thus much safer. The oxide becomes a hazard if it is applied incorrectly. Some of it will dissolve over a period of hours and days if acidic foods and liquids, such as cider, are stored in pottery containers, resulting in dissolved Pb^{2+}, up to hundreds or even thousands of parts per million, in the food:

$$PbO(s) + 2\,H^+(aq) \longrightarrow Pb^{2+}(aq) + H_2O$$

Indeed, lead-glazed dishes are still a major source of dietary lead, especially, but not exclusively, in developing countries. The leaching of lead from glazed ceramics used to prepare food is one of the major sources of the element for children in Mexico, where lead contamination is a leading public health problem.

Various salts of lead have been used as pigments for millenia, since they give stable, brilliant colors. *Lead chromate*, $PbCrO_4$, is the yellow pigment used in paints applied to school buses and for the yellow stripes on roads. *Red lead*, Pb_3O_4, is used in corrosion-resistant paints and has a bright red color. It was used in great quantities in the past to produce a rust-resistant surface coating for iron and steel. *Lead acetate* is often used in preparations to color gray hair, since the Pb^{2+} ion of this soluble salt will combine with the SH group in hair proteins to give a dark color.

Lead pigments have been used to produce the colors used in glossy magazines and food wrappers. In past centuries, lead salts were used as coloring agents in various foods. *White lead*, $Pb_3(CO_3)_2(OH)_2$, was used extensively until the middle of the twentieth century as a major component of white indoor paint. Since it was more durable than unleaded paint, it was often used on surfaces subject to punishment, such as kitchen cabinets and window trim. However, when the paint peels or chips off, small children may eat the paint flecks, since Pb^{2+} has a sweet taste. Persons who renovate old homes should ensure that dust from layers of old paint is properly contained. Children in inner-city slums, in which old coats of paint continue to peel, are often found to have elevated blood levels of lead. In indoor paint white lead has now been replaced by the pigment *titanium dioxide*, TiO_2. Although now banned from use in indoor paints (since 1978 in the United States), lead pigments continue to be used in exterior paints, with the result that soil around houses may eventually become contaminated. Some of this lead-contaminated soil may be ingested by small children because of its sweet taste. The use of *lead arsenate*, $Pb_3(AsO_4)_2$, as a pesticide was formerly another source of Pb^{2+} to soil.

An additional source of sweet-tasting lead-containing dust was the surface of some types of PVC miniblinds that had lead incorporated as a stabilizer in the plastic and underwent partial decomposition from exposure to UV in sunlight. Lead is used as a stabilizer in a variety of other PVC products as well, including children's toys.

Lead dust, originally soil containing tiny particles of lead compounds, is now the biggest source of lead for children in U.S. inner cities. The lead is collected from individually small but numerous contributions from many sources already mentioned: paint flakes, ceramics, plastics, gasoline, recycling plants, and even lead salts used in hair-coloring preparations.

Green Chemistry: Replacement of Lead in Electrodeposition Coatings

Sheet metal surfaces made of steel undergo corrosion very rapidly unless they are covered with a protective coating. Since the 1960s a technique called *electrodeposition* has competed with spray painting for coating steel. In 1976 the first automobiles were treated by electrodeposition. In this technique the surface to be treated is dipped in a bath, with the surface acting as a cathode or anode, and the coating is deposited electrophoretically. Electrodeposition has many advantages over spray painting, including

- lower air pollution, due to decreased solvent emissions,

- better corrosion protection, due to better coverage of poorly accessible areas,

- reduced waste, due to high transfer efficiency, and

- more uniform coating thickness.

Virtually all primer coats for automobiles are done by this method. Red lead, Pb_3O_4, mentioned previously, offers significant corrosion resistance and primer coats use large amounts of this material. Although lead has been banned from residential paints, the demand for corrosion resistance has resulted in exemptions from environmental regulations regarding lead in automobile paints.

PPG Industries discovered that *yttrium oxide* serves as an excellent replacement for lead as a corrosion inhibitor and won a Presidential Green Chemistry Challenge Award in 2001. On a weight basis, yttrium oxide is twice as effective in inhibiting corrosion as red lead but only 1/120th as toxic. An additional consideration is the pretreatment process used to assist in adhesion and corrosion resistance prior to the application of the electrocoat. The use of yttrium reduces the amount of nickel and eliminates chromium from metal pretreatments, compared to the lead process. It is estimated that employing yttrium in automobile electrodeposition will eliminate not only the use of 1 million pounds of lead but also 25,000 lb of chromium and 50,000 lb of nickel on an annual basis.

Dissolution of Otherwise Insoluble Lead Salts

The presence of significant concentrations of lead in natural waters seems paradoxical, given that both its sulfide, PbS, and its carbonate, $PbCO_3$, are highly insoluble in water:

$$PbS(s) \rightleftharpoons Pb^{2+} + S^{2-} \qquad K_{sp} = 8.4 \times 10^{-28}$$

$$PbCO_3(s) \rightleftharpoons Pb^{2+} + CO_3^{2-} \qquad K_{sp} = 1.5 \times 10^{-13}$$

In both salts, however, the anions behave as fairly strong bases. Thus both of the dissolution reactions are followed by the reaction of the anions with water:

$$S^{2-} + H_2O \rightleftharpoons HS^- + OH^-$$

$$CO_3^{2-} + H_2O \rightleftharpoons HCO_3^- + OH^-$$

Because these reactions reduce the concentrations of the original anions produced by dissolution of the PbS or $PbCO_3$, the position of equilibrium in the original reactions shifts to the right, thereby dissolving more of the salt, analogous with the process involving $CaCO_3$ that we analyzed in Chapter 9. Thus the solubilities of PbS and $PbCO_3$ in water are substantially increased by the reaction of the anion with water (see Additional Problem 2).

If highly acidic water comes into contact with minerals such as PbS, the "insoluble" solid dissolves to a much greater extent than in neutral waters. This occurs because the sulfide ion initially produced is subsequently converted almost entirely to *bisulfide ion*, HS^-, which in turn is converted by the acid to dissolved *hydrogen sulfide* gas, H_2S, since both S^{2-} and HS^- act as bases in the presence of acid:

$$S^{2-} + H^+ \rightleftharpoons HS^- \qquad K = 1/K_a(HS^-) = 7.7 \times 10^{12}$$

$$HS^- + H^+ \rightleftharpoons H_2S \qquad K' = 1/K_a(H_2S) = 1.0 \times 10^7$$

When these two reactions are added to that for the dissolution of PbS into Pb^{2+} and S^{2-}, the overall reaction is seen to be

$$PbS(s) + 2\,H^+ \rightleftharpoons Pb^{2+} + H_2S(aq)$$

Since the equilibrium constant $K_{overall}$ for an overall process that is the sum of three others is the product of their equilibrium constants, in this case $K_{overall} = K_{sp}KK' = 6.5 \times 10^{-8}$. By application of the law of mass action to this reaction, we find for the expression for the equilibrium constant in terms of concentrations

$$K_{overall} = [Pb^{2+}]\,[H_2S]/[H^+]^2$$

Under conditions in which no significant amount of hydrogen sulfide gas is vaporized, but which are sufficiently acidic that almost all the sulfur exists as H_2S rather than as S^{2-} or HS^-, the stoichiometry of the reaction allows us to write that $[Pb^{2+}] = [H_2S]$. By substitution of this relationship into the above equation, we obtain

$$[Pb^{2+}]^2 = 6.5 \times 10^{-8}\,[H^+]^2 \quad \text{or} \quad [Pb^{2+}] = 2.5 \times 10^{-4}\,[H^+]$$

Thus the solubility of PbS increases linearly with the H^+ concentration in acidic water. At pH = 4, the solubility of PbS and the concentration of Pb^{2+} ion in water is calculated to be 2.5×10^{-8} M, whereas at pH = 2, the solubility is 2.5×10^{-6} M. We conclude that dangerous concentrations of lead ion can occur in highly acidic bodies of water that are in contact with "insoluble" lead minerals.

PROBLEM 11-8

By calculations similar to those for PbS, deduce the relationship between the solubility of mercuric sulfide, HgS ($K_{sp} = 3.0 \times 10^{-53}$), and the hydrogen ion concentration in acidic water. Is the solubility of HgS increased by exposure to acid?

Ionic 4+ Lead in Automobile Batteries

In highly oxidizing environments, lead can form the 4+ ion, one form of Pb(IV). Thus the oxide PbO_2, written in ionic form as $Pb^{4+}(O^{2-})_2$, exists, as do the mixed oxides Pb_2O_3 and Pb_3O_4, which are combinations of Pb(II) as PbO and Pb(IV) as PbO_2. The mixed oxide Pb_3O_4, known as red lead, has been widely used as a paint pigment, especially as a coating for iron since it forms a surface layer that prevents rusting.

The elemental lead and the lead oxide PbO_2 employed as the two electrodes in storage batteries in almost all vehicles now constitute the major use of the element. Storage batteries that are not recycled constitute the main source of lead in municipal waste. The majority of used lead storage batteries, however, are recycled for their lead content. During the recycling operation lead can be released into the environment if careful controls are not maintained. Indeed, such recycling operations often constitute urban hot spots of lead emission into the surrounding communities. Although lead recycling operations in the United States are carried out under strict control, that is not necessarily the case in developing countries, where batteries are often shipped for recycling.

Tetravalent Organic Lead Compounds as Gasoline Additives

Whereas the compounds of Pb(II) are ionic, most Pb(IV) compounds are covalent molecules rather than ionic compounds of Pb^{4+}. In this respect, tetravalent lead is similar to the corresponding forms of the other elements (C, Si, Ge, Sn) in its group of the periodic table.

Commercially and environmentally, the most important covalent compounds of lead(IV) are tetraalkyl componds, PbR_4, especially those in which R is the methyl group, CH_3, and the ethyl group, CH_2CH_3, namely, **tetramethyllead,** $Pb(CH_3)_4$, and **tetraethyllead,** $Pb(C_2H_5)_4$. In the past, both compounds found widespread use as additives in leaded gasoline. As discussed in Chapter 6, this practice has now been phased out in North America and in many other developed countries, except in aviation fuel, for which no acceptable substitute has yet been found.

When these additives are used in gasoline, the atoms of lead that are liberated by the combustion of the tetraalkyl compounds must be removed before they can form metallic deposits and damage the vehicle's engine. In order to convert the combustion products into volatile forms that can leave the engine in the exhaust gases, small quantities of *ethylene dibromide* and *ethylene dichloride* are also added to the leaded gasoline. As a result, the lead is removed from the engine and enters the atmosphere from the tailpipe as a mixture of the mixed dihalide $PbBrCl$ and the dihalides $PbBr_2$ and $PbCl_2$. Subsequently, under the influence of sunlight, these compounds form PbO. The lead oxide exists in particulate form as an aerosol in the atmosphere for hours or days. Consequently, not all of it is deposited in the immediate surroundings of the roadway. It can thus enter the food chain at more distant sites if it is deposited on vegetables or on fields used by grazing animals. Furthermore, a small fraction of the ethylene dihalides are converted into dioxins and furans and enter the environment in these forms.

Since tetraalkyl lead compounds are volatile, they evaporate to some extent from gasoline and enter the environment in gaseous form. They are not water-soluble, but they are readily absorbed through the skin. In the human liver PbR_4 molecules are converted into the more toxic PbR_3^+ ions, which are neurotoxins because they can cross the blood–brain barrier. In substantial doses, these organic compounds of lead cause symptoms that mimic psychosis. It is not clear what the effects, if any, may be of chronic low-level exposure. At very high exposures, tetraalkyllead compounds are fatal, as was discovered many years ago when several employees of the companies that originally produced these compounds died.

Environmental Lead from Leaded Gasoline

In contrast to mercury, little or no methylation of inorganic lead occurs in nature. Thus almost all the tetraalkylated lead in the environment probably originated from leaded gasoline.

A high proportion of environmental lead in many parts of the world comes from that emitted from vehicles and occurs in the environment mainly in inorganic form. The conversion to nonleaded gasoline in North America and Europe, the initial impetus for which was the interference of lead in exhaust gases with the proper functioning of catalytic converters, has had the welcome side effect of greatly decreasing the average amount of lead ingested by urban inhabitants. Indeed, the noted environmentalist Barry Commoner has called the elimination of lead from gasoline "one of the [few] environmental success stories." European scientists have recently been able to trace the rise and fall of atmospheric alkylated lead levels by analyzing different vintages of a French red wine (*Chateauneuf-du-Pape*) that used grapes grown near two busy auto routes. They found that the concentration of *trimethyllead*—the degradation product of the tetramethyl compound—rose steadily to a maximum in the mid-1970s, followed by a steady decline to about one-tenth of the peak concentration by the early 1990s as the compound was phased out of gasoline. This pattern of usage is consistent with the variation in U.S. consumption of lead for use in gasoline, plotted in Figure 11-3. The plot shows a sharp rise from 1930 to 1970, followed by an even sharper decline thereafter.

In many countries of the world the use of leaded gasoline continues. In these areas, the air is the major source of lead ingested by humans, as it was in the past in North America and Europe. For example, in Mexico airborne lead from vehicular emissions is a major source of the element in the blood of children. Some of the gasoline-based lead enters the body directly from inhaled air and some enters indirectly from food into which lead has been incorporated. Airborne lead oxide eventually settles on soil, water, fruits, or leafy vegetables, thereby entering the food chain, since soluble lead is absorbed by plants. Microorganisms do bioconcentrate lead, but in contrast to mercury, lead does not undergo biomagnification in the food chain.

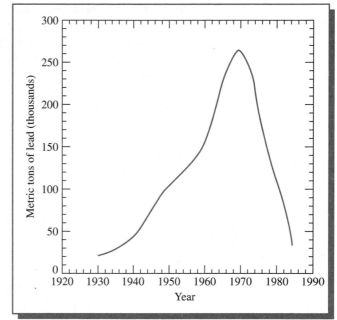

FIGURE 11-3 The historical consumption of lead in gasoline in the United States. [Source: C. E. Dunlop et al., "Past Leaded Gasoline Emissions as a Nonpoint Source Tracer in Riparian Systems," *Environmental Science and Technology* 34 (2000): 1211.]

Lead's Effects on Human Reproduction and Intelligence

Most lead ingested by humans is present initially in the blood, but that amount eventually reaches a plateau. Any excess enters the soft tissues, including the organs, particularly the brain. Eventually lead is deposited in bone, where it replaces calcium, because Pb^{2+} and Ca^{2+} ions are similar in size. Indeed, lead absorption by the body increases in persons with a calcium (or iron) deficiency and is much higher in children than in adults. A study in Mexico indicated that pregnant women can decrease the lead levels in their blood—and presumably in the blood of their developing fetus—by taking calcium supplements.

At high levels, inorganic lead (Pb^{2+}) is a general metabolic poison. The toxicity of lead is proportional to the amount present in the soft tissues, not to that in blood or bone. Lead remains in human bones for decades; thus it can accumulate in the body. The dissolving of bone, as occurs with old age or illness such as osteoarthritis and advanced periodontal disease or in times of stress such as pregnancy and menopause, results in the remobilization of bone-stored lead back into the bloodstream where it can produce toxic effects. There is some evidence that excess lead can lead to the deterioration of bones in adults. Recently a correlation has been found between periodontal bone loss and blood lead levels in U.S. adults, particularly in those who smoke. Children exposed to environmental lead also have more dental cavities.

The humans most at risk from low levels of Pb^{2+} are fetuses and children under the age of about seven years. Both these groups are more sensitive to lead than are adults, partly because they absorb a greater percentage of dietary lead and their brains are growing rapidly. The metal readily crosses the placenta and thus is passed from mother to unborn child. Because of the immaturity of the fetus's blood–brain barrier, there is little to prevent the entry of lead into its brain. Indeed, in the past women who worked in the lead industry suffered higher-than-average rates of miscarriages and stillbirths. In addition, lead is transferred postnatally from the mother in her breast milk and/or from the tap water used to prepare formula for bottle-fed babies.

The principal risk to children from lead is interference with the normal development of their brains. A number of studies have found small but consistent and significant neuropsychological impairment in young children due to environmental lead absorbed either before or after birth. Lead appears to have deleterious effects on children's behavior and attentiveness, and possibly also on their IQs. This is illustrated in Figure 11-4, where a mental development score is plotted as a function of age for groups of young children differentiated by the amount of lead in their umbilical cords at birth. A study of children in a lead-smelting community (Port Pirie) in Australia indicates that children with a blood lead level of 300 ppb had an average IQ 4 to 5 points lower than those whose level was 100 ppb. This is consistent with other studies that indicate that there is an IQ deficit of about 2–3 points for each 100-ppb increase in blood lead. A very recent study of U.S. children found an overall IQ difference of about 5 points per 100 ppb, but the ratio increased to about 7 points per 100 ppb at very low levels.

A survey in 1976–1980 of American children aged from 6 months to 5 years found that about 4% of them had blood lead levels in excess of 300 ppb

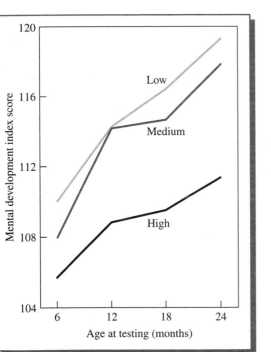

FIGURE 11-4 The effect of prenatal exposure to lead on the mental development of infants. Lead exposure is measured by its concentration in the blood of the child's umbilical cord. [Source: D. Bellinger et al., "Longitudinal Analyses of Prenatal and Postnatal Lead Exposure and Early Cognitive Development," *New England Journal of Medicine* 316 (1987): 1037–1043. Reprinted by permission of the *New England Journal of Medicine*.]

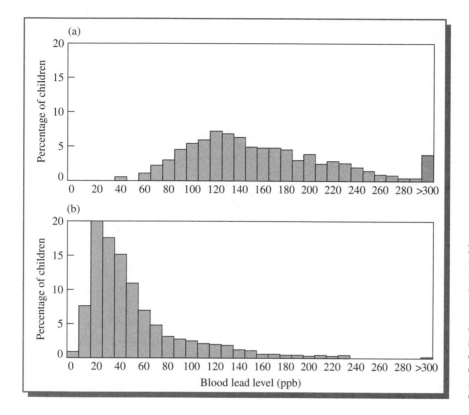

FIGURE 11-5 The distribution of blood lead levels in U.S. children aged 1–5 years (a) in 1976–1980 and (b) in 1988–1991. [Source: R. A. Goyer, "Results of Lead Research: Prenatal Exposure and Neurological Consequences," *Environmental Health Perspectives* 104 (1996): 1050–1054.]

and that an additional 20% had levels over 200 ppb (see Figure 11-5a). These concentrations represent two of the cutoffs that had been proposed in the past as "safe" levels, but it appears that there may be no level at which lead does not produce a deleterious effect (i.e., there is no threshold) in young and unborn children. A second survey, in 1988–1991, indicated that blood lead levels in U.S. children had fallen substantially; less than 9% of those aged 1–5 years had blood lead levels greater than 100 ppb (see Figure 11-5b).

In summary, on an atom-for-atom basis, lead is not as dangerous as mercury. However, the general population is exposed to lead from a greater variety of sources and generally at higher levels than those associated with mercury. Overall, more people are adversely affected by lead, though on average to a lesser extent, than those individuals exposed to mercury. Both metals are more toxic in the form of their organic compounds than as the simple inorganic cations. In terms of its environmental concentration, lead is much closer—within a factor of 10—to the level at which overt signs of poisoning become manifest than is any other substance, including mercury. Thus it is appropriate that society continue to take steps to further reduce human exposure to lead.

PROBLEM 11-9

The concentrations of lead in blood samples are often reported in units of micrograms of Pb per deciliter of blood or in micromoles of lead per liter of blood. Calculate the value of the concentration in these units of a blood sample containing 60 ppb lead, assuming that the density of blood is 1.0 g/mL.

Cadmium

Cadmium, Cd, belongs to the same subgroup of the periodic table as zinc and mercury, but it is more similar to zinc. Like zinc, the only common ion of cadmium is the 2+ species. In contrast to mercury, cadmium's compounds with simple anions such as chloride are ionic salts rather than covalent molecules.

Environmental Sources of Cadmium

Most cadmium is produced as a by-product of zinc smelting, since the two metals usually occur together. Some environmental contamination by cadmium often occurs in the areas surrounding zinc, lead, and copper smelters. As is the case for the other heavy metals, burning coal introduces cadmium into the environment. The disposal by incineration of waste materials that contain cadmium is also an important source of the metal in the environment.

An important use of cadmium is as an electrode in rechargeable *nicad* (nickel–cadmium) batteries used in calculators and similar devices. When current is drawn from the battery, the solid metal cadmium electrode partially disintegrates to form insoluble *cadmium hydroxide*, $Cd(OH)_2$, by incorporating hydroxide ions from the medium in which it is in contact. When the battery is being recharged, the solid hydroxide, which was deposited on the metal electrode, is converted back to cadmium metal:

$$Cd(s) + 2\,OH^- \rightleftharpoons Cd(OH)_2(s) + 2\,e^-$$

Each nicad battery contains about 5 g of cadmium, much of which is volatilized and released into the environment if the spent batteries are incinerated in garbage. The metallic cadmium tends to condense onto the smallest particles in the incinerator smokestream, which are precisely the ones that are difficult to capture by pollution-control devices placed in the gas stack. In order to avoid releasing airborne cadmium into the environment upon combustion, some municipalities now require nicad batteries to be separated from other garbage. The recycling of metals from such batteries has also begun in some areas. However, some American states and European countries are taking steps to make the use of nicad batteries illegal because of the potential environmental contamination by cadmium. Battery manufacturers hope to replace nicad batteries soon with ones that do not contain cadmium.

In ionic form, the main use of cadmium is as a pigment. Because the color of *cadmium sulfide*, CdS, depends on the size of the particles, cadmium pigments of many hues can be prepared. Both CdS and CdSe have been used extensively to color plastics. CdSe is also used in photovoltaic devices (such as photoelectric cells) and in TV screens. For several centuries painters have used cadmium sulfide pigments in paints to produce brilliant yellow colors and oppose any ban on them, since at present there are no suitable replacements. Van Gogh could not have painted his famous *Sunflowers* canvas without cadmium yellows, although it is speculated that cadmium poisoning may have contributed to the painter's anguished mental state.

Cadmium is released into the environment during the incineration of plastics and other materials that contain it as a pigment or as a stabilizer. Its release to the atmosphere also occurs when cadmium-plated steel is recycled, since the element is fairly volatile when heated (its boiling point is 765°C).

Human Intake of Cadmium

The MCL for cadmium in drinking water is 5 ppb in the United States. Although Cd^{2+} is rather soluble in water, unless sulfide ions are also present to precipitate the metal as CdS, humans usually receive only a small proportion of their cadmium directly from drinking water or from air, except for individuals who live near mines and smelters, particularly those that process zinc.

Smokers are also exposed to cadmium that is absorbed from soil and irrigation water by tobacco leaves and released into the smokestream when a cigarette is burned. Heavy smokers have approximately double the cadmium intake of nonsmokers from all sources.

Owing to its similarity to zinc, plants absorb cadmium from irrigation water. The use on agricultural fields of phosphate fertilizers, which contain ionic cadmium as a natural contaminant, and of sewage sludge contaminated with cadmium from industrial releases increases the cadmium level in soil and subsequently in plants grown in it. In the future, cadmium may be removed from phosphate fertilizer before it is sold to the consumer (see also Chapter 12). Soil also receives cadmium from atmospheric deposition. Since cadmium uptake in plants increases with decreasing soil pH, one effect of acid rain is to increase cadmium levels in food.

For most of us, the greatest proportion of our exposure to cadmium comes from our food supply. Seafood and organ meats, particularly kidneys, have higher levels than do most other foods. However, most of the cadmium in the diet usually comes from potatoes, wheat, rice, and other grains, since most people consume so much more of them than of seafood and kidneys. An exception is the Inuit people in Canada's Northwest Territories; a prized component of their diet is caribou kidneys, organs that are highly contaminated by cadmium that has reached the Arctic regions on the wind from industrial regions in Europe and North America.

Historically, all episodes of serious cadmium contamination are due to pollution from nonferrous mining and smelting. The most acute environmental problem involving cadmium occurred in the Jintsu River Valley region of Japan, where rice for local consumption was grown with irrigation water drawn from a river that was chronically contaminated with dissolved cadmium from a zinc mining and smelting operation upstream. Hundreds of people in this area, particularly older women who had borne many children and who had poor diets, contracted a degenerative bone disease called *itai-itai,* or "ouch-ouch," so named because it caused severe pain in the joints. In this disease, some of the Ca^{2+} ions in the bones are replaced by Cd^{2+} ions since they have the same charge and virtually the same size. The bones slowly become porous and can subsequently fracture and collapse. The intake of cadmium by itai-itai sufferers was estimated at about 600 μg/day, which is about 10 times the average intake of North Americans.

Protection Against Low Levels of Cadmium

Cadmium is acutely toxic: the lethal dose is about 1 g. Humans are protected against chronic exposure to low levels of cadmium by the sulfur-rich protein **metallothionein,** the usual function of which is the regulation of zinc metabolism. Because it has many sulfhydryl groups, metallothionein can complex almost all ingested Cd^{2+}, and the complex is subsequently eliminated in the urine. If the amount of cadmium absorbed by the body exceeds the capacity of metallothionein to complex it, the metal is stored in the liver and kidneys. Indeed, there is evidence that chronic exposure to cadmium eventually leads to an increased chance of acquiring kidney diseases.

The average cadmium burden in humans is increasing. Although cadmium is not biomagnified, it is a cumulative poison since, if not eliminated quickly (by metallothionein, as discussed), its lifetime in the body is several decades. The areas at greatest risk from cadmium are Japan and central Europe; in both regions the pollution of the soil by cadmium is particularly high due to contamination from industrial operations. Rice grown in many areas of Japan is often contaminated with rather high cadmium levels. As a consequence, the dietary intake of cadmium by residents of Japan is substantially greater than for peoples of other developed countries. Indeed, in Japan the average daily amount of cadmium ingested is beginning to approach the maximum level recommended by health authorities, though this limit has a large built-in safety factor relative to levels at which health effects would occur.

Arsenic

Arsenic is not actually a metal; it is a metalloid—its properties are intermediate between those of metals and nonmetals. However, for convenience we discuss it in this chapter.

Arsenic compounds such as the oxide As_2O_3, *white arsenic*, were the poison of choice for murder and suicide from ancient times through the Middle Ages. In the seventeenth century arsenic was believed in some European societies to be not only a poison but also a magical substance that was a cure for certain ailments, including impotence, and a prophylactic against the plague. Indeed, arsenic compounds have been used therapeutically for 2000 years, and even today about 50 Chinese medicines contain the element. There are small background levels of arsenic in many foods, and a trace amount of this element apparently is essential to good human health.

Arsenic(III) Versus Arsenic(V) Toxicity

Arsenic occurs in the same group of the periodic table as phosphorus, so it also has an s^2p^3 electron configuration in its valence shell. Loss of all three p electrons gives the $3+$ ion, whereas sharing of the three electrons gives trivalent arsenic; collectively, these two forms are designated As(III). Alternatively, loss of all five valence-shell electrons gives the $5+$ ion, and sharing them all gives pentavalent arsenic; collectively, these two forms are designated As(V). Overall, arsenic acts much as does phosphorus, although it has more of a tendency to form ionic rather than covalent bonds, since it is more metal-like. Due to the similarity in properties, arsenic compounds coexist with those of phosphorus in nature. Consequently, arsenic often contaminates phosphate deposits and commercial phosphates.

Arsenic's lethal effect when consumed in an acute dose is due to gastrointestinal damage, resulting in severe vomiting and diarrhea. Inorganic As(III) is more toxic than As(V), although the latter is converted by reduction to As(III) in the human body. It is thought that the greater toxicity of As(III) is due to its ability to be retained longer in the body by becoming bound to sulfhydryl groups in one of a number of different enzymes. Due to the subsequent inactivity of the enzymes, energy production in the cell declines and the cell is damaged. Once arsenic is methylated in the liver, it does not bind tightly to enzymes and hence is largely detoxified.

Anthropogenic Sources of Arsenic to the Environment

Anthropogenic environmental sources of arsenic stem

- from the continuing use of its compounds as pesticides,

- from its unintended release during the mining and smelting of gold, lead, copper, and nickel, in whose ores it commonly occurs (the leachate from abandoned gold mines of previous decades and centuries still sometimes being a significant source of arsenic pollution in water systems),

- from the production of iron and steel, and

- from the combustion of coal, of which it is a contaminant.

The arsenic present in raw coal can become a serious pollutant around areas where the fossil fuel is burned. This is particularly serious when the coal is burned in small, unventilated stoves rather than in large power plants. In these cases, which occur in regions of some developing countries, the arsenic not only becomes an indoor air pollutant but also contaminates the food and water stored indoors. This problem is particularly acute in the Guizhou province of China, where arsenic levels in the coal are extraordinarily high, above 1% in some cases. High levels of arsenic in coal are also found in some areas of India. By contrast, arsenic in U.S. coal averages about 22 ppm, and most coal worldwide has arsenic levels of less than 5 ppm.

Arsenic compounds found widespread use as pesticides before the modern era of organic chemicals. Although its use in these applications has decreased, arsenic contamination remains an environmental problem in some areas of the world. The common arsenic-based pesticides include the insecticide lead arsenate, $Pb_3(AsO_4)_2$, and the herbicide *calcium arsenate*, $Ca_3(AsO_4)_2$, both of which contain As(V) in the form of the **arsenate ion,** AsO_4^{3-}. The herbicides *sodium arsenite*, Na_3AsO_3, and *Paris Green*, $Cu_3(AsO_3)_2$, both contain As(III) in the form of the **arsenite ion,** AsO_3^{3-}. Some of the methylated derivatives of arsenic acids (to be discussed) are also used as herbicides. The environmental consequences of using another heavy metal, tin, in a pesticide are discussed in Box 11-1.

Since the 1970s arsenic has been used in the form of the compound **chromated copper arsenate,** CCA, to pressure-treat lumber to prevent rot and termite damage. Unfortunately, some of the arsenic leaches out of the wood over time. About 90% of the current industrial use of arsenic in the United States is in wood preservatives. U.S. producers of CCA-treated

BOX 11-1 | Organotin Compounds

Although inorganic compounds of tin (Sn) are relatively nontoxic, the bonding of one or more carbon chains to the metal results in substances that are toxic. Such organotin compounds have some common uses, such as additives to stabilize PVC plastics and as fungicides to preserve wood, and therefore are of environmental concern.

Tin forms a series of compounds of general formula R_3SnX, which are molecular substances though often shown in formulas as if they were ionic, i.e., (R_3Sn^+) (X^-), where R is a hydrocarbon group and X is a monatomic anion; corresponding compounds such as $(R_3Sn)_2O$ also occur. All these compounds are toxic to mammals when R is a very short alkyl chain; maximum toxicity occurs when R is the ethyl group, C_2H_5, and decreases progressively with increasing chain length.

For fungi the greatest toxic activity is attained when each hydrocarbon chain has four carbons in an unbranched chain, that is, when R is the n-butyl group, $—CH_2CH_2CH_2CH_3$ (or simply $n\text{-}C_4H_9$). *Tributyltin oxide*, $(R_3Sn)_2O$ where $R = n\text{-}C_4H_9$, and the corresponding fluoride have both been used as fungicides; commonly they are incorporated as antifouling agents in the paint applied to docks, to the hulls of boats, to lobster pots, and to fishing nets, etc. to prevent the accumulation of slimy marine organisms such as the larvae of barnacles. In recent years tributyltin has been incorporated into polymeric coatings for boat hulls; a thin layer of the compound subsequently forms around the hull. The tin compounds replaced copper(1) oxide, Cu_2O, in such applications since their effectiveness lasts longer than a single season.

Unfortunately, some of the tributyltin compound leaches into the surface waters in contact with the coatings or paint, particularly in harbors where the boats are moored, and subsequently enters the food chain via the microorganisms that live near the surface. This can lead to sterility or death for fish and some types of oysters and clams that feed on these microorganisms. Some countries have restricted the use of tributyltin compounds to large ships. Thus, although the concentration of tributyltin has decreased in the waters of small harbors and marinas, the pollutant still tends to concentrate in marine coastal regions due to its use on large vessels. Scientists are worried that the presence of tributyltin compounds in these waters could affect fish reproduction.

For this reason, the International Maritime Organization banned new applications of tributyltin to ships of any size starting in 2003 and requires that this material be removed from all old applications by 2008. Ironically, the triazine herbicide added to copper-based antifoulant paints that were introduced to replace those based upon tributyltin degrades only slowly in water and has now begun to accumulate there.

Higher organisms have enzymes that break down tributyltin fairly rapidly, so it is not very toxic to humans. However, most humans now have detectable levels of tributyltin in their blood.

wood phased out the use of the arsenic compound at the end of 2003 for wood destined for residential structures such as decks, picnic tables, fences, and playground equipment. CCA is discussed in further detail later in the Chromium section. The U.S. EPA has already banned arsenic in all other pesticides.

Arsenic in Drinking Water

Arsenic—much of it from natural sources—is one of the most serious environmental health hazards. The presence of significant levels of arsenic, As, in drinking water supplies is a significant and controversial environmental issue. Although arsenic has been used for millennia as an acute poison, the major health problem stemming from its presence at low levels in drinking water is cancer. Drinking arsenic-contaminated water has also been linked to diabetes and cardiovascular disease, perhaps by disrupting a hormonal process associated with both conditions. Natural levels of arsenic can be quite high, and it is more common for health problems to arise from this source than from anthropogenic arsenic.

Arsenic is carcinogenic in humans. Lung cancer results from the inhalation of arsenic and probably also from its ingestion. Cancers of the lung, bladder, and skin, and perhaps also of the kidney, arise from ingested arsenic, including that in water. There is evidence that smoking and simultaneous exposure to environmental arsenic act **synergistically** in causing lung cancer, i.e., their effect taken together is greater than the sum of their individual effects if each acted independently.

Drinking water, especially that derived from groundwater, is a major source of arsenic for many people. Although anthropogenic uses of arsenic can result in contamination of water, by far the greatest problems occur with contamination by natural processes. Groundwater in several parts of the world is highly contaminated with arsenic. Unfortunately, the arsenic is tasteless, odorless, and invisible, so it is not easily detected.

Major problems from high arsenic levels occur in the Bengal Delta, with the result that many people—up to 40 million—in Bangladesh and in the West Bengal region of India drink arsenic-laced water. The World Health Organization (WHO) has called this the "largest mass poisoning of a population in history." The problem is due to the creation of tens of millions of tube wells that mine groundwater that was previously inaccessible. The concrete tube wells extend 20 m (60 ft) or more into the ground. Ironically, the wells were constructed by UNICEF in the 1970s and early 1980s in an otherwise highly successful project to eliminate diarrhea, cholera, and other waterborne diseases and to reduce the high child mortality rate caused by use of microbially unsafe water from streams, ponds, and shallowly dug wells used in the past. About half the wells—affecting about 50 million people in Bangladesh—produce water with arsenic levels that exceed, in some cases by a factor of more than 50, the 10-ppb WHO guideline for drinking water. Generally, the deeper the well beyond about 20 m, the lower the concentration of arsenic. Several million people living in the Bengal Delta region will probably contract skin disorders from drinking arsenic-laced groundwater if remedial action is not taken; a fraction of them will also suffer from the more serious ailment of *arsenicosis*, which can cause cancer of the skin, bladder, kidneys, and lungs. Skin lesions appear after 5–15 years of exposure to high levels of arsenic in drinking water. A very large number of residents of West Bengal, India, already have skin lesions that may develop into skin cancer because they consumed arsenic-laced groundwater from underground wells.

Recent evidence from Bangladesh shows that the detrimental effects of water from the tube wells may be due not solely to arsenic but also to elevated levels of manganese—which intensifies arsenic poisoning—and perhaps in a few locations also to high levels of lead and chromium in the water. However, it has been established that rice and vegetables grown in Bangladesh using irrigation water from tube wells are also contaminated by arsenic, and this may be the dominant source of the element for some people.

The origin of the dissolved arsenic in the water in Bangladesh is controversial. Normally the element is coprecipitated with and adsorbed on the surface of iron(II) oxides in the soil. However, the arsenic, along with the iron, dissolves when insoluble Fe(III) is reduced by natural organic carbon to the more soluble Fe(II) state. Indeed, the higher the concentration of dissolved iron, the higher the arsenic concentration in the water. The controversy centers around whether the dominant process is the natural one in which buried peat acts as the reducing agent as it has been doing for millennia, or whether the release has been greatly accelerated in recent years as an indirect effect of annually lowering the water table by extracting massive amounts of water for crop irrigation. In the latter mechanism, the subsequent recharge of the depleted aquifer transports carbon in the water drawn down from the surface, resulting in further reduction of iron oxides and solubilization of the arsenic. Reduction of arsenic from the +5 state in which it exists when adsorbed to the iron mineral to the more soluble +3 state is also believed to help release the element into the water.

Arsenic-contaminated drinking water is also a major problem in Chile, Argentina, Mexico, Nepal, Vietnam, Taiwan, China, and other areas of India as well. Indeed, 8% of the deaths of Chilean adults over 30 are attributable to arsenic poisoning. In a study of residents of Taiwan who were exposed to high levels of the element in their well water, a relationship between arsenic exposure and skin cancer incidence has been established. As in Bangladesh, the arsenic problems began only when people began to drink groundwater, which was touted as being purer than surface water, since the latter is often contaminated by sewage.

Drinking Water Standards for Arsenic

Drinking water, especially groundwater, is a major source of arsenic for most people. The global average inorganic arsenic content of drinking water is about 2.5 ppb. The European Union has followed the World Health Organization in setting 10 ppb as the acceptable limit for arsenic in drinking water. Canada has set its drinking water standard for arsenic at 25 ppb, and Australia at 7 ppb. The standard in many developing countries is still 50 ppb, which is not now considered to be protective.

In the last days (2000) of the Clinton administration, the MCL for arsenic in U.S. drinking water was lowered from 50 ppb to 10 ppb. Although the Bush administration at first withdrew this regulation, it later concluded that it was warranted. As a result, the 10-ppb limit became law in February 2002, with the compliance date set for 2006.

One of the difficulties in setting a standard for arsenic levels in drinking water is determining the manner in which the element operates as a carcinogen. For carcinogens that induce cancer directly—by damaging DNA—the assumption is made that no amount of exposure to the substance is safe, since the risk rises from zero in direct proportion to exposure. However, there is some evidence that arsenic does not act directly but indirectly, by inducing cell damage and regrowth or by inhibiting repair of DNA damage caused by other carcinogens such as UV light or tobacco smoke. For carcinogens that act indirectly there can be a threshold, a level below which the substance can be considered safe and not to cause damage.

Assuming that no threshold exists, linear extrapolations of human cancer incidence from populations that were exposed to high levels of arsenic leads to the conclusion that there is a 1-in-1000 lifetime risk of dying from cancer induced by normal background levels of arsenic. This estimate makes arsenic almost equal to environmental tobacco smoke and radon exposure as an environmental carcinogen. Drinking water over a lifetime at the 50-ppb level, the old U.S. standard, would cause bladder or lung cancer in about 1% of the population, a much greater risk than continuously consuming any other water-based contaminant at its MCL. About 57 million Americans currently drink water containing more than 1 ppb (i.e., $1\mu g/L$) of arsenic (Figure 11-6). Most of these water systems lie in the western and midwestern states and use groundwater with naturally occurring arsenic. Some environmentalists argue that the arsenic standard should be lowered still further, to 3 ppb, at

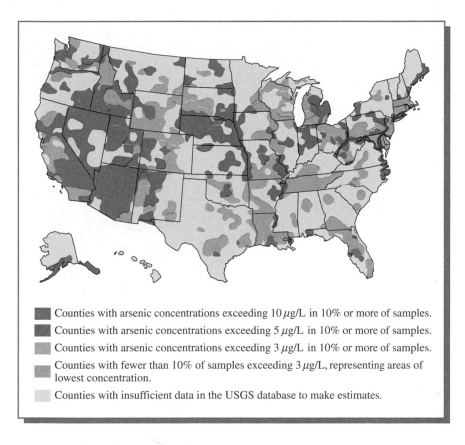

Counties with arsenic concentrations exceeding 10 μg/L in 10% or more of samples.

Counties with arsenic concentrations exceeding 5 μg/L in 10% or more of samples.

Counties with arsenic concentrations exceeding 3 μg/L in 10% or more of samples.

Counties with fewer than 10% of samples exceeding 3 μg/L, representing areas of lowest concentration.

Counties with insufficient data in the USGS database to make estimates.

FIGURE 11-6 Average arsenic concentrations in U.S. drinking water. [Source: "Pressure to Set Controversial Arsenic Standard Increases," *Environmental Science and Technology* 34 (2000): 208A.]

which the cancer risk is still 1 per thousand, whereas at 10 ppb it is about 3 per thousand.

Some scientists are not convinced that these estimates of cancer risk are at all realistic, since the extrapolation of the cancer incidence from high arsenic levels to the low environmental concentrations may not be valid if arsenic acts indirectly as a carcinogen. It will be difficult to resolve this issue by analyzing cancer trends in different parts of the United States, however, since the fraction of bladder and lung cancers caused by arsenic is still a small percentage of the total for these diseases.

One argument against making the arsenic standard even as low as 10 ppb is that it will force some small-scale suppliers of drinking water to shut down, since they cannot afford the cost of equipment to remove the element. Such a shutdown might lead their consumers to turn to water supplies that are even less safe in other respects. Indeed, lowering the standard to 10 ppb is estimated to cost users of small water utilities, i.e., many people in rural areas, several hundred dollars a year, whereas it will cost users of large facilities only a few dollars annually.

Removal of Arsenic from Water

At present the most widely used process for removing arsenic is to pass the drinking water over activated *alumina* (aluminum oxide), onto the surface of which the arsenic is adsorbed. Arsenic cannot be removed from water by cation exchange, since it occurs as an anion. However, anion exchange and reverse osmosis are both promising ways to remove arsenic from drinking water, although the latter process is expensive.

Like calcium and magnesium, arsenic can be removed from drinking water at large treatment facilities by precipitating it in the form of one of its insoluble salts. The arsenic in surface water normally exists as the arsenate ion, AsO_4^{3-}. The salt formed by the ferric ion, Fe^{3+}, and arsenate is insoluble, so soluble *ferric chloride*, $FeCl_3$, is dissolved in the water, and the precipitated *ferric arsenate* is filtered from the resulting mixture:

$$Fe^{3+} + AsO_4^{3-} \longrightarrow FeAsO_4(s)$$

Arsenic in groundwater often exists as As(III) since reducing conditions occur underground and it must be oxidized to As(V) before this process can used.

Steady State of Arsenic in Water

A fairly realistic model for the mass balance of arsenic in a typical large water body, in this case Lake Ontario, is shown in Figure 11-7. The arsenic in the lake is in a steady state: as much enters the lake as leaves it each year. The

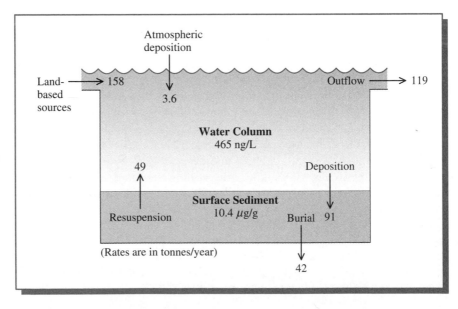

FIGURE 11-7 Steady-state model mass-balance diagram for arsenic in Lake Ontario. [Source: Adapted from S. Thompson et al., "A Modeling Strategy for Planning the Virtual Elimination of Persistent Toxic Chemicals from the Great Lakes," *Journal of Great Lakes Research* 25 (1999): 814.]

lake receives 161 tonnes of As per year, almost all of it from river and lake flows that originate in land-based sources and the rest from the atmosphere, mainly in the form of arsenic dissolved in rain and snow. About three-quarters of the annual input quantity leaves the lake by outflow (to the St. Lawrence River). The other quarter corresponds to the net amount deposited into the surface sediment, after correction for the arsenic redissolved in the water column from this source. Over time, the sediment arsenic, with a concentration of about 10 ppm, becomes buried. The net input and output of arsenic for Lake Ontario are equal, so it is in a steady state, and the concentration of the element in the water, about 0.5 ppb, remains constant with time.

Arsenic in Organic and Other Molecular Forms

The common environmental organic forms of arsenic are not simple methyl derivatives, as with mercury and lead; they are water-soluble acids that can be excreted by the body and thus are less toxic than some inorganic forms. In particular, arsenic occurs most commonly in water, as the As(V) acid H_3AsO_4 or one of its deprotonated forms; the completely deprotonated form is the arsenate ion. Like its phosphorus analog, the structure of the acid is $(OH)_3AsO$. Biological methylation in the environment by methylcobalamin initially involves the replacement of one or more —OH groups of the acid by —CH_3 groups. Monomethylation by the human liver and kidneys converts most but not all ingested inorganic arsenic to $(CH_3)(OH)_2AsO$ and then to the corresponding dimethyl acid, which are then readily excreted.

Although much daily exposure to arsenic by North American adults is due to food intake, especially meat and seafood, much of the arsenic in these sources occurs in the organic form and is therefore readily excreted. In seafood the common forms of arsenic are either the $(CH_3)_4As^+$ itself or the ion with one methyl group replaced by —CH_2CH_2OH or —CH_2COOH. The organic forms of arsenic in seafood are probably noncarcinogenic and are much less toxic than inorganic forms, as illustrated in dramatic fashion by their high LD_{50} values, of the order of thousands of milligrams per kilogram, compared to those for inorganic arsenic, which are about 1% of these values (see Table 11-2).

In contrast, neutral As(III) compounds such as *arsine*, AsH_3, and *trimethylarsine*, $As(CH_3)_3$, are the most toxic forms of arsenic. Curiously, the trimethyl compound is produced by the reaction, under humid conditions, of molds in wallpaper paste with the arsenic-containing green pigment $CuHAsO_3$ in wallpaper. Instances of mysterious illnesses and even of human "death by wallpaper" due to chronic exposure to the $As(CH_3)_3$ gas released into rooms by this mechanism have been reported. Some historians believe Napoleon was fatally poisoned by the trimethylarsine emitted from the wallpaper in his chronically damp house on the island of St. Helena, where he

TABLE 11-2	LD$_{50}$ Values for Some Common Forms of Arsenic	
Name	Formula	LD$_{50}$ (mg/kg)
Arsenous acid	H_3AsO_3	14
Arsenic acid	H_3AsO_4	20
Methylarsonic acid	$CH_3AsO(OH)_2$	700–1800
Dimethylarsonic acid	$(CH_3)_2AsO(OH)$	700–1800
Arsenocholine	$(CH_3)_3As^+CH_2CH_2OH$	6500
Arsenobetaine	$(CH_3)_3As^+CH_2COO^-$	>10,000

Source: X. C. Le, "Arsenic Speciation in the Environment," *Canadian Chemical News* (September 1999): 18.

lived in exile. There have also been episodes of human poisoning from gaseous arsine that was accidentally generated and released when aqueous solutions of As(III) as $HAsO_2$ came into contact with an easily oxidized metal such as aluminum or zinc and the arsenic was reduced:

$$2\,Al(s) + HAsO_2 + 6\,H^+ \longrightarrow 2\,Al^{3+} + AsH_3 + 2\,H_2O$$

Chromium

Chromium normally occurs in the form of inorganic ions. Its common oxidation states are +3 and +6, i.e., Cr(III) and Cr(VI), also known as *trivalent* and *hexavalent* chromium.

Under oxidizing, i.e., aerobic, conditions chromium exists in the (VI) state, usually as the **chromate ion,** CrO_4^{2-}, although under even slightly acidic conditions this oxyanion is protonated to $HCrO_4^-$.

$$H^+ + CrO_4^{2-} \rightleftharpoons HCrO_4^-$$

The oxyanions of chromium(VI) are highly soluble in water. Both the Cr(VI) ions mentioned are yellow and impart a yellowish tinge to water even at chromium levels as low as 1 ppm. (At high concentrations not encountered in the environment, chromate dimerizes to give the orange *dichromate ion,* $Cr_2O_7^{2-}$, familiar as a strong oxidizing agent in the laboratory.)

Under reducing, i.e., anaerobic, conditions chromium exists in the (III) state. In aqueous solution this state occurs as the +3 ion, i.e., Cr^{3+}. However, the aqueous solubility of this ion is not high, and Cr(III) is often precipitated as its hydroxide, $Cr(OH)_3$, under alkaline, neutral, or even slightly acidic conditions:

$$Cr^{3+} + 3\,OH^- \rightleftharpoons Cr(OH)_3(s)$$

Thus, whether chromium occurs as an ion dissolved in water or as a precipitate depends on whether the aqueous environment is oxidizing or reducing. The difference is important, since hexavalent Cr(VI) is toxic and a suspected carcinogen, whereas trivalent Cr(III) is much less toxic and even acts as a trace nutrient. Chromate ion, CrO_4^{2-}, readily enters biological cells, apparently because of its structural similarity to *sulfate ion*, SO_4^{2-}. Inside the cell it can oxidize DNA and RNA bases. Because hexavalent chromium is more toxic, more soluble, and more mobile than trivalent chromium, it is considered to pose a greater health risk. (The term *hexavalent chromium* was made famous a few years ago in the movie *Erin Brockovich*, the story of how a legal assistant battled successfully against pollution of local groundwater by this substance.)

Chromium Contamination of Water

Chromium is widely used for electroplating, corrosion protection, and leather tanning. In tanning, Cr(III) binds to protein in animal skin to form leather that is resistant to water, heat, and bacteria. As a consequence of industrial emissions, chromium is a common water pollutant, especially of groundwater under areas with metal-plating industries. It is also the second most abundant inorganic contaminant of groundwater under hazardous waste sites. The MCL for total chromium in U.S. drinking water is 100 ppb.

Most dissolved heavy metals can be removed from wastewater by simply increasing the pH, since their hydroxides are insoluble. However, Cr(VI) does not precipitate out at any pH. Owing to the low solubility and hence the low mobility of Cr(III), however, the usual way to extract chromium(VI) from water is to first use a reducing agent to convert Cr(VI) to Cr(III):

$$CrO_4^{2-} + 3\,e^- + 8\,H^+ \rightleftharpoons Cr^{3+} + 4\,H_2O$$
$$\text{(soluble)} \qquad\qquad\qquad \text{(insoluble)}$$

Reducing agents commonly employed for this conversion are gaseous SO_2 or a solution of *sodium sulfite*, Na_2SO_3. In addition, reducing the Cr(VI) to Cr(III) by adding iron in the form of Fe(II) and then adding base to precipitate Cr(III) is a common practice in purifying Cr-contaminated wastewater. Fine-grained elemental iron placed in permeable underground walls positioned in the path of flowing polluted groundwater is another application of this technique. The iron reduces the chromium, and then as Fe^{3+} it forms a complex insoluble Fe(III)–Cr(III) compound. This reduction process can occur spontaneously in soils with, e.g., Fe^{2+} or organic carbon as the reducing agent. Hexavalent chromium is quite mobile in soils, since it is not strongly absorbed by many types of soil.

However, it can be reduced to the less-mobile trivalent form by the humic substances in soils that are rich in organic matter.

The Wood Preservative CCA

Another potentially significant source of chromium to the environment stems from its presence in chromated copper arsenate (CCA), a widely used wood preservative. This substance is used to protect wooden structures, such as residential docks, destined to be used in aquatic environments. Not only chromium but also arsenic and copper leach from the structures into the water over time. For environmental and human health reasons, CCA largely replaced organic preservatives such as *creosote* and *pentachlorophenol*, mentioned in Chapter 8.

The amount of CCA forced into the wood is almost 10% of the mass of the lumber. The chromium used is initially hexavalent. However, during a period of *fixation*, which lasts for several weeks after treatment, almost all the Cr(VI) is reduced to Cr(III) by reaction with carbon in the wood. This process produces insoluble complexes that are slow to leach from the treated wood over its lifetime, since the copper and chromium at least are bound to the wood. Leaching of heavy metals from the wood is very slow by a few months after treatment, with more copper and arsenic than chromium being lost.

Green Chemistry: Removing the Arsenic and Chromium from Pressure-Treated Wood

Wood that is used for exterior construction decays in three to five years unless it is treated with pesticides that prevent destruction from termites, fungi, and other wood-destroying agents. Most of the preserved exterior wood that is presently used is commonly called *pressure-treated wood*. Pressure-treated wood is found in over 50% of homes in the United States. It is also used in decks, fences, retaining walls, piers, docks, wooden bridges, picnic tables, and playground equipment, and lasts from 10 to 20 times longer than untreated wood. Treatment of wood results in the conservation of millions of trees each year and limits the use of scarce woods that contain natural preservatives, such as redwoods.

Pressure-treated wood is produced by placing the wood in a horizontal cylinder and evacuating the cylinder, which draws out much of the moisture from the wood cells. An aqueous preservative solution is then pumped into the cylinder and the pressure is raised, forcing the preservative solution into the wood cells. In the United States the preservative solution used in 95% of pressure-treated wood is the chromated copper arsenate (CCA) discussed in the previous section.

Although the percentages vary, the most common formulation for the preservative solution is 35.3% CrO_3, 19.6% CuO, and 45.1% As_2O_3. Treatment with CCA results in wood with concentrations of copper, chromium, and arsenic of 1000–5000 mg/kg. In 2001, 7 billion board feet of pressure-treated wood (enough to build 450,000 homes) was produced, utilizing 60 million kilograms of CCA. The CCA contained 20 million kilograms of hexavalent chromium and 18 million kilograms of arsenic. A single 12-ft long 2×6 board contains about 27 g of arsenic, enough to kill more than 200 adults.

Although the preservatives are "locked" into the wood, health officials and environmentalists have long been concerned with the potential for leaching of arsenic and chromium from pressure-treated wood and the ingestion of these elements by infants and children from direct contact with the wood. Studies of the soils beneath decks made of pressure-treated wood gave copper, chromium, and arsenic concentrations averaging 75, 43, and 76 mg/kg, while control soils averaged 17, 20, and 4 mg/kg. Studies also indicate that measurable amounts of arsenic can be dislodged from the surfaces of pressure-treated wood by direct contact.

The U.S. EPA announced recently that, because of the environmental and human health concerns associated with CCA, wood producers have voluntarily ceased production (as of December 31, 2003) of CCA-treated wood intended for residential use, either as construction materials or for play structures, picnic tables, decks, fencing, patios, etc. Chemical Specialties, Inc. (CSI) in 1996 introduced a new wood preservative called *Preserve* to replace CCA, for which they earned a Presidential Green Chemistry Challenge Award in 2002. Preserve is formulated with an alkaline quaternary (ACQ) wood preservative. The active ingredients in the preparation are copper and a *quaternary ammonium salt*, $R_4N^+Cl^-$ (either didecyl dimethyl ammonium chloride or alkyl dimethyl benzyl ammonium chloride). According to the World Health Organization, none of these ingredients are mammalian or human carcinogens.

Because there are no environmental and health concerns for ACQ, under the EPA system, ACQ is registered as a nonrestricted pesticide for treatment of wood products. Analogous formulations of copper and ACQ are used as algaecides and fungicides in lakes, rivers, and streams, as well as fish hatcheries and potable water supplies. Quaternary ammonium salts are also used as surfactants in typical household and industrial detergents and disinfectants and, unlike arsenic, they have low toxicity to mammals. It is also noteworthy that the copper used in the ACQ formulations is obtained from scrap copper. ACQ-treated wood not only eliminates the cancer and toxicity concerns associated with CCA, but it offers the advantages of simplified disposal of treated wood and elimination of hazardous waste generation at the approximately 450 treatment sites across the United States.

Review Questions

1. What is a *sulfhydryl group*, and how does it interact biochemically with heavy metals? How does the interaction affect processes in the body?

2. What principle underlies the use of chelates for heavy-metal poisoning?

3. Do heavy metals bioconcentrate? Do any biomagnify?

4. What are some important sources of airborne mercury?

5. Is the liquid or the vapor of mercury more toxic? Describe the mechanism by which mercury vapor affects the human body.

6. What is an *amalgam*? Give two examples and explain how they are used.

7. Explain how the *chlor-alkali process* leads to the release of mercury to the environment.

8. Name two uses for mercury in batteries.

9. Write the formulas for the methylmercury ion, for two of its common molecular forms, and for dimethylmercury. What is the principal source of exposure of humans to methylmercury?

10. Explain why mercury vapor and methylmercury compounds are much more toxic than other forms of the element.

11. What is meant by *Minamata disease*? Explain its symptoms and how it first arose.

12. List several uses for organic compounds of mercury. Which ones have been phased out?

13. What are the two common ionic forms of lead?

14. Explain how lead can dissolve — e.g., in canned fruit juice — even though it is insoluble in mineral acids.

15. Explain why lead contamination of drinking water by lead pipes is less common in hard-water areas than in soft-water areas.

16. Why are lead compounds used in paints? Why were mercury compounds used in paints?

17. Explain why heavy-metal compounds such as PbS and $PbCO_3$ become much more soluble in acid water.

18. In what forms does lead exist in the lead storage battery?

19. What are the formulas and names of the two organic compounds of lead that were used as gasoline additives? What was their function? What organic compounds were also added?

20. Discuss the toxicity of lead, especially with respect to its neurological effects. Which sub-groups of the population are at particular risk from lead?

21. What are the main sources of cadmium in the environment?

22. Explain how *nicad* batteries operate. What other uses are made of cadmium?

23. What is the main source of cadmium to humans?

24. Describe what is meant by *itai-itai* disease and relate where it arose and why.

25. What is *metallothionein*? What is its significance with respect to cadmium in the body?

26. What are some uses of arsenic that result in contamination of the environment?

27. What organic compounds of arsenic are of environmental significance? Why is arsenic in organic acid forms not very toxic to humans?

28. What are the main health concerns about arsenic in drinking water? Why is the drinking water in many regions of Bangladesh heavily polluted with arsenic?

29. Describe how arsenic can be removed from water.

30. What are the two important oxidation states of chromium? Which one is the more toxic?

31. Explain how Cr(VI) can be removed from wastewater.

32. What is CCA? Name two toxic heavy metals it contains.

33. Complete the chart shown in outline below:

Element	Common ionic forms	Common organo-metallic forms	Most toxic forms
Mercury			
Lead			
Cadmium			
Arsenic			
Chromium			

Green Chemistry Questions

1. The replacement of lead with yttrium in electrodeposition coatings won PPG a Presidential Green Chemistry Challenge Award.

(a) Which of the three focus areas (see the Introduction to Green Chemistry) for these awards does this award best fit into?

(b) List one of the twelve principles of green chemistry (see the Introduction to Green Chemistry) that are addressed by the green chemistry developed by PPG.

2. What environmental advantages does the use of yttrium oxide have over the use of lead oxide in electrodeposition coatings?

3. What environmental advantages does electrodeposition offer over spray painting?

4. The removal of arsenic and chromium from pressure-treated wood won Chemical Specialties, Inc. a Presidential Green Chemistry Challenge Award.

(a) Which of the three focus areas (see the Introduction to Green Chemistry) for these awards does this award best fit into?

(b) List one of the twelve principles of green chemistry (see the Introduction to Green Chemistry) that are addressed by the chemistry developed by Chemical Specialties, Inc.

Additional Problems

1. A man who weighs 50 kg eats 1 kg of fish a day. If the fish contains the legal limit of 0.5 ppm of methylmercury, and assuming that this substance is evenly distributed within his body, calculate the steady-state concentration of methylmercury that will result. By comparison of this concentration to that of the fish, decide whether biomagnification is occurring in the transfer of methylmercury from the fish to the man. Would your answer to the latter question differ if he ate only 0.2 kg of fish a day?

2. By adding the dissolution reaction for PbS(s) to that reaction of S^{2-} with water, determine the overall reaction when PbS dissolves and most of the resulting sulfide ion reacts with water. Calculate the solubility of PbS with and without the subsequent reaction of sulfide, given that for HS^-, $K_a = 1.3 \times 10^{-13}$.

3. **(a)** Fit approximately an exponential decay curve to the distribution of blood lead levels among children, based on the portion of the curve in Figure 11-5b from 20 ppb (the zero point of the function) to higher levels. By integration, determine the total percentage of children with levels in excess of 100 ppb that your function predicts. **(b)** What does the fact that the curve in Figure 11-5b does not continue to rise as the blood lead level comes close to zero tell us about the background level of lead in the environment?

4. Although (for the sake of simplicity) many authors often state that the predominant form of arsenic in water is the arsenate ion, the fact is that, since the AsO_4^{3-} ion is basic, the forms $HAsO_4^{2-}$, $H_2AsO_4^-$, and H_3AsO_4 will all be present. Given that for H_3AsO_4 the successive acid dissociation constants are 6.3×10^{-3}, 1.3×10^{-7}, and 3.2×10^{-12}, deduce the predominant form of arsenic in waters of pH = 4, 6, 8, and 10.

5. The object of this problem is to estimate the mass of lead that would have been deposited annually on each square meter of land near a typical, busy, six-lane freeway from the lead compounds emitted by cars using the roadway. Use reasonable estimates for the number of cars passing a point each day and their average mileage per liter or gallon of gasoline. Assume that the gasoline contained about 1 g Pb/gal or 0.2 g/L, and make the approximation that half the lead was evenly deposited within 1000 m on each side of the freeway.

6. How does the phenomenon of acid rain indirectly affect the risk to human health from mercury, lead, and cadmium?

7. The 2+ ions of mercury, lead, and cadmium each form a series of complexes by attaching to

themselves in successive equilibrium reactions up to four chloride ions. Deduce the formulas for the species for one of these metals, including the net charges on the complexes. Would the complexes with three or four chlorines be more likely to be found in fresh water or in seawater?

8. By reference to textbooks in your library or to websites on heavy metals and/or water pollution, determine why copper is also considered to be toxic, and find out what types of organisms are at risk from elevated levels in the environment. Does speciation affect the toxicity of copper?

9. Based on the material in this chapter, write a paragraph supporting your choice of which of the five metals you consider still requires the most regulatory action for environmental control.

10. The vapor pressure of mercury at 20°C is 0.001201 torr. Imagine a chemistry laboratory somewhere that has been in use for decades and has had so many mercury spills over the years that it has accumulated sufficient liquid mercury in various areas, e.g., in cracks in the floor and in other crevices, that the Hg liquid–vapor equilibrium has been established. (This is truly a nightmare scenario!) What is the concentration of Hg vapor in this laboratory room in milligrams per cubic meter? Compare this to the threshold limit value of 0.05 mg/m^3 established by the American Conference of Governmental Industrial Hygienists for safe exposure based on a 40-hour work week. [Note: 1 atm = 760 torr.]

11. In the chapter it is mentioned that both Pb^{2+} and Cd^{2+} are similar in size to Ca^{2+} and thus can become incorporated into bones by replacing Ca^{2+} ions. Explain why there is a size similarity among these three ions, despite their very different atomic masses and the fact that the metals are found in three different rows of the periodic table.

12. Lena Ma, a chemist at the University of Florida, Gainesville, has recently reported studies of a particular fern species with an incredible *hyperaffinity* for arsenic. For example, after two weeks, ferns grown in soils containing 6, 500, and 1500 ppm As were found to contain 755, 7849, and 15,861 ppm As in their fronds. Determine the enrichment factor in each case, and discuss the trend in terms of soil As content. Suggest a potential use for this fern. Finally, suggest two reasons why a fern might have adapted such an affinity for arsenic.

Further Reading

1. S. Krishnamurthy, "Biomethylation and Environmental Transport of Metals," *Journal of Chemical Education* 69 (1992): 347.

2. T. W. Clarkson, "The Three Modern Faces of Mercury," *Environmental Health Perspectives* 110, supplement 1 (2002): 11.

3. G. J. Myers and P. W. Davidson, "Does Methylmercury Have a Role in Causing Developmental Disabilities in Children?" *Environmental Health Perspectives* 108, supplement 3 (2002): 413.

4. C. Hanish, "Where Is Mercury Deposition Coming From," *Environmental Science and Technology* (1 April 1998): 716A.

5. T. W. Clarkson, "Mercury: Major Issues in Environmental Health," *Environmental Health Perspectives* 100 (1992): 31.

6. R. Hoffmann, "Winning Gold," *American Scientist* 82 (1994): 15.

7. H. W. Mielke, "Lead in the Inner Cities," *American Scientist* 87 (1999): 62.

8. R. A. Goyer, "Results of Lead Research: Prenatal Exposure and Neurological Conse- quences," *Environmental Health Perspectives* 104 (1996): 1050.

9. P. A. Baghurst et al., "Environmental Expo- sure to Lead and Children's Intelligence at the Age of Seven Years," *New England Journal of Medicine* 327 (1992): 1279; "Exposure to Environmental Lead and Visual-Motor Integration at Age 7 Years: The Port Pirie Cohort Study," *Epidemiology* 6 (1995): 104.

10. R. L. Canfield et al., "Intellectual Impairment in Children with Blood Lead Concentrations Below 10 μg per Deciliter," *New England Journal of Medicine* 348 (2003): 1517.

11. O. Beattie and J. Geiger, *Frozen in Time: The Fate of the Franklin Expedition* (Vancouver, Canada: Greystone Books, Douglas and McIntyre, Ltd., 2000).

12. T. A. Tsuda et al., "Inorganic Arsenic: A Dangerous Enigma for Mankind," *Applied Organometallic Chemistry* 6 (1992): 309.

13. A. H. Smith et al., "Cancer Risks from Arsenic in Drinking Water," *Environmental Health Perspectives* 97 (1992): 259.

14. F. Pearce, "Arsenic's Fatal Legacy Grows," *New Scientist* (9 August 2003): 4.

15. A. Lykknes and L. Kvittingen, "Arsenic: Not So Evil After All?" *Journal of Chemical Education* 80 (2003): 497.

16. J. A. Hingston et al., "Leaching of Chromated Copper Arsenate Wood Preservatives: A Review," *Environmental Pollution* 111 (2001): 53.

Websites of Interest

Log on to www.whfreeman.com/envchem3e/ and click on Chapter 11.

Environmental Instrumental Analysis 4	Inductively Coupled Plasma Determination of Lead

The analysis of heavy metals in environmental samples is now routinely accomplished by the spectroscopic method discussed in this box.

Anthropogenic lead contamination in the environment in North America and Europe has been decreasing since tetraethyl- and tetramethyllead were phased out of gasolines over the last two decades. Presently, major sources of environmental lead are residual lead in aerosol particles and dust near roadways; smelting ash; lead pipes in plumbing (*plumbum* is Latin for lead); battery manufacture, recycling, and disposal; cigarette smoke; and old lead-based paints. Like some of the other toxic metals under discussion in this book, lead can be identified quickly and accurately using an atomic emission technique called **inductively coupled plasma spectroscopy** (ICP).

Each element has a unique atomic structure with electrons in well-defined energy levels. The movement of electrons between these levels, which requires the absorption or emission of energy, is also well defined, and therein lies the key to atomic emission spectroscopy. If the atoms in a sample are excited using a very-high-energy source—such as a flame, spark, or plasma—many of the atoms' electrons will be raised to higher energy levels. Almost immediately these excited-state electrons will relax by returning to the ground state, but that return is accompanied by the *emission* of a photon whose energy corresponds to the difference between the excited-state and ground-state energy levels. And just as the energy of the promotion is

well defined, which means that only specific energies can be absorbed by a particular atom, the energy released by this relaxation—and the photon containing that energy—is very specific. Since photon energy is strictly related to wavelength (see Chapter 1), a means of elemental detection can be based on detecting the light emitted from a sample after the atoms in it are excited by some means: that light is characteristic of the atoms excited in the sample.

In the case of ICP, the excitation source is a very-high-temperature plasma. Light emitted from sample atoms injected into the plasma is collected by lenses and mirrors and focused onto the diffraction grafting. This grating separates individual wavelengths (as does a prism) and focuses the light on a photomultiplier tube (PMT), which converts the light into electronic signals. The wavelengths of the light are specific to the elements in the plasma that emitted the photons, and the intensity of the light as measured by the PMT reflects the concentration of that element in the sample. Atomic identity and amount are the two parameters of emission spectroscopy that must be determined if it is to be used as an analytical tool in environmental analysis.

The components of an ICP spectrometer depend on which of two basic designs are used. The first is called a *sequential spectrometer* and the second a *simultaneous spectrometer*. The first design uses only one PMT (detector) and requires a process of scanning through the emission wavelengths to determine multiple elements in one sample. This scanning process is usually accomplished by very exacting rotation of the diffraction grating in order to

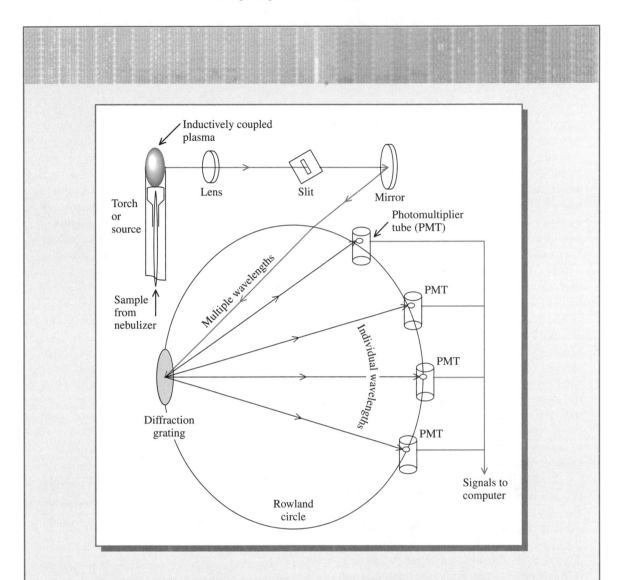

separate the light emissions. The rotation is carefully controlled by a computer so that the signal generated by the PMT can be correlated with the wavelength of light that falls on it; i.e., the computer knows by the grating position—which it is controlling—which wavelength is striking the PMT. Sequential ICPs require time to scan and sample the light for each element's emission, but the cost of a sequential instrument is modest compared to the second, more powerful, spectrometer design, the simultaneous ICP spectrometer.

Simultaneous spectrometers have a similar configuration to the simpler sequential instru-

(continued on p. 564)

Inductively Coupled Plasma Determination of Lead (continued)

ments—they both have a plasma source (also called a torch) and a monochromator to separate the light into individual wavelength components. However, in the simultaneous instrument, the spatially separate wavelength beams are simultaneously focused on many PMTs—one PMT for each wavelength to be detected instead of a single PMT as in the sequential instrument. In this way, many elemental emission wavelengths can be detected at one time without having to scan between them, and therefore many elements can be determined in the same sample simultaneously (see the diagram on page 563). These multiple PMTs are arranged along a curve called the Rowland circle. The instrument's manufacturer places the appropriate PMTs at specific locations along the Rowland circle at points that correspond to the place where light of a known wavelength will strike. And once again the wavelength of the light identifies the element emitting it and its intensity conveys the concentration of that element that was injected into the plasma.

Street dust is a measure, to some degree, of the lead to which people in cities are exposed by inhalation. It is thought that most of the lead in urban dust originated in the past from particulates released by the combustion of leaded gasoline. With that in mind, a recent comparison has been made of the lead measured in dust samples taken from the streets of Manchester, England, in 1997 and a similar study in 1975. Samples of dust, dirt, and soil were scraped off streets with a spatula and analyzed by atomic absorption spectrometry (AAS) and by a modified ICP technique. In AAS, a high-temperature flame or hot carbon tube is used to atomize the element of interest (lead in our case) in a dust sample that has been dissolved in an acid solution. Once the lead atoms have been atomized (converted to elemental Pb) and vaporized by the high temperature, a source lamp with a specific wavelength is shone through the vaporized atoms. Since only specific energies can be absorbed by Pb atoms (because the gaps between the atomic energy levels are quantized and specific to that element), the amount of absorption of the source lamp's light can be used as a measure of the amount of lead in the beam. A PMT on the opposite side of the sample atoms from the source lamp records a decrease in its light intensity when the sample is introduced. The larger the absorption, the more lead atoms are preszent in the sample.

Category	1975 ppm Pb (no. of samples)	1997 ppm Pb (no. of samples)
> 100 cars/hour	1001 ± 40 (180)	577 ± 53 (17)
< 10 cars/hour	933 ± 186 (53)	536 ± 93 (13)
Playgrounds, parks, gardens	1014 ± 206 (49)	572 ± 77 (47)

The table shows abridged results from both studies. Average lead amounts are reported for three different sampling categories: high traffic, low traffic, and areas where children play.

Reference: S. M. Nageotte and J. P. Day, "Lead Concentrations and Isotope Ratios in Street Dust Determined by Electrothermal Atomic Absorption Spectrometry and Inductively Coupled Mass Spectrometry," Analyst 123 (1998): 59–62.

Ion Chromatography of Environmentally Significant Anions

The quantitative determination of levels of environmentally important ions, such as those discussed in the preceding chapters, can be accomplished using chromatographic methods described in this box.

The need to determine the prevalence of common anions like phosphate (PO_4^{3-}), nitrate (NO_3^-), or fluoride (F^-) isn't immediately clear. The biospheric significance of these ubiquitous ions is not as obvious as is that of PCBs, for example, or pesticides, or toxic metals like mercury or cadmium. The reason that these ionic components are important is that they can give an indication of the relative reduction/oxidation potential in an aqueous sample taken from an environment such as a stagnant lake (PO_4^{3-}), or of the contamination of groundwater from fertilizer runoff (NO_3^-), or whether municipal water supplies need to be supplemented with fluoride (F^-) for the health of children's teeth. Although these charged ions can be detected by widely available ultraviolet detectors common in most high-performance liquid chromatographic systems, a more sensitive means of detection involves ionic conductivity. This chromatographic method is called *ion chromatography with ionic conductivity detection*. Although cations can also be separated by ion chromatography (IC), only anionic separations will be discussed here.

The heart of the separation process in an ion chromatograph is a short column (10–15 cm) packed with small-diameter particles called *ion-exchange resins*. These are often made of a styrene/divinylbenzene polymer or microparticles of silica coated with compounds containing an anionic functional group such as a quaternary amine, $-N(CH_3)_3^+OH^-$, or a primary amine, $-NH_3^+OH^-$, when they are to be used for anion separation. Ion-exchange resins with sulfonate, $-SO_3^-H^+$, or carbonate, $-COO^-H^+$, functional groups are used for cationic separation.

The actual process of chromatographic separation occurs after a sample containing analyte anions (and their associated cations) are injected into the chromatographic column. With gas chromatography (see Environmental Instrumental Analysis Box 2), the mobile phase is an inert gas that does not chemically interact with the chromatographic surface. The moble phase in ion chromatography, on the other hand, is a solution of cations and anions with a carefully controlled pH; buffers are often used. This complex mixture of mobile-phase ions—carefully chosen for each group of analyte ions to be separated—interacts with the analyte ions and the functional groups of the column's chromatographic surface. That interaction takes many forms depending on a number of variables but might be best described in the most common ion chromatography as a competition of the mobile-phase anions and the analyte anions for chromatographic sites on the packing material (the charged functional groups such as $-N(CH_3)_3^+$ or $-SO_3^-$). This competition yields different overall travel times for each of the analytes as they pass down the column; some are retained longer than others. (The overall down-column movement is provided by pumping of the mobile phase by an

(continued on p. 566)

Ion Chromatography of Environmentally Significant Anions *(continued)*

external liquid pump.) Different analyte travel times—as in gas chromatography—translate into different exit (or retention) times for each anion in the original mixture. The result is chromatographic separation.

The process for anionic ion chromatographic retention by ion-exchange resins can be represented by the equation

$$RN(CH_3)_3{}^+HCO_3{}^-(s) + \textbf{anion}^-(aq) \longrightarrow$$
$$RN(CH_3)_3{}^+\, \textbf{anion}^-(s) + HCO_3{}^-(aq)$$

In this equation the term **anion**$^-$ represents any of the analyte anions mentioned above. When a test sample is injected onto the column, this anion is quickly retained by complexation with the stationary phase near the head of the column. The next step in the chromatographic process takes place as a mobile phase with a carefully controlled amount of an anionic ion such as bicarbonate, $HCO_3{}^-$, is pumped through the column. The presence of the bicarbonate anion in the mobile phase forces the equilibrium in the above equation to the left and the retained analyte anion is freed and moves down the column in the flowing mobile phase. As the analyte moves along, it repeatedly undergoes this same process of retention and movement (or exchange between the stationary and mobile phase). Most importantly, different analyte anions (e.g., fluoride, phosphate, or chloride) undergo this exchange process to differing degrees and therefore travel at different overall rates during their time in the IC column. The result is that different analytes exit the chromatographic column at different times, i.e., chromatographic separation has taken place.

The task of detecting analyte anions in the presence of the anions always present in the mobile phase is by no means a trivial one. Since both kinds of anions—analytes and mobile phase—conduct electricity, using an ordinary conductivity cell as a detector at the end of the IC column is normally not practical. The problem is especially difficult because, in order to get adequate separation of some important anions, the mobile phase often has to have high ionic content to displace the analyte anions from the chromatographic surface, something that is obviously required for separation. Therefore, most of the ionic conductivity passing through the detector is attributable to the mobile-phase ions and not the analyte—an unworkable situation when one is trying to detect the analyte anions by their conductivity.

An ingenious solution to this problem is called *conductivity suppression* or *eluent (mobile-phase) suppression*. This technology converts the mobile-phase anions from an easily dissociated ionic form to a (soluble) molecular form that does not strongly influence the signal produced by the conductivity detector. The suppression module is placed after the chromatographic column, but before the conductivity detector. In an anion-exchange system, the suppression module might carry out the reaction

$$Na^+(aq) + HCO_3{}^-(aq) + \textbf{resin}^-H^+(s) \longrightarrow$$
$$\textbf{resin}^-Na^+(s) + H_2CO_3(aq)$$

Here **resin**$^-$**H**$^+$ represents a *cation*-exchange resin that will exchange cations—instead of anions as in the chromatographic column described. This process basically prevents (or suppresses) the mobile phase's anions from contributing to the conductivity by converting current-conducting bicarbonate anion into relatively undissociated H_2CO_3. Therefore, the conductivity detector's signal is based almost completely on the passage of analyte anions through the detector cell. (These anions are not affected by the cation ionexchange resin.) This results in lower detection limits for the analytes of interest and a more stable baseline (less noise and drift) than a similar system without eluent suppression.

The figure below is a schematic of an ion chromatographic system, showing the injector, the chromatographic column, the eluent suppressor module, the conductivity detector, and the processes that occur at each step.

The figure in the next column is an example of the kind of chromatogram that a system of this type would generate. The anions detected are fluoride, chloride, phosphate, and nitrate. As with all chromatograms, detector signal intensity is plotted versus time.

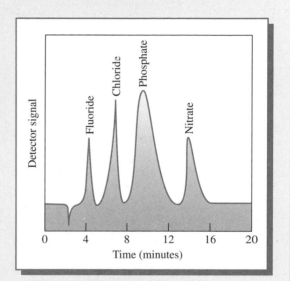

Reference: P. Janos and P. Aczel, "Ion Chromatographic Separation of Selenite and Selenate Using a Polyanionic Eluent," *Journal of Chromatography A* 749 (1996): 115–122.

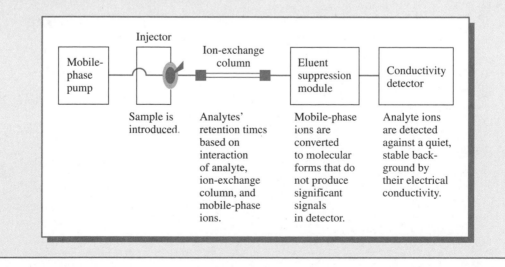

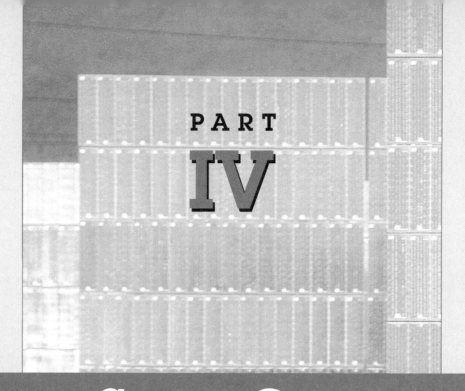

PART

IV

SOME OTHER ENVIRONMENTAL CONCERNS

CHAPTER 12

Hazardous and Municipal Wastes, and the Contamination of Soils and Sediments

In this chapter we turn our attention to the environmental aspects of the solid state—particularly of soil and of the sediments of natural water systems—and of ways that polluted soils and sediments can be remediated. A closely related issue is the nature and disposal of concentrated wastes of all kinds, including both domestic garbage and hazardous waste, and of their possible recycling.

The material in this chapter has been arranged in the order of generally increasing toxicity and hazard. Thus we begin with the least toxic substances—domestic and commercial garbage—and consider its disposal by landfilling, incineration, or recycling. We then consider soils and sediments and their contamination by chemicals. Finally, we look at hazardous wastes and some of the high-technology methods that are being developed to dispose of them.

A tractor moving shredded paper in the warehouse of a paper-recycling plant. The paper can be reused or recycled in several different ways. (Digital Vision)

Domestic and Commercial Garbage: Its Disposal and Minimization

The great majority of the material that we discard and that must be disposed of is not hazardous but is simply *garbage* or *refuse*. The greatest single constituent of this **solid waste** (defined as waste that is collected and transported by a means other than water) is construction and demolition debris, almost all of which is either reused or is eventually buried in the ground. The second largest volume of waste is that generated by the commercial and industrial sectors, followed by the domestic waste generated by residences. Typically, a North American generates about 2 kg of domestic and commercial waste a day, twice as much as the average European. In these discussions we will not consider the much larger amounts of waste generated by the petroleum industries, by agriculture, as ashes from power plants, or as sewage, which was discussed in Chapter 10.

A breakdown by the type of solid waste typically generated in countries at various levels of economic development is shown in Figure 12-1. Notice that the fraction of the waste that is vegetable matter declines as the level of economic development rises. The opposite is true of paper, which in industrialized countries is the largest single component of waste and dominates commercial-sector waste. Historically, the largest component of paper waste was newspapers; now the volume of paper packaging is similar. The amount of packaging has grown, in part, because so many goods now are produced far from their ultimate destination and must be transported safely over long distances. Plastics, glass, and metals each account for about one-tenth of the volume of solid waste in developed countries, whereas organic matter (food waste) accounts for about twice this value. These proportions would differ significantly in areas that collect materials for recycling or composting: the glass and metals components would be much smaller.

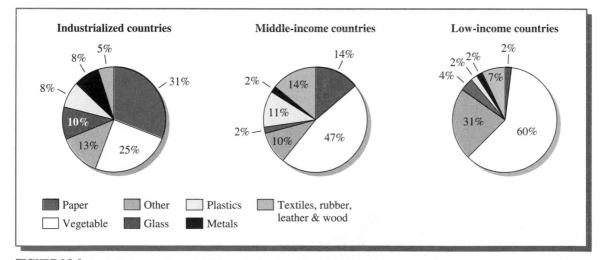

FIGURE 12-1 Typical composition of solid waste for countries at different levels of economic development. [Source: "Waste and the Environment," *The Economist* (29 May 1993): 5 (Environment Survey section).]

Burying Garbage in Landfills

The main method used for disposal of **municipal solid waste,** MSW, is to place it in a **landfill** (also variously called a garbage dump or a rubbish tip), which is a large hole in the ground that usually is covered with soil and/or clay after it is filled. For example, 85–90% of domestic and commercial waste is currently landfilled in the United Kingdom, about 6% is incinerated, and the same fraction is recycled or reused; similar figures apply to many municipalities in North America. Landfilling dominates the disposal methods because its direct costs are substantially lower than disposal by any other means.

In the past, landfills were often simply large holes in the ground that had been created by mineral extraction—especially old sand or gravel pits. In many instances they leaked and contaminated the aquifers that lay beneath them; this was especially true for landfills that used former sand pits, since water easily percolates through sand. These landfills were not designed, controlled, or supervised, and they accepted many types of wastes, including hazardous ones.

Modern municipal landfills are much more elaborately designed and engineered, often accept no hazardous waste, and have their sites selected to minimize impact on the environment. The components of a typical modern landfill are illustrated in Figure 12-2.

In a **sanitary landfill** the MSW is compacted in layers (to reduce its volume) and is covered with about 20 cm (8 in.) of soil at the conclusion of each day's operations. Thus the landfill consists of many adjacent *cells*, each corresponding to a day's waste (Figure 12-2). After one layer of cells is completed, another is begun, and the process is continued until the hole is filled. Usually, the landfill is eventually capped by a meter or so of soil, or preferably clay, a material that is fairly impervious to rain. A *geomembrane* made of plastic may be added on top as a liner instead of the clay or over it. The system recommended by the U.S. EPA is illustrated in Figure 12-3.

During the time that municipal wastes in a landfill are decomposing—aerobically at first, then anaerobically after a few months or a year—water from precipitation, liquid from the waste itself, and groundwater that seeps into the landfill all percolate through the garbage, producing a liquid called **leachate.** This liquid contains dissolved, suspended, and microbial contaminants extracted from the solid

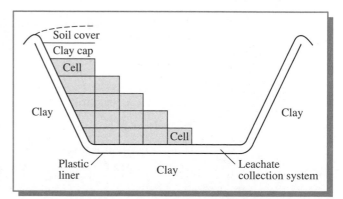

FIGURE 12-2 Components of a modern landfill (in the process of being filled).

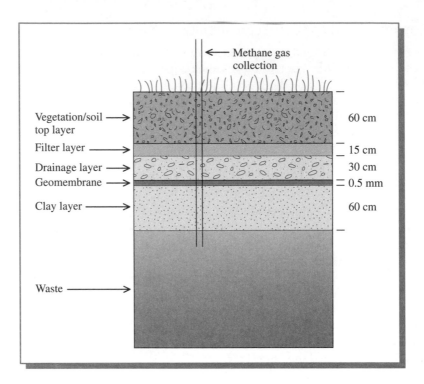

FIGURE 12-3 Landfill design with cover system recommended by the U.S. EPA.

waste. Leachate volume is relatively high for the first few years after a site is covered. Typically, leachate contains

- volatile organic acids such as *acetic acid* and various longer-chain fatty acids,

- bacteria,

- heavy metals, usually in low concentration (those of most concern in leachate are lead and cadmium), and

- salts of common inorganic ions such as Ca^{2+}.

The **micropollutants** present even in MSW leachate include common volatile organic compounds such as *toluene* and *dichloromethane*.

Stages in the Decomposition of Garbage in a Landfill

There are three stages of decomposition in a municipal landfill. Operating landfills still receiving garbage undergo all three stages simultaneously in different regions or depths. In practice, only food and yard waste biodegrade.

Rubber, plastics, and much of the paper content of garbage are very slow to degrade.

• In the first, short, **aerobic stage,** oxygen is available to the waste and it oxidizes organic materials to CO_2 and water with the release of heat. The internal temperature can rise to 70–80°C, since the reactions are exothermic. The carbon dioxide released from the organic matter as it decomposes makes the leachate acidic, thereby further facilitating its ability to leach metals encountered in the waste. Since the majority of biodegradable material is *cellulose,* whose empirical formula is approximately CH_2O, we can approximate this phase of the reaction as

$$CH_2O + O_2 \longrightarrow CO_2 + H_2O$$

Some organic matter is partially oxidized to aldehydes, ketones, and alcohols, which give fresh waste its characteristic sweet smell.

• In the second, **anaerobic acid phase,** the process of *acidic fermentation* occurs, generating *ammonia, hydrogen,* and **carbon dioxide** gases and large quantities of partially degraded organic compounds, especially organic acids. The pH of the leachate in this phase is 5.5–6.5 and is chemically aggressive. Other organic and inorganic substances dissolve in this leachate due to its acidity. Again, carbon dioxide is released, but no **methane.** This phase of the reaction can be approximated by the reaction

$$2 CH_2O \longrightarrow CH_3COOH$$

although longer-chain fatty acids, which subsequently decompose into acetic acid, are formed initially, as is hydrogen gas.

In this phase the leachate has a high oxygen demand (see BOD and COD, Chapter 9), as well as relatively high heavy-metal concentrations. Anaerobic decomposition produces volatile carboxylic acids and esters, which dissolve in the water present. The sickly sweet smell that emanates from landfills during this phase is due to these esters and to thioesters.

• The third, **anaerobic**—or **methanogenic**—**stage** starts about six months to a year after coverage and can continue for very long periods of time. Anaerobic bacteria work slowly to decompose the organic acids and hydrogen that were produced in the second stage. Since the organic acids are consumed in the process, the pH rises to about 7 or 8, and the leachate becomes less reactive. The main products of this stage are carbon dioxide and methane. To a first approximation, the overall reaction here is

$$CH_3COOH \longrightarrow CH_4 + CO_2$$

Methane generation usually continues for a decade or two and then drops off relatively quickly. Some methane is also formed when hydrogen gas combines with carbon dioxide. Much lower BOD values and a smaller volume are associated with landfills in this phase. Because the leachate is not acidic in this phase, the heavy-metal concentrations drop since these substances are not as soluble at higher pH.

Often the methane gas produced by a landfill is vented to the atmosphere by being directed into wells or gravel-packed seams in the landfill. In some municipalities the methane gas is burned as it is released through vents (see Figure 12-3) rather than being released into the air. This treatment of the methane is especially desirable, given that the greenhouse gas potential of CH_4 is much greater than that of the CO_2 produced by its combustion (Chapter 4). The heat produced from the combustion of this gas can be used for practical purposes.

PROBLEM 12-1

Calculate the volume of methane gas, at 15°C and 1.0 atm pressure, that is released annually by the anaerobic decomposition of 1 kg of garbage, assuming the latter is 20% biodegradable organic in nature and that decomposition occurs evenly over a 20-year period. [Hint: Add together the equations for the two anaerobic stages of decomposition.]

Leachate from a Landfill

Engineering is needed to control the leachate from a landfill. Otherwise the liquid can flow out at the bottom of the landfill and percolate through porous soil to contaminate the groundwater below it. Alternatively, if the soil under the landfill is nonporous, the leachate can build up and gradually overflow the site (the overflowing bathtub effect), possibly contaminating nearby surface waters.

The typical components used to control the leachate consist of

• a **leachate collection and removal system,** followed by treatment of the liquid. Often the potential effect of leachate on groundwater is monitored by digging and testing several wells in the vicinity.

• a **liner** placed around the walls and bottom of the landfill. The liner material is either synthetic (e.g., a plastic such as 2-mm-thick high-density polyethylene) or natural (e.g., compacted clay). The material chosen is impervious to water and will largely prevent the leakage of the contaminated leachate into the groundwater, especially if and when the collection system fails due to clogging, etc. Since 1991 new landfills in the United States must have at least six layers of protection between the garbage and the underlying groundwater! Liners have been developed that consist of *bentonite clay* — which is an excellent sealant and efficiently binds heavy metals, preventing their migration out of the landfill — sandwiched between two layers of a plastic such as polypropylene.

Leachate treatment systems must address all the liquid's major components. The treatment of leachate, usually done at a sewage treatment plant, is accomplished by aerobic degradation to rapidly decrease the BOD, sometimes using advanced oxidation methods that employ ozone (Chapter 10). In the past, collected leachate was often simply returned to the top of the landfill, since, during its second percolation through the waste, much of its organic content would be biologically degraded; however, this practice is now discouraged in the United States.

Incineration of Garbage

Besides landfilling, the most common way to dispose of wastes, particularly organic and biological ones, is by **incineration:** the oxidation by controlled burning of materials to simple, mineralized products such as carbon dioxide and water. The primary incentive in the incineration of municipal solid waste is to substantially reduce the *volume* of material that must be landfilled. In the case of toxic or hazardous substances, an even more important goal is to eliminate the toxic threat from the material. Incineration of hospital wastes is done to sterilize them as well as to reduce their volume.

Many municipalities throughout the world burn domestic garbage in incinerators. For example, Japan and Denmark burn more than half their domestic waste, but the practice is banned in some areas. The combustible components of the garbage, such as paper, plastics, and wood, provide the fuel for the fire. The most common domestic MSW incinerators are one-stage **mass burn** units; the two-stage **modular** type is more modern. In the latter, wastes are placed in the primary chamber and burn at a temperature of about 760°C. The gases and airborne particles that result from the first stage are then burned more completely, at temperatures in excess of 870°C, in the secondary combustion chamber. The quantity of waste gases that must later be controlled is greatly reduced in the two-stage units compared to the one-stage unit, although the gases are further heated as they exit the one-stage unit to produce more complete combustion. In some incinerators an attempt is made to recover some of the heat of the combustion processes and to convert it to steam, hot water, or even electricity.

The output from municipal incinerators includes not only the final gases but also solid residues that amount to about one-third of the initial weight of the garbage. **Bottom ash** is the noncombustible material that collects at the bottom of the incinerator. **Fly ash** is the finely divided solid matter that is usually trapped by environmental pollution controls in the stack to prevent it from being released into the outside air. Much of the ash consists of the inorganic constituents of the waste, which form solids rather than gases even when fully oxidized. Although fly ash accounts for only 10–25% of the total ash mass, it is generally the more toxic component, since heavy metals and dioxins and furans readily condense onto its small particles. The low density and small particle character of the ash make inadvertent dispersal into the environment a significant risk. Of particular concern are heavy metals in the ash, which could potentially be leached from it and pollute nearby surface water and groundwater. For many years it was common for incinerator ash to be taken to a hazardous waste landfill. Techniques such as the addition of an adhesive or melting and vitrification have now been developed to solidify ash into a leach-resistant material that need not be classified as hazardous waste. In some countries such as Denmark and the Netherlands the ash is mainly recycled into asphalt.

The main environmental concern about incineration is the air pollution that it generates, consisting of both gases and particulates. The emission controls on MSW incinerators can control a large fraction, but not all, of the toxic substances emitted into the air from the combustion process. About half the capital costs of new incinerators is spent on air pollution control equipment. Typically the controls include a **baghouse filter,** which is made of woven fabric and is used to filter particulates, especially those with diameters over 0.5 μm, from the flow of output gas. Periodically the bags are shaken or the air flow is reversed to collect the fly ash. Also typical is a **gas scrubber,** which is a stream of liquid or solid that is passed through the gas stream, removing some particles and gases. If the liquid stream consists of *lime* and water [$Ca(OH)_2$], or if the solid stream consists of lime, acid gases such as HCl and SO_2 are efficiently removed since they are neutralized to salts by the lime. Heavy metals are also captured by the alkaline environment, since they form insoluble hydroxides. In some modern installations, *nitrogen oxides* are removed by spraying ammonia or *urea* into the hot exhaust gases (recall the chemistry explained in Chapter 2). In another new technology used in garbage incinerators, activated charcoal or lignite coke powder is blown into the exhaust gases, which are subsequently filtered by baghouse; much of the dioxin, furan, and mercury content of the exhaust gases is removed, since these components adsorb onto the charcoal or coke surface.

Although public concern has centered on emissions from *hazardous waste* incinerators (to be discussed later in the chapter), several U.S. surveys in the 1990s indicated that much more dioxin and furan emissions emanate from medical waste and municipal waste incinerators than from hazardous waste ones, although cement kiln units used for hazardous wastes (discussed in a later section) also make a significant contribution. There are more than 1000 medical waste incinerators in the United States. Emissions to the air from incinerators are most likely to happen during start-up and when equipment fails. Because medical waste and backyard barrel garbage incinerators operate in a start-and-stop mode, they tend to produce more airborne pollutants per unit mass of incinerated waste than do larger incinerators. Overall, municipal, backyard, and medical incinerators are believed to be a major anthropogenic source of both mercury and dioxins/furans in the U.S. environment and a moderately important source of cadmium and lead.

Green Chemistry: Polyaspartate—A Biodegradable Antiscalant and Dispersing Agent

In pipes, boilers, water cooling systems, and other devices that handle water, scale buildup (Figure 12-4) tends to reduce water flow and heat transfer, thereby lowering efficiency. In addition, scale may lead to corrosion and damage of these devices. Scale is generally the result of the precipitation of insoluble compounds such as *calcium carbonate, calcium sulfate,* and *barium sulfate.* Compounds called *antiscalants* or *dispersants* are employed to prevent the buildup of scale. Whereas antiscalants prevent the formation of scale, dispersants allow its formation but maintain the scale in a state of suspension so that it can simply be washed away.

One of the most commonly used antiscalants and dispersants is the polyanion **polyacrylate** or PAC:

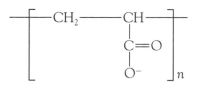

Short chains of this polymer act as antiscalants while longer chains act as dispersants. The anionic *carboxylate groups,* —COO^-, of PAC are able to form complexes with the cations (such as calcium and barium) normally found in scale, thus preventing the formation of scale or dispersing it. Globally, several hundred million kilograms of PAC are produced each year, a significant portion of which is used as a dispersant or antiscalant. Although PAC is nontoxic, it is nonvolatile and does not degrade in the environment. When used for water treatment, it builds up in lakes and streams, or at best it must be removed in wastewater treatment plants as a sludge and then landfilled.

To prevent this environmental burden, biodegradable antiscalants and dispersants such as **polyasparate** have been developed. Polyaspartate can be used to replace PAC, but because it undergoes biodegradation to innocuous products

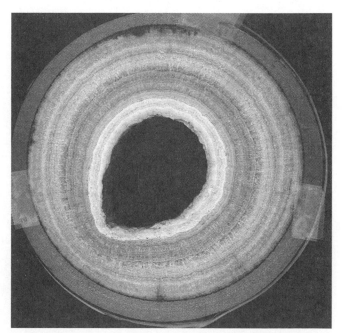

FIGURE 12-4 Scale buildup in a water pipe. (Ward Lopes)

(such as carbon dioxide and water), it eliminates the need for removal in wastewater treatment plants and disposal in landfills.

Although the performance of polyaspartate is comparable to that of PAC, its price was formerly prohibitive. The Donlar Corporation developed a new synthesis of polyaspartate that lowered the cost of the polymer so that it was competitive with PAC. For this accomplishment Donlar won a Presidential Green Chemistry Challenge Award in 1996. Donlar's synthesis (Figure 12-5) begins by heating *aspartic acid* (a naturally occurring amino acid) to produce *polysuccinimide*, followed by basic hydrolysis to produce polyaspartate. This straightforward synthesis is not only economically desirable but also environmentally sound. The first step simply requires heat and yields only water as a by-product, while the second step uses water under basic conditions to produce the desired product. The product of this synthesis is generally called **thermal polyaspartate** (TPA) because of the heat used in the synthesis. Polyaspartate can also be used in fertilizers (to enhance uptake of nutrients) and detergents (as builders).

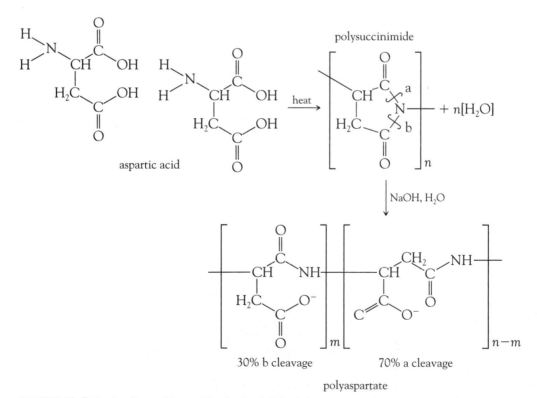

FIGURE 12-5 Donlar Corporation synthesis of polyaspartate.

The Recycling of Household and Commercial Waste

In the past few decades there has been mounting pressure in developed countries to reduce the amount of material discarded as waste after a single use. The incentives here are to conserve the natural resources, including energy, from which the materials are produced and to reduce the volume of material that must be buried as garbage, incinerated, etc. The **four Rs** of such waste management philosophies are

- to **reduce** the amount of materials used (sometimes called *source reduction*);

- to **reuse** materials once they are formulated;

- to **recycle** materials to recover components that can be refabricated; and

- to **recover** the energy content of the materials if they cannot be used in any other way.

These principles can be and are applied to all types of wastes, including hazardous ones, but in the following discussions we concentrate on their application to domestic materials, particularly in regard to recycling.

A distinction is often made between **preconsumer recycling,** which involves the use of waste generated during a manufacturing process, and **postconsumer recycling,** which involves the reuse of materials that have been recovered from domestic and commercial consumers. The postconsumer items most often collected for recycling are

- paper (especially newspapers and cardboard),

- aluminum (especially beverage cans),

- steel (especially food cans), and

- plastic and glass containers.

The labor, energy, and pollution costs associated with collecting the materials, sorting them, and transporting them to facilities where they can be reused must be considered in any analysis of recycling. In addition, historically the demand for recycled materials in these categories has been unstable, with prices swinging wildly in response to changes in supply and demand. For these and similar reasons, the recycling of paper, glass, and plastics usually needs to be justified on noneconomic and nonenergy grounds—such as savings in landfill space. The most economically viable forms of recycling materials usually involve a minimum of chemical reprocessing and correspond more closely to **reuse**—e.g., using newspapers to make paperboard or insulation, and reusing glass and plastic containers.

Debates still rage both in the popular press and in the scientific literature as to whether or not recycling is a worthwhile activity.

The Recycling of Metals and Glass

From the viewpoints of both economics and energy conservation, the recycling of metals makes sense. Virgin metal must be obtained by reduction of the oxidized form of the element found in nature. The reduction process requires energy that does not need to be expended again when the metallic form of the element is recycled.

Consider the reduction of aluminum and iron from the oxide ores. By definition, the enthalpies of these processes equal the negative of their enthalpies of formation:

$$Al_2O_3 \longrightarrow 2\,Al + \tfrac{3}{2}\,O_2 \qquad \Delta H° = -\Delta H_f° \,(Al_2O_3)$$

$$= +1676 \text{ kJ/mol oxide}$$

$$= +31 \text{ kJ/g metal}$$

Recycling aluminum cans saves 95% of the energy that is needed to produce Al metal from bauxite ore. Since the energy required for the aluminum reduction must come in the form of electricity, and this energy accounts for about 25% of the cost of its production, it makes good economic sense to recycle this metal, and, in fact, most aluminum cans are recycled. Recycling steel cans saves about two-thirds of the energy required to produce them from iron ore.

> **PROBLEM 12-2**
>
> The enthalpy of formation of the principal ore of iron, Fe_2O_3, is -824 kJ/mol. Calculate the enthalpy of the reaction in which 1.00 g of metallic iron is formed from the ore. Given your result, would you expect the price that recycling operators are willing to pay for scrap iron per kilogram to be greater or less than that for scrap aluminum?

In the case of paper, glass, and plastics, there is no significant change in the average oxidation state of the principal materials during their transformation from inexpensive raw material components—wood, sand and lime, and oil, respectively—to finished products; thus there are no great energy savings when they are recycled.

Modern, low-polluting, and energy-efficient electric furnaces cannot handle as high a proportion of used glass as can their more polluting, more energy-consuming counterparts that use fossil fuels. Consequently, the recycling of too much glass can result in producing more pollution and using more energy than would otherwise be the case!

The Recycling of Paper

People in developed countries throw away more paper than any of the other components of municipal solid waste (see Figure 12-1), and it seems an obvious material to recycle. However, the production of virgin paper, for example, uses only about one-quarter more energy than recycling of old paper. The transportation of waste paper to recycling mills and the deinking process itself are heavy consumers of energy. Notwithstanding these considerations, tremendous quantities of paper, especially newsprint, are currently recycled. Indeed, corrugated paper products are the most intensely recycled material in North America.

The first step in recycling paper is mechanical dispersal into its component fibers in water. Then it is cleaned to remove nonfibrous contaminants, followed by treatment with *sodium hydroxide* or *sodium carbonate* to deink it. A detergent is added to help disperse the pigment, and the ink particles are removed by washing or flotation on air bubbles, which rise to the top. The resulting deinked stock is usually less white than virgin fiber, so the two types are often blended. If necessary, the whiteness of the recycled stock can be improved by bleaching, usually with *peroxides* and *hydrosulfites*. The used ink, which is recovered in a sludge with some pulp fibers, is later pressed to remove water and then can be burned to produce steam for use in mill operations, or it can be treated to detoxify it. In general, the use and release into the environment of materials such as *chlorine* or other bleaching agents, acids, and organic solvents is significantly less with the production of recycled paper than with the creation of the virgin material.

There is a limit to the number of times that paper can be recycled, however, since with each cycle the pulp fibers become progressively shorter and so lose some of their integrity. Newsprint can be recycled back into newsprint about six to eight times.

From 1985 to 2000 almost $20 billion was spent by the paper industry in the United States on technology and capital investment required to recycle paper; the ultimate goal was to recycle about half of all paper used in the country. By the mid-1990s, almost 40% of paper in the United States and more than half of that in some western European countries was being recycled. Food boxes and egg cartons are some of the products usually made from recycled pulp fiber.

Perhaps a more clever use of waste paper in the future will be its conversion to fuel *ethanol*, as discussed in Chapter 6. Paper can be incinerated directly to recover its energy content, reducing the amount of fossil fuels burned in power stations. According to an analysis by Britain's *Centre for Environmental Technology*, recycling paper is environmentally superior to landfilling it but is actually inferior to burning it for its fuel value when all factors are taken into consideration.

The Recycling of Tires

Another consumer commodity that presents a waste-management headache is vehicle tires. In North America, about one 10-kg rubber tire per person per year on average is discarded; thus about one-third of a *billion* tires are added to the supply of approximately 3 billion tires presently stored in mountainous piles, awaiting ultimate disposal! Because the tires are made primarily from oil and consequently are flammable, tire fires in these huge piles are not uncommon and produce tremendous amounts of smoke, *carbon monoxide,* and toxins such as **PAHs** and **dioxins.** The fires are difficult to extinguish because of air pockets in and between the tires.

There have been efforts to use tires either as fuel or as a filler for asphalt, but currently such applications consume only about 10% of the tires that are discarded annually. Some used tires are also utilized for their rubber content, to produce landscaping products.

A number of attempts at commercially reprocessing shredded scrap tires by **pyrolysis**—the thermal degradation of a material in the absence of oxygen—have been made. The resulting products are low-grade gaseous and liquid fuels and a **char** containing minerals and a low-grade version of the material called **carbon black,** which can be further treated and converted into *activated carbon.* It may eventually be possible to convert the liquid component into high-grade char, thereby making the process economically profitable. The "rubber" in tires consists of about 62% of a hydrocarbon polymer and 31% of carbon black—added to strengthen the tires and reduce wear—so there is a ready market for the latter. Using the liquid component as a fuel is problematic because of its high content of aromatic hydrocarbons.

The Recycling of Plastics

One of the triumphs of industrial chemistry in the twentieth century was the development of a wide variety of useful **plastics.** All plastics are composed at the molecular level of polymeric organic molecules, very long units of matter in which a short structural unit is repeated over and over again. All the raw materials (except chlorine) from which the plastics are currently made are obtained from crude oil.

Conceptually, the simplest organic polymer is **polyethylene** (or **polyethene**), the molecules of which are composed of many thousands of —CH_2— units bonded together:

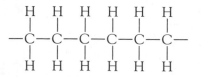

This polymer is prepared by combining many molecules of *ethylene* (ethene) (hence the name) and is an example of an **addition polymer.** Depending on exactly how the polymerization takes place, either **low density polyethylene (LDPE)** (the plastic given the recycling designation number 4) or **high density polyethylene (HDPE)** (the cloudy white or opaque plastic given the designation number 2) is formed.

There are several other addition polymers similar to polyethylene in which one (or more) of the four hydrogen atoms in each ethylenic —CH₂—CH₂— unit is replaced by a group or atom X, giving the polymer

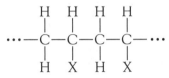

If the X attached to every second carbon atom in each chain is chlorine, then the clear (or blue-tinted) polymer **polyvinyl chloride (PVC)** (recycling number 3) is obtained. If the substituent X is a methyl group, we have **polypropylene** (recycling number 5), and if it is a benzene ring we have **polystyrene** (number 6). The plastics formed from all these polymers are used extensively in packaging, as indicated by the original uses listed in Table 12-1.

The other plastic that is commonly recycled (given the number 1) is the clear plastic **poly(ethylene terephthalate) (PET).** Its structure is a chain of two CH₂ units alternating with a unit of the organic molecule *terephthalic acid.* PET is used in the form of film (for magnetic tape as well as photographic film), fiber, and molded resin (e.g., plastic bottles).

In the last quarter of the twentieth century plastics became the symbol of a throwaway society, since much of the product—especially that used in packaging—was designed to be used once and then discarded. Many environmentalists believed that waste plastic was a major culprit in the garbage crisis. Indeed, molded plastics take up a greater percentage of volume in landfills than their percentage by mass because their densities are low, although they are eventually compressed by the weight of the materials piled on them, as well as by compacting machinery before they are placed in the landfill. Plastics are the second most common constituent of municipal garbage, following paper and cardboard by quite a margin. The per capita annual use of plastics in North America is approximately 30 kg.

For a number of reasons, including the fact that landfills, especially throughout Europe, are reaching their capacity and that many citizens in developed countries are opposed to their incineration, many plastics are now collected from consumers and recycled. As of the mid-1990s, over 80% of the mass of plastics recycled in the United States consists of PET and

TABLE 12-1	Commonly Recycled Plastics		
Plastic recycling number	Acronym and name of plastic	Original use examples	Recycle use examples
1	PET Poly(ethylene terephthalate)	Beverage bottles; food & cleanser bottles; drugstore product containers	Carpet fibers; fiber-fill insulation; non-food containers.
2	HDPE High-density polyethylene	Milk, juice, & water bottles; margarine tubs; crinkly grocery bags	Oil & soap bottles; trash cans; grocery bags; drain pipes
3	PVC (or V) Polyvinyl chloride	Food, water, & chemical bottles; food wraps; blister packs; construction material	Drainage pipes; flooring tiles; traffic cones
4	LDPE Low-density polyethylene	Flexible bags for trash, milk, & groceries; flexible wraps & containers	Bags for trash & groceries; irrigation pipes; oil bottles
5	PP Polypropylene	Handles, bottle caps, lids, wraps, & bottles; food tubs	Auto parts; fibers; pails; trash containers
6	PS Polystyrene	Foam cups & packaging; disposable cutlery; furniture; appliances	Insulation; toys; trays; packaging "peanuts"
7	Other	Various	Plastic "timber": posts, fencing, & pallets

HDPE, in approximately equal amounts, with LDPE the only other one of significance. Some countries, such as Sweden and Germany, have made manufacturers legally responsible for the collection and recycling of the packaging used in their products.

There is little doubt that the public in many developed countries has embraced recyling of plastics. As of the late 1990s about half the urban communities in the United States had curbside recycling programs that included plastics. However, the recycling rate for PET soda bottles in 1996 had fallen to 34%, considerably less than the 45% rate achieved in 1994.

There has been much resistance to plastics recyling in some quarters, including many in the plastics industry. Their argument is that virgin plastics are a low-cost product that is made from a relatively low-cost raw material (oil). The input energy for making plastics is very small compared to that used for making aluminum or steel from its raw materials. The cost of cleaning used plastic and converting it back into its monomers so that it can be polymerized again is substantial, compared to the current cost of oil. Some executives in the plastics industry believe that the best disposal method for plastics is simply to burn them and use the heat energy provided, especially given the fact that there is little objection by the public to simply burning most (over three-quarters) of the oil produced — in vehicles, domestic furnaces, and power plants. In addition, experiments indicate that the presence of plastics makes the other materials in domestic garbage burn more cleanly and reduces the need for supplemental fossil fuel to be added. Although plastics account for less than 10% of the mass of garbage, they make up more than one-third of its energy content.

Environmentalists counter these arguments by pointing out that if environmental impacts were to be included in determining the cost of virgin materials, recycled plastic would be the cheaper choice. Also, the combustion of some plastics, notably PVC, produces dioxins and furans and releases *hydrogen chloride* gas.

Ways of Recycling Plastics

There are one physical and three chemical ways to recycle plastics:

1. Reprocess the plastic (a physical process) by remelting or reshaping. Usually the plastics are washed, shredded, and ground up, so that clean, new products can then be made.

2. Depolymerize the plastic to its component monomers by a chemical or thermal process so that it can be polymerized again.

3. Transform the plastic chemically into a low-quality substance from which other materials can be made.

4. Burn the plastic to obtain energy (energy recycling).

Examples of the *reprocessing* option include the production of carpet fibers from recycled PET, of plastic trash cans, grocery bags, etc. from recycled HDPE, and of CD cases and office accessories such as trays and rulers from recycled polystyrene. Further examples of reprocessing are listed for each category of packaging plastic in the last column of Table 12-1.

The *depolymerization* option can be employed with PET and other polymers of the —A—B—A—B—A—B— type, in which units of types A and B alternate in the structure. These **condensation polymers** are produced by combining small molecules that contain A and B units. During the polymerization process, A and B form the polymer and the remaining parts of the molecules combine. For example, in the production of PET, the molecule *methanol*, CH_3OH, is formed from the OH unit of one component and the CH_3 of the other. In the chemical *depolymerization* process, a catalyst and heat are applied to a mixture of methanol and the plastic to *reverse* the polymerization process and recover the original components:

$$CH_3—A—CH_3 + HO—B—OH + CH_3—A—CH_3 + HO—B—OH + \cdots$$

$$\xrightleftharpoons[\text{depolymerization}]{\text{polymerization}} CH_3—A—B—A—B—A \cdots —A—B—A—B—OH + \text{many } CH_3OH$$

For PET, B is —CH$_2$—CH$_2$— and A is

Physical recycling of PET is currently more economically viable than chemical reprocessing.

One of the difficulties in depolymerization of plastics is the fact that organic and inorganic compounds are often added to the original polymer to modify the physical properties of the plastic, such as its flexibility, and these must be removed before the monomers can be reused.

For many addition polymers it is difficult to devise a process by which the original monomers can be re-formed. For example, the monomer yield for the thermal depolymerization of polystyrene is about 40%, but it is close to zero for polyethylene because the chain will be broken at random positions and will not exclusively produce two-carbon units.

Examples of the *transformation* option are:

- *Reductive* processes such as the production of synthetic crude oil by hydrogenation of plastics or by heating them to a high temperature to "crack" the polymer molecules, a process that can be used even with mixed plastics. The pyrolysis of polyethylene to monomers that can be converted to lubricants has also been proposed.

- *Oxidative* processes such as the gasification of plastics by adding oxygen and steam to produce synthesis gas (a mixture of hydrogen and carbon monoxide, discussed in Chapter 6).

Green Chemistry: Development of Recyclable Carpeting

In the United States over 4.5 billion pounds of carpet are landfilled annually. This translates into 800 million square yards of carpet, enough to cover 160,000 football fields, or one-quarter of the entire state of Rhode Island! Carpeting is not biodegradable, takes up valuable and rapidly declining landfill space, and is ultimately made from petroleum, a nonrenewable resource. As landfill fees escalate—estimates indicate they will double every five years—and the costs of transporting, installing, and replacing carpeting rise, the demand for an economical and environmentally responsible alternative to landfilling of this product has increased.

The two major components of carpeting are the backing and the face fiber. Since the 1970s, polyvinyl chloride (PVC) has been the material of choice for carpet backing. Environmental and health concerns about PVC include *vinyl chloride* (the monomer used to produce PVC) and *phthalate* plasticizers. Vinyl chloride, a known carcinogen, is volatile. Some believe that this compound outgases from the polymer, but others argue that the high temperatures used to process PVC should eliminate virtually all of the volatile monomer. Phthalates, which are added to PVC to make it more flexible, migrate out of the polymer and have become widely dispersed in the environment. The growing concern over the effects of phthalates on the human reproductive system was discussed in Chapter 8. In addition, as also discussed in Chapter 8, when PVC burns it produces toxic by-products including dioxins, furans, and hydrochloric acid.

Recycling is now commonplace when it comes to paper, glass, and plastic. However, most of us do not think of recycling carpeting. In the United States only 4% of carpeting is recycled. Part of the reason for this is that the PVC backing interferes with the recycling process. Shaw Industries won a Presidential Green Chemistry Challenge Award in 2003 for its development of a new type of carpeting that allows for *closed-loop recycling*, i.e., recycling of used carpeting back into carpeting with little loss of material. This new type of carpeting, known as *EcoWorx*, employs polyolefin backing and nylon-6 fiber. In addition to lending itself to recycling, the polyolefin backing has low toxicity and eliminates the significant environmental and health concerns related to PVC.

The process of recycling EcoWorx carpeting begins with grinding the used carpeting and separating the heavier particles (the polyolefin backing) from the lighter ones (the nylon-6 fibers) with a stream of air (a process know as elutriation). The polyolefin particles can then be reused in the extrusion process to produce new backing. The nylon-6 is depolymerized to its monomer (caprolactam) and repolymerized to virgin nylon-6 fibers. The fibers are then used to form new carpeting, completing the cycle. Both the backing and the fibers of EcoWorx carpeting can be used again and again to form new carpeting.

There are several additional environmental advantages of EcoWorx, as well as economic advantages:

• Recycling requires no additional petroleum feedstocks; this benefits not only the environment, but also the economic bottom line.

• Recycling reduces the amount of landfill space needed.

- The polyolefin-backed carpeting is 40% lighter in weight than carpeting backed with PVC. More carpeting can be shipped by truck within weight limits, thus lowering fuel consumption, cost, and pollution.

- The use of polyolefins eliminates the energy-intensive heating process required for PVC. Again, this results in lowering fuel consumption, cost, and pollution.

- In carpet backing (including PVC backing), significant amounts of inorganic fillers are used to provide loft and bulk. Traditionally, virgin *calcium carbonate* was employed for this purpose. EcoWorx contains 60% class C fly ash (a waste by-product from the burning of lignite or sub-bituminous coal) as a filler. Using fly ash as a filler utilizes an unwanted by-product and precludes the use of a virgin chemical.

Returning goods for recycling is often a significant barrier to recycling. To overcome this difficulty, Shaw developed a system for returning the carpeting at the end of its useful life, at no cost to the consumer.

Life Cycle Assessments

One technique used in minimizing the production of wastes and in pollution prevention is the **life cycle assessment** (or **analysis**), LCA—an accounting of all the inputs and outputs in a product's life, from raw material extraction to final disposal. This cradle-to-grave analysis for a product can be used to identify the types and magnitudes of environmental impacts that the product (or process) has, including both the natural resources used and the pollution produced.

The results of a life cycle assessment can be used in two ways:

- to identify opportunities within the life cycle to minimize the overall environmental burden of a product and

- to compare two or more alternative products to determine which is the more environment-friendly.

An example of the first use is in the production of motor vehicles; the life cycle and the most important inputs and outputs are illustrated in Figure 12-6. In designing new cars, life cycle assessments are used to help minimize pollution while maintaining economic viability. The analysis is particularly useful for identifying new environmental burdens that would arise if others are decreased. For example, vehicles could be made much lighter by increased use of plastics; however, the types of plastics used are difficult to recycle and they would increase the eventual burden of solid waste in landfills.

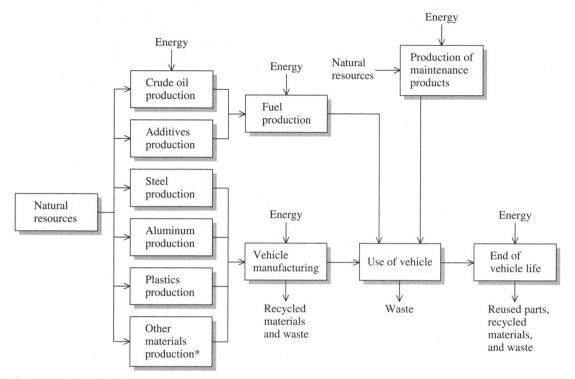

*For example, rubber, lead, glass, paints, and coolants.

FIGURE 12-6 Important input and output components in life cycle assessments of motor vehicles. [Source: M. Freemantle, "Total Life-Cycle Analysis Harnessed to Generate 'Greener' Automobiles," *Chemical and Engineering News* (27 November 1995): 25.]

Soils and Sediments

The contamination of soil by wastes is not solely a phenomenon of modern times. In Roman times metal ores were mined and the ores smelted, polluting the surrounding countryside with waste from the mines. The production of materials and chemicals in Europe even at the start of the Industrial Revolution produced substantial pollution. However, the extent of contamination and the hazard from discarded materials expanded greatly in the last century, particularly in the period after World War II.

We begin this section by discussing the nature of soil and of sediments.

Basic Soil Chemistry

Most soils are composed mainly of small particles of weathered rock that are **silicate minerals.** At the atomic level, these minerals consist of polymeric inorganic structures in which the fundamental unit is a silicon atom surrounded tetrahedrally by four oxygen atoms. Since these oxygen atoms are

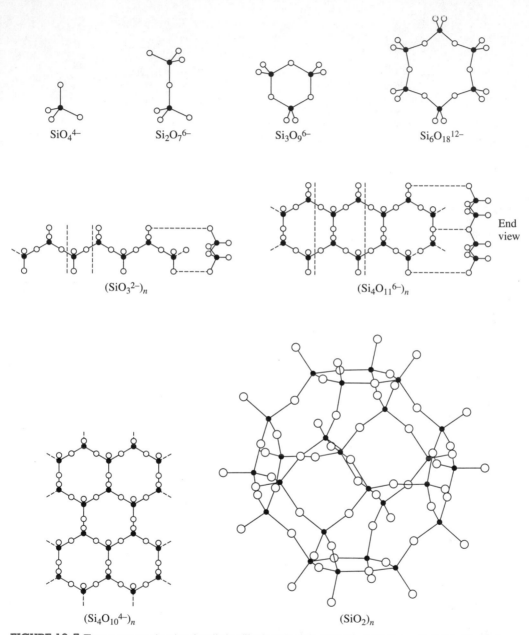

SiO_4^{4-} $Si_2O_7^{6-}$ $Si_3O_9^{6-}$ $Si_6O_{18}^{12-}$

$(SiO_3^{2-})_n$ $(Si_4O_{11}^{6-})_n$ End view

$(Si_4O_{10}^{4-})_n$ $(SiO_2)_n$

FIGURE 12-7 The common structural units in silicate minerals. [Source: R. W. Raiswell, P. Brimblecombe, D. L. Dent, and P. S. Liss, *Environmental Chemistry* (London: Edward Arnold Publishers, 1980).]

in turn each bonded to another silicon, etc., the resulting structure is an extended network. There are many variations on the silicate structural theme. Some networks have exactly twice as many oxygens (formally, O^{2-}) as silicons (formally, Si^{4+}) and correspond to electrically neutral SiO_2 polymers. In others, some of the tetrahedral sites are occupied by aluminum ions, Al^{3+}; the extra negative charge in these networks is neutralized by the presence of other cations such as H^+, Na^+, K^+, Mg^{2+}, Ca^{2+}, and Fe^{2+}. Some common silicon–oxygen structural units are illustrated in Figure 12-7.

Over time, the weathering of the silicate minerals from rocks can involve chemical reactions with water and acids in which ion substitution occurs. Eventually, these reactions yield substances that are important examples of a class of soil materials known as **clay minerals.** A mineral having a particle size less than about 2 μm is by definition a component of the clay fraction of soil.

In addition to clay, there are several other soil types; the definition of each type depends on particle size, as indicated in Figure 12-8. Notice the factor-of-10 increase in size with each transition in type: the upper boundary for silt is 10 times that for clay, that for fine sand is 10 times that for silt, etc. Because the particle size of sand is large, it has a relatively low density and water runs through it easily. In contrast, soils composed of clay are dense and have poor drainage and aeration. The best agricultural soils have a combination of soil types present. The range of element content for major and minor elements in the mineral component of soil is given in Table 12-2.

Clay particles act as colloids in water. Because the clay particles are much smaller than those of sand or silt, their total surface area per gram is thousands of times larger. Consequently, most important processes in soil occur on the surface of colloidal clay particles.

Particles of clay possess an outer layer of cations that are bound electrostatically to an electrically charged inner layer, as illustrated in Figure 12-9. The most common cations in soil are H^+, Na^+, K^+, Mg^{2+}, and Ca^{2+}. Depending on the concentration of cations in the water surrounding the clay particle, the cations on the particle are capable of being exchanged for them. For example, in water rich in potassium ions but poor in other ions,

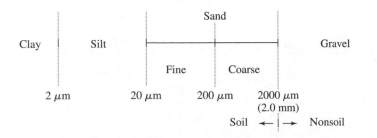

FIGURE 12-8 The soil particle size classifications system of the International Society of Soil Science. [Source: G. W. vanLoon and S. J. Duffy, *Environmental Chemistry*, (Oxford: Oxford University Press, 2000).]

TABLE 12-2	Elements Content of the Mineral Components of Soils		
Major elements (%)		**Minor elements (ppm)**	
Si	30–45	Zn	10–250
Al	2.4–7.4	Cu	5–15
Fe	1.2–4.3	Ni	20–30
Ti	0.3–0.7	Mn	~400
Ca	0.01–3.9	Co	1–20
Mg	0.01–1.6	Cr	10–50
K	0.2–2.5	Pb	1–50
Na	Trace–1.5	As	1–20

Source: G. W. vanLoon and S.J. Duffy, *Environmental Chemistry* (Oxford: Oxford University Press, 2000).

K^+ ions will displace the ions bound to the surface of the clay particle (see Figure 12-9). If, on the other hand, the soil is acidic—i.e., rich in H^+ ions—the metal ions on the surface will be displaced by H^+ ions and the metal ions will enter the aqueous phase. Generally, the greater the positive charge on a cation, the more strongly it binds to the particle. Heavy metals dissolved in soil water are often adsorbed onto the surface of clay particles.

In addition to minerals, the other important components of soil are organic matter, water, and air. The proportion of each component varies greatly from one soil type to another. The organic matter, which gives soil its dark color, is primarily a material called **humus.** Humus is derived principally from photosynthetic plants, some components of which (such as cellulose and hemicellulose) have already been decomposed by organisms that live in the soil. The undecomposed plant material in humus is mainly

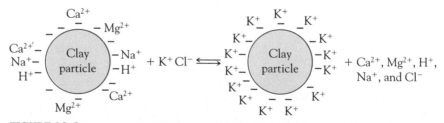

FIGURE 12-9 Ion-exchange equilibria on the surface of a clay particle. The addition of K^+ ions to the soil water displaces the exchange equilibria to the right, whereas removal of K^+ ions from solution displaces it to the left. [Source: R.W. Raiswell, P. Brimblecombe, D. L. Dent, and P. S. Liss, *Environmental Chemistry* (London: Edward Arnold Publishers, 1980).]

protein and lignin, both polymeric substances that are largely insoluble in water. A significant amount of the carbon in lignin exists in the form of six-membered aromatic benzene rings connected by chains of carbon and oxygen atoms (see Figure 8-4). Much of the organic matter in soil also consists of colloidal particles.

As a result of the partial oxidation of some of the lignin, many of the resulting polymeric strands contain *carboxylic acid* groups, —COOH. This dark-colored portion of humus consists of **humic** and **fulvic acids** and is soluble in alkaline solutions due to the presence of the acid groups. Humic acid is *insoluble* in acid solution, whereas fulvic acid is *soluble*. The humic acid is less soluble in acid than the fulvic not only because its molecular weight is much greater (100 to 1000 times greater), but also because its oxygen content is lower, so there are fewer —OH groups per carbon to form hydrogen bonds with the water. The acid groups are often adsorbed onto the surfaces of the clay minerals, to an extent dependent on the distribution of surface charge on the particles. Humic and fulvic acids form colloids that are hydrophilic, whereas those of clay are hydrophobic.

Since some of the carbon in the original plant material is transformed to carbon dioxide and thus lost as a gas to the surroundings, humus has relatively more nitrogen than the original plants; its other main components are carbon, oxygen, and hydrogen. Generally speaking, nonpolar organic molecules are more strongly attracted to the organic matter in soil than to the surfaces of particles derived from minerals.

PROBLEM 12-3

The percentage composition of a typical fulvic acid is 50.7% carbon by mass, 45.1% oxygen, and 4.22% hydrogen. Derive the (simplest) empirical formula for the substance.

The Acidity and Cation-Exchange Capacity of Soil

If the soil at the surface contains minerals with elements in a reduced state, oxidation by atmospheric oxygen can produce acid. An example is the oxidation of sulfur in *pyrite*, discussed as acid mine drainage in Chapter 9. Acid rain, of course, provides another source of acidity in certain areas (Chapter 2).

Quantitatively, the ability to exchange cations is expressed as the soil's **cation-exchange capacity,** CEC, which is defined as the quantity of cations that are reversibly adsorbed per unit mass of the (dry) material. The quantity of cations is given as the number of moles of positive charge (usually expressed as centimoles or millimoles) and the soil mass is usually taken to be 100 g or 1.00 kg. Typical values of the CEC for common clay minerals range from 1 to 150 centimoles per kilogram (cmol/kg). The CEC values are determined in large part by the surface area per gram of the mineral. CEC values for the organic component of soil are high, due to the large number of —COOH groups that can bind to and exchange cations; e.g., the CEC of peat can be as high as 400 cmol/kg.

PROBLEM 12-4

The CEC for a soil sample is found to be 20 cmol/kg. What is the CEC value for this sample in units of millimoles per 100 grams?

Biologically, the exchange of cations by soils is the mechanism by which the roots of plants take up metal ions such as potassium, calcium, and magnesium. Although the roots release hydrogen ions to the soil in exchange for the metal ions, this is not the main reason why soil in which plants grow is often somewhat acidic. Most of the acidity is due to metabolic processes involving the roots and microorganisms in the soil, which result in the production of *carbonic acid*, H_2CO_3, and of weak organic acids.

Rainwater that is acidic releases base cations from soil particles by exchanging them for H^+ ions. The acidity of water flowing through the soil stays low for this reason. However, once the base cations have been exhausted, aluminum ions are released, as discussed in Chapter 2. It is now known that in the past, base cations from dust particles, especially those containing calcium and magnesium carbonates, neutralized some of the acidity in precipitation, but along with *sulfur dioxide*, their emission from industrial sources has been curtailed in recent decades.

The pH of soil can vary over a significant range for a variety of reasons. For example, soils in areas of low rainfall but high concentrations of the soluble salt sodium carbonate, Na_2CO_3, become alkaline due to the (hydrolysis) reaction of the **carbonate ion,** CO_3^{2-}, with water, as discussed in Chapter 9.

Soils that are too alkaline for agricultural purposes can be remediated either by the addition of elemental sulfur, which releases hydrogen ions as it is oxidized by bacteria to *sulfate ion*, or by the addition of the sulfate salt of a metal such as iron(III) or aluminum, which reacts with the soil's water to extract hydroxide ions and thereby release hydrogen ions:

$$2\,S(s) + 3\,O_2 + 2\,H_2O \longrightarrow 4\,H^+ + 2\,SO_4^{2-}$$

$$Fe^{3+} + 3\,H_2O \longrightarrow Fe(OH)_3(s) + 3\,H^+$$

The pH of water present in the soil is determined by the concentration of hydrogen and hydroxide ions. However, soil has **reserve acidity** due to the large number of hydrogen atoms in the —COOH and —OH groups in the organic fraction and on the cation-exchange sites on minerals that are occupied by H^+ ions. In other words, soils act as weak acids, retaining their H^+ ions in a bound condition until acted upon by bases. Thus the pH of soil tends to be buffered against large increases in pH, since this bound hydrogen ion can be slowly released into the aqueous phase. In the process of **liming,** which is the addition of salts such as calcium carbonate to soil, carbonate ions neutralize acids present in the uppermost soil zones, producing carbon dioxide and water. Once this process has occurred, calcium ions can replace hydrogen ions in the organic matter or clays. The additional carbonate ions that enter the aqueous phase combine with the newly released H^+ ions, again to produce weakly acidic carbonic acid, which dissociates into carbon dioxide gas and water. Thus liming is a procedure by which the pH of a soil can be raised somewhat, and it is the practical method by which acidic soils can be remediated.

Soil Salinity

In hot, dry climates, salts and alkalinity tend to accumulate since there is little rainfall to leach ions from the soil. In contrast to other climates, the net movement of water in arid climates is upward rather than downward: water evaporation and loss by transpiration of plants exceeds rainfall. The salts that accompany the upward migration of water remain at or near the surface when the water has escaped. Salt accumulation at the surface also occurs in semiarid regions due to the use of poor-quality irrigation water, whose salt content remains after the water has evaporated.

Ions are also liberated at the surface of soil in the weathering of otherwise insoluble minerals. A simple example is the reaction of *olivine:*

$$Mg_2SiO_4 + 4\,H_2O \longrightarrow Si(OH)_4 + 2\,Mg^{2+} + 4\,OH^-$$

Additional hydroxide ion is produced when the *silicate ion,* SiO_4^{4-}, produced by mineral weathering reacts as a strong base with water.

As a general rule, hydrolysis of silicate minerals at the surface produces cations and hydroxide ions. In nonarid climates the hydroxide is neutralized by acids that are naturally produced in the soil (see later section), but this does not occur in arid climates. There is very little organic matter in the soils of arid areas. The hydroxide reacts with atmospheric carbon dioxide that dissolves in water to produce *bicarbonate* and carbonate ions:

$$OH^- + CO_2 \longrightarrow HCO_3^-$$
$$HCO_3^- \longrightarrow H^+ + CO_3^{2-}$$

Consequently, bicarbonate and carbonate salts accumulate in arid soils. If the predominant cations in the soil are calcium and magnesium, most of the carbonate ion will be locked away as their insoluble carbonate salts. However, if the predominant cations are sodium and potassium, the soil when moist will have a high pH, since the carbonate salts of these ions are soluble and the free CO_3^{2-} will act as a base:

$$CO_3^{2-} + H_2O \rightleftharpoons HCO_3^- + OH^-$$

Increasing soil salinity is a major problem in Australia, especially in regions where wheat and other shallow-rooted crops have replaced natural, long-rooted vegetation. This replacement, plus irrigation of crops including rice and cotton, has resulted in a rise of the water table and, with it, the salt that was formerly deep in the soil.

Sediments

Sediments are the layers of mineral and organic particles, often fine-grained, that are found at the bottom of natural water bodies such as lakes, rivers, and oceans. The ratio of minerals to organic matter in sediments varies substantially, depending on location. Sediments are of great environmental importance because they are sinks for many chemicals, especially heavy metals and organic compounds such as PAHs and pesticides, from which they can be transferred to organisms that inhabit this region. Thus the protection of sediment quality is a component of overall water management.

The map in Figure 12-10 shows watersheds in the continental United States where sediments are sufficiently contaminated to pose environmental

FIGURE 12-10 U.S. watersheds where contaminated sediments may pose environmental risks. [Source: U.S. EPA, in B. Hileman, "EPA Finds 7% of Watersheds Have Polluted Sediments," *Chemical and Engineering News* (26 January 1998): 27.]

risks. According to a U.S. EPA report, 7% of all watersheds pose a risk to people who eat fish from them and to the fish and wildlife themselves. The two pollutants found at high levels most frequently at contaminated sites are **PCBs** and mercury, although **DDT** (and its metabolites) and PAHs were also found at high concentrations at many sampling stations.

The transfer of hydrophobic organic pollutants to organisms may proceed by intermediate transfer to **pore water,** which is the water present in the microscopic pores that exists within the sediment material. Organic chemicals equilibrate between being adsorbed on the solid particles and dissolved in the pore water. For this reason, pore water is often tested for toxicity in determining sediment contamination levels.

Just as the total concentration of organic material in sediments may not be a good measure of the amounts that are biologically available, the same is true for the levels of heavy-metal ions present. Different sediments with the same total concentration of the ions of a heavy metal can vary by a factor of at least 10 in terms of the toxicities to organisms arising from the metal. This variation occurs principally because *sulfides* in the sediments control the availability of the metals. Recall from Chapter 11 that the toxic heavy metals mercury, cadmium, and lead form the very insoluble sulfide salts HgS, CdS, and PbS; analogous insoluble salts are formed by nickel and many other metals. If the concentration of sulfide ions *exceeds* the total of that of the metals, virtually all the metal ions will be tied up as insoluble sulfide salts and will be unavailable biologically at normal pH values. However, if the sulfide concentration is *less* than that of the metals, the difference is biologically available. The sulfide ion that is available to complex with metals is the amount that will dissolve in cold aqueous acid and is termed **acid volatile sulfide,** AVS. Industrially polluted sediments may have AVS concentrations of hundreds of micromoles of sulfur per gram, whereas uncontaminated sediments from oxidizing environments can have values as low as 0.01 μmol/g.

The Binding of Heavy Metals to Soils and Sediments

The ultimate sink for heavy metals, and for many toxic organic compounds as well, is deposition and burial in soils and sediments. Heavy metals often accumulate in the top layer of the soil and are therefore accessible for uptake by the roots of crops. For these reasons it is important to know the nature of these systems and how they function.

Humic materials have a great affinity for heavy-metal cations and extract them from the water that passes through by the process of ion exchange. The binding of metal cations occurs largely by the formation of complexes with the metal ions by —COOH groups in the humic and fulvic acids. For example, for fulvic acids the most important interactions probably involve a —COOH group and an —OH group on adjacent carbons of a benzene ring in the polymer structure, where the heavy-metal M^{2+} ion replaces two H^+ ions:

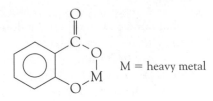

Humic acids normally yield water-insoluble complexes, whereas those of smaller fulvic acids are water-soluble.

PROBLEM 12-5

Draw the structure that would be expected if a dipositive metal ion M^{2+} were to be bound to two —COO^- groups on adjacent carbons in a benzene ring.

Heavy metals are retained by soil in three ways:

- by adsorption onto the surfaces of mineral particles,

- by complexation by humic substances in organic particles, and

- by precipitation reactions.

The precipitation processes for mercury and cadmium ions involve the formation of the insoluble sulfides HgS and CdS when the free ions in solution encounter *sulfide ion*, S^{2-}. Significant concentrations of aqueous sulfide ion occur near lake bottoms in summer months when the water is usually oxygen-depleted, as discussed in Chapter 9. However, the total concentration of mercury in soil water can exceed the limits set by the solubility product of HgS because some of the mercury will take the form of the moderately soluble molecular compound $Hg(OH)_2$ and does not participate in the equilibrium with the sulfide.

In acidic soils the concentration of Cd^{2+} can be substantial, since this ion adsorbs only weakly onto clays and other particulate materials. Above a pH of 7, however, Cd^{2+} precipitates as the sulfide, carbonate, or phosphate. Thus the liming of soil to increase its pH is an effective way of tying up cadmium ion and thereby preventing its uptake by plants.

Like many other chemicals, heavy-metal ions are often adsorbed onto the surfaces of particulates, especially organic ones that are suspended in water, rather than simply being dissolved in water as free ions or as complexes with soluble biomolecules such as fulvic acids. The particles eventually settle to the bottom of lakes and are buried when other sediments accumulate on top of them. This burial represents an important sink for many water pollutants and is a mechanism by which the water is cleansed. Before they are covered by subsequent layers of sediments, however, freshly deposited matter at the bottom of a body of water can recontaminate the water above it by desorption of the chemicals, since adsorption and desorption establish an equilibrium. Furthermore, the adsorbed pollutants can enter the food web if the particles are consumed by bottom-growing and -feeding organisms.

Although mercury in the form of Hg^{2+} is firmly bound to sediments and does not readily redissolve into water, environmental problems have arisen in several bodies of water due to the conversion of the metal into *methylmercury* and its subsequent release into the aquatic food web. The overall cycling of mercury species between air, water, and sediments was illustrated in Figure 11-1.

As previously discussed, anaerobic bacteria methylate the *mercuric ion* to form $Hg(CH_3)_2$ and CH_3HgX, which then rapidly desorb from sediment particles and dissolve in water, thereby entering the food web. Although the levels of methylmercury dissolved in water can be extremely low (of the order of hundredths of a part per trillion), a biomagnification factor of 10^8 results in ppm-range concentrations in the flesh of some fish. The devastating consequences of methylmercury poisoning have already been described (Chapter 11).

Excavation and analysis of the sediments at the bottom of a body of water can yield a historical record of contamination by various substances. For example, the curves in Figure 12-11 show the levels of mercury and of lead in the sediments of the harbor in Halifax, Nova Scotia, as a function of depth and, therefore, of year. For decades raw sewage has been dumped

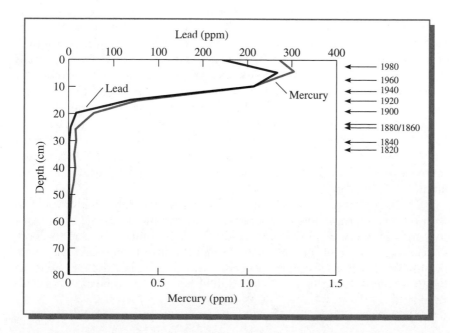

FIGURE 12-11 Lead and mercury concentrations in the sediments of Halifax Harbor versus depth (and therefore year of deposit). [Source: D. E. Buckley, "Environmental Geochemistry in Halifax Harbour," *WAT on Earth* (1992): 5.]

into this harbor; consequently its sediments are a historical record of the levels of pollutants in sewage. The metal pollution peaked about 1970, having begun to increase dramatically about 1900. These trends are also typical of heavy-metal levels in other water bodies, such as the Great Lakes; the characteristic decreases of the past few decades are due to the imposition of pollution controls.

In summary, then, both soils and sediments act as vast sinks and reservoirs in the containment of heavy metals.

Mine Tailings

In modern times many minerals (and in some cases fossil fuels) are extracted from much, much larger quantities of rock (or sand, etc.) than formerly, since the remaining supplies of the minerals occur in dilute form. This practice produces huge quantities of unwanted crushed rock in the form of dry, coarse-grained waste that must be disposed of, usually in slag heaps or landfills close to the mine. Eventually these waste piles are covered with soil and vegetation.

A more important environmental problem arises from disposal of the **tailings** of mining processing—fine-grained slurries that are more mineral-rich and often contain chemicals such as *cyanide* that were used to extract or process the ore. The toxic components of tailings are a potential source of pollution to local surface water, groundwater, and soil.

To prevent their dispersal into the environment, tailings are usually deposited as slurries in dams constructed on-site for the purpose. To prevent leakage into the soil, a clay or geomembranic liner is usually incorporated at the landfill's base. Over time, the solid settles to the bottom of the dam and the water evaporates or is drained off, i.e., the tailings are dewatered, although this can take a long time to achieve. In some cases, toxic materials such as heavy metals, nitrates, or excess acidity is removed by treatment of the tailings. Eventually, to establish a vegetated cover, organic matter and fertilizer must be added, since the dried tailings themselves contain little or no organic matter and consequently are sterile as well as hostile to plants.

If the crushed rock or tailings contain *iron pyrites*, exposure to oxygen will produce *sulfuric acid* by the series of reactions discussed as acid mine drainage in Chapter 9. The acidity can be neutralized by the continuous addition of limestone.

The most important environmental problem associated with a tailings dam is its potential failure, resulting in a catastrophic discharge of the tailings into a waterway and/or onto land. The failure might be due to flood, earthquake, or simply the loss of stability of the dam to pressure over time. A number of such incidents have occurred within the last decade—e.g., in Spain—with devastating results to wildlife, fish, and, in some cases, agricultural land.

An alternative to the storage of tailings on land is disposal in the deep ocean by using pipelines reaching down 100 m or more. Lack of oxygen there will slow the process of oxidation, and within a few years the tailings will be covered by other debris. Some biologists are dubious about this plan, because the environment for bottom-dwelling organisms will suffer. Heavy-metal contamination of fish has occurred in areas where this means of disposal is practiced. Due to the fine-grained nature of the tailings, dispersal over wide areas of the ocean floor occurs.

The Remediation of Contaminated Soil

Even areas thought to be rather pristine can have localized areas of contaminated soil. For example, the large-scale forestry industry in New Zealand has resulted in the contamination of several hundred sites where lumber is treated with the preservative *pentachlorophenol*, PCP (Chapter 8). Furan and dioxin contaminants of the PCP are also found in the soils. All three contaminant types have now leached into the groundwater at some of the sites and have begun to bioacccumulate in the food chain.

PROBLEM 12-6

Predict the predominant dioxin congener that would contaminate PCP. Is it very toxic? [Hint: The production of dioxin congeners from chlorophenols was discussed in Chapter 8.]

Contaminated soil is found most often not only near waste disposal sites and chemical plants, but also near pipelines and gasoline stations. The three main types of technologies currently available for the remediation of contaminated sites are

- containment or immobilization,

- mobilization, and

- destruction.

In general, the technologies can be applied **in situ,** i.e., in the place of contamination, or **ex situ,** i.e., after removing the contaminated matter to another location. Owing to the costs and risks, e.g., air pollution arising from excavation, in situ processes usually are preferred.

Among the techniques associated with in situ **containment** (i.e., the isolation of wastes from the environment) are capping of the contaminated site, especially with clay, and/or the imposition of cut-off walls of low permeability that prevent the lateral spread of contaminants. Ex situ containment would be the placing of the excavated soil in a special landfill. **Immobilization** techniques include solidification and stabilization, and are especially useful for inorganic wastes, which tend to be difficult to treat by other methods. A concentrated waste can be solidified by reaction with *portland cement*, for example, or by entombing the wastes in molten glass in the process of **vitrification.** By these techniques, the solubility and mobility of the contaminants are reduced.

Mobilization techniques are mainly accomplished in situ and include soil washing and the extraction of contaminant vapor from soil for highly volatile, water-insoluble contaminants such as gasoline. Heating of the soil to increase the rate of evaporation and air injection wells are sometimes used in conjunction with **soil vapor extraction,** in which the contaminants are removed by drilling wells in the soil and appling vacuum extraction. As indicated in Table 12-3, this technique is the most frequently used innovative technology at *Superfund* sites (Box 12-1) in the United States. A related technology is **thermal desorption,** in which wastes are heated to cause volatile organic compounds to vaporize. Both soil vapor extraction and thermal desorption are useful to remediate both volatile and semivolatile organic compounds, the latter including many PAHs.

In situ **soil washing** is accomplished by injecting fluids through wells into subsurface soil and collecting them in other wells. The fluid can simply be water, which will remove water-soluble constituents, or an aqueous solution that is acidic or basic in order to remove basic and acidic contaminants, respectively. Other options in soil washing include the use of solutions containing chelating agents such as *EDTA* (see Chapter 11) to remove metals, and oxidizing agents to oxidize and thereby solubilize previously insoluble species. Sometimes the washing solution uses **surfactants,** or *surf*ace-*act*ive agents. These are substances such as detergents that possess both hydrophobic and hydrophilic components within the same molecule

TABLE 12-3	Common Innovative Remediation Technologies in Projects at U.S. Superfund Sites (as of 1996)		
Technology	Sites in design or installation	Sites operational or completed	Total number
Soil vapor extraction	69	70	139
Thermal desorption	22	28	50
Bioremediation (ex situ)	24	19	43
Bioremediation (in situ)	14	12	26
In situ flushing	9	7	16
Soil washing	8	1	9
Solvent extraction	4	1	5

Source of data: U.S. EPA, *Innovative Treatment Technologies: Annual Status Report,* 8th ed., 1996.

BOX 12-1 | The Superfund Program

In 1980 the federal government of the United States established a program now known as *Superfund* to clean up abandoned and illegal toxic waste dumps, since dangerous chemicals from many such sites were polluting groundwater. The cleanup costs are shared by chemical companies, the current and past owners of the sites, and the government. Many billions of dollars have already been spent on remediation, and many billions more will eventually be required. Progress in the cleanups has been rather slow on account of the litigation involved and the huge amounts of money at stake. Many decades will pass before even the highest-priority sites are all cleaned up.

The Superfund program is administered by the Environmental Protection Agency, which has identified nearly 1,300 waste sites having such serious potential to cause harm to humans and the environment that they have been placed on a National Priorities List. As of 1994, priority sites were still being added faster than they were being removed by completion of remediation. New Jersey, Pennsylvania, and California have the greatest number of priority sites.

By the late 1990s, the EPA had finished cleanup work at 300 sites, begun work at more than 700 others, and conducted emergency removal of materials at more than 3,000 additional locations. In all, over 30,000 sites have been identified as potentially in need of cleanup.

The most common contaminants at the Superfund sites are the heavy metals lead, cadmium, and mercury, and the organic compounds benzene, toluene, ethylbenzene, and trichloroethylene.

and therefore can increase the mobilization of hydrophobic contaminants into the aqueous phase. **Biosurfactants** produced by microbes have recently been discovered that can selectively remove certain heavy metals such as cadmium from soil. Currently soil washing and flushing are the most common innovative technologies used to remove metals at Superfund sites.

The containment, mobilization, and immobilization techniques by themselves do not result in the *elimination* of the hazardous contaminants. **Destruction** techniques, principally incineration and bioremediation, do result in permanent elimination because they chemically or biochemically transform the contaminants. Organic contaminants in soil can be oxidized (mineralized) by feeding the excavated soil into the combustion chamber of an incinerator, or by using incineration or one of the specialty oxidation techniques to be discussed later to treat the substances that have been extracted from the soil. Bioremediation uses the metabolic activities of microorganisms to destroy toxic contaminants and is also discussed in detail later.

Electrochemical techniques are sometimes used to remediate soil. Placing electrodes in the contaminated ground and applying a dc voltage

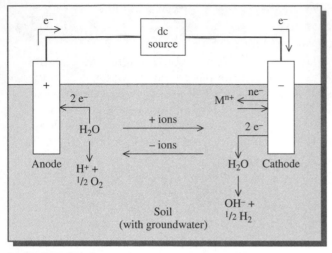

FIGURE 12-12 Electrochemical remediation of metal-contaminated soil.

between them results in ion transport within the soil: the ions travel within the groundwater electrolyte. If heavy-metal ions are dissolved in the electrolyte water, they will eventually move to the (negatively charged) cathode and be deposited on it. Indeed, other metal ions tend to be desorbed from their positions on negatively charged clay surfaces in the process, since hydrogen ion is released at the anode (as water is electrolyzed) and subsequently migrates in the groundwater toward the cathode (see Figure 12-12). Recall that heavy-metal ions are much more soluble in an acidic than in a neutral or alkaline environment.

In situ chemical oxidation can often be used to remediate soils (and groundwater) contaminated with chlorinated solvents and/or with *BTEX* (Chapter 6). The oxidizing agent is injected directly by means of a well into the underground waste, and may or may not be extracted at the other side of the contaminated zone. Typically, a salt of *permanganate ion*, MnO_4^-, is used for *TCE, PCE,* and *MTBE* deposits, whereas *ozone* or *hydrogen peroxide* is used for BTEX and PAHs or, in some cases, for the C_2 chlorinated solvents as well. The MnO_2 product of oxidation by permanganate is a natural constituent of soil. Hydrogen peroxide is often supplemented with ferrous salts (together called *Fenton's Reagent*) to create hydroxyl radical by a reaction described in Chapter 10.

The Analysis and Remediation of Contaminated Sediments

We now realize that many river and lake sediments are highly contaminated by heavy metals and/or toxic organic compounds and that such sediments act as sources to recontaminate the water that flows above them.

One way to determine the extent of contamination of a sediment is to analyze a sample of it for the total amounts of lead, mercury, and other heavy metals that are present. However, this technique fails to distinguish between toxic materials already present in either their toxic form or in a form that can be resolubilized into the water, and those that are firmly bound to sediment particles and unlikely to become resolubilized. Thus a more meaningful test involves extracting from a sediment sample the substances that are soluble in water or in a weakly acidic solution and analyzing the resulting liquid. In this way the permanently bound and therefore inactive toxic agents can be left out of the account. Finally, the effect of sediments on organisms that usually dwell in or on them can be determined by adding the organisms to a sediment sample and observing whether they survive and reproduce normally.

Several types of remediation have been used for highly contaminated sediments. The simplest solution is often to simply cover the contaminated sediments with clean soil or sediment, thereby placing a barrier between the contaminants and the water system. In other instances the contaminated sediments are dredged from the bottom of the water body to a depth below which the contaminant concentration is acceptable. If the sediments are high in organic content and inorganic nutrients, they are often used to enhance soil used for nonagricultural purposes. In some cases the sediment can be used for cropland provided that its heavy metals and other contaminants will not enter the growing food. Cadmium is usually the heavy metal of greatest concern in such sediments; if the pH of the resulting soil is 6.5 or greater, most of the cadmium will not be soluble, so a higher total concentration is often tolerated.

Several chemical and biological methods of decontaminating sediments are in use. For example, treatment with calcium carbonate or lime increases the pH of the sediments and thereby immobilizes the heavy metals. In some situations, contaminated sediments are simply covered with a chemically active solid, such as limestone (calcium carbonate), *gypsum* (calcium sulfate), *iron(III) sulfate*, or *activated carbon*, that gradually detoxifies the sediments. In other cases, the sediments are first dredged from the bottom of the water body and then treated. Heavy metals are often removed by acidifying the sediments or treating them with a chelating agent; in both cases the heavy metals become water-soluble and leach from the solid. For organic contaminants, extraction of toxic substances using solvents and destruction by either heat treatment of the solid or the introduction of microorganisms that consume them are the main options. The cleaned sediments can then be returned to the water body or spread on land. These techniques for removing metals and organics from sediments are also often useful on contaminated soils.

Bioremediation of Wastes and Soil

Recall from Chapter 10 that *bioremediation* involves the use of living organisms, especially microorganisms, to degrade environmental wastes. It is a technology that is experiencing rapid growth, especially in collaboration with genetic engineering, which is used to develop strains of microbes with the ability to deal with specific pollutants. Bioremediation is used particularly for the remediation of waste sites and soils contaminated with semivolatile organic compounds such as PAHs. It is a popular method for use at Superfund sites (see Table 12-3).

Bioremediation exploits the ability of microorganisms, especially bacteria and fungi, to degrade many types of wastes, usually to simpler and less toxic substances. Indeed, for many years it was thought that microorganisms could and would eventually biodegrade *all* organic substances, including all pollutants, that entered natural waters or the soil. The discovery that some compounds, chloroorganics especially, were resistant to rapid biodegradation was responsible for correcting that misconception. Substances resistant to biodegradation are termed **recalcitrant** or **biorefractory.** In addition, other substances, including many organic compounds, biodegrade only partially; they are transformed instead into other organic compounds, some of which may be biorefractory and/or even more toxic than the original substances. An example of the latter phenomenon is the potential conversion of the once widely used solvent *1,1,1-trichloroethane* (methyl chloroform, now banned as an ozone-depleting substance, as discussed in Chapter 1) into carcinogenic *vinyl chloride*, $CHCl = CH_2$, by a combination of abiotic and microbial steps.

If a bioremediation technique is to operate effectively, several conditions must be fulfilled:

• the waste must be susceptible to biological degradation and in a physical form that is susceptible to microbes;

• the appropriate microbes must be available; and

• the environmental conditions, such as pH, temperature, and oxygen level, must be appropriate.

An example of biodegradation is the degradation of aromatic hydrocarbons by soil microorganisms when land is contaminated by gasoline or oil. The largest bioremediation project in history was the treatment of some of the oil spilled by the *Exxon Valdez* tanker in Alaska in 1989. The bioremediation consisted of adding nitrogen-containing fertilizer to more than 100 km of the shoreline that had been contaminated, thereby stimulating the growth of indigenous microorganisms, including those that could degrade hydrocarbons. Both surface and subsurface oil was biodegraded in this operation. Some of the aromatic components in crude oil in marine spills become more susceptible to biodegradation once they are photooxidized by sunlight into more polar species.

In contexts such as contaminated soils, the biodegradation of PAHs is slow since they are strongly adsorbed onto soil particles and are not readily released into the aqueous phase where biodegradation could occur. PAH-contaminated soils are especially prevalent at gasworks sites used in the 1850–1950 period for the production of "town gas" from coal or oil. The pollution here is mainly in the form of deposits of **tars,** which are waste products that are high-molecular-weight organic liquids denser than water that are mixed with the soil and contain high levels of both BTEX and PAHs. Unfortunately, groundwater that comes into contact with the tar can become contaminated if some of its more soluble constituents such as benzene and naphthalene dissolve, although most of the tar is insoluble in water. The other common soil contaminants at gasworks sites are *phenols* and *cyanide*.

Bioremediation processes take place under either aerobic or anaerobic conditions. In the **aerobic treatment** of wastes, aerobic bacteria and fungi that utilize oxygen are employed; chemically, the processes are oxidations, as the microorganisms use the wastes as food sources. In some of the aerobic soil bioremediation procedures, oxygen-saturated water is pumped through the solid to ensure that O_2 availability remains high. For example, about 85,000 tonnes of soil contaminated by gasoline, oil, and grease from a fuel plant in Toronto were recently decontaminated by first enclosing it in plastic and then pumping air, water, and fertilizer into it to encourage the population of aerobic bacteria to multiply and devour the hydrocarbon pollutants. The bioremediation process took only three months.

There are many examples of wastes that can be degraded by anaerobic microorganisms, although usually the most rapid and complete biodegradations are obtained with aerobic microorganisms. The anaerobic process usually works best when there are some oxygen atoms in the organic wastes themselves. In general, the process corresponds to the *fermentation* discussed in Chapter 9, in which biomass with an approximate empirical formula of CH_2O decomposes ultimately to methane and carbon dioxide:

$$2\ CH_2O \longrightarrow CH_4 + CO_2$$

An advantage of anaerobic biodegradation is its production of *hydrogen sulfide*, which in situ precipitates heavy-metal ions as the corresponding sulfides.

Several strategies used in bioremediation are based on the fact that microorganisms evolve quickly—due to their short reproductive cycle—and they develop the ability to use the food source at hand, even if it is chemical wastes. One remediation strategy is to isolate the most efficient degrading biomicroorgansims flourishing at a contaminated site, grow a large population of them in the laboratory, and finally return the enhanced population to the site. Another strategy is to introduce microorganisms that were found at other sites that are useful in degrading a particular type of waste, rather than waiting for them to evolve. Unfortunately, microorganisms adapted to one environment may not be capable of surviving in another if additional hostile contaminants are present. A third strategy is to encourage an increase in the population of indigenous microorganisms at the site by adding nutrients to the wastes and by ensuring that the acidity and moisture levels are optimum.

Rather than waiting for microorganisms to evolve spontaneously, an alternative strategy is to use genetic engineering to develop microbes specifically designed to attack common organic pollutants. However, regulatory authorities have thus far been reluctant to allow genetically modified organisms to be released into the environment, since public opposition to such a move would be substantial.

In addition to bacteria, **white-rot fungi** can be used in biodegradation. This species protects itself from pollutants by degrading them outside the cell wall by secreting enzymes there that catalyze the production of *hydroxyl radicals* and other reactive chemicals. Since hydroxyl radical in particular is quite nonspecific about which substances it oxidizes, the fungi are useful in degrading mixtures of waste, including various chlorinated substances such as DDT and *2,4,5-T*, as well as PAHs.

Bioremediation of Organochlorine Contamination

It has been discovered that PCBs in sediments undergo some biodegradation. PCB molecules with relatively few chlorine atoms undergo oxidative aerobic biodegradation by a variety of microorganisms. For the reaction to begin, a pair of nonsubstituted carbons—one ortho to the point of connection between the rings and one meta site next to it—must be available on one of the benzene rings. After 2,3 hydroxylation at these two sites, the 1,2 carbon–carbon bonds to the ortho carbons split in sequence, thereby destroying the second ring and producing compounds that readily degrade (see Figure 12-13).

PROBLEM 12-7

Deduce the chlorine substitution positions on the benzene ring in the benzoic acid that results from the aerobic degradation of 2,3',5-trichlorobiphenyl.

Although PCB molecules that are heavily substituted by chlorines will not undergo this process, since they are unlikely to have adjacent unsubstituted carbons, they will instead undergo **anaerobic degradation,** as will perchlorinated organic compounds such as TCE and *HCB* (Chapter 7). In the absence of oxygen, anaerobic microorganisms facilitate the removal of chlorine atoms and their replacement by hydrogen atoms,

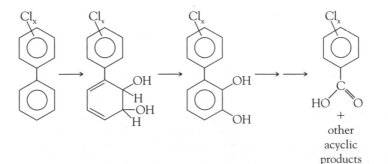

FIGURE 12-13 Example of the aerobic degradation of PCB molecules.

apparently by a reductive dechlorination mechanism that initially involves the addition of an electron to the molecule. In the case of PCBs, this reductive dechlorination occurs most readily with meta and para chlorines. Apparently steric effects block the ortho position from being attacked in most anaerobic mechanisms. Thus the products of anaerobic treatment here are ortho-substituted congeners, ultimately 2-chloro-biphenyl and 2,2′-dichlorobiphenyl especially.

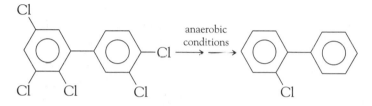

Since dioxin-like toxicity of PCBs requires several meta and para chlorines, the anaerobic degradation process significantly reduces the health risk from PCB contamination. Of course once adjacent ortho and meta sites without chlorine are available, aerobic microorganisms—if available—could degrade the biphenyl structure, as already discussed. Figure 12-14 illustrates the change in composition of a commercial PCB sample after it has resided for some time in sediment that contains anaerobic bacteria.

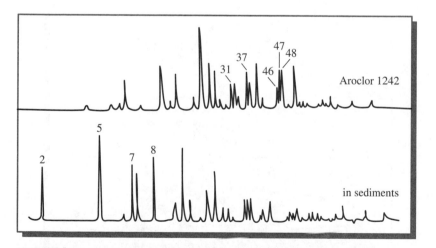

FIGURE 12-14 Concentrations profiles (chromatograms) for various PCB congeners of commercial Aroclor 1242 before and after biodegradation (different scales) in a sediment. The major components of the peaks correspond to the following positions of chlorine substitution: 2 = 2-chlorobiphenyl; 5 = 2,2′ and some 2,6; 7 = 2,3′; 8 = 2,4′ and some 2,3; 31 = 2,2′, 5,5′; 46 = 2,4,4′, 5; 47 = 2,3′, 4′,5; 48 = 2,3′, 4,4′. Notice that biodegradation produces ortho-substituted congeners that are present in low concentrations or absent in the original sample. [Source: D. A. Abramowicz and D. R. Olson, *CHEMTECH* (July 1995): 36–40.]

Phytoremediation of Soils and Sediments

The technique of **phytoremediation,** the use of vegetation for the in situ decontamination of soils and sediments of heavy metals and organic pollutants, is an emerging technology. As illustrated in Figure 12-15, plants can remediate pollutants by three mechanisms:

- the direct uptake of contaminants and their accumulation in the plant tissue (phytoextraction),

- the release into the soil of oxygen and biochemical substances such as enzymes that stimulate the biodegradation of pollutants, and

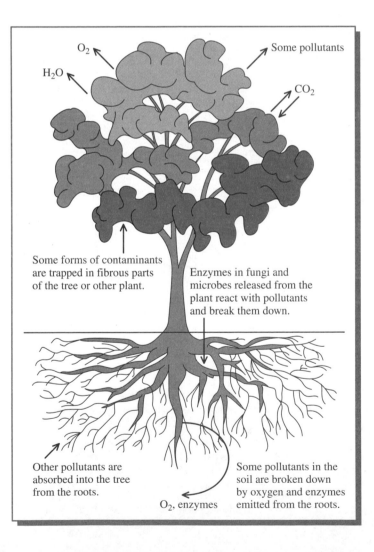

FIGURE 12-15
Mechanisms of phytoremediation by a plant.

- the enhancement of biodegradation by fungi and microbes located at the root–soil interface.

Advantages of phytoremediation include its relatively low cost, aesthetic benefits, and nonobtrusive nature.

Certain plants are **hyperaccumulators** of metals, i.e., they are able to absorb much higher than average levels of these contaminants (by a factor of at least 10–100, to yield a contaminant concentration of 0.1% or more) through their roots and to concentrate them much more than do normal plants. This ability probably evolved over long periods of time as the plants grew on natural soils that contained high concentrations of pollutants, especially heavy metals. In bioremediation these plants are grown on contaminated sites and then harvested and burned. In some instances, the resulting ash is so concentrated in metal that it can be mined!

Phytoremediation is an attractive technique because metals are often difficult to extract with other technologies since their concentration is usually so small. For example, the shrub called the *Alpine pennycress* has the ability to hyperaccumulate cadmium, zinc, and nickel. Scientists are experimenting with various types of plants that can extract lead from soils. One difficulty with phytoremediation is that hyperaccumulators usually are plants that are slow to grow and therefore slow to accumulate metals. However, fast-growing poplar trees show promise of being efficient phytoremediators. One recently developed type of hybrid poplar efficiently absorbs TCE from hazardous waste sites and from groundwater. In general, there is a need to harvest the plants before they lose their leaves or begin to decay, so that contaminants do not become dispersed or return to the soil.

Plants can efficiently take up organic substances that are moderately hydrophobic, with log K_{ow} values (Chapter 7) from about 0.5 to 3, a range that includes BTEX components and some chlorinated solvents. Substances that are more hydrophobic than the upper limit of this range bind so strongly to roots that they are not easily taken up within the plant. Once taken up, the plant may store the substance by transformation within its lignin component or it may metabolize it and release the products into the air.

Substances that plants release into the soil include chelating ligands and enzymes; by complexing a metal, the former can decrease its toxicity, and the latter in some cases can biodegrade pollutants. For example, it has been found that the plant-derived enzyme *dehalogenase* can degrade TCE. Plants also release oxygen at the roots, thereby facilitating aerobic transformations. As previously discussed, the fungi that exist in symbiotic association with a plant also have enzymes that can assist in the degradation of organic contaminants in soil.

Bioremediation in general and phytoremediation in particular are rapidly emerging technologies. The long-term potential for these techniques to be used at many sites that require decontamination is apparent.

Experiments at several test sites have shown that phytoremediation can be used successfully to degrade petroleum products in soils. The number of Superfund sites applying bioremediation technology is shown in Table 12-3.

Hazardous Wastes

In this section, we consider the nature of various types of hazardous wastes and discuss how individual samples of such wastes can be destroyed as an alternative to simply dumping them and thereby deferring the problem to a later date.

Currently there are more than 50,000 hazardous waste sites and perhaps 300,000 leaking underground storage tanks in the United States alone. The Superfund program of the U.S. EPA was created to remediate waste sites; the eventual cost is estimated to be $31 billion (see Box 12-1).

The Nature of Hazardous Wastes

A substance can be said to be a hazard if it poses a danger to the environment, especially to living things. Thus **hazardous wastes** are substances that have been discarded or designated as waste and that pose a danger. Most of the hazardous wastes with which we shall deal are commercial chemical substances or by-products from their manufacture; biological materials are not considered here.

In Chapter 8 we paid a great deal of attention to substances that were **toxic,** i.e., they threaten the health of an organism when they enter its body. Other common types of hazardous materials include those that are

- **ignitable** and burn readily and easily;

- **corrosive** because their acid or base character allows them to easily corrode other materials;

- **reactive** in senses not covered by ignition or corrosion, i.e., by explosion; and

- **radioactive.**

Some waste materials are hazardous in more than a single category.

The Management of Hazardous Wastes

There are four strategies in the management of hazardous waste. In order of decreasing preference, they are:

- **Source reduction:** The deliberate minimization, through process planning, of hazardous waste generation in the first place. The green chemistry cases presented throughout this text provide many examples of this strategy.

• **Recycling and reuse:** The use in a different process as raw materials, whether by the same company or a different one, of hazardous wastes generated in a process.

• **Treatment:** The use of any physical, chemical, biological, or thermal process—including incineration—that reduces or eliminates the hazard from the waste. Examples of such technology are discussed in subsequent sections.

• **Disposal:** Burial of the nonliquid waste in a properly designed landfill. In the past, liquid hazardous wastes were often injected into deep underground wells.

Landfills that are specially designed to accommodate hazardous wastes have several characteristics in addition to those discussed for sanitary landfills. The locations of such landfills should be

• in an area with clay or silt soil, to provide an additional barrier to leachate dispersal, and

• away from groundwater sources.

Often the hazardous wastes in such landfills are grouped according to their physical and chemical characteristics so that incompatible materials are not placed near each other.

Toxic Substances

As mentioned, toxic wastes are those that can cause a deterioration in the health of humans or other organisms when they enter a living body. Their characteristics, and many examples, were discussed in Chapters 8 and 11 particularly, and will not be reviewed in detail here. Those of main concern are heavy metals, organochlorine pesticides, organic solvents, and PCBs.

As an example of the magnitude of the problem of toxic substance waste management, consider the PCBs that are still used in the capacitors in the ballasts tubes of fluorescent light fixtures. Within the sealed capacitor container is a thick, gel-like liquid of concentrated PCB oil that is absorbed in several layers of paper. A typical capacitor contains about 20 g—about a tablespoon—of liquid PCB. Although each ballast does not contain much PCB, the number of these lighting fixtures in use in the developed world is huge. Thus the ultimate collection and disposal of PCBs from these sources will be a task requiring many years and many dollars. Disposal methods for toxic organic compounds are discussed later in this chapter.

Incineration of Toxic Waste

Incinerators that deal with hazardous waste are often more elaborate than those that burn municipal waste because it is important that the material be more completely destroyed and that emissions be more tightly controlled. In some cases the waste (e.g., PCBs) will not ignite on its own and must be added to an existing fire fueled by other wastes or by supplemental fuel such as natural gas or petroleum liquids. Modern facilities employ very hot flames, ensure that there is sufficient oxygen in the combustion zone, and keep the waste compounds in the combustion region long enough to ensure that their **destruction and removal efficiency,** DRE, is essentially complete, i.e., >99.9999%, called "six nines." The presence of carbon monoxide at a concentration greater than 100 ppm in the gas emissions is often used as an indicator of incomplete combustion.

About 3 million tonnes of hazardous wastes are burned annually in the United States, although this is only 2% of the amount generated. Three-quarters of the hazardous waste is dealt with by aqueous treatment, and 12% is disposed of on land or injected into deep wells.

The two most common forms of toxic waste incinerators are the rotary kiln and the liquid injection types. The **rotary kiln incinerator** can accept wastes of all types, including inert solids such as soil and sludges. The wastes are fed into a long (>20 m) cylinder that is inclined at a slight angle (about 5° from the horizontal) away from the entrance end and slowly rotates so that unburned material is continually exposed to the oxidizing conditions of 650–1100°C; see Figure 12-16. Over a period of about an hour the waste makes its way down the cylinder and is largely combusted. The hot exit gases from the kiln are sent to a secondary (nonrotating) combustion chamber equipped with a burner in which the temperature is 950–1200°C. The gas stays in the chamber for at least two seconds so that destruction of organic molecules is essentially complete. In some installations, liquid wastes can be fed directly into this chamber as the fuel. The gases exiting the secondary chamber are rapidly cooled to about 230°C by an evaporating water spray (in some cases with heat recovery), since they would otherwise destroy the air pollution equipment that they next enter.

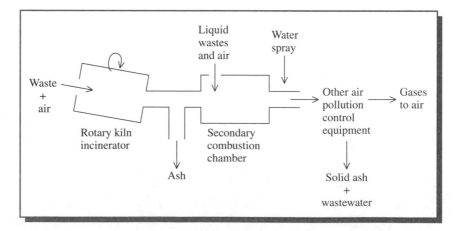

FIGURE 12-16
Schematic diagram of the components of a rotary kiln incinerator, including air pollution equipment.

Rotary kiln and other types of incinerators usually employ the same series of steps to purify the exhaust of particulates and of acid gases before it is released into the air as do garbage incinerators, i.e., a gas scrubber and a baghouse filter.

Cement kilns are a special type of large rotary kiln used to prepare cement from limestone, sand, clay, and shale. Very high temperatures of 1700°C or more are generated in cement kilns in order to drive off the carbon dioxide from limestone, $CaCO_3$, in the formation of *lime*, CaO. In addition to hotter combustion temperatures (compared to incinerators), wastes in cement kilns are burned with the fuel right in the flame; the residence time of the material in the kiln is also longer. Liquid hazardous wastes are sometimes used as part of the fuel (up to 40%) for these units, the remainder being a fossil fuel, usually coal. Recently, techniques have been developed that allow kilns to handle sludges and solids. Cement kilns burn more hazardous waste (about a million tonnes a year) than do commercial incinerators in the United States, although even more waste is incinerated on-site by chemical industries.

In the **liquid injection incinerator**—a vertical or horizontal cylinder—pumpable liquid wastes are first dispersed into a fine mist of small droplets. The finer these droplets, the more complete their subsequent combustion, which occurs at about 1600°C with a waste residence time of a second or two. A fuel or some "rich" (easily combustible) waste is used to produce the high temperatures of the combustion. In contrast to the rotary kiln, only a single combustion chamber is used, although in some modern versions a secondary input of air is introduced to improve oxygen distribution and generate more complete combustion. The exit gases can be passed through a spray dryer to neutralize and remove acid gases, followed by a baghouse filter to remove particulates, before being released into the outside air.

The incineration of hazardous waste has garnered much attention from environmentalists and some of the general public because of the potential release of toxic substances—particularly from the stack into the air—resulting from the operation of these units. Of special concern are the organic **products of incomplete combustion,** PICs, that have been found both in gases and adsorbed on particles emitted from incinerators. The PICs must be formed in the post-flame region because they could not survive the temperatures of the flame. Some of the most prevalent PICs are methane and benzene.

In their research to understand the production of these pollutants, scientists and engineers have discovered that reactions can occur downstream of the flame in "quench" zones and in pollution control devices, where temperatures fall below 600°C. Both gas-phase and surface-catalyzed processes apparently occur. For example, trace amounts of various dioxins and furans form at 200–400°C on fly ash and soot surfaces where the processes may be catalyzed by transition metal ions. A temperature of about 400°C is optimal for dioxin formation; below this the reaction is slow and above it they are quickly decomposed. It has not been firmly established whether the dioxins and furans result from the coupling on surfaces of precursor compounds such as chlorobenzenes and chlorophenols, or from the so-called de novo synthesis involving chlorine-free furan- or dioxin-like structures reacting with inorganic chlorides.

Concern has also been expressed about increased emissions that could occur when the incinerator is being closed down and during accidents or power failures, when lower temperatures would result for some time, since much greater quantities of dioxins and furans could presumably form under such conditions. **Fugitive emissions,** which include emissions from valves, minor ruptures, incidental spills, etc. are also a concern. The dust emitted from cement kiln incinerators has been found to contain toxic metals and some PICs. In fact, the health risk from toxic metal ion emissions (usually as oxides or chlorides) from the hazardous waste incinerators is found to exceed that from the toxic organics. As in the case of municipal incinerators, the solid residue from hazardous waste units can amount to one-third of the original waste volume and contains traces of toxic materials, as does the wastewater from the scrubber units.

Concerns about incineration have spurred the development of other technologies for disposing of hazardous waste. In **molten salt combustion** wastes are heated to about 900°C and destroyed by being mixed with molten sodium carbonate. The spent carbonate salt contains $NaCl$, $NaOH$, and various metals from the combusted waste, which can be recovered so that the $NaCO_3$ can be reused. No acidic gas is evolved, since it reacts within the salt. In **fluidized bed incinerators** a solid material such as limestone, sand, or alumina is suspended in air (fluidized) by means of a jet of air, and the wastes are combusted in the fluid at about 900°C. A secondary combustion chamber completes the oxidation of the exhaust gases. **Plasma incinerators** can achieve temperatures of 10,000°C by passing a strong electrical current through an inert gas such as argon. The plasma consists of a mixture of electrons and positive ions, including nuclei, and can successfully decompose compounds, producing much lower emissions than traditional incinerators. In such a **thermal** or **hot** plasma, all the particles travel at high speeds and are thermally hot.

Supercritical Fluids

The use of **supercritical fluids** is another modern alternative to incineration. The supercritical state of matter is produced when gases or liquids are subjected to very high pressures, and in some cases to elevated temperatures. At pressures and temperatures at or beyond the **critical point,** separate gaseous and liquid phases of a substance no longer exist. Under these conditions, only the supercritical state, with properties that lie between those of a gas and those of a liquid, exists. For example, for water, the critical pressure is 218 atm (22.1 megapascals) and the critical temperature is 374°C, as illustrated in the phase diagram in Figure 12-17. Depending on exactly how much pressure is applied, the physical properties of the supercritical fluid vary between those of a gas (at relatively lower pressures) and those of a liquid (at higher pressures); the variation of properties with pressure or temperature is particularly acute near the critical point. Thus the density of supercritical water can vary over a considerable range, depending upon how much pressure (beyond 218 atm) is applied. Other substances that readily

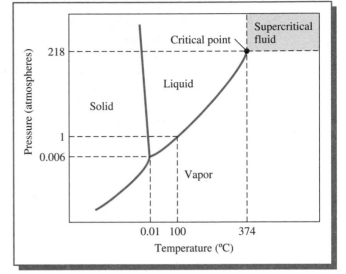

FIGURE 12-17 Phase diagram for water (not to scale). Notice the region for the supercritical state (shaded light green), which exists at temperatures and pressures beyond the critical point.

form useful supercritical fluids are carbon dioxide (which is used for many extractions in the food industry, including the decaffeination of coffee beans), xenon, and argon. See Table 12-4 for their critical temperatures and pressures.

One rapidly developing innovative technology for destruction of organic wastes and hazardous materials such as phenols is **supercritical water oxidation** (SCWO). Initially, the organic wastes to be destroyed are either dissolved in aqueous solution or suspended in water. The liquid is then subjected to very high pressure and a temperature in the 400–600°C range so that the water lies beyond its critical conditions and so is a supercritical fluid. The solubility characteristics of supercritical water differ markedly from those of normal liquid water: most organic substances become much *more* soluble and many ionic substances become much *less* soluble. Similarly, and also because very high pressures are applied, O_2 is much more soluble in supercritical than in liquid water.

TABLE 12-4	Supercritical Fluid Characteristics	
Substance	Critical temperature (°C)	Critical pressure (atm)
Water	374.1	217.7
Carbon dioxide	31.3	72.9
Argon	150.9	48.0
Xenon	16.6	58.4

At the elevated temperatures associated with supercritical water, the dissolved organic materials are readily oxidized by the ample amounts of O_2 that are pumped into and dissolve in the fluid. Hydrogen peroxide may be added to generate hydroxyl radicals, which initiate even faster oxidation. Because materials diffuse much more rapidly in the supercritical state than in liquids, the reaction is generally complete within seconds or minutes. One practical problem with the SCWO method is that insoluble inorganic salts that are formed in the reactions can corrode the high-pressure equipment and so shorten its lifetime. This corrosion problem can be solved by designing the reactor so that there are no zones where salts can build up.

The advantages of this technology include the rapidity of the destruction reactions and the lack of the gaseous NO_X by-products that are characteristic of gas-phase combustion. The required pressure and temperature conditions are readily accessible with available high-pressure equipment. There are some intermediate products of oxidation—mainly organic acids and alcohols, and perhaps also some dioxins and furans—formed with the SCWO process, however, which give rise to concerns about the toxicity of the effluent from the process. In the variant of this technology in which a catalyst is used, the percentage conversion to fully oxidized products is increased and the amount of intermediates that persist is decreased.

In the **wet air oxidation** process, temperatures (typically 120–320°C) and pressures lower than those required to achieve supercritical conditions for water are used to efficiently oxidize aqueous wastes (often catalytically). The oxidation is efficient because the amount of oxygen that dissolves at the enhanced pressures favors the reaction. The process is generally much slower than in supercritical water, requiring about an hour. The method is cheaper to operate, however, than the relatively expensive SCWO technology.

Supercritical carbon dioxide has been used to extract organic contaminants such as the gasoline additive MTBE from polluted water. After extraction, the pressure can be lowered, at which point the carbon dioxide becomes a gas, leaving behind the liquid contaminants to be incinerated or otherwise oxidized. Similarly, supercritical fluids could be used to extract contaminants such as PCBs and DDT from soils and sediments.

Nonoxidative Processes

All the processes just described have employed oxidation as the means of destroying the hazardous wastes. However, a closed-loop **chemical reduction process** has been devised with no uncontrolled emissions, using a *reducing* rather than an oxidizing atmosphere to destroy hazardous wastes. One advantage to the absence of oxygen is that there is no opportunity for the incidental formation of dioxins and furans. The reducing atmosphere is achieved by using hydrogen gas at about 850°C as the substance with which the preheated mist of wastes reacts. The carbon in the wastes is converted to methane (and some temporarily to other hydrocarbons such as benzene, which are subsequently hydrogenated to produce additional methane). The oxygen, nitrogen, sulfur, and chlorine are converted into their hydrides. The process is actually enhanced by the presence of water, which under these reaction conditions can act as a reducing agent and form additional hydrogen by the water-gas shift reaction with methane (see Chapter 6). PAH formation, which is characteristic of other processes at high temperatures in the absence of air, is suppressed by maintaining the hydrogen content at more than 50%. The gas output is cooled and scrubbed to remove particulates. The hydrocarbon output from the process is subsequently burned to provide heat for the system; thus there are no direct emissions to the atmosphere.

PROBLEM 12-8

Construct and balance chemical equations for the destruction of the PCB molecule with the formula $C_{12}H_6Cl_4$ (a) by combustion with oxygen to yield CO_2, H_2O, and HCl and (b) by hydrogenation to yield methane and HCl.

Chemical dechlorination methods for the treatment of chlorine-containing organic wastes, especially PCBs from transformers, have been developed and used in various parts of the world, though they are rather expensive to operate. The basic idea is to substitute a hydrogen atom or some other nonhalogen group for the covalently bound chlorine atoms on the molecules, thereby detoxifying them. The mostly dechlorinated wastes can then be incinerated or disposed of in landfills. The commonly used reagent for this purpose is MOR, the alkali metal (M = sodium or potassium) salt of a polymeric alcohol. In the reaction, an —OR group replaces each of the chlorines, which depart as the salt MCl:

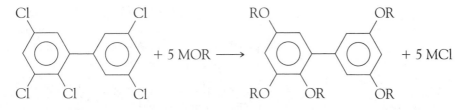

The process is carried out at temperatures above 120°C in the presence of *potassium hydroxide* and occurs most efficiently for highly chlorinated PCBs. An alternative dechlorination method is the reaction of the PCBs with dispersions of metallic sodium to give sodium chloride and a polymer containing many biphenyl units joined together.

Review Questions

1. Define the term *solid waste* and name its five largest categories for developed countries.

2. Describe the components and steps in the creation of a *sanitary landfill*.

3. Describe the three stages of waste decomposition that occur in a sanitary landfill, including the products of each stage. Are all stages equal in production or consumption of acidity?

4. Define the term *leachate*, explain how this substance arises, and list several of its common components. How can leachate be controlled and how is it treated?

5. Explain the difference between the two common types of MSW incinerators.

6. What is the difference between *bottom ash* and *fly ash* in an incinerator? Describe some of the air pollution control devices found on incinerators.

7. What is meant by the *four Rs* in waste management?

8. Why can the recycling of metals often be justified by economics alone?

9. Describe the processes by which paper and rubber tires can be recycled.

10. What are the common types of consumer packaging plastics that can be recycled? What four ways are used to recycle plastics?

11. What are some of the arguments for and against the recycling of plastics?

12. What is meant by a *life cycle assessment* and what are the two main uses for LCAs?

13. Describe the main inorganic constituents of soil. How do clay, sand, and silt particles differ in size?

14. What are the names and origins of the principal organic constituents of soil? Are both types of acids soluble in base? in acid?

15. What is meant by a soil's *cation-exchange capacity*? What are its common units?

16. What is meant by a soil's *reserve acidity*? How does it arise?

17. Describe several methods, including chemical equations, by which soils that are too acid or too alkaline can be treated.

18. Describe the processes by which soil in arid areas becomes salty and alkaline.

19. What are meant by the terms *sediments* and *pore water*?

20. By what three ways are heavy metals bound to sediments?

21. How can mercury stored in sediments be solubilized and enter the food chain?

22. Describe how mine tailings are usually stored and how this represents a potential environmental problem.

23. How can sediments contaminated by heavy metals be remediated so they can be used on agricultural fields?

24. Describe two ways by which contaminated sediments can be treated without removing the sediments themselves.

25. List the three categories of technologies commonly used to remediate contaminated soils. Give examples of each.

26. List three conditions that must be fulfilled if bioremediation of soil is to be successful.

27. Describe the two ways in which PCBs in sediments are bioremediated.

28. Define *phytoremediation* and list the three mechanisms by which it can operate. What are *hyperaccumulators*?

29. Define the term *hazardous waste*. What are the five common types?

30. List, in order of decreasing desirability, the four strategies used in the management of hazardous waste.

31. Name and describe the three common types of incinerators used to destroy hazardous waste. What does *DRE* stand for and how is it defined?

32. Define the term *PIC* and describe how they are formed.

33. Explain the advantages and disadvantages of supercritical water oxidation for the destruction of hazardous wastes.

34. How is the chemical reduction process carried out? What advantages does it have over oxidation methods?

Green Chemistry Questions

1. What are the environmental advantages of using polyaspartate as an antiscalant/dispersant versus polyacrylate?

2. The synthesis of polyaspartate as developed by Donlar won a Presidential Green Chemistry Challenge Award.

(a) Which of the three focus areas (see the Introduction to Green Chemistry) for these awards does this award best fit into?

(b) List at least three of the twelve principles of green chemistry (see the Introduction to Green Chemistry) that are addressed by the chemistry developed by Donlar Corporation.

3. The development of recyclable carpeting by Shaw Industries won a Presidential Green Chemistry Challenge Award.

(a) Which of the three focus areas (see the Introduction to Green Chemistry) for these awards does this award best fit into?

(b) List three of the twelve principles of green chemistry (see the Introduction to Green Chemistry) that are addressed by the chemistry developed by Shaw Industries.

4. What are the environmental advantages of using polyolefin-backed nylon fiber carpeting in place of PVC-backed carpeting?

Additional Problems

1. Consider a small city of 300,000 people in the northern United States or southern Canada or central Europe, and suppose that its residents on average produce 2 kg/day of MSW, about one-quarter of which will decompose anaerobically to release methane and carbon dioxide evenly over a 10-year period. Calculate how many homes could be heated by burning the methane from the landfill, given that residential requirements are about 10^8 kJ/year in that climate zone. Consult Chapter 6 for data on methane combustion energetics.

2. By reference to the general solubility rules for sulfides found in introductory chemistry textbooks, deduce which metals would *not* have their sediment availabilities determined by AVS. Which of these metals would occur in the form of insoluble carbonates instead, and thus be biologically unavailable in marine environments?

3. The acid volatile sulfide concentration in a sediment is found to be 10 μmol/g. The majority of the sulfide is present in the form of insoluble FeS, since the Fe^{2+} concentration in the sediment is 450 $\mu g/g$. What mass of mercury in the form of Hg^{2+} can be tied up by the remaining sulfide in 1 tonne of such sediment? Note that you may assume that a negligible amount of iron is not tied up as FeS.

4. For the PCB congener 2,4,4′,5-tetrachlorobiphenyl, deduce **(a)** the chlorine substitution pattern on the benzoic acid that results from its aerobic degradation and **(b)** the various PCB congeners with one or two chlorines that could result from its anaerobic degradation.

5. Generate lists of the aspects of production, distribution, and disposal that you would employ in a life cycle assessment done for the purpose of deciding which is the more environmentally friendly container for beer: glass bottles or aluminum cans. Do you think the conclusions of such an analysis would depend significantly on the extent of recycling of the containers?

6. As mentioned in the chapter, waste with a high cellulose content, such as paper, wood scraps, and corn husks, can be converted to ethanol for use as a fuel. One way this can be done is to first perform an acid hydrolysis of the cellulose to convert it into glucose. This is then followed by fermentation of the glucose to produce ethanol. Write a balanced equation for the fermentation of glucose, $C_6H_{12}O_6$, into ethanol, C_2H_5OH. Given that the pyranose monomer unit in cellulose has the formula $C_6H_{10}O_5$ (i.e., glucose−H_2O), determine the volume of ethanol that could be produced in this way from 1.00 tonne of hardwood scraps (46% cellulose), assuming 100% conversion in both steps.

7. One way to reduce the lifetime of plastic wastes is to make them photodegradable, so that some of the bonds in the polymers will break on absorption of light. When designing a photodegradable plastic, what range of sunlight wavelengths would it be best for the plastic to absorb? What is the main limitation of the breakdown of these plastics? If you were to design a photodegradable plastic that would break down on absorption of 300-nm light, what would be the maximum bond energy of the bonds to be cleaved?

8. Removal of toxic wastes from soils and sediments can be performed by desorption/extraction using an appropriate solvent. Traditionally, chlorinated solvents such as dichloromethane have been used, because of the high solubility of polychlorinated contaminates in such solvents. More recently, supercritical fluids, such as supercritical CO_2, have been used for such extraction, as discussed in the chapter. What are the advantages of using supercritical CO_2, compared to dichloromethane? [W. Zhou et al., "Desorption of Polychlorinated Biphenyls from Contaminated St. Lawrence River Sediments with Supercritical Fluids," *Industrial and Engineering Chemistry Research* 43 (2004): 397] have reported successful extraction of PCBs from St. Lawrence River sediments, using pure and mixed-solvent supercritical CO_2. After 60 minutes extraction of a sediment sample containing 2200 ppm PCBs, the sediment was found to contain ≈50 ppm PCB using supercritical CO_2, and less than 5 ppm PCB using supercritical CO_2 with 5 mol% added methanol. Calculate the extraction efficiency in each case. Suggest reasons for the difference in the two efficiencies.

Further Reading

1. M. B. McBride, *Environmental Chemistry of Soils* (New York: Oxford University Press, 1994).

2. P. S. Phillips and N. P. Freestone, "Managing Waste—The Role of Landfill," *Education in Chemistry* (January 1997): 11.

3. F. Pearce, "Burn Me," *New Scientist* (22 November 1997): 31.

4. B. Piasecki et al., "Is Combustion of Plastics Desirable?" *American Scientist* 86 (1998): 364.

5. K. Tuppurainen et al., "Formation of PCDDs and PCDFs in Municipal Waste Incineration and Its Inhibition Mechanisms: A Review," *Chemosphere* 36 (1998): 1493.

6. T. E. McKone and S. K. Hammond, "Managing the Health Impacts of Waste Incineration," *Environmental Science and Technology* (1 September 2000): 380A.

7. R. Brown et al., "Bioremediation," *Pollution Engineering* (October 1999): 26.

8. M. E. Watanabe, "Phytoremediation on the Brink of Commercialization," *Environmental Science and Technology* 31 (1997): 182A.

9. P. B. A. Kumar et al., "Phytoextraction: The Use of Plants to Remove Heavy Metals from Soils," *Environmental Science and Technology* 29 (1995): 1232.

10. D. A. Wolfe et al., "The Fate of the Oil Spilled from the *Exxon Valdez*," *Environmental Science and Technology* 28 (1994): 561A–568A.

11. F. Pearce, "Tails of Woe," *New Scientist* (11 November 2000): 46.

12. M. L. Hitchman et al., "Disposal Methods for Chlorinated Aromatic Waste," *Chemical Society Review* (1995): 423.

13. D. Simonsson, "Electrochemistry for a Cleaner Environment," *Chemical Society Reviews* 26 (1997): 181.

14. D. Amarante, "Applying in Situ Chemical Oxidation," *Pollution Engineering* (February 2000): 40.

15. H. Black, "The Hottest Thing in Remediation," *Environmental Health Perspectives* 110 (2002): A146.

Websites of Interest

Log on to www.whfreeman.com/envchem3e/ and click on Chapter 12.

13

Radioactivity, Radon, and Nuclear Energy

In all the previous chapters we have been concerned with chemicals and chemical processes. We finish the book by considering *nuclear* processes and how they affect the environment, our health, and our energy supply. These concerns all center on the effects of radioactivity, and it is with this topic that we begin. This allows us to discuss **radon,** the most important radioactive indoor air pollutant, and the topic of depleted uranium. We then switch to nuclear energy and explore the ways in which electricity can be produced from it and the environmental consequences of the radioactive waste that these processes generate.

The San Onofre nuclear power generating station at San Diego, California. All current nuclear energy is generated by fission, though fusion plants may be viable in the future. (Corbis Images)

Radioactivity and Radon Gas

The Nature of Radioactivity

Although most atomic nuclei are stable indefinitely, some are not. The unstable, or **radioactive,** nuclei spontaneously decompose by emitting a small particle that is very fast moving and therefore carries with it a great deal of energy. In some types of nuclear decomposition processes, atoms are converted from those of one element to those of another as a consequence of this emission. Very heavy elements are particularly prone to this type of decomposition, which occurs by the emission of a small particle. The nuclei produced by emission of the particle may or may not themselves be radioactive; if they are, they will undergo another decomposition at a later time.

Recall from introductory chemistry that the **mass number** is the *number* of heavy particles—protons and neutrons—and not the actual mass of the nucleus. An **alpha (α) particle** is a radioactively emitted particle that has a charge of $+2$ and a mass number of 4—it has two neutrons and two protons—and it is identical to a common *helium* nucleus. Thus an α particle is written as $^4_2$He, where 4 is its mass number and 2 refers to its nuclear charge (i.e., number of protons). The nucleus that remains behind after an atom has lost an α particle has a nuclear charge that is 2 units *less* than the original, and it is 4 units *lighter*. For example, when a $^{226}_{88}$Ra (radium-226) nucleus emits an α particle, the resulting nucleus has a mass number of $226 - 4 = 222$ units and a nuclear charge of $88 - 2 = 86$; this is a wholly new element that is an isotope of the element radon. The process can be written as a *nuclear reaction:*

$$^{226}_{88}\text{Ra} \longrightarrow\ ^{222}_{86}\text{Rn} + {}^4_2\text{He}$$

Notice that both the total mass number and the total nuclear charge individually balance in such equations.

A **beta (β) particle** is an electron. It is formed when a neutron splits into a proton and an electron in the nucleus. Since the proton remains behind in the nucleus when the electron leaves it, the nuclear charge (or atomic number) *increases* by 1 unit (you may imagine this effect as "subtracting a negative particle"). There is no change in mass number of the nucleus, since the total number of neutrons plus protons remains the same. For example, when an atom of the lead isotope $^{214}_{82}$Pb (lead-214) decays radioactively by the emission of a β particle, the nuclear charge of the product is $82 + 1 = 83$, corresponding to the element bismuth; the mass number remains 214:

$$^{214}_{82}\text{Pb} \longrightarrow\ ^{214}_{83}\text{Bi} + {}^{\ 0}_{-1}\text{e}$$

Notice that the symbol $^{\ 0}_{-1}$e used here for the electron shows its mass number (zero) and its charge; in the equation the total mass numbers and nuclear charge numbers each balance.

One other important type of radioactivity is the emission of a **gamma (γ) particle** (also called a *ray*) by a nucleus. This is a huge amount of energy concentrated in one photon and possesses no particle mass. Neither the nuclear mass number nor the nuclear charge changes when a γ particle is emitted. The emission of γ ray often accompanies the emission of an α or β particle from a radioactive nucleus. The properties of all three types of nuclear radiation are summarized in Table 13-1.

TABLE 13-1	Summary of Small Particles Produced by Radioactivity		
Particle symbol and name	Chemical symbol	Comment	Effect on nucleus of particle emission
α (alpha)	^4_2He	Nucleus of a helium atom	Atomic no. reduced by 2
β (beta)	$^0_{-1}\text{e}$	Fast-moving electron	Atomic no. increased by 1
γ (gamma)	None	High-energy photon	None

PROBLEM 13-1

Deduce the nature of the species that belongs in the blank for each of the following nuclear reactions:

(a) $^{222}_{86}\text{Rn} \longrightarrow \, ^4_2\text{He} +$ _____

(b) $^{214}_{83}\text{Bi} \longrightarrow \beta +$ _____

(c) $^{214}_{-}\text{Po} \longrightarrow \, ^{214}_{-}\text{Pb} +$ _____

(d)   _____ $\longrightarrow \, ^{234}_{90}\text{Th} + \alpha$

The Health Effects of Ionizing Radiation

The α and β particles that are produced in the radioactive decay of a nucleus are not in themselves harmful chemicals, since they are simply the nucleus of a helium atom and an electron. However, they are ejected from the nucleus with an incredible amount of energy of motion. When this energy is absorbed by the matter encountered by the particle, it often ionizes atoms or molecules; for that reason, it is called **ionizing radiation,** or just *radiation*. This radiation is potentially dangerous if we absorb it, since the molecular components of our bodies can be ionized or otherwise damaged.

Although α and β particles are energetic, they cannot travel far within the human body, since they lose more and more of their energy—and consequently slow down—as they collide with more and more atoms.

Alpha particles can travel only a few thousandths of a centimeter within the body, so they are not penetrating. This is true because they are relatively massive, and when they interact with matter they slow down, capture electrons from it, and are converted into harmless atoms of helium gas. If an α particle is emitted outside the body, it will usually be absorbed in the air or by the layer of dead skin, so it will do you no harm. However, inhaled or ingested radioactive atoms can cause serious internal damage when they emit

α particles. The damage is particularly severe with α particles since their energy is concentrated in a small area of absorption located within about 0.05 mm of the point of emission. In their interaction with matter, α particles are highly damaging—the most highly damaging of all particles—since they can knock atoms out of molecules or ions out of crystal sites. If the molecules affected are DNA or its associated enzymes, cancer can result.

Beta particles move much faster than α particles since they are much lighter and can travel about 1 m in air or about 3 cm in water or biological tissue before losing their excess energy. Like α particles, they can cause considerable damage to cells if they are emitted from particles that have been inhaled or ingested and the radioactive nucleus is consequently close to the cell when it decays.

Gamma rays easily pass through concrete walls—and our skin. A few centimeters of lead are required to shield us from γ rays. Gamma particles are the most penetrating and therefore the most damaging of the three, traveling a few dozen centimeters into our bodies or even right through them. They are generally the most dangerous type of radioactivity, since they can penetrate matter efficiently and do not have to be inhaled or ingested. Although they can pass through our bodies, γ rays lose some of their energy in the process, and cells can be damaged by this transferred energy, since it can ionize molecules. Ionized DNA and protein molecules cannot carry out their normal functions, potentially resulting in radiation sickness and cancer.

The ions produced by radiation when their energy is transferred to molecules are free radicals, and hence they are highly reactive (Chapters 2, 3, and 12). For example, a water molecule can be ionized by an α, β, or γ ray or by an X-ray. The resulting H_2O^+ free-radical ion subsequently dissociates into a hydrogen ion and the *hydroxyl free radical*, OH:

$$H_2O + radiation \longrightarrow e^- + H_2O^+$$

$$H_2O^+ \longrightarrow H^+ + OH$$

If the affected water molecule is contained in a cell, the hydroxyl radical can engage in harmful reactions with biological molecules in the cell, such as DNA and proteins. In some cases, radiation damage is sufficient to kill cells of living organisms. This is the basis of food irradiation, where the death of microorganisms helps prevent subsequent spoilage of the food.

If human beings are exposed to substantial, though sublethal, amounts of ionizing radiation, they can develop *radiation sickness*. The earliest effects of this malady to be observed occur in tissues containing cells that divide rapidly, because damage to the cell's DNA or protein can affect cell division. Such rapidly dividing cells are found in bone marrow, where white blood cells are produced, and in the lining of the stomach. It is not surprising, then, to find that early symptoms of radiation sickness include nausea and a drop in white blood cell count. Children are more susceptible to radiation than adults because their tissues involve more cell division. On the other hand, radiation can be effectively used to kill cancer cells since they are dividing rapidly. Unfortunately, radiation therapy cannot be completely selective in terms of the cells it affects, so it has side effects such as nausea.

Long-term effects from radiation may show up in genetic damage, because chromosomes may have undergone damage or their DNA may have mutated. Such damage may lead to cancer in the person exposed or to effects in her or his offspring if the changes occurred in the ovaries or testes.

Quantifying the Amount of Radiation Energy Absorbed

The amount of radiation absorbed by the human body is measured in **rad** (radiation absorbed dose) units, where 1 rad is the quantity of radiation that deposits 0.01 joule of energy to 1 kilogram of body tissue. The rad is not a particularly useful quantity, however, since the damage inflicted by 1 rad of α particles is 10 to 20 times greater than that inflicted by 1 rad of β particles or γ rays. The scale of absorbed radiation that incorporates this biological effectiveness factor is the **rem** (roentgen equivalent man). A more modern unit than the rem is the sievert, Sv, which equals 100 rem.

On average, we each receive about 0.3 rem, i.e., 300 mrem or 3000 μSv, of radiation annually. The origin on average is about

- 55% from radon in indoor and outdoor air;

- 8% from cosmic rays from outer space;

- 8% from rocks and soil;

- 11% from natural radioactive isotopes (e.g., ^{40}K, ^{14}C) of elements that are present in our own bodies; and

- 18% from anthropogenic sources, chiefly medical X-rays.

The average contribution from nuclear power production is negligible at present.

An *acute* exposure of more than 25 rem results in a measurable decrease in a person's white blood cell count; over 100 rem produces nausea and hair loss; and an exposure of over 500 rem results in a 50% chance of death within a few weeks.

Radioactive Nucleus Decay

The radioactive decay of the atoms in an isotope sample does not occur all at once. For example, in a sample of **uranium-238,** ^{238}U, just large enough to be visible, there are about 10^{24} atoms. Only about 10^7 of the ^{238}U nuclei in the sample decompose in a given second, so it requires billions of years for the decomposition process to be complete for the sample as a whole.

Since all radioactive nuclei disintegrate by processes that are kinetically first-order, it is convenient to express the decomposition rate as the time period required for half the nuclei in a sample to disintegrate—its **half-life,** $t_{1/2}$. (This is the same property that was used in earlier chapters to discuss the decomposition of substances by chemical reactions.) For example, the half-life of ^{238}U is about 4.5 billion years. Thus about half of this isotope of uranium existing when the Earth was formed (about 4.5 billion years ago) has now disintegrated; half the *remaining* ^{238}U, amounting to one-quarter of the original, will disintegrate over the next 4.5 billion years, leaving one-quarter of the original still intact. After three half-lives have passed, only one-eighth of the original will remain, and only one-sixteenth will be there after four half-lives.

PROBLEM 13-2

Recall from introductory chemistry that for first-order processes, the fraction F of reactant left after time t of reaction has elapsed is given by $F = e^{-kt}$, where k is the rate constant. Calculate the time required for 99% of a sample of ^{238}U to disintegrate. [Hint: The relationship between k and $t_{1/2}$ is $k = 0.693/t_{1/2}$.]

Radon from the Uranium-238 Decay Sequence

Many rocks and granite soils contain uranium, so the radioactive decay process takes place under our feet each day, with radon gas being one of its unwelcome products. Each $^{238}_{92}U$ nucleus eventually emits an α particle, and an atom of the *thorium* isotope $^{234}_{90}Th$ is formed:

$$^{238}_{92}U \longrightarrow {}^{234}_{90}Th + {}^4_2He$$

This is the first of 14 sequential radioactive decay processes that a ^{238}U nucleus undergoes, as illustrated in Figure 13-1a. The last of these reactions produces $^{206}_{82}Pb$, a nonradioactive (stable) isotope of *lead;* therefore the sequence stops. The concentration of each member of the series can be determined using steady-state principles (see Box 13-1).

Of particular interest is the portion of the 14-step sequence of ^{238}U radioactive decay that involves radon, since this element is the only one, other than the helium produced from the α particles, that is gaseous and therefore is mobile. Details concerning this portion of the radioactive decay series are shown in Figure 13-1b. The immediate precursor of the radon is radium-226, which has a half-life of 1,600 years and decays by emission of an α particle:

$$^{226}_{88}Ra \longrightarrow {}^{222}_{86}Rn + {}^4_2He$$

The ^{222}Rn isotope has a half-life of 3.8 days, which is long enough for it to diffuse through the solid rock or soil in which it is initially formed. Most radon escapes directly into outdoor air when the surface of the Earth where it appears

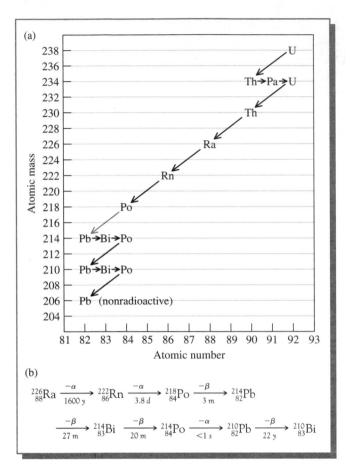

FIGURE 13-1 (a) The ^{238}U radioactive decay series. (b) The radium–radon portion of the ^{238}U radioactive decay series. The symbol above the arrow signifies the type of particle (α or β) emitted during the transition. The time period indicated below an arrow is the half-life of the unstable isotope.

is not covered, e.g., by a building. The very small background concentration of radon in air that this produces nevertheless yields about half of our exposure to radioactivity, as listed above. Although the radon decays in a few days, it is constantly replaced by the decay of more radium.

Some scientists have pointed out that radon gas accumulates to unhealthy levels in caves, including some that are often used for recreational purposes. However, it is in certain homes that radon becomes an important indoor air pollutant. Most radon that seeps into homes comes from the top meter of soil below and around the foundation; radon produced much deeper than this will probably decay to a nongaseous and therefore immobile element before it reaches the surface. Loose, sandy soil allows the maximum diffusion of radon gas, whereas frozen, compacted, or clay soil inhibits its flow. Radon enters the basements of homes through holes and cracks in their concrete foundations. The intake is increased significantly if the air pressure in the basement is low. The material used to construct the homes and water from artesian wells are other potential sources of radon in homes. Groundwater systems serving up to a few hundred people often have radon levels almost 10 times those of surface waters. When well water is heated and exposed to air, as occurs when it exits from a showerhead, radon is released to the air. However, radon from water usually represents only a small fraction of that arising from soil, although it represents a greater health hazard than that contributed from water disinfection by-products and other dissolved chemicals.

Measuring the Rate of Disintegration and Health Threat from Environmental Radiation

The rate of radioactive disintegration in a sample of matter is measured in **curies**, Ci, where 1 Ci equals 3.7×10^{10} disintegrations per second. Environmental regulations are usually expressed in the number of

BOX 13-1 | Steady-State Analysis of the Radioactive Decay Series

The various members, except the first and last, of the 14-step radioactive series of Figure 13-1a in a body of undisturbed uranium ore are each in a steady state (Chapter 1) with respect to their concentrations. Since all nuclear disintegration processes are kinetically first-order, the rate of production of each species, C, is proportional to the concentration of B, the species that precedes it in the chain $A \rightarrow B \rightarrow C \rightarrow D \rightarrow \cdots$. The rate of disintegration of C is proportional to its own concentration. Thus

$$d\,[C]/dt = k_p\,[B] - k_d\,[C] = 0$$

at steady state

so

$$[C]_{ss}/[B]_{ss} = k_p/k_d$$

Since each first-order rate constant is inversely proportional to the half-life period for the process, it follows that

$$[C]_{ss}/[B]_{ss} = t_C/t_B$$

where t_C and t_B correspond to the half-life periods for decay by disintegration of C and B. Thus the steady-state concentration of any species along the series relative to the one that precedes it is directly proportional to its half-life decay period and inversely proportional

to the half-life decay period of the species that produces it. For example, the half-life of ^{222}Rn is 3.8 days and that of the ^{226}Ra that produces it is 1600 years or 1.5×10^5 times as long, so the steady-state ratio of radon to radium is $1/1.5 \times 10^5$ or 6.5×10^{-6}. In qualitative terms, the radon concentration never builds up close to that of the radium since it decays so quickly after it is formed.

PROBLEM 1

Show that the ratio of the steady-state concentration of C relative to the first member, A, of the radioactive series $A \rightarrow B \rightarrow C \rightarrow D \rightarrow \cdots$ is equal to the ratio of their half-lives for disintegration, and thus that the steady-state concentration of each member of the series is directly proportional to its disintegration half-life.

PROBLEM 2

Using the equations developed in Box 10-3, deduce how long it will take for radon-22 to reach 80% of its steady-state concentration after a sample of radium-226 is purified of contamination by any other substance. [Hint: Recall that for first-order reactions, $t_{0.5} = 0.693/k$.]

picocuries, pCi. Thus 1 pCi $= 10^{-12}$ Ci, or 0.037 disintegrations per second. The U.S. EPA has set 4 pCi per liter of air as the maximum radon level in houses.

We can calculate the amount of energy from radiation that is absorbed by a person's lungs in a year if he or she breathes air containing radon at the 4 pCi/L level, since the energy of each α particle emitted by a radon atom is known by measurement to be 9.0×10^{-13} J. Since 4 pCi/L is equivalent to $4 \times 10^{-12} \times 3.7 \times 10^{10} = 0.15$ disintegrations per liter per second, and since there are $60 \times 60 \times 24 \times 365$ sec in a year, the total annual number of disintegrations in 1 L of air is 4.7×10^6. Thus the total amount of energy liberated annually in the process is $4.7 \times 10^6 \times 9.0 \times 10^{-13}$ J $= 4.2 \times 10^{-6}$ J. If we assume that all this energy is absorbed by a person's lung tissues (rather than by the air in the lungs), that the lung volume is about 1 L, and that the lung mass is about 3 kg, then since 1 rad = 0.01 J/kg, the energy absorbed is 1.4×10^{-4} rad or 0.14 mrad. Using a factor of 10 to convert rads to rems for α-particle radiation, we find that the annual radiation dose is about 1.4 mrem, i.e., about 0.5% of background exposure.

The Daughters of Radon

Radon, the heaviest member of the noble gas group, is chemically inert under ambient conditions and remains a monatomic gas. As such, it becomes part of the air that we breathe when it enters our homes. Because of its inertness, physical state, and low solubility in body fluids, radon *itself* does not pose much of a danger; the chance that it will disintegrate during the short time it is present in our lungs is small and, as discussed above, the range of α particles in air before they lose most of their energy is less than 10 cm.

The danger arises instead from the radioactivity of the next three elements in the disintegration sequence of radon—namely, *polonium*, *lead*, and *bismuth* (see Figure 13-1b). These *descendants* are termed **daughters** of radon, which in turn is called the *parent* element. In macroscopic amounts, these particular daughter elements are solids, and when formed in the air from radon they all quickly adhere to dust particles. Some dust particles adhere to lung surfaces when inhaled, and it is under these conditions that the elements pose a health threat. In particular, both the ^{218}Po, which is formed directly from ^{226}Rn, and the ^{214}Po, which is formed later in the sequence (Figure 13-1b), emit energetic α particles that can cause radiation damage to the bronchial cells near which the dust particles reside. This damage can eventually lead to lung cancer. Indeed, as will be discussed, radon (or rather its daughters) is the second leading cause of such cancers, although it follows smoking by a wide margin.

Although some radon daughters in the sequence disintegrate by β-particle emission, the deleterious health effects of these particles are considered negligible because the α particles carry much more energy and, as discussed, it is the disruption of cell molecules by the burst of high energy that initiates cancer.

Notice that the sequence (see Figure 13-1b) of radon decay to ^{210}Pb formation takes less than a week on average. In contrast, disintegration of ^{210}Pb to ^{210}Bi has a half-life of 22 years, and, in fact, most of the lead will have been cleared from the body before this process occurs.

Measuring the Health Danger from Radon and Its Daughters

The greatest exposure to α particles from radon disintegration is experienced by miners who work in poorly ventilated underground uranium mines. Their rate of lung cancer is indeed higher than that of the general public, even after corrections to the data have been made for the effects of smoking. From statistical data relating their excess incidence of lung cancer to their cumulative level of exposure to radiation, a mathematical relationship between cancer incidence and radon exposure has been developed. Scientists have extrapolated this relationship to determine the risk to the general population from the generally lower levels of radon to which the public is exposed.

Based on linear extrapolation, a 1998 analysis by the National Research Council concluded that in the mid-1990s, about 22,000 lung cancer deaths per year (the actual range is 3,000–33,000), corresponding to about 14% of the U.S. total, were associated with exposure to radon in indoor air. Almost 90% of these radon victims were smokers, because radon and cigarette smoke are synergistic in causing the disease. Nevertheless, radon exposure accounted for about one-quarter of the lung cancer deaths for nonsmokers. An analysis for western Germany predicted that about 2,000 deaths there annually will be due to indoor radon, and the same number is predicted for the United Kingdom.

Those skeptical of using uranium miners' data point out that the calculated estimates of radon-caused lung cancer may be too high, since miners work in much dustier conditions than are found in homes and their breathing during hard labor is much deeper than normal. Consequently, there is a much greater chance that radon daughters will find their way deep into the lungs of miners in comparison with the general population. The miners' exposure to arsenic and diesel exhaust may also contribute to an increase in the lung cancer rate that would have been counted as due to radon.

In order to establish whether or not radon gas buildup in homes causes lung cancer, several epidemiological studies were undertaken in the 1990s. These analyses, one from Sweden, one from Canada, and one from the United States, reached contradictory conclusions about the risk of radon to householders. In the Swedish report the rate of lung cancer in nonsmokers and especially in smokers was found to increase with increasing levels of radon in their homes. The Canadian study focused on residents of Winnipeg, Manitoba, which has the highest average radon levels in Canada; no linkage between radon levels and lung cancer incidence was found. The U.S. study, conducted among nonsmoking women in Missouri, found little evidence for a trend of increasing lung cancer with increasing indoor radon concentration.

The environmental radon problem has received the greatest attention in the United States, where there are currently programs to test the air in the basements of a large number of homes for significantly elevated levels of the gas. Once radon is identified, the owners can then alter the air circulation patterns to reduce radon levels in living areas, thereby reducing the additional risk of contracting lung cancer. It has been pointed out, however, that the normally high mobility of the U.S. population means that, on average, a given individual who happens to live in a house with a high level of radon will be there for only a few years and likely spend most of his or her life in a house with a lower level (since only about 7% of houses have high levels). Consequently, the estimate of increased lung cancer death mentioned at the beginning of this section is likely much too high.

Depleted Uranium

Depleted uranium is what remains of natural uranium once most of the ^{235}U isotope, used in power reactors and for nuclear weapons, and the ^{234}U have been extracted from it. Indeed, about 200 kg of depleted uranium is produced for every kilogram of the highly enriched element, since uranium is only 0.7% ^{238}U. Because uranium is so dense (70% denser than lead), it is useful for making armor-piercing weapons, especially projectiles used against tanks. When the shells hit hard targets, the uranium ignites and the combustion creates clouds of dust containing *uranium oxide*. This dust settles and contaminates the soil in the area, but before that, it can be inhaled by people in the vicinity. Since most of the ^{235}U has been extracted from the uranium (for use in bombs and power production), and since ^{238}U has such a very long half-life, depleted uranium is less radioactive in terms of α-particle emissions (by almost half) than the naturally occurring element. However, the β emission from daughters of ^{238}U is still present. Concerns have been expressed about the effect of the residual radioactivity from depleted uranium on troops and civilians exposed to it during wartime.

Dirty Bombs

A **dirty bomb** is a conventional, chemical-based explosive mixed with radioactive material that would be distributed over a wide area as a result of the bomb's explosion. Dirty bombs could be made by terrorists, using radioactive material stolen from either hospitals or research institutes, or purchased on the black market from supplies originating in the former Soviet Union or other countries that have undergone disruption with a consequent loss of security at their nuclear energy facilities. Although the actual danger to human health from a dispersal of dirty bomb materials into the environment may well be quite small, the fear it would generate in the public and the expense of decontamination of wide areas would be substantial.

A similar terrorist threat would be the deliberate crashing of hijacked aircraft into nuclear power plants or into containers of nuclear fuel. However, most nuclear energy industry executives deny that the power plants or containers could be breached, and radioactivity released, by such an event.

Nuclear Energy

Although most of the energy we use originates as heat generated by the combustion of carbon-containing fuels, heat in commercial quantities can also be produced indirectly when certain processes involving atomic nuclei occur; this power source is called **nuclear energy,** and it is used mainly to produce electricity. Since nuclear forces are much stronger than chemical bond forces, the energy released per atom in nuclear reactions is immense compared to that obtained in combustion reactions. One of the attractions of nuclear power is that it does not generate *carbon dioxide* or other greenhouse gases during its operation. Some policy-makers have promoted expansion of nuclear power as a way to combat global warming in the future.

There are two processes by which energy is obtained from atomic nuclei: fission and fusion.

- In **fission,** the collision of certain types of heavy nuclei (all of which have many neutrons and protons) with a neutron results in the splitting of the nucleus into two similarly sized fragments. Since the separated fragments are more stable energetically than was the original heavy nucleus, energy is released by the process.

- The combination of two very light nuclei to form one combined nucleus is called **fusion.** It also results in the release of huge amounts of energy, since the combined nucleus is more stable than the original, lighter ones.

Fission Reactors

Currently there are more than 400 fission-based nuclear power plants in operation in more than 30 countries in the world, generating 16% of global electricity needs and 2% of the total commercial energy produced. Global production more than tripled between 1980 and 2000. The fraction of electricity produced by nuclear energy in various countries is listed in Table 13-2.

The most economically useful example of fission, and the one mainly used by power plants, is induced by the collision of a ^{235}U nucleus with a neutron. The combination of these two particles is unstable. When it decomposes, the products vary but are typically a nucleus of *barium*, ^{142}Ba, one of *krypton*, ^{91}Kr, and three neutrons:

$$^{1}_{0}n + ^{235}_{92}U \longrightarrow ^{142}_{56}Ba + ^{91}_{36}Kr + 3\, ^{1}_{0}n$$

$$\text{\textcircled{n}} + \left(^{235}_{92}U\right) \longrightarrow \left(^{142}_{56}Ba\right) + \left(^{91}_{36}Kr\right) + \begin{array}{c} \text{\textcircled{n}} \\ \text{\textcircled{n}} \\ \text{\textcircled{n}} \end{array}$$

TABLE 13-2	Proportion of Electricity Generated by Nuclear Power	
Country	Proportion
France	76%
South Korea	41%
Sweden	39%
Japan	34%
United Kingdom	22%
United States	20%
Canada	14%
Russia	12%
Netherlands	4%

Not all the uranium nuclei that absorb a neutron form exactly the same products, but the process always produces two nuclei of about these sizes, together with several neutrons.

The two new nuclei produced in fission reactions are very fast-moving, as are the neutrons. It is heat energy from this excess kinetic energy that is used to produce electrical power. Indeed, the generation of electricity by nuclear energy and by the burning of fossil fuels both involve using the energy source to produce steam, which is then used to turn large turbines that produce the electricity.

An average of about three neutrons are produced per ^{235}U nucleus that reacts; one of these neutrons can produce the fission of another ^{235}U nucleus, and so on, yielding a **chain reaction.** In atomic bombs, the extra neutrons are used to induce a very rapid fission of all the uranium that is constrained to stay in a small volume, so energy is released explosively. In contrast, the energy is released gradually in a nuclear power reactor by ensuring that, on average, only one neutron released from each ^{235}U fission event initiates the fission of another nucleus. The extra neutrons produced in fission are absorbed by the **control rods** in a nuclear power reactor. These are bars made from neutron-absorbing elements such as *cadmium*. The position of the rods in the reactor can be varied to control the rate of fission.

The neutrons that leave ^{235}U nuclei upon fission are too fast-moving to be efficiently absorbed by other nuclei and cause further fission, so they must be slowed down if they are to be useful. This is accomplished by the **moderator,** which, depending upon the type of reactor, can be regular water, heavy water (i.e., water enriched in deuterium), or *graphite*. The **coolant** material that is used to carry off the heat energy produced by fission in most reactor types is water, but gaseous carbon dioxide is used in some. Ultimately, high-pressure, high-temperature steam is created—just as in a fossil-fuel power generating plant—and used to turn turbines and create electricity. The operating efficiency of nuclear power reactors is significantly lower than fossil-fuel ones since the core of the nuclear reactor, from which heat is drawn, is only a little above 300°C, whereas that of fossil-fuel plants may be as high as 550°C. The various steps in the production of electricity by a fission-based nuclear power plant are illustrated in Figure 13-2.

PROBLEM 13-3

Calculate the maximum percentage efficiency in converting heat at 300°C into electricity according to the second law of thermodynamics (see Chapter 6), assuming the cooling water is at 17°C. Repeat the calculation for 550°C heat. Do your calculated values lie above the average efficiencies of 30% and 40% observed for nuclear and coal-fired power plants, respectively?

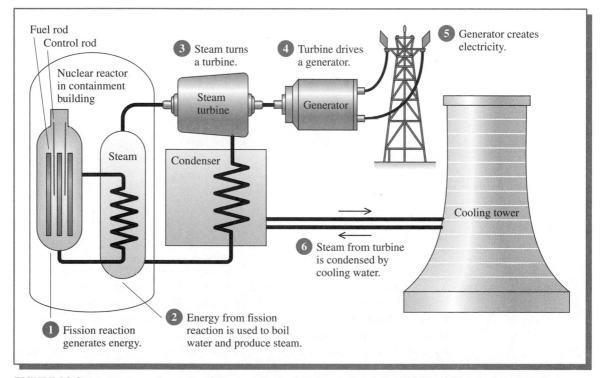

FIGURE 13-2 Step-by-step schematic diagram of the production of electrical power by a nuclear fission reactor.

The uranium in reactors is contained in a series of enclosed bars called **fuel rods.** When the uranium fuel in a rod is "spent," i.e., when its ^{235}U content is too low for it to be useful as a fuel, since the number of neutrons produced per second is low, it is removed from the reactor.

The only naturally occurring uranium isotope that can undergo fission is ^{235}U, which constitutes only 0.7% of the native element. The remaining uranium is ^{238}U (99.3% natural abundance). A neutron produced by the fission of a ^{235}U nucleus can be absorbed on collision by a ^{238}U nucleus. The resulting ^{239}U nucleus is radioactive and emits a β particle, as does the heavy product (^{239}Np) of this process; consequently a nucleus of **plutonium,** ^{239}Pu, is produced:

$$\mathrm{^1_0n} + \mathrm{^{238}_{92}U} \longrightarrow \mathrm{^{239}_{92}U} \xrightarrow{-\beta} \mathrm{^{239}_{93}Np} \xrightarrow{-\beta} \mathrm{^{239}_{94}Pu}$$

$$\textcircled{n} + \left(\mathrm{^{238}_{92}U}\right) \longrightarrow \left(\mathrm{^{239}_{92}U}\right) \longrightarrow \beta + \left(\mathrm{^{239}_{93}Np}\right) \longrightarrow \beta + \left(\mathrm{^{239}_{94}Pu}\right)$$

Thus, ^{239}Pu is produced as a by-product of the operation of nuclear power reactors.

Unfortunately, both the pair of nuclei into which the ^{235}U nucleus fissions and the ^{239}Pu by-product are highly radioactive substances. As a consequence, the material in spent fuel rods is much more radioactive than was the original uranium. Most of the common fission products of uranium emit both β and γ rays; for example,

$$\mathrm{^{142}_{56}Ba} \longrightarrow \beta + \mathrm{^{142}_{57}La}$$

After the barium decays (half-life of 11 min), the product, a lanthanum isotope, also decays by β emission. The spent fuel rods, in which heat is still being produced due to radioactive decay, are immersed in ponds of water for a few months until they have cooled down.

Although many of the radioactive products of ^{235}U fission have short half-lives and therefore decay rapidly, a few have much longer half-lives. After 10 years, most of the radioactivity in spent fuel is due to *strontium-90*, ^{90}Sr (half-life of 29 years), and to *cesium-137*, ^{137}Cs (half-life of 30 years). The dispersal of radioactive strontium and cesium into the environment would constitute a serious environmental problem. Ions of both metals can be readily incorporated into the body because strontium and cesium readily replace chemically similar elements that are integral parts of animal bodies, including humans. Strontium, a Group 2 metal, concentrates in the bones and teeth, replacing ions of *calcium*, also a Group 2 metal that forms +2 ions. Ions of *cesium*, a Group 1 metal, can replace those of *potassium*, also from Group 1 and also +1 in charge, which is widespread in all body cells.

Once these elements are in place in the body, their radioactive disintegration produces β particles that can damage the cells in which they are present or near. Radioactive waste from the spent fuel rods of nuclear power plants must therefore be carefully monitored and will eventually have to be deposited in a secure environment from which it cannot escape.

Environmental Problems of Uranium Mining and Refining

Several of the steps in the *nuclear fuel cycle*, illustrated in Figure 13-3, generate environmental waste.

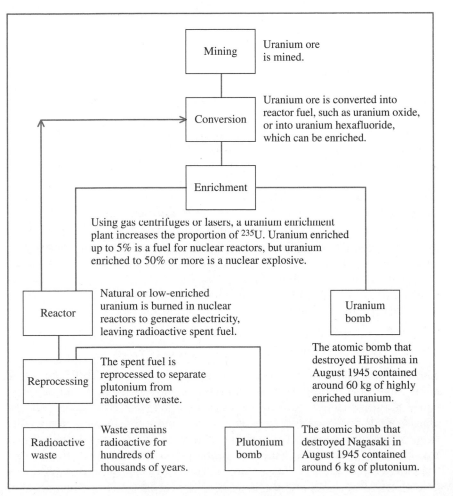

FIGURE 13-3 The nuclear fuel cycle. [Source: R. Edwards, "A Struggle for Nuclear Power," *New Scientist* (22 March 2003): 8.]

Major suppliers and processors of uranium ore are Canada, Australia, Russia, Niger, Ukraine, and Kazakstan. During the mining of uranium, contamination of the environment by radioactive substances commonly occurs. Since naturally occurring uranium slowly decays into other substances that are also radioactive, uranium ore contains a variety of radioactive elements. As a result, the very large volume of waste material that remains after the uranium is chemically extracted from the ore is itself radioactive.

The radioactive material that was immobilized in the original rock ore is in the form of liquid and powder tailings after mining. Gaseous emissions also escape from the wastes. As in other mining operations, the liquid tailings are normally held in special ponds until the solids separate. Pollution of the local groundwater can occur if these ponds leak or overflow. In addition, when the solid tailings are exposed to the weather and are partially dissolved by rainfall, they can contaminate local water supplies.

The use of the solid tailings as landfill on which buildings are constructed can also lead to problems, since the radon produced by the radioactive decay of the radium in the tailings is quite mobile. As previously noted, radon is a particular hazard for uranium ore miners, since the radioactive gas is always present in the ore and is released into the air of the mine. Indeed, the incidence of lung cancer among such miners was particularly high until mine ventilation was improved to permit more frequent changes of air and thus a more efficient clearing out of accumulated radon gas.

In most but not all nuclear power reactors (the Canadian *CANDU* system being the main exception), the uranium fuel must be enriched in the fissionable ^{235}U isotope; its abundance is increased to 3.0% from the 0.7% in the naturally occurring element. The extent of enrichment required for use in bombs is much greater. Uranium sufficiently enriched for this purpose is called *weapons-grade* material and is 90% or more ^{235}U. Enrichment is a very expensive, energy-intensive process since it requires physical rather than chemical means of separation, given the fact that all isotopes of a given element behave identically chemically. For separation, the uranium is temporarily converted to the gaseous compound *uranium hexafluoride*, UF_6.

The Future of Fission-Based Nuclear Power

The view of nuclear power from fission in the public eye in North America and many European countries shifted from positive to negative in recent decades, partly as a consequence of the accidents at the *Three Mile Island* power plant in Harrisburg, Pennsylvania, in 1979 and at the Chernobyl, Ukraine, power station in 1986. No new reactors have been ordered in the United States since the Three Mile Island incident, and several power plants in both the United States and Canada have been shut down. The last new plant to open in Great Britain started in 1995, and there are no plans for new ones. Some observers believe that the nuclear power industry could eventually revive as oil and gas supplies dwindle and as carbon dioxide emission restrictions become more stringent. Currently, the only new nuclear power plants being built are in countries where the price of electricity is regulated by the government, since combined-cycle (cogeneration) natural-gas-fired power plants are more efficient and cheaper than nuclear power at present. The main advantages and disadvantages of nuclear power production are summarized in Table 13-3.

The Catastrophe at Chernobyl

A more realistic fear than a nuclear power plant running out of control and blowing up like an atomic bomb is that the highly radioactive fission products contained within operating fuel rods could be spread into the surrounding countryside if a nonnuclear explosion occurs in a power plant. This in fact did occur at one of the nuclear power plants in Chernobyl, Ukraine, in 1985. Engineers at the plant lost control of the reactor during a routine test by overriding the plant's safety mechanisms and withdrew most of the control rods from the reactor core. As a consequence, the reactor overheated and a fire broke out in the graphite moderator. This subsequently produced a huge explosion, which blew off the heavy plate covering the building, and several hundred million curies of radioactivity were released into the air and spread over a wide area. Several dozen people, mainly plant operators and firefighters, died immediately from high doses of radiation. The dispersed radioactivity was mainly concentrated in the noble gas isotopes ^{131}Xe and ^{85}Kr, in ^{131}I, and in the cesium isotopes ^{134}Cs and ^{137}Cs, all of which are fission products.

TABLE 13-3	Inherent Advantages and Disadvantages of Nuclear Power Production from Fission	
Advantages	**Disadvantages**
Minimal air and water pollution	Production of radioactive wastes that require special handling
Efficient use of fuel resources	Possibility of an accident producing serious human health problems
Relatively low operating cost	Long-term storage of waste and decommissioning of plants will likely involve high costs
	Requirement of an international security system to prevent diversion of nuclear materials to weapons use

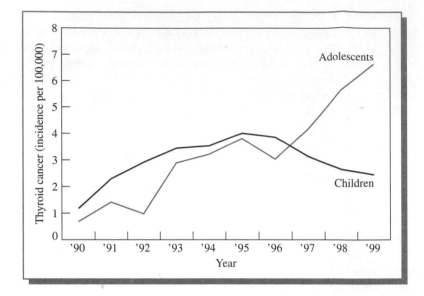

FIGURE 13-4 Incidence of thyroid cancer in adolescents (>14 years old) and in children (<14 years old) in the Chernobyl region of Ukraine. [Source: R. Stone, "Living in the Shadow of Chernobyl," *Science* 292 (2001): 420].

The major chronic health consequence to the public of the explosion at Chernobyl has been the great increase in thyroid cancer among children in the area. A plot of thyroid cancer incidence (Figure 13-4) shows that while the incidence in young children has peaked, that in adolescents is still rising. Evidence suggests that the increase in thyroid cancer is restricted to those already born when the explosion occurred. Thankfully, thyroid cancer has a high cure rate.

The thyroid cancer was initiated presumably by β radiation from the radioactive iodine, ^{131}I, which has a half-life of only 8 days. The human body concentrates iodine in the thyroid gland. Indeed, tablets of *potassium iodide*, KI, designed to flood the body with iodine and thereby dilute the radioactive form, were distributed to the inhabitants around Chernobyl, but only after a week's delay following the explosion.

There have also been increases in birth defects among children born in the areas around Chernobyl that were heavily contaminated by radioactivity. However, fewer additional cases of leukemia have been observed than were expected from the amount of ^{137}Cs expelled from the reactor. Cesium-137 is the most prevalent radioactive isotope from the explosion that is still in the soil of the area.

The Accident at Three Mile Island

The other nuclear power plant accident was much less serious than the one at Chernobyl. It occurred in 1979 at the Three Mile Island plant in

Pennsylvania. The problem originated with failures in the water-based cooling systems in the reactor. Although control rods were inserted into the core and the fission process stopped, the reactor kept heating. The source of the heat was the radioactive decay by γ-ray emission, not of uranium, but of fission products that had naturally built up over time in the fuel rods. Because of mechanical malfunction and operator error, the core became partially uncovered, with the result that it heated even further ($>2200°C$) and half of it melted. Hydrogen and oxygen gases were produced from the decomposition of the superheated water.

Fortunately, there was no large explosion at Three Mile Island: the containment facility was not breached. In contrast to the Chernobyl accident, the liquid and solid radioactive materials that escaped from the reactor at Three Mile Island did not escape from the containment building; the release of ^{131}I was about a million times less. Significant amounts of radioactive noble gas isotopes ^{131}Xe and ^{85}Kr did escape into the air but, as at Chernobyl, they would have been quickly diluted in the atmosphere.

Plutonium and the Problem of Nuclear Waste

The plutonium-239 isotope that is produced during uranium fission is an α-particle emitter and has a long half-life, 24,000 years. After 1000 years the main sources of radioactivity from spent fuel rods will be plutonium and other very heavy elements, since the medium-sized nuclei produced in fission, having much shorter half-lives than 1000 years, will have largely decayed by that time. Thus the long-term radioactivity of the spent fuel rods can be greatly reduced by chemically removing the very heavy elements from it.

PROBLEM 13-4

Given that the half-life of ^{239}Pu is 24,000 years, how many years will it take for the level of radioactivity from plutonium in a sample to decrease to 1/128 (i.e., about 1%) of its original value?

The ^{239}Pu that forms in fuel rods is itself fissionable, and once its concentration in the rods becomes high enough, some of it also undergoes fission and contributes to the power output of the reactor. The plutonium that accumulates over time in fuel rods can later be chemically removed from the spent fuel by **reprocessing.** In this procedure, the fuel rods are dissolved in acid; the plutonium is then separated from other metallic elements present in the solution by exploiting the differences in the solubility of their salts. Since reprocessing uses chemical procedures, it is much less energy-intensive than the isotope separation processes. However, the liquid that remains after the valuable heavy elements have been removed is still highly radioactive due to the presence of fission products. No method of disposal for this waste has yet been approved or implemented; it is still stored in metal tanks. Unfortunately, many of the older tanks in which the liquid is stored have begun to leak at installations such as Hanford, in Washington State, where plutonium for nuclear weapons was manufactured by reprocessing. Reprocessing is a costly process, in part because the radioactive waste it produces must be stored.

Spent fuel rods from civilian power reactors in a number of countries have been reprocessed in France and the United Kingdom, and to a smaller extent in Russia and Japan, though not in the United States or Canada. This reprocessing has resulted in a buildup of hundred of tonnes of plutonium. Comparable quantities of the element are also available from the dismantling of nuclear weapons by the United States and Russia (see Box 13-2).

Aside from questions of health and safety, a major problem associated with the handling of plutonium from both civilian and military sources involves the security measures needed to prevent the material from slipping into the hands of terrorists and rogue governments who wish to fashion their own bombs. Only a few kilograms of weapons-grade plutonium, which consists of 93% or more of ^{239}Pu, is required to make an atomic bomb. A somewhat larger

---

BOX 13-2 | Radioactive Contamination by Plutonium Production

Plutonium has been deliberately produced for more than 50 years to provide the readily fissionable material needed for nuclear weapons. In the United States, most plutonium production and processing were carried out at Hanford, Washington. Because of the huge quantities of radioactive waste that were produced, stored, and disposed of at this facility, the surrounding environment is now so heavily polluted that it has been called the "dirtiest place on Earth." About 190,000 m^3 of highly radioactive solid waste and 760 million liters of moderately radioactive liquid waste and toxic chemicals were deposited in the ground at this site. As much as a metric ton of plutonium may be contained within the masses of solid wastes buried there. The cleanup of the wastes will cost between 50 and 200 billion dollars and will not be completed earlier than the year 2020. One proposed—though expensive—technique for immobilizing the radioactive wastes involves passing a strong electrical current through the contaminated soil for a period of days; the electricity would fuse the soil and sand into a glassy rock from which the contaminants could not escape.

Plutonium was used as the actual explosive material in some atomic bombs, and as a "trigger" for hydrogen (fusion) bombs, forcing the reactants together and thereby initiating the thermonuclear explosion. Consequently, about 100 metric tons of plutonium must be removed from nuclear weapons as tens of thousands of them are dismantled by the United States and Russia over the next few decades.

quantity of *reactor-grade* plutonium, which contains more of the other isotopes of plutonium and similar elements, is needed for a bomb. The world's current stockpile of plutonium exceeds 1000 metric tons and continues to grow.

Breeder reactors are nuclear power reactors that are designed specifically to maximize the production of by-product plutonium; such reactors actually produce more fissionable material than they consume. These special reactors start with a small amount of ^{238}Pu (initially obtained from conventional uranium-fueled reactors) and produce more of it from ^{238}U. The reactors are sodium-cooled and use fast neutrons, rather than those moderated to produce fission. The fast neutrons are more likely to be captured by ^{238}U and produce its daughters, which quickly decay to plutonium. There have been operational problems with these very technologically sophisticated reactors, and the programs have been abandoned in the United States, the United Kingdom, Germany, and France. Russia and Japan still operate demonstrator breeder reactors, and both India and China are building demonstration units. Overall, the future of breeder reactors is in doubt.

Two methods have been proposed to dispose of excess plutonium:

• To mix it with other highly radioactive liquid waste and then to **vitrify** the mixture into durable glass logs that would be buried far underground in metal canisters, as will be discussed. **Vitrification** chemically bonds liquid waste into a stable and durable borosilicate glass in which fission product oxides make up about 20% of the mass.

• To convert it to *plutonium dioxide,* PuO_2, and mix it with *uranium oxide* to produce a **mixed oxide fuel,** MOX (containing a few percent plutonium), that could be used in existing nuclear power plants. Indeed, some MOX is now employed in reactors in France, Germany, and Switzerland, and in the future will probably be used in the United States, Canada, Belgium, and Japan. However, producing MOX fuel is much more expensive at present than producing low-enriched uranium, so the incentive to use it is not economics. The issue of using mixed oxide fuel is very controversial in Great Britain, where one-third of the world's nonmilitary plutonium is presently stockpiled.

Nuclear energy has been used to generate electricity now for many decades and constitutes an appreciable energy source in some countries (see Table 13-1). However, there is still no consensus among scientists and policymakers concerning the best procedure for the long-term storage of the radioactive wastes generated by these plants. Initially, spent fuel rods are simply stored above ground—often under cooled water—for several years or decades until the level of radioactivity has been reduced and the rods have therefore cooled somewhat. At this stage, the rods can be transferred to dry storage, e.g., in concrete canisters. If the fuel rods are reprocessed to remove the plutonium, the remaining highly radioactive waste is subsequently resolidified.

The radiation from plutonium is weak and becomes a concern only if the material is ingested or inhaled, since its radiation (α particles) cannot pass through dead layers of skin or clothing or indeed through even a few centimeters of air. Elemental, metallic plutonium cannot be easily absorbed by the body. However, when exposed to air, plutonium forms the oxide PuO_2, a powdery dust that disperses readily and can be inhaled. Even microscopic amounts of plutonium oxide lodged in the lungs can induce lung cancer.

Whether or not the plutonium is removed, most nuclear waste disposal plans assume that the solid material would be encapsulated and immobilized in a glass or ceramic form and then buried far below the Earth's surface. The container for this ultimate disposal would likely be made of a metal, such as *copper* or *titanium*, that is highly resistant to corrosion. The canisters are designed to last for several hundred years at least before leakage could occur; by that time the level of radioactivity would have declined substantially, though almost all of the plutonium would remain. In Sweden, canisters are being designed to last for 100,000 years, at which time the level of radioactivity inside would be no greater than that of uranium ore.

Canisters will ultimately be buried in vaults 300–1000 m below the surface. The geological features of the burial sites should include high stability (from disruption by earthquakes or volcanic activity) and low permeability to assure minimal interactions with groundwater and with the biosphere. Deep geological disposal is the only method by which safety requirements can be met without burdening future generations with monitoring and management responsibilities. However, some governments do not accept the idea of *permanent* disposal of such wastes. They want to be able to recover the plutonium from spent fuel rods if it turns out to be needed in the future for nuclear power.

Originally, the U.S. plan for the storage of high-level nuclear waste was to use about 10,000 stainless-steel containers, each weighing several tonnes. However, a nickel–chromium–molybdenum alloy of steel *(C-22)* is now proposed for this role since it is more resistant to corrosion. The United States plans to use a depository to be created 300 m below the surface at Yucca Mountain, Nevada. In contrast to the plans of other countries, the depository will be in the unsaturated zone (see Chapter 10), 300 m above the water table, where air is still present in the soil. Conditions in the unsaturated zone are oxidizing. The original idea was to have a dry repository, since water would have been the main agent for release and transport of radioactive nuclei. However, some evidence now exists that there is rapid transport of water through the area even at this depth. Consequently, a so-called drip shield of titanium would be placed over each canister to keep water away from it. Other countries, including Canada, plan to use deeper sites, which will be in the saturated zone so that ambient conditions will be reducing, and to use copper canisters.

After the engineered barriers have eventually failed, the release of radioactive nuclei will depend on the chemical durability of the fuel. Since the uranium in the spent nuclear fuel occurs as U(IV) in the form of UO_2, even small amounts of moisture will result in its oxidation to U(VI), as UO_2^{2+}, which is quite soluble and mobile, at a much faster rate than would occur under reducing conditions. The presence of Fe^{2+}, resulting from the rusting of the steel canister, would reverse the oxidation and precipitate the uranium oxide of any U(VI) that came into contact with it.

Fusion Reactors

The optimum energetic stability per nuclear particle (protons and neutrons) occurs for nuclei of intermediate size, such as iron. That is why the fission of a heavy nucleus into two fragments of intermediate size releases energy. Similarly, the fusion of two very light nuclei to produce a heavier one also releases substantial quantities of energy. Indeed, fusion reactions are the source of the energy in stars, including our own Sun, and in hydrogen bombs.

Unfortunately, fusion reactions all have extremely large activation energies due to the huge electrostatic repulsion that exists between the positively charged nuclei when they are brought very close together, which must happen before fusion can occur. Consequently, it is difficult to initiate and sustain a controlled fusion reaction that provides more energy than it consumes.

The fusion reactions that have the greatest potential as producers of useful commercial energy involve the nuclei of the heavier isotopes of hydrogen, namely *deuterium*, ^2H, and *tritium*, ^3H. Note that two different sets of products can be produced from these sample reactions:

$$\underset{\text{deuterium}}{^2_1H} + \underset{\text{deuterium}}{^2_1H} \begin{cases} \nearrow\ ^3_2He + ^1_0n \\ \text{or} \\ \searrow\ ^3_1H + ^1_1H \end{cases}$$

The energy that is released when one of these reactions occurs is about 4×10^8 kJ/mol, which is about 1 million times the energy produced in a typical exothermic *chemical* reaction. An abundant supply of deuterium is available, since it is a nonradioactive, naturally occurring isotope (constituting 0.015% of hydrogen) and thus is a natural component of all water.

A somewhat lower activation energy is required for the reaction of deuterium with tritium:

$$\underset{\text{deuterium}}{^2_1H} + \underset{\text{tritium}}{^3_1H} \rightarrow\ ^4_2He + ^1_0n$$

However, because tritium is a radioactive element (a β emitter) with a short half-life (12 years), it is not a significant component of naturally occurring hydrogen and would have to be synthesized by the fission of the relatively scarce element *lithium*.

The environmental consequences of generating electrical power from fusion reactors should be less serious than those associated with fission reactors. The only radioactive waste produced *directly* in quantity would be tritium, although the neutrons emitted in the process could produce radioactive substances when they are absorbed by other atoms. Although the β particle that tritium emits is not sufficiently energetic to penetrate the outer layer of human skin, tritium is nevertheless dangerous since biological systems incorporate it as readily as they do normal hydrogen (^{1}H or ^{2}H) by inhalation, absorption through the skin, or ingestion of water or food. Currently, tritium in drinking water (some of which results from artificial sources) constitutes the source of about 3% of our exposure to radioactivity.

About 80% of the energy emitted in the deuterium–tritium reaction is associated with the neutrons. The energy, captured by using neutron-heated coolants, would be used to create superheated steam to drive turbines. Unfortunately, the intense neutron bombardment will cause severe degradation of the structure used to confine the fusion reactants and will generate large amounts of radioactive nuclei. These problems would be largely overcome if so-called *advanced fuels* were to be used as reactants. Thus the fusion reaction of deuterium with helium-3 ($^3_2$He) releases only a few percent of its reaction energy as neutrons (the products of the dominant process being protons and ^{4}He nuclei), and that of two ^{3}He nuclei (to produce two protons and ^{4}He) is essentially neutron-free. These processes could operate with much higher energy conversion efficiencies but require even higher initiation temperatures and confinement conditions. The other serious problem is the lack of ^{3}He on Earth: the Moon is the best source for this material!

PROBLEM 13-5

Write and balance the two fusion reactions involving ^{3}He mentioned above.

The Energy Released in Nuclear Processes

The energy, E, released in fission and fusion processes comes from the conversion of a tiny fraction, m, of the masses of the atoms and other particles involved. According to Einstein's famous equation, the energy released is

$$E = mc^2$$

where c is the speed of light.

For example, in the conversion of one mole of deuterium atoms (mass 2.0140 g) and one mole of tritium atoms (mass 3.01605 g) into one mole of ^{4}He atoms (mass 4.00260 g) and one mole of neutrons (mass 1.008665 g), a total of 0.0188 g of matter is lost by conversion to energy. Since $m = 0.0188 \times 10^{-3}$ kg and $c = 2.99792 \times 10^8$ m/s, the energy released is

$$E = mc^2 = (0.0188 \times 10^{-3} \text{ kg}) \times (2.99792 \times 10^8 \text{ m/s})^2$$
$$= 1.69 \times 10^{12} \text{ kg m}^2/\text{s}^2$$
$$= 1.69 \times 10^{12} \text{ J}$$

This energy, 1690 million kilojoules per mole, is about 7 million times the energy released when the same quantity of hydrogen is burned in oxygen to produce water.

PROBLEM 13-6

Calculate the amount of energy released in the fission process described earlier in the chapter in which uranium-235 and a neutron are fissioned to barium-142, krypton-91, and 3 neutrons. Assume 1 mol of ^{235}U reacts and that the atomic masses of ^{235}U, ^{142}Ba, ^{91}Kr, and a neutron are, respectively, 235.044, 141.926, 91.923, and 1.008665. Is the energy released much greater than, much less than, or about the same per nucleus as that for the fusion reaction between deuterium and tritium?

Review Questions

1. What is the particulate nature of the radioactive emissions α and β particles? What is a γ ray?

2. Why are α particles dangerous to health only if ingested or inhaled?

3. What is meant by the terms *rad* and *rem*?

4. Explain the origin of radon gas in buildings.

5. Explain what is meant by the term *daughters of radon*. Why they are more dangerous to health than radon itself?

6. What is *depleted uranium*? Is it radioactive at all?

7. Explain what is meant by a *dirty bomb*.

8. Define *fission* and write the reaction in which a ^{235}U nucleus is fissioned into typical products.

9. What are the functions in a fission power reactor of (a) the moderator and (b) the coolant?

10. Explain why the spent fuel rods from fission reactors are more radioactive than the initial fuel.

11. Why would radioactive strontium and cesium be particularly harmful to human health?

12. Write the nuclear reaction that produces plutonium from ^{238}U in a fission reactor.

13. Describe why the mining of uranium ore often pollutes the local environment.

14. What is a *breeder reactor*? Why is it useful to breed fissionable fuel? What is meant by *reprocessing* and why is it done?

15. Describe the two main methods that have been proposed to dispose of excess plutonium.

16. Define *fusion*, and give two examples of fusion processes (i.e., reactions) that may be used in power reactors of the future.

17. Describe the nature of any radioactive by-products of the operation of fusion reactors. What damage could the neutrons do?

Additional Problems

1. Another possible radioactive decay mechanism involves the emission of a positron, a particle with the same mass as an electron but the opposite (i.e., positive) charge. The net nuclear effect of positron emission is the change of a proton into a neutron. Explain how the result of positron emission differs from β emission in terms of the periodic table. Deduce the symbol for a positron to be used in balanced nuclear equations, by analogy with the symbol $_{-1}^{0}e$ used for a β particle (the symbol e for an electron is also used here to represent a positron). Predict the decay products and write the balanced equation for the positron decay of the radioactive isotopes $_{11}^{22}Na$ and $_{7}^{13}N$.

2.   Tablets of KI were distributed to people living around Chernobyl a week after the nuclear reactor explosion occurred, to help flush the radioactive ^{131}I isotope out of their bodies. What percentage of the ^{131}I released by the explosion would have remained by that time? How long would it have taken for 99% of the released ^{131}I to undergo radioactive decay? Use the kinetics equation for the first-order decay of a species A as a function of time: $[A] = [A]_o\, e^{-kt}$, where k is the first-order rate constant for the decay. (The rate constant can be obtained from the value of the half-life given in the chapter: $t_{1/2} = 0.693/k$.)

3.   One physical property that differs for $^{235}UF_6$ and $^{238}UF_6$ is the rate of effusion of these gases through porous membranes. The lighter $^{235}UF_6$ will effuse faster, resulting in enrichment of this desired isotope after passing through the membrane. The effusion rate of gases can be described by Graham's law, which states that the effusion rate of a particle is inversely proportional to the square of its mass. Use this law to determine relative effusion rates of $^{235}UF_6$ and $^{238}UF_6$. Based on your result, explain why it is necessary in practice to perform this process through hundreds of membrane barriers to achieve useful enrichment.

Further Reading

1.   P. K. Hopke et al., "Health Risks Due to Radon in Drinking Water," *Environmental Science and Technology* 34 (2000): 921.

2.   J. H. Lubin et al., "Lung Cancer in Radon-Exposed Miners and Estimation of Risk from Indoor Exposure," *Journal of the National Cancer Institute* 87 (1995): 817.

3.   K. E. Warner, P. N. Courant, and D. Mendez, "Effects of Residential Mobility on Individual Versus Population Risk of Radon-Related Lung Cancer," *Environmental Health Perspectives* 103 (1995): 1144.

4.   R. Stone, "Living in the Shadow of Chernobyl," *Science* 292 (2001): 420.

5.   N. Zevos, "Radioactivity, Radiation, and the Chemistry of Nuclear Waste," *Journal of Chemical Education* 79 (2002): 692.

6.   R. C. Ewing, "Less Geology in the Geological Disposal of Nuclear Waste," *Science* 286 (1999): 415.

7.   K. D. Crowley and J. F. Ahearne, "Managing the Environmental Legacy of U.S. Nuclear-Weapons Production," *American Scientist* 90 (2002): 514.

8.   R. Matthews, "Here Comes the Sun," *New Scientist* (9 February 2002): 36.

Websites of Interest

Log on to www.whfreeman.com/envchem3e/ and click on Chapter 13.

APPENDIX
BACKGROUND ORGANIC CHEMISTRY

In Chapters 7 and 8, the most important environmental problems caused by toxic organic chemicals are discussed in detail. In this appendix we provide some necessary background in organic chemistry for those students whose previous education has not included this subject.

The organic compounds of interest environmentally are mostly electrically neutral molecules containing covalent bonds. Stable compounds of this type inevitably involve the formation of *four* bonds by carbon; in carbon-centered free radicals, they form three bonds. Conceptually at least, chemists view all organic chemicals as "derived" from those simple organic compounds that contain only carbon and hydrogen, that is, **hydrocarbons.** We shall follow this convention and divide our discussion into several sections, most of which deal with specific types of hydrocarbons.

Alkanes

The simplest hydrocarbons are those that contain strings of carbon atoms, each one singly bonded to its closest neighboring carbon atom(s) and to several hydrogen atoms. Such hydrocarbons are called **alkanes,** of which the simplest are *methane*, CH_4; *ethane*, C_2H_6; and *propane*, C_3H_8. Commercial supplies of all three are readily available from natural gas wells. Structural formulas for these three alkanes are:

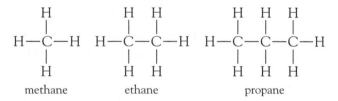

methane   ethane   propane

For convenience, chemists often write down the formulas for such species by gathering together in one unit all the hydrogens bonded to a given carbon and displaying only the carbon–carbon bonds; thus ethane is represented as $CH_3—CH_3$ and propane as $CH_3—CH_2—CH_3$. In another common representation called a **condensed formula,** the C—C single bonds are not shown; rather the formula lists each carbon and the atoms attached to it. For example, the condensed formula for ethane is CH_3CH_3. Each carbon atom in an alkane molecule forms four equiangular single bonds, so the geometry about each carbon is tetrahedral. Thus all alkanes are three-dimensional, nonplanar molecules, even though, for the sake of clarity, their

TABLE 1	Some of the Simple Unbranched Alkanes		
Molecular formula	Name	Condensed formula	Boiling point (°C)
CH_4	Methane	CH_4	−164
C_2H_6	Ethane	CH_3CH_3	−89
C_3H_8	Propane	$CH_3CH_2CH_3$	−42
C_4H_{10}	Butane	$CH_3CH_2CH_2CH_3$	−0.5
C_5H_{12}	Pentane	$CH_3(CH_2)_3CH_3$	36
C_6H_{14}	Hexane	$CH_3(CH_2)_4CH_3$	69
C_7H_{16}	Heptane	$CH_3(CH_2)_5CH_3$	98
C_8H_{18}	Octane	$CH_3(CH_2)_6CH_3$	126
C_9H_{20}	Nonane	$CH_3(CH_2)_7CH_3$	151
$C_{10}H_{22}$	Decane	$CH_3(CH_2)_8CH_3$	174
$C_{11}H_{24}$	Undecane	$CH_3(CH_2)_9CH_3$	196
$C_{12}H_{26}$	Dodecane	$CH_3(CH_2)_{10}CH_3$	216

structural formulas seem to represent them as planar molecules involving bond angles of 90° and 180°.

Table 1 lists the alkanes having one through twelve carbon atoms in a continuous string; more complex alkanes contain branches, as we shall see shortly. The alkane with four carbon atoms is called *butane* and is a gas; longer alkanes are liquids or solids under ordinary conditions. When five or more carbons are present in the alkane, the Latin-based abbreviation for that number (*pent* for 5, *hex* for 6, *hept* for 7, *oct* for 8, etc.) is employed as the prefix to the ending *-ane* in its name. For example, the molecule $CH_3—CH_2—CH_2—CH_2—CH_3$, which can be written more simply as $CH_3(CH_2)_3CH_3$, is called *pentane* since it has 5 carbon atoms; its formula is C_5H_{12}. When all the carbons lie in one continuous chain (without branches), the molecule is said to be *straight-chained* or *unbranched,* and often the prefix *n-* is added to the name; thus the pentane molecule described is called *n-pentane*.

The constituent atoms of many organic compounds may be "reshuffled" in chemical reactions to yield new structures—the ingredient atoms remain exactly the same, but the way in which they are linked together (their "order of linkage") can be altered. And so we get distinctly different compounds. These sets of different compounds with the same molecular formula, but different structures, are **structural isomers;** there are other types of

isomerism, too, which we need not consider here. Sometimes the difference between isomers is slight—a matter of small differences in physical
properties; sometimes it is enormous—the biological activity of two isomers
may differ profoundly. Alkanes with four or more carbon atoms have isomers in which the chain of carbon atoms is branched: not all the carbons
are part of an unbroken path of bonded atoms. An example is the isomer of
n-pentane illustrated below:

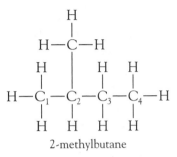

2-methylbutane

In naming such alkanes and other organic molecules, the short chains
of carbon atoms that comprise the branches are assigned group names ending in *-yl*, which are derived by deleting the *-ane* ending from the name of
the alkane hydrocarbon that has the same length (in terms of linked carbon
atoms). Thus the CH_3— group is called the *methyl* group, CH_3CH_2— is
called *ethyl*, etc. The names of these groups are listed as prefixes to the name
for the longest continuous chain of carbon atoms, and each is preceded by a
number that indicates the carbon atom of the chain to which the group is
attached. For example, the molecule shown above is called *2-methylbutane*,
since butane is the alkane consisting of four carbons in an unbranched
chain, at the second carbon atom of which the —CH_3 group is bonded.

There are compounds in which one or more of the hydrogen atoms in
hydrocarbons such as the alkanes have been replaced by another atom such
as fluorine, chlorine, or bromine. Things that can be substituted for hydrogen atoms are called **substituents.** Some simple substituted hydrocarbons
are encountered in Chapter 1; examples include the substituted methanes
CF_2Cl_2 (dichlorodifluoromethane), CHF_2Cl (chlorodifluoromethane), and
CF_3Br (bromotrifluoromethane), and the substituted ethane CHF_2—CH_2F,
which is called 1,1,2-trifluoroethane, where the numbers refer to the carbon
numbers to which the fluorines are bonded; its structure is:

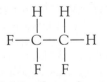

1,1,2-trifluoroethane

PROBLEM 1

Write out the structural formula and the condensed formula for each of the following alkanes: (a) *n*-pentane; (b) 3-ethylhexane; (c) 2,3-dimethylbutane.

Alkenes and Their Chlorinated Derivatives

In some organic molecules, one or more pairs of the carbon atoms are joined by double bonds; since each carbon atom forms a total of four bonds, there are only two additional bonds formed by such carbon atoms. The simplest hydrocarbon of this type is a colorless gas called *ethene*, usually known by its older name *ethylene*:

ethene (ethylene)

Notice that the actual planar geometry of this molecule, with bond angles of about 120° around each carbon, can be shown in the structural formula. Condensed formulas normally show the double bond: $CH_2\!=\!CH_2$ or $H_2C\!=\!CH_2$.

A $C\!=\!C$ bond can be a part of a longer sequence of carbon atoms that are joined together by other single or double or triple CC bonds. For example, *propene* is a three-carbon chain with one adjacent pair of carbons joined by a double bond:

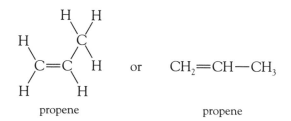

propene   propene

The name for a hydrocarbon chain containing a $C\!=\!C$ bond is the same as that used for the alkane of the same length, except that the *-ane* ending of the alkane is replaced by *-ene*. The molecule is numbered such that the $C\!=\!C$ unit is part of the continuous chain, and such that the $C\!=\!C$ unit is at the lower-numbered end of the chain. Collectively, hydrocarbons containing $C\!=\!C$ bonds are called **alkenes.** If there are two $C\!=\!C$ bonds in a hydrocarbon, the prefix *di-* is placed before the *-ene* ending; thus the hydrocarbon below is called *1,3-pentadiene*:

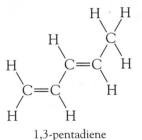

1,3-pentadiene

The numbers preceding the name are those assigned to the first carbon atom that participates in each of the double bonds. The alternative numbering scheme, that is, assigning the CH_3 carbon on the right to be #1, is not used since the first double bond would then start at carbon #2 and the name would be *2,4-pentadiene*; thus the first double bond would not have the lowest possible number.

In some derivatives of ethene, one or more of its hydrogen atoms have been replaced by chlorine atoms. The chloroethenes, like ethene itself, are planar molecules. The simplest example is $CH_2{=}CHCl$, called *chloroethene* but known in the chemical industry as *vinyl chloride*; it is produced in huge quantities since the common plastic material polyvinyl chloride (PVC) is subsequently prepared from it.

A number is usually placed in front of the name of the substituent to indicate the specific carbon atom to which it is bonded; thus $Cl_2C{=}CH_2$ is called *1,1-dichloroethene* to distinguish it from *1,2-dichloroethene*, $CHCl{=}CHCl$. The molecule *1,1,2-trichloroethene*, $CCl_2{=}CHCl$, is a liquid solvent that has found extensive uses. Note that the prefix numbers in the compound name here are superfluous since it has no isomers and there is no need to distinguish one isomer from another. This compound is usually referred to by its traditional name *trichloroethylene*. The structural formulas of a few substitution products of ethene are:

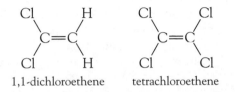

1,1-dichloroethene   tetrachloroethene

The liquid compound *tetrachloroethene*, $CCl_2{=}CCl_2$, finds use on a large scale as the dry-cleaning solvent used commercially to remove grease spots and other stains on clothing. The prefix 1,1,2,2- is not used as part of its name since it is superfluous (no other arrangements of chlorine being possible). Note that when all the hydrogens in a molecule have been replaced by a given atom or group, the prefix *per* can be used instead of the actual number; thus tetrachloroethylene is also called *perchloroethylene*, giving rise to its nickname "perc."

PROBLEM 2

Write structural formulas for each of the following: (a) 1,1-dichloropropene, (b) perchloropropene, (c) 2-butene.

PROBLEM 3

Determine the correct name for each of the following: (a) $CHCl_2CHCl_2$, (b) $CH_3—CH_2—CH=CH_2$, (c) $CH_2=CH—CH=CH_2$.

Symbolic Representations of Carbon Networks

Organic molecules often contain extensive networks of carbon atoms. Chemists find it convenient to construct shorthand visual representations of such molecules using a symbolic system of lines that indicate only the position of the *bonds* (not including bonds to hydrogen atoms), rather than writing out a structure in which the C and H atoms are shown explicitly. To indicate the presence of a carbon atom, a "kink" is shown in the chain's representation. For example, the molecule *n*-butane can be represented as (a) or (b) below:

$$CH_3—CH_2—CH_2—CH_3$$

(a)   (b)

The hydrogen atoms are not shown at all in the "stripped-down" version; the number of them at any carbon atom can be deduced by subtracting from 4 the number of bonds to that carbon that are displayed explicitly. Thus in the representation below for 2-chloropropane, carbons #1 and #3 must possess 3 hydrogens, since they are shown as forming one other bond, whereas carbon #2 has one hydrogen, since it is shown as forming three other bonds:

2-chloropropane

PROBLEM 4

Write out the full structural formulas for each of the following molecules:

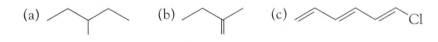

(a)   (b)   (c)

PROBLEM 5

Draw symbolic ("kinky") bond diagrams for each of the following molecules:

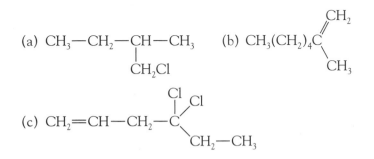

(a) $CH_3-CH_2-CH-CH_3$ (b) $CH_3(CH_2)_4C\overset{CH_2}{\underset{CH_3}{\diagup}}$
                   |
                CH_2Cl

(c) $CH_2=CH-CH_2-C\overset{Cl}{\underset{CH_2-CH_3}{\diagup}}\diagdown^{Cl}$

Common Functional Groups

In addition to being replaced by simple single-atom substituents like Cl and F, the hydrogen atoms in alkanes and alkenes can be replaced by more complex "attachments" called **functional groups**—these are typically headed by oxygen or by nitrogen atoms. The common functional groups are listed in Table 2. The simplest such polyatomic group is —O—H, usually simply shown as —OH; it is called the **hydroxyl group.** Compounds that correspond to alkanes or alkenes with the hydrogen of one C—H bond replaced by an —OH group are called **alcohols.** Familiar examples are methyl alcohol or *methanol* (also called wood alcohol), and ethyl alcohol or *ethanol* (grain alcohol):

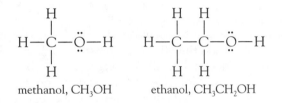

methanol, CH_3OH ethanol, CH_3CH_2OH

The use of alcohols as fuels is discussed in Chapter 6.

TABLE 2	Some Common Functional Groups
Name of compound type	**Functional group**
Chloride	$-Cl$
Fluoride	$-F$
Alcohol	$-OH$
Ether	$-O-$
Aldehyde	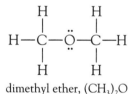
Carboxylic acid	
Amine	

Compounds called **ethers** contain an oxygen atom connected on both sides to a carbon atom or chain:

$$H-\underset{\underset{H}{|}}{\overset{\overset{H}{|}}{C}}-\overset{..}{\underset{..}{O}}-\underset{\underset{H}{|}}{\overset{\overset{H}{|}}{C}}-H$$

dimethyl ether, $(CH_3)_2O$

In more formal names for such compounds, the $-OCH_3$ group is known as *methoxy*, and the $-OCH_2-CH_3$ group is known as *ethoxy*, so that dimethyl ether would be named *methoxymethane*. The use of ethers as gasoline fuel additives is discussed in Chapter 6.

There are organic compounds analogous to alcohols and ethers in which sulfur occurs in the position otherwise occupied by oxygen. The prefix *thio-* is used to denote this substitution; thus we have thioalcohols, or just **thiols,** such as CH_3SH, and **thioethers,** such as CH_3-S-CH_3.

As discussed in Chapter 2, carbon–oxygen double bonds are found in some organic molecules. Molecules that contain the $H-C=O$ group bonded to hydrogen or to a carbon are known as **aldehydes;** the important examples encountered in polluted air are *formaldehyde*, $H_2C=O$, and *acetaldehyde*, $CH_3C(H)=O$. (Atoms or groups shown inside parentheses are bonded to the preceding carbon but do not themselves participate in the bond displayed next in the formula.)

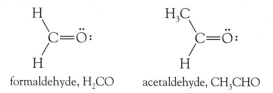

formaldehyde, H_2CO acetaldehyde, CH_3CHO

If the C=O group is connected to an —OH group, the system is called a **carboxylic acid;** examples are *formic acid* and *acetic acid:*

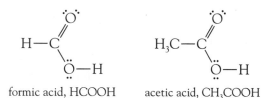

formic acid, HCOOH   acetic acid, CH_3COOH

Groups headed by nitrogen atoms are known as **amino groups;** they are found attached to carbon chains in some organic molecules. Compounds in which the amino group is bonded to a hydrocarbon chain are called **amines.** Note that nitrogen atoms form a total of three bonds, some (or all) of which can be directed to carbons. Two examples are:

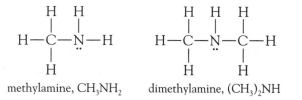

methylamine, CH_3NH_2 dimethylamine, $(CH_3)_2NH$

Molecules that contain *both* the carboxylic acid group and an amino group are called **amino acids.** They are the most important groups in proteins, which include the enzymes that accelerate specific biological reactions. The amino acid called *cysteine* is illustrated below; notice that it contains a thiol group as well as an amino and a carboxylic acid group.

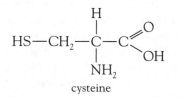

cysteine

Alcohols, acids, and amines that contain short carbon chains are quite soluble in water. The reason is that molecules of these three types contain

O—H or N—H bonds, which possess a hydrogen atom that is partially depleted of electron density by the highly electronegative atom (O or N) to which it is bonded. The partial positive charge δ^+ of the hydrogen is attracted to regions of unbonded electron density—lone pairs—on atoms of adjacent molecules:

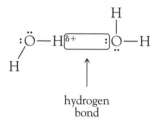

hydrogen
bond

Such interactions are called **hydrogen bonds;** the forces holding the two atoms together—and therefore also holding together the two molecules to which the atoms belong—are not nearly as strong as those of a regular bond within a molecule, but they are much stronger than the forces that operate between molecules in hydrocarbons. Water molecules illustrate this situation. They stick together because each hydrogen atom is hydrogen-bonded to the lone pair of the H_2O molecule closest to it. The attraction between H_2O molecules due to these interactions results in a relatively high boiling point for liquid water, much higher than anticipated for a molecule of its mass. For a (nonionic) substance to be freely soluble in water, these secondary bonds between adjacent water molecules must be replaced by similar interactions between the substance and the water molecules. Consequently, molecules that contain N—H or O—H bonds and a short chain of carbons are soluble in water because the hydrogen bonds they form with H_2O molecules replace those that are broken when the substance is incorporated into the liquid.

Hydrogen atoms bonded to carbon cannot form hydrogen bonds with water molecules, since the carbon is not sufficiently electronegative to produce much of a positive charge on a hydrogen atom bonded to it. In addition, there are no lone pairs of electrons on the carbon atoms. Consequently, there is no driving force that can disrupt the extensive network of hydrogen bonding within liquid water in order to incorporate a large number of molecules of hydrocarbons or chlorinated organic molecules. In both hydrocarbon and chlorinated organic molecules all the hydrogens are bonded to carbon, and for this reason, such molecules are not very soluble in water. Even molecules with one O—H or N—H group and many carbon atoms are insoluble in water, since their overall character is dominated by the large number of carbons. The forces of attraction that do exist between organic molecules that contain no hydrogen-bonding capacity are quite nonspecific and nondirectional; consequently, different hydrocarbons are quite soluble in each other, and organochlorine molecules are soluble in hydrocarbons. We can restate the familiar generalization that "like dissolves like": compounds tend to dissolve in other substances having the same types of intermolecular interactions.

Write the structural formulas and symbolic diagrams for each of the following: (a) ethyl alcohol, (b) ethylamine, (c) acetic acid (the carboxylic acid with a methyl group bonded to the carbon).

Rings of Carbon Atoms

Networks of carbon atoms exist as rings in many organic molecules. The most common rings are those that contain five, six, or seven carbon atoms. Molecules containing rings are named by placing the prefix *cyclo* in front of the usual name for the carbon chain of that length. Thus a ring of six carbons, all joined by single C—C bonds, is called *cyclohexane*. The molecule shown at the right below is called *methylcyclopentane*.

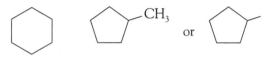

cyclohexane   methylcyclopentane

Write out both simple and symbolic bond diagrams for the following molecules: (a) cyclopropane, (b) chlorocyclobutane, (c) any isomer of dimethylcyclohexane.

Benzene

One of the most common and most stable organic structural units is the **benzene ring,** which is a planar hexagon of six carbon atoms. In the parent hydrocarbon, it also contains six hydrogen atoms, one bonded to each carbon and lying in the C_6 plane:

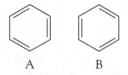

A   B

Each carbon in C_6H_6 is bonded to two carbons and to one hydrogen, so in order to form four bonds it must be doubly bonded to one of its neighboring carbons. The two ways of achieving this result are shown in the so-called Kekulé structures (A and B at the bottom of the previous page). In fact, benzene molecules adopt neither of these two forms, each of which would have alternating short $C=C$ and long $C-C$ bonds; rather they exist in an averaged "resonance" structure in which all CC bonds have the same intermediate length. (The term *resonance* here alludes to the mathematics of the bond description and is commonly interpreted as "blend" or "hybrid.") This result is represented by the structure shown below, with the hexagon containing an enclosed circle to represent the three double bonds; often, however, just one of the Kekulé structures is shown, it being understood, at least by chemists, that no actual alternation of bonds is meant.

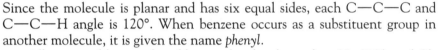

Since the molecule is planar and has six equal sides, each $C-C-C$ and $C-C-H$ angle is 120°. When benzene occurs as a substituent group in another molecule, it is given the name *phenyl*.

Alkenes readily react by addition of molecules such as H_2, HCl, and Cl_2 across the double bonds—that is, with one atom attaching itself to each of the two carbons on the double bond—and thereby convert the $C=C$ units to single bonds. For example, the addition of HCl to ethene produces chloroethane. The corresponding reactions do *not* readily occur to the bonds in benzene or its derivatives. Benzene can be hydrogenated, i.e., hydrogen can be added to its double bonds, but only under rather extreme conditions. This difference in the behavior of benzene compared to alkenes is an example of the special stability of a six-membered ring containing three sets of alternating double and single bonds. The electrons of the bonds interact with each other in a manner that makes the molecule energetically much more stable than would be expected from adding up single- and double-bond energies appropriate to alkanes and alkenes. The extra stability disappears if even one of the three double bonds is hydrogenated or otherwise added to by other molecules. Thus the six-membered benzene ring is a unit of great inherent stability and survives intact in media that would destroy other $C=C$ bonds. Benzene and other molecules that possess this extra stability are said to be **aromatic** systems.

An exception to the rule that benzene does not add to its double bonds occurs when the attacking atom or molecule is a free radical such as a chlorine atom; recall that such species possess an unpaired electron. Thus, whereas molecular chlorine, Cl_2, itself does not add to the double bonds in benzene, a single chlorine atom does. In fact, as early as 1825, Michael Faraday found that chlorine gas would react with benzene if the reaction mixture was exposed to strong light, which we now realize splits the Cl_2

molecules into free Cl atoms. Once such a reaction starts, it continues until all the carbon atoms have added one chlorine atom, and 1,2,3,4,5,6-hexachlorocyclohexane is produced:

$$C_6H_6 + 3\ Cl_2 \xrightarrow{\text{UV light}} C_6H_6Cl_6$$

Chlorinated Benzenes

Although benzene does not readily undergo addition reactions with molecules or ions that are not free radicals, it does participate in **substitution** reactions: one of its hydrogen atoms can be replaced by a group such as methyl, hydroxyl, etc., that forms a single bond. Of particular interest is substitution by chlorine, since many compounds of environmental concern contain chlorine-substituted benzene rings. When benzene is reacted with chlorine gas in the presence of a catalyst such as iron(III) chloride, $FeCl_3$, one of the hydrogens (explicitly drawn on the ring below for clarity) is replaced by chlorine, and HCl gas is released:

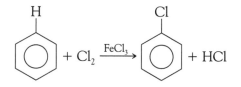

Notice that the aromatic C_6 ring survives intact. Notice also that in these structures the lone pairs on the chlorine atoms are not shown, as they were in the previous structures for alcohols and amines. Chemists use both types of structures—those that show the lone pairs and those that do not—for such compounds.

    If the reaction is allowed to continue, i.e., if excess Cl_2 is available, one or more hydrogen atoms of the chlorobenzene molecules will in turn be replaced by chlorine. There are three isomeric dichlorobenzenes, all of which could in principle be produced in such a reaction:

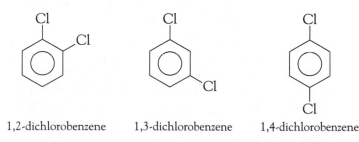

1,2-dichlorobenzene   1,3-dichlorobenzene   1,4-dichlorobenzene

The numbering scheme begins at one of the "substituted" carbons; the direction of numbering around the ring is chosen to yield the smallest possible number for the second substituent. In older nomenclature, 1,2-disubstituted benzene is called the *ortho* substituted isomer, 1,3-disubstitution calls for the prefix *meta*, and the 1,4 isomer is termed *para*. Thus the compound 1,4-dichlorobenzene shown at the bottom, right, on the previous page is also called *para*-dichlorobenzene or *p*-dichlorobenzene. When using the Kekulé structure for benzene in which double and single bonds are displayed, it is important to remember that the choice of a structure (A or B on page AP-11) for the positions of the double bonds is an arbitrary one. This has the consequence that there are only three, not five, isomeric dichlorobenzenes. For example, 1,6-dichlorobenzene is not different from the 1,2 isomer; they represent the same molecule viewed from different perspectives. (To avoid any such complications, many chemists use only the circle-containing hexagon symbol shown on page AP-12.)

PROBLEM 8

Deduce the structures and names for the three chemically different trichlorobenzenes.

Review Questions

1. What is the name of the hydrocarbon $CH_3(CH_2)_4CH_3$? Is it an alkane? Draw its structural formula.

2. Draw the structural formula for 3-ethylheptane.

3. What would be the name for the substituent group $CH_3CH_2CH_2CH_2—$?

4. Draw structural and condensed formulas for (a) trichloroethene and (b) 1,1-difluoroethene.

5. What is the main use for the compound $CCl_2{=}CCl_2$? What two names are used for this compound?

6. Draw structural formulas for each of the following: methyl alcohol, methylamine, formaldehyde, and formic acid.

7. Draw the structural formulas and symbolic representations for cyclopentane and for cyclopentene.

8. Explain what the term *hydrogen bonding* means. Explain why a short-chain alcohol such as methanol is soluble in water whereas a long-chain one such as octanol is not.

9. Is the six-membered ring system in benzene particularly stable or unstable?

10. Do molecules such as H_2 readily add to benzene? Do free radicals such as atomic hydrogen readily add to benzene?

11. What is another name for perchlorobenzene? Do you expect it to be more soluble in aqueous or in hydrocarbon media?

Further Reading

For more extensive background in organic chemistry, consult a modern introductory textbook such as K. P. C. Vollhardt and N. E. Schore, *Organic* *Chemistry*, 4th ed. (New York: W. H. Freeman and Company, 2003).

ANSWERS TO SELECTED ODD-NUMBERED PROBLEMS

CHAPTER 1

1-1 (a) 427 kJ/mol; junction of UV-B with UV-C

(b) 299 kJ/mol; junction of UV with visible region

(c) 160 kJ/mol; junction of visible and infrared regions

(d) 29.9 kJ/mol; beginning of thermal IR region

1-3 390.7 nm; 127.5 nm

1-5 307 nm

1-7 No

1-9 $O* + H_2O \longrightarrow 2\,OH$

1-11 491 nm; visible

1-13 The dichloroperoxide reaction would become relatively more important.

1-15 The reaction is highly endothermic and is therefore too slow to be significant.

1-17 (a) I: $F + O_3 \longrightarrow FO + O_2$

$FO + O \longrightarrow F + O_2$

II: $2 \times [F + O_3 \longrightarrow FO + O_2]$

$2\,FO \longrightarrow [FOOF] \longrightarrow 2\,F + O_2$

(b) $F + O_3 \longrightarrow FO + O_2$

$FO + O_3 \longrightarrow F + 2\,O_2$

net $2\,O_3 \longrightarrow 3\,O_2$

(c) $X + O_3 \longrightarrow XO + O_2$

$XO + O_3 \longrightarrow X + 2\,O_2$

net $2\,O_3 \longrightarrow 3\,O_2$

1-19 Presumably they have efficient tropospheric sinks and short atmospheric lifetimes.

Box 1-2, Problem 1 $2\,O_3 + \text{sunlight} \longrightarrow 3\,O_2$

Box 1-3, Problem 1 (a) CF_2Cl_2 (b) $C_2F_3Cl_3$ (c) $C_2HF_3Cl_2$ (d) $C_2H_2F_4$

Green Chemistry Questions

1. (a) 2

(b) 1, 4

3. No, not if the carbon dioxide is a waste by-product from another process.

5. Harpin is applied to the plant, which elicits the plant's own natural defenses.

Additional Problems

1. (b) 4×10^{15} grams

3. Net is $O_2 + \text{UV photon} \longrightarrow 2\,O$

Each O reacts as $O + O_2 \longrightarrow O_3$

Overall $3\,O_2 + \text{UV photon} \longrightarrow 2\,O_3$

5. $k_1[O_3]/k_2[O]$

7. $ClONO_2 + \text{photon} \longrightarrow Cl + NO_3$

$NO_3 + \text{photon} \longrightarrow NO + O_2$

$Cl + O_3 \longrightarrow ClO + O_2$

$NO + O_3 \longrightarrow NO_2 + O_2$

$ClO + NO_2 \longrightarrow ClONO_2$

net $2\,O_3 + 2\,\text{photons} \longrightarrow 3\,O_2$

9. In each cycling of chlorine and bromine, a fraction of the halogen not already bound as BrCl forms this molecule, so eventually all the halogen becomes tied up as BrCl.

11. (a) 140

(b) To distinguish between 2 isomers.

CHAPTER 2

2-1 1.2×10^{-14} M; 0.30 ppt

2-3 4×10^{-8} atm; 4×10^{-16} atm; the reaction does not remain in equilibrium as the gas quickly cools.

2-5 O_3 is reduced from 160 to 80 ppm; yes.

2-7 $2\,CO(NH_2)_2 + 4\,NO + O_2 \longrightarrow$
$4\,N_2 + 2\,CO_2 + 4\,H_2O$

2-9 $2\,NaOH + SO_2 \longrightarrow Na_2SO_3 + H_2O$

$Ca(OH)_2$ is the substance

$Na_2SO_3 + Ca(OH)_2 \longrightarrow$
$CaSO_3 + 2\,NaOH$

Net process is $SO_2 + Ca(OH)_2 \longrightarrow$
$CaSO_3 + H_2O$

2-11 $NO_3^- + 10 H^+ + 8 e^- \longrightarrow$
$$NH_4^+ + 3 H_2O$$

2-13 $PM_{0.1}$; PM_∞ or PM_{100}

2-15   3:1 ratio; area is bigger with smaller particles

Box 2-1, Problem 1   (a) 0.032 ppm
                    (b) 7.9×10^{11} molecules/cm³
                    (c) 1.3×10^{-9} M
       Problem 3   (a) 9.3×10^{11} molecules/cm³
                    (b) 74.2 $\mu g/m^3$

Green Chemistry Questions

1. (a) Both are volatile and flammable hydrocarbons (VOCs); vapors would contribute to pollution of the troposphere; flammability would reduce worker safety.
   (b) It is a VOC and would contribute to tropospheric pollution, but it is not flammable.
   (c) It is a greenhouse gas, but waste carbon dioxide is used and recycled.

Additional Problems

1. 4×10^{10} molecules/cm³ sec; 4×10^5 molecules/cm³ sec; that with O_3
3. (a) $O_3 + 2 KI + H_2O \longrightarrow I_2 + O_2 + 2 KOH$
   (b) 120 ppb
5. The outer two decline in height and the center one increases.
7. 1.0%
9. 0.0015 g and 0.0016 g
11. 2.65×10^5 L; 1080 kg
13. Yes, 250 $\mu g/m^3$ is well above the detection threshold of 123 $\mu g/m^3$.

CHAPTER 3

3-1   CFCs have no H or multiple bonds, so they don't react with OH or light and thus don't oxidize. No, since CH_2Cl_2 has H atoms and so will react with OH.

3-3   Quite endothermic

3-5   If a C-bonded H is abstracted by OH, the C-centered radical H_2COH will lose the hydroxyl H to O_2 abstraction, producing

formaldehyde. If instead OH abstracts the OH hydrogen initially, the resulting C-centered H_3CO radical loses H to O_2 abstraction, producing formaldehyde.

3-7   Overall: $H_2CO + 2 NO + 2 O_2 + $ sunlight
$$\longrightarrow CO_2 + H_2O + 2 NO_2$$

3-9   Overall: $CH_3(H)CO + 7 O_2 + 7 NO + $
       sunlight $\longrightarrow 2 CO_2 + 7 NO_2 + 4 OH$

3-11   Same as for Problem 3-7

3-13 CH_3OOH

3-15   3.88

3-17   b, c, and d

3-19   (a) $BrO + O \longrightarrow Br + O_2$
       (b) $BrO + ClO \longrightarrow Br + Cl + O_2$
       (c) $2 BrO \longrightarrow BrOOBr$
       (d) $BrO + UV \longrightarrow Br + O$

Additional Problems

1.	$CO + OH$	$\longrightarrow HOCO$
	$HOCO + O_2$	$\longrightarrow HOO + CO_2$
	$HOO + NO$	$\longrightarrow OH + NO_2$
	NO_2	$\xrightarrow{UV} NO + O$
	$O + O_2$	$\longrightarrow O_3$

Overall: $CO + 2 O_2 \xrightarrow{UV} CO_2 + O_3$

3.	$CO + OH$	$\longrightarrow HOCO$
	$HOCO + O_2$	$\longrightarrow CO_2 + HOO$
	$HOO + O_3$	$\longrightarrow OH + 2 O_2$

5. NO_2: $\cdot \ddot{O} - \ddot{N} = \ddot{O}$

     HNO_2: $H - \ddot{O} - \ddot{N} = \ddot{O}$

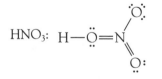

     HNO_3:

CHAPTER 4

4-1 H_2, Cl_2 cannot; all the others can.

4-3   (a) 8.0 kJ/mol; 1.3×10^{-20} J/molecule; 1.5%
       (b) 28.1 kJ/mol; 4.67×10^{-20} J/molecule; 5.3%

4-5   0.41 ppm; 2800 Gt
4-7 2.5×10^{15} g/year
4-9   483 Tg
4-11   No; yes; no

Green Chemistry Questions
1.   (a) 2
       (b) 1, 3, 4, 5, 6

Additional Problems
1.   (a) Symmetric and antisymmetric stretch
            and bending vibrations for both
       (b) Only the SO_2 symmetric stretch will
            contribute much.
       (c) Short atmospheric lifetimes
3.   0.48
5.   0.785
7.   21%; warming is not linearly related to
       concentration, and relative humidity could
       be different at 18°C.
9.   23.5 and 12.1 ppm
11.   1180 L; 13,500 L; 23,500 L

CHAPTER 5

5-1 $t = 0.69/k$; 17, 23, 46, and 69 years; 115
        years; 4.6%
5-3   0.00112, 0.00152, and 0.00254 mol CO_2/kJ
5-5   2.274 tonnes

Green Chemistry Questions
1.   (a) 2
       (b) 1, 5, 7, 9, 10
3.   Growing crops requires fertilizers and pesti-
       cides. Energy is needed to plant, cultivate, and
       harvest; to produce, transport, and apply fertil-
       izers and pesticides; to make and run tractors;
       and to transport seeds, biomass, monomers,
       and polymers. Use of land to produce crops for
       chemicals also removes land that could be
       used to produce food and feed.

Additional Problems
5.   1.4%

CHAPTER 6

6-1   68%
6-3   387°C
6-5   (a) 22.7 kJ/g for methanol, 29.7 kJ/g
            for ethanol
       (b) 17.9 kJ/mL for methanol, 23.5 kJ/mL
            for ethanol
       Ethanol > methanol on both a volume
       and mass basis, but both are inferior to
       methane and to a lesser extent to gasoline
       on a mass basis.
6-7   3:2; 20%
6-11 $H_2 + CO_3^{2-} \longrightarrow CO_2 + H_2O + 2\ e^-$
        $2\ CO_2 + 4\ e^- + O_2 \longrightarrow CO_3^{2-}$
        $2\ H_2 + O_2 \longrightarrow 2\ H_2O$
6-13   4:1
6-15   419 nm; visible light

Additional Problems
1. 5.5×10^{24} J; 0.007%
3.   One-third of the CO
       $C + 4/3\ H_2O \longrightarrow 2/3\ CH_3OH + 1/3\ CO_2$
7.   0.0011 EJ

CHAPTER 7

7-1   (a) 4.0×10^{-5} ppm; 0.040 ppb
       (b) 3.0 μg/L
7-3   5.0 ppm
7-5   (a) Yes
       (b) No
7-7   0.00010 mg/kg/day; 0.0055 mg

Green Chemistry Questions
1.   (a) 3
       (b) 4
3.   (a) A pesticide must meet one or more of the
            following requirements. It
            (1) reduces pesticide risks to human
                 health;
            (2) reduces pesticide risks to nontarget
                 organisms;

(3) reduces the potential for contamination of valued environmental resources;

(4) broadens adoption of IPM or makes it more effective.

(b) 1 and 2

5. (a) 3

(b) 1, 4

7. 1, 2, and 3

Additional Problems

1. About 12 g

3. 2.2×10^{-11} mol/m^3; 1.1×10^{-7} mol/m^3; 1.1×10^{-12} mol/m^3

0.22 mol; 0.77 mol; 2×10^{-8} mol

5. 0.070 ppb

CHAPTER 8

8-1 No, the same.

Yes

The unique ones are 1,2; 1,3; 1,4; 1,6; 1,7; 1,8; 1,9; 2,3; 2,7; 2,8.

8-3 (a) 2,3,5,6- and 2,3,4,5-tetrachlorophenols

(b) Two 2,3,5,6-tetrachlorophenols produce the 1,2,4,6,7,9-HCDD.

Two 2,3,4,5-tetrachlorophenols produce the 1,2,3,6,7,8-HCDD.

Two 2,3,4,6-tetrachlorophenols produce the 1,2,3,6,8,9-HCDD, the 1,2,4,6,7,9-HCDD, and the 1,2,3,6,7,8-HCDD.

8-5 1.87×10^5 molecules

8-7 2,3; 2,4; 2,5; 2,6; 3,4; 3,5; 2,2′; 2,3′; 2,4′; 2,5′; 2,6′; 3,3′; 3,4′; 3,5′; 4,4′

With rotation, 2,2′ and 2,6′ interconvert, as do 2,3′ and 2,5′, as well as 3,3′ and 3,5′.

8-9 1,2; 1,3; 1,4; 2,3; 2,4; 3,4; 1,6; 1,7; 1,8; 1,9; 2,6; 2,7; 2,8; 3,6; 3,7; 4,6 dichlorodibenzofurans

8-11 28.6 pg

Green Chemistry Questions

1. (a) 2

(b) 1, 4, and 6

Additional Problems

1. 2.0×10^{-12} mol/m^3; 1.5×10^{-10} mol/m^3; 4.9×10^{-5} mol/m^3

3. Octachloro, and 1,2,3,4,6,7,9- and 1,2,3,4,6,7,8-heptachlorodibenzo-p-dioxins, and 1,2,3,6,8,9- and 1,2,3,6,7,8- and 1,2,4,6,7,9-hexachlorodibenzo-p-dioxins

7. (a) 11 years

(b) 22.51 ppm; 1989; yes

(c) 10.7 ppm; 1989

CHAPTER 9

9-5 16 mg/L

9-7 (a) 1.6×10^{-35}

(b) 1.6×10^{77}

9-9 10^{141}

9-11 7.9×10^{-3} M; 6.3×10^{-20} M; 8.32 and 2.51

9-13 5.5×10^{-3} M; 0.001%; no

9-15 5.5×10^{-6} M; 5.5×10^{-7} M; 5.5×10^{-8} M

9-17 2.1×10^6 and 220 at pH $= 4$; 66 and 7.0×10^{-3} at pH $= 8.5$

9-19 2.6×10^{-4} M

9-21 51 mg CaCO$_3$ per L; slightly greater

9-23 3.6

Green Chemistry Questions

1. It removes the outermost layer of the raw cotton fiber, known as the cuticle, which makes the fiber wetable for bleaching and dyeing.

3. (a) 2

(b) 1, 3, 4, 6

Additional Problems

1. 2.4 ppm; 6.3 mg/L

3. 3.2×10^{-6} M; 3.9×10^{-7} M; yes

9. 6.1 mg/L; yes

CHAPTER 10

10-1 $Ca(HCO_3)_2 + Ca(OH)_2 \longrightarrow 2\ CaCO_3 + H_2O$

1:1 ratio

10-3 $+1$; OH$^-$; NH$_3$ + HOCl

10-5　11 ppm nitrogen; slightly less stringent

10-7　$Fe \longrightarrow Fe^{2+}$

$2 H_2O + 2 e^- \longrightarrow H_2 + 2 OH^-$

$Fe + 2 H_2O \longrightarrow Fe^{2+} + 2 OH^- + H_2$

10-9　Because vinyl chloride is carcinogenic

10-11　0.42 gram

10-13　High energy costs

10-15　NO, NO_2, NO_2^-, NO_3^-

10-17　214.3 kJ/mol; 558 nm; 45%

Green Chemistry Questions

1. It is biodegradable and its synthesis is performed under mild conditions, uses only water as a solvent, and excess ammonia is recycled.

Additional Problems

1. 0.79; 0.54; 0.27; 0.10
5. An O–O bond, as in $HOSO_2OO^-$; -1, so it needs to gain an electron to become -2
7. (a) $CH_3CCl_3 + e^- \longrightarrow CH_3CCl_2 + Cl^-$
   (b) $CH_3CCl_3 + OH \longrightarrow CH_3CCl_2 + H_2O$
11. 8.8×10^{-4} atm
13. 27.4 atm

CHAPTER 11

11-1　$M^{2+} + H_2S \longrightarrow MS + 2 H^+$

$M^{2+} + 2 R{-}S{-}H \longrightarrow$

$\qquad\qquad R{-}S{-}M{-}S{-}R + 2 H^+$

11-3　0.051 gram

11-5　0.5 mg; 2.0×10^5 grams

11-7　2×10^{-5} grams

11-9　6.0 micrograms per deciliter; 0.29 micromole per liter

Green Chemistry Questions

1. (a) 3
   (b) 4
3. Lower air pollution, better corrosion protection, and reduced waste

Additional Problems

1. Yes; yes, the answer changes
3. (a) $P = 36.6\, e^{-0.03B}$ (approximately); 61%

(b) There are almost no lead-free environments.

7. $MCl^+, MCl_2, MCl_3^-, MCl_4^{2-}$; seawater.

CHAPTER 12

12-1　3.9 L

12-3　$C_3H_3O_2$

12-5

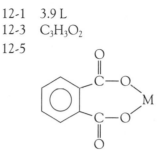

12-7　2,5-dichlorobenzoic acid

Green Chemistry Questions

1. It is biodegradable and as a result it places less of a load on the environment. The Donlar synthesis of polyaspartate is only two steps and requires only heat and hydrolysis with sodium hydroxide.
3. (a) 1
   (b) 1, 5, 6

Additional Problems

1. About 8000 homes
3. 1.14 kilograms
7. UV-B; need sunlight exposure to decompose; 399 kJ/mol

CHAPTER 13

13-3　48%; 60%; yes, substantially above

13-5　$^2_1H + ^3_2He \longrightarrow ^4_2He + ^1_1p$ (or ^1_1H)

$2\,^3_2He \longrightarrow ^4_2He + 2\,^1_1p$ (or $2\,^1_1H$)

Additional Problems

1. ^0_1e; $^{22}_{11}Na \longrightarrow ^{22}_{10}Ne + ^0_1e$; $^{13}_7N \longrightarrow ^{13}_6C + ^0_1e$
3. The effusion rate of $^{235}UF_6$ is 0.43% faster than that of $^{238}UF_6$. Since very little enrichment occurs in a single step, it is necessary to repeat the process hundreds of times to achieve suitable enrichment

APPENDIX

1. a)

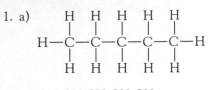

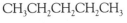

CH₃CH₂CH₂CH₂CH₃

b)

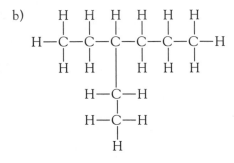

(CH₃CH₂)₂CHCH₂CH₃

c)

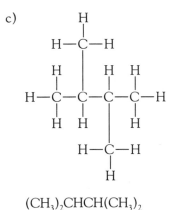

(CH₃)₂CHCH(CH₃)₂

3. a) 1, 1, 1, 2-tetrachloroethane
    b) 1-butene
    c) 1, 4-butadiene

5. a)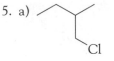

b)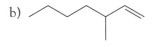

c)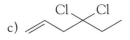

7. a)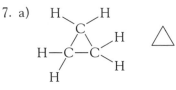

b)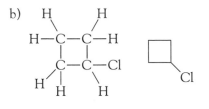

c)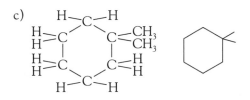

INDEX

Note: Page numbers followed by f, t, and b indicate figures, tables, and boxed material, respectively.